ELEMENTS

OF

ANALYTICAL MECHANICS,

BY

W. H. C. BARTLETT, LL. D.,

PROFESSOR OF NATURAL AND EXPERIMENTAL PHILOSOPHY IN THE UNITED STATES MILITARY ACADEMY AT WEST POINT

AND

AUTHOR OF ELEMENTS OF SYNTHETICAL MECHANICS, ACOUSTICS, OPTICS, AND SPHERICAL ASTRONOMY.

EIGHTH EDITION, REVISED, CORRECTED, AND ENLARGED.

NEW YORK:

A. S. BARNES & CO., 111 & 113 WILLIAM STREET,

(CORNER OF JOHN STREET.)

SOLD BY BOOKSELLERS, GENERALLY, THROUGHOUT THE UNITED STATES.

1866.

G. W. WOOD,
Printer,
John-street, cor. Dutch.

TO

COLONEL SYLVANUS THAYER,

OF THE CORPS OF ENGINEERS, AND LATE SUPERINTENDENT OF THE

UNITED STATES MILITARY ACADEMY,

This Work

IS MOST RESPECTFULLY AND AFFECTIONATELY DEDICATED,

IN GRATITUDE FOR THE PRIVILEGES

ITS AUTHOR HAS ENJOYED UNDER A SYSTEM OF INSTRUCTION

AND GOVERNMENT WHICH GAVE VITALITY TO

THE ACADEMY,

AND OF WHICH HE IS THE FATHER.

*

PREFACE.

It is now six years since the publication of the first edition of the present work. During this interval, it has been corrected and amended according to the suggestions of daily experience in its use as a text-book. It now appears with an additional part, under the head, Mechanics of Molecules; and this completes—in so far as he may have succeeded in its execution—the design of the author to give to the classes committed to his instruction, in the Military Academy, what has appeared to him a proper elementary basis for a systematic study of the laws of matter. The subject is the action of forces upon bodies,—the source of all physical phenomena—and of which the sole and sufficient foundation is the comprehensive fact, that all action is ever accompanied by an equal, contrary, and simultaneous reaction. Neither can have precedence of the other in the order of time, and from this comes that character of permanence, in the midst of endless variety, apparent in the order of nature. A mathematical formula which shall express the laws of this antagonism will contain the whole subject; and whatever of specialty may mark our perceptions of a particular instance, will be found to have its origin in corresponding peculiarities of physical condition, distance, place, and time, which are the elements of this formula. Its discussion constitutes the study of Mechanics. All phenomena in which bodies have a part are its legitimate subjects, and no form of matter under extraneous influences is exempt from its scrutiny. It embraces alike, in their reciprocal action, the gigantic and distant orbs of the celestial regions, and the proximate atoms of the ethereal atmosphere which pervades all space and establishes an unbroken continuity upon which its Divine Architect and Author may impress the power of His will at a single point and be felt everywhere. Astronomy, terrestrial physics, and chemistry are but its specialties; it classifies all of human knowledge that relates to inert matter into groups of phenomena, of which the rationale is in a common principle; and in the hands of those gifted with the priceless boon of a copious mathematics, it is a key to external nature.

The order of treatment is indicated by the heads of MECHANICS OF SOLIDS, of FLUIDS, and of MOLECULES,—an order suggested by differences of physical constitution.

The author would acknowledge his obligation to the works of many eminent writers, and particularly to those of M. Lagrange, M. Poisson, M. Couchey, M. Fresnel, M. Lamé, Sir William R. Hamilton, the Rev. Baden Powell, Mr. Airy, Mr. Pratt, and Mr. A. Smith.

WEST POINT, 1858.

CONTENTS.

INTRODUCTION.

PART I.

MECHANICS OF SOLIDS.

PART II.

MECHANICS OF FLUIDS.

PART III.

MECHANICS OF MOLECULES.

PART IV.

APPLICATIONS TO SIMPLE MACHINES, PUMPS, &c.

TABLES.

The Greek Alphabet is here inserted to aid those who are not already familiar with it, in reading the parts of the text in which its letters occur.

Letters.	Names.	Letters.	Names
Α α	Alpha	Ν ν	Nu
Β β ϐ	Bēta	Ξ ξ	Xi
Γ γ ſ	Gamma	Ο ο	Omicron
Δ δ	Delta	Π ϖ π	Pi
Ε ε	Epsilon	Ρ ρ ϱ	Rho
Ζ ζ ζ	Zēta	Σ σ ς	Sigma
Η η	Eta	Τ τ ⁊	Tau
Θ ϑ θ	Thēta	Υ υ	Upsilon
Ι ι	Iōta	Φ φ	Phi
Κ κ	Kappa	Χ χ	Chi
Λ λ	Lambda	Ψ ψ	Psi
Μ μ	Mu	Ω ω	Omega

ELEMENTS

OF

ANALYTICAL MECHANICS.

INTRODUCTION.

PHYSICAL SCIENCE.

§ 1.—The term *Nature*, is employed to signify all the bodies of the universe, collectively.

Of the existence of bodies, we are rendered conscious by the impressions they make upon the mind through the senses.

The condition of every body is subject to a variety of changes. These changes are brought about by agents external to the bodies themselves; and to investigate nature with reference to these changes and their causes, is the object of *Physical Science.*

PHYSICAL PROPERTIES.

§ 2.—*Physical Properties,* are those external signs by which the existence of bodies are made known to us through the medium of the senses. These properties are either *primary* or *secondary.*

PRIMARY PROPERTIES.

§ 3.—A *Primary Property* is that without which the existence of the body cannot be conceived. There are two of these—*Extension* and *Impenetrability.*

Extension is that by which every body occupies a limited portion of space. From it the body derives its figure and volume.

Impenetrability is that which prevents two bodies from occupying the same space at the same time. It determines a body's identity.

A body, then, is any thing which has extension and impenetrability.

SECONDARY PROPERTIES.

§ 4.—*Secondary Properties* are those which are not necessary to a conception of a body's existence, though all bodies may, and indeed do, possess them in a greater or less degree. They are *Compressibility*, *Expansibility*, *Porosity*, *Divisibility*, and *Elasticity*.

1.—*Compressibility* is that property by which a body may be made to occupy a smaller, and *expansibility* that by which it may be made to occupy a larger space, without, in either case, altering the quantity of its matter.

2.—*Porosity* is that property by which a body does not fill all the space within its exterior boundary, but leaves holes or *pores* between its elements.

In many cases the pores are visible to the naked eye; in others they are only seen by the aid of the microscope; and when so minute as to elude the power of this instrument, their existence may be inferred from experiment. Sponge, cork, wood, bread, &c., are bodies whose pores are obvious to unassisted vision. The human skin appears full of them, when viewed with the magnifying glass. The pores of one body are filled with some other body, and the pores of this with a third, as in the case of a sponge containing water, and the water containing air, and so on till we come to the most subtle of substances, *ether*, which pervades all bodies and all space.

3.—*Divisibility* is that property in consequence of which, by various mechanical means, such as beating, pounding, grinding, &c., a body may be reduced to fragments, homogeneous to each other, and to the entire mass.

By the aid of mathematical processes, the mind may be led to admit the infinite divisibility of bodies, though their practical division, by mechanical means, is subject to limitation. Many examples, however, prove that this process may be carried to an incredible extent. Nature furnishes numerous instances of objects, whose existence can only be detected by means of the most acute senses, assisted by the most powerful artificial aids.

Mechanical subdivisions for purposes connected with the arts are exemplified in the grinding of corn, the pulverizing of sulphur, charcoal and saltpetre, for the manufacture of gunpowder; and Homœopathy affords a remarkable instance of the extended application of this property of bodies.

In common gold lace, a silver thread is covered with gold so attenuated, that the quantity on a foot of thread weighs less than $\frac{1}{6000}$ of a grain. An inch of such thread will therefore contain $\frac{1}{72000}$ of a grain of gold; and if the inch be divided into 100 equal parts, each of which would be distinctly visible, the quantity of the precious metal in each of such pieces would be $\frac{1}{7200000}$ of a grain. One of these particles examined through a microscope of a magnifying power equal to 500, will appear 500 times as long, and the gold covering it will be visible, having been divided into 3,600,000,000 parts, each of which exhibits all the characteristics of this metal.

Dyes are likewise susceptible of an incredible divisibility. With 1 grain of blue carmine, 10 lbs. of water may be tinged blue. These 10 lbs. of water contain about 617,000 drops. Supposing that 100 particles of carmine are required in each drop to produce a uniform tint, it follows that this one grain of carmine has been subdivided 62 millions of times.

According to Biot, the thread by which a spider suspends herself is composed of more than 5000 single threads.

Our blood, which appears like a uniform red mass, consists of small red globules swimming in a transparent fluid called serum. The diameter of one of these globules does not exceed the 4000th part of an inch: whence it follows that one drop of blood, such as would hang from the point of a needle, contains at least one million of these globules.

But more surprising than all, is the microcosm of organized nature in the Infusoria. Of these creatures, which for the most part we can see only by the aid of the microscope, there exist many species so small that millions piled on each other would not equal a single grain of sand, and thousands might swim at once through the eye of the finest needle. The coats-of-mail and shells of these animalcules exist in such prodigious quantities, that extensive strata of rocks, as, for instance, the smooth slate near Bilin, in Bohemia, consist almost entirely of them. By microscopic measurements, 1 cubic line of this slate contains about 23 millions, and 1 cubic inch about 41,000 millions of these animals. As a cubic inch of this slate weighs 220 grains, 187 millions of these shells must go to a grain, each of which would consequently weigh about the $\frac{1}{187}$ millionth part of a grain. Conceive further, that each of these animalcules, as microscopic investigation has proved, has its limbs, entrails, &c., the possibility vanishes of our forming the most remote conception of the dimensions of these organic forms.

In cases where the finest instruments are unable to give the least aid in estimating the minuteness of bodies,—in other words, when bodies evade the perception of our sight and touch,—our olfactory nerves frequently detect the presence of matter in the atmosphere, of which no chemical analysis could afford us the slightest intimation.

Thus, for instance, a single grain of musk diffuses in a large and airy room a powerful scent, that frequently lasts for years;

and papers laid near musk will make a voyage to the East Indies and back without losing the smell. Imagine how many particles of musk must radiate from such a body every second, in order to render the scent perceptible in all directions.

4.—*Elasticity* is that property by which a body resumes of itself its figure and dimensions, when these have been changed or altered by any extraneous cause. Different bodies possess this property in very different degrees, and retain it with very unequal tenacity. Glass, tempered steel, ivory and whalebone, are among the more elastic solids. All fluids are highly elastic.

REST, MOTION, FORCE.

§ 5.—The state of a body by which it continues in the same place, is called *rest;* that by which it passes from one place to another, is called *motion;* and whatever changes the state of a body or the elements of a body, *with respect to rest or motion*, is called *force*. The existence of force is inferred from the changes, with respect to rest or motion, which all bodies and their internal elements are found to be continually undergoing. Its nature, or in what it consists, is unknown.

CONSTITUTION OF BODIES.

§ 6.—Several hypotheses have been proposed to explain the constitution of a body, and the mode of its formation. The most remarkable of these was by *Boscovich*, about the middle of the last century. According to this eminent philosopher:

1. All matter consists of indivisible and inextended *atoms*.

2. These atoms are endowed with attractive and repulsive forces, varying both in intensity and direction by a change of distance, so that at one distance two atoms attract one another, and at another distance they repel.

3. This law of variation is the same in all atoms. It is, therefore, mutual; for the distance of atom a from atom b, being the same as that of b from a, if a attract b, b must attract a with *precisely* an equal force.

4. At all considerable or *sensible* distances, these mutual forces are attractive and sensibly proportional to the square of the distance inversely. It is the attraction called *gravitation*.

5. At the small and insensible distances in which sensible contact is observed, and which do not exceed the 1000th or 1500th part of an inch, there are many alternations of attraction and repulsion, according as the distance of the atoms is changed. Consequently, there are many situations within this narrow limit, in which two atoms neither attract nor repel.

6. The force which is exerted between two atoms when their distance is diminished without end, and is just vanishing, is an insuperable repulsion, so that no force whatever can press two atoms into mathematical contact.

Such, according to *Boscovich*, is the constitution of a material atom and the *whole* of its constitution, and the immediate efficient cause of *all* its properties.

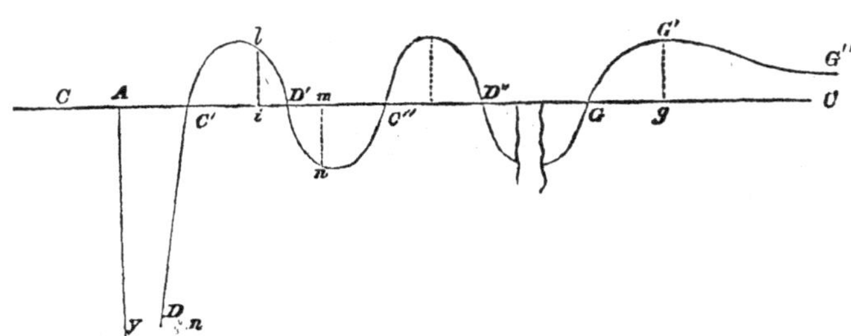

Boscovich represents his law of atomical action by what may be called an exponential curve. Let the distance of two atoms be estimated on the line $A\ C'\ C$, A being the situation of one of them, while the other is placed anywhere on this line. When placed at i, for example, we may suppose that it is attracted by

A, with a certain intensity. We represent this intensity by the length of the line *i l*, perpendicular to *A C*, and express the direction of the force, namely, from *i* to *A*, because it is attractive, by placing *i l* above the axis *A C*. Should the atom be at *m*, and be repelled by *A*, we express the intensity of repulsion by *m n*, and its direction from *m* towards *G* by placing *m n* below the axis.

This may be supposed for every point on the axis, and a curve drawn through the extremities of all the perpendicular ordinates, will be the exponential curve or scale of force.

As there are supposed a great many alternations of attractions and repulsions, the curve must consist of many branches lying on opposite sides of the axis, and must therefore cross it at C', D', C'', D'', &c., and at *G*. All these are supposed to be contained within a very small fraction of an inch.

Beyond this distance, which terminates at *G*, the force is always attractive, and is called the force of *gravitation*, the maximum intensity of which occurs at *g*, and is expressed by the length of the ordinate $G'g$. Further on, the ordinates are sensibly proportional to the square of their distances from *A*, inversely. The branch $G' G''$ has the line *A C*, therefore, for its asymptote.

Within the limit $A\ C'$ there is repulsion, which becomes infinite, when the distance from *A* is zero; whence the branch $C' D''$ has the perpendiculiar axis, *A y*, for its asymptote.

An atom being placed at *G*, and then disturbed so as to move it in the direction towards *A*, will be repelled, the ordinates of the curve being below the axis; if disturbed so as to move it from *A*, it will be attracted, the corresponding ordinates being above the axis. The point *G* is therefore a position in which the atom is neither attracted nor repelled, and to which it will tend to return when slightly removed in either direction, and is called the *limit of gravitation.*

If the atom be at C', or C'', &c., and be moved ever so little

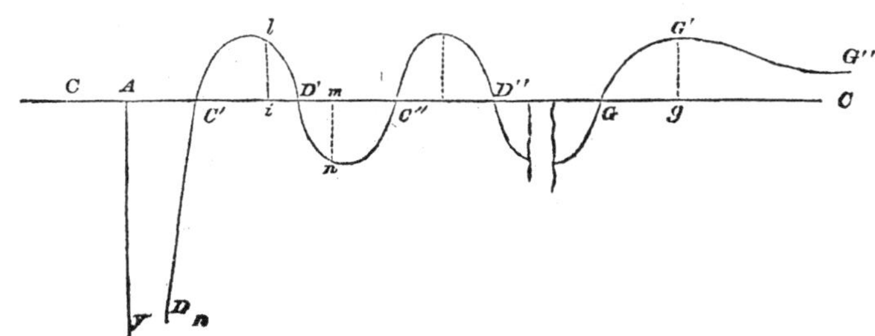

towards A, it will be repelled, and when the disturbing cause is removed, will fly back; if moved from A, it will be attracted and return. Hence C', C'', &c., are positions similar to G, and are called *limits of cohesion*, C' being called the *last limit of cohesion.* An atom situated at any one of these points will, with that at A, constitute a *permanent molecule* of the simplest kind.

On the contrary, if an *atom* be placed at D', or D'', &c., and be then slightly disturbed in the direction either from or towards A, the action of the atom at A will cause it to recede still further from its first position, till it reaches a limit of cohesion. The points D', D'', &c., are also positions of indifference, in which the atom will be neither attracted nor repelled by that at A, but they differ from C', C'', G, &c., in this, that an atom being ever so little removed from one of them has no disposition to return to it again; these points are called *limits of dissolution.* An atom situated in one of them cannot, therefore, constitute, with that at A, a permanent molecule, but the slightest disturbance will destroy it.

It is easy to infer, from what has been said, how three, four, &c., atoms may combine to form molecules of different orders of complexity, and how these again may be arranged so as by their action upon each other to form particles. Our limits will not permit us to dwell upon these points, but we cannot dismiss the subject without suggesting one of its most interesting consequences.

According to the highest authority, the SUN and other heavenly bodies have been formed by the gradual subsidence of a vast *nebula* towards its centre. Its molecules, forced by their gravitating action within their neutral limits, are in a state of tension, which is the more intense as the accumulation is greater; and the molecular agitations in the sun, caused by the successive depositions at its surface, make this body, in consequence of its vast size, a perpetual fountain of that incessant stream of ethereal waves which constitute the essence of *light* and *heat*. The internal heat of the earth has the same explanation. All bodies would appear self-luminous were the acuteness of our sense of sight increased beyond its present limit in the same proportion that the sun exceeds them in size. The sun far transcends all the other bodies of our system in regard to heat and light, and is in a state of *incandescence*, because of the mode of its formation and of its vastly greater dimensions.

2.—The molecular forces, here considered, are the effective causes which determine a body to be a solid, liquid, or gas. If the attractions prevail over the repulsions, the body is *solid;* if these antagonistic forces be equal, it is *liquid;* and if the repulsions prevail over the attractions, it is a *gas*.

3.—The molecular forces may so act upon the elements of dissimilar bodies as to cause a new combination or union of their atoms. This may also produce a separation between the combined atoms or molecules, in such manner as to entirely change the individual properties of the bodies. Such efforts of the molecular forces are called *chemical action;* and the disposition to exert these efforts, *chemical affinity*.

4.—Beyond the last limit of gravitation, atoms attract each other: hence, all the atoms of one body attract each atom of another, and vice versa: thus giving rise to attractions between bodies of sensible magnitudes through sensible distances. The intensities of these attractions are proportional to the number of atoms in the attracting body directly, and to the square of the distance between the bodies inversely.

5.—The term *universal gravitation* is applied to this force, when it is intended to express the action of the heavenly bodies on each other; and that of *terrestrial gravitation* or simple *gravity*, when we wish to express the action of the earth upon the bodies forming with itself one whole. The force is always of the same kind, however, and varies in intensity only by reason of a difference in the number of atoms and their distances. Its effect is always to generate motion, when the bodies are free to move.

Gravity, then, is a property common to all terrestrial bodies, since they constantly exhibit a tendency to approach the earth and its centre. In consequence of this tendency all bodies possess weight, and, unless supported, fall to the surface of the earth; and if prevented by any other bodies from doing so, they exert a pressure on these latter.

This is one of the most important properties of terrestrial bodies, and the cause of many phenomena, of which a fuller account will be given hereafter.

DENSITY.

§ 7.—*Density* is a term employed to express the greater or less proximity of a body's atoms. The relative densities of different bodies must, therefore, be proportional to the number of atoms they contain under equal volumes. The weights of bodies being proportional to the number of their atoms, the density of any body is measured by the quotient arising from dividing its weight by the weight of an equal volume of some other body, assumed as a standard, and whose density is regarded as unity. The density of pure water, at the temperature of 38°,75 Fahrenheit, is assumed as the unit, the water possessing its maximum density at that temperature.

MASS.

§ 8.—The *mass* of a body is the quantity of matter it contains; and this being proportional to its weight, the mass of a body may be measured by the quotient arising from dividing its weight by the weight of some other body assumed as the unit of mass. A cubic foot of distilled water, at its maximum density, may be assumed as the unit of mass. And, therefore, a body whose mass is expressed by any number, say 20, will contain twenty times the matter contained in a cubic foot of distilled water at its greatest density.

If the mass of a body be denoted by M, its weight by W, and that of a unit of mass by g, then will

$$W = M \cdot g \quad \ldots\ldots\ldots \quad (1)$$

If V denote the body's volume, and D its density; then will

$$M = V \cdot D \quad \ldots\ldots\ldots \quad (2)$$

and by substitution above,

$$W = V \cdot D \cdot g \quad \ldots\ldots\ldots \quad (3)$$

The masses of bodies are so constantly in view in discussing and applying physical principles, as to make it important to understand well the method of getting their numerical values. Equation (1) may be written

$$\frac{W}{g} = M,$$

and in which W and g must be expressed in terms of the same unit. But g may have two values, very different in kind. It may be expressed in pounds, or any other unit of weight, or in feet, or any other unit of length. In the first case, the body assumed as the unit of mass and that whose mass is desired, are simply weighed, and the ratio of the weights taken. In the second case, the body assumed as the unit of mass is permitted to fall in vacuo, and the velocity its own weight can generate in it, in one second of time, ascertained.

A cubic foot of pure water, at its maximum density, weighs 62,3791 pounds avoirdupois, and the measure of a body's mass is given by

$$M = \frac{W}{62{,}3791};$$

in which W must be expressed in avoirdupois pounds.

The velocity which the weight of a body can impress upon itself, in one second of time, on the parallel of 45°, is 32,1801 feet; and the measure for a body's mass is given by

$$M = \frac{W}{32{,}1801},$$

in which W may be expressed, as before, in pounds; but in this case the cubic foot of water ceases to be the unit of mass, and in its stead we take so much of the water, or of any other body, as will weigh 32,1801 pounds, as the unit of mass. In this latter case, any body which weighs one pound will be $\frac{1}{32{,}1801}$ of the unit of mass.

Had the pound been made greater than it is in the proportion of 62,3791 to 32,1801, or the foot less in the proportion of 32,1801 to 62,3791, then would the same number have expressed both the pounds avoirdupois in the weight of a cubic foot of pure water at the standard temperature, and the number of feet in the velocity this weight could generate in the same cubic foot in one second of time.

UNORGANIZED AND ORGANIZED BODIES.

§ 9.—All bodies connected with the earth are distributed into two classes, viz.: *Unorganized* and *Organized.*

The unorganized class embrace all minerals, as metals, stones, earths, alkalies, water, air, and the like.

The organized class include all animals and vegetables.

The *unorganized bodies* form the lower class, and are, so to speak, the substratum of the organized. They are acted upon solely by influences external to themselves, and have nothing that can properly be called life. They have no definite or periodical duration.

Organized bodies are more or less perfect individuals, possessing organs adapted to the performance of certain functions. They possess vitality, and are continually appropriating to themselves unorganized bodies, changing their properties, and, by this process, increasing their bulk. They possess the faculty of reproduction. They retain only for a limited time the vital principle, and, when life is extinct, they sink into the class of unorganized bodies.

HEAVENLY BODIES.

§ 10.—The *Heavenly Bodies* form a distinct class. In the changes they bring about within themselves, they resemble organized bodies; and may, in one sense, be said to possess organs. Those of our earth are its continents, oceans, and atmosphere. The researches of Geology furnish the most ample evidence of vast changes having taken place in the earth. It now supports and nourishes both the animal and vegetable kingdoms. There was a time when neither of these existed upon it. It was once all fluid, from excessive heat; it is now incrusted with an indurated envelope of many miles in thickness, inclosing a molten, liquid mass. In many places its continents are being elevated, while in others they are being depressed; corresponding changes are taking place in the shores of the ocean, and the climates of the same zones are undergoing modifications. What is true of our earth, is doubtless equally true of the other heavenly bodies.

NATURAL PHILOSOPHY.

§ 11.—*Natural Philosophy* is a name given to that branch of physical science which treats of the *general* properties of *unorganized bodies;* the changes they undergo without affecting their internal constitution; the causes of these changes, and the laws which govern both the causes and changes.

THERMOTICS, ACOUSTICS, OPTICS, ELECTRICS.

§ 12.—Natural Philosophy embraces the subjects of *Thermotics*, *Acoustics*, *Optics*, and *Electrics*. The first treats of *heat*, the second of *sound*, the third of *light*, and the fourth of *electricity*. The subject of *magnetics* is omitted here, because it is now merged into that of electrics. The phenomena which appertain to these different heads, all have a common source in the action of forces upon bodies—the nature of the bodies, the kind and mode of action of the forces, and the sense employed to excite the mind to a perception of the effects, constituting the main distinction.

Natural Philosophy relies upon observation and experiment for its data. From these we deduce the varied information we have acquired about bodies; by the former we notice any changes that transpire in the condition or relations of any body, as they spontaneously arise without interference on our part; whereas, in the performance of an experiment, we purposely alter the natural arrangement of things, to bring about some particular condition we desire. To accomplish this, we make use of appliances called *philosophical apparatus*, the proper use and application of which, it is the office of *Experimental Philosophy* to teach.

If we notice that in winter water becomes converted into ice, we are said to make an observation; if, by means of freezing mixtures or evaporation, we cause water to freeze, we are then said to perform an experiment.

These observations and experiments are next subjected to calculation, from which are deduced what are called *the laws of nature*, or *the rules that like causes will invariably produce like results*. To express these laws with the greatest possible brevity, mathematical formulas are used. When it is not practicable to represent them with mathematical precision, we are content with inferences and assumptions based on analogies, or with probable *hypotheses*, as the means for explanation and further deductions.

A hypothesis gains in probability the more nearly it accords with the ordinary course of nature, the more numerous the observations and experiments on which it is founded, and the more simple the explanation it offers of the phenomena for which it is intended to account.

CHEMISTRY.

§ 13.—*Chemistry* treats of the *individual* properties of unorganized bodies, by which, as regards their constitution, they may be distinguished from one another. It also investigates the transformations that take place in the interior of these bodies, and by which their substance is altered and remodelled; and, lastly, it detects and classifies the laws that regulate chemical changes.

NATURAL HISTORY.

§ 14.—*Natural History* treats of the organized bodies. It comprises three divisions, viz.: *Anatomy*, which is concerned with the dissections of plants and animals; *Vegetable and Animal Chemistry*, which investigates their internal constitution; and *Physiology*, which explains the objects and offices of their various organs.

ASTRONOMY.

§ 15.—*Astronomy* teaches the knowledge of the heavenly bodies. It consists of two branches—Physical and Spherical Astronomy. The former treats of the constitution and physical condition of the heavenly bodies, their mutual influences and actions on each other, and, generally, seeks to explain the causes of celestial phenomena; the latter is concerned with the appearances, magnitudes, distances, arrangements, and motions of these bodies. All measurements upon them are made from stations on the earth, and by instruments that give the sides and angles of spherical triangles projected upon the concave of the celestial vault; and hence the name.

GEOLOGY, PHYSICAL GEOGRAPHY, METEOROLOGY.

§ 16.—*Geology*, *Physical Geography*, and *Meteorology*, are strictly branches of Physical Astronomy. The first teaches a knowledge of the structure and history of the earth's crust; the second treats of the nature and character of its surface; and the third is concerned with the phenomena of the atmosphere and climate.

PHYSICS.

§ 17.—Natural Philosophy, Chemistry, Natural History, and Astronomy, are but branches of the more general subject called *Physics*—a science so vast in its range as to embrace whatever is known and can be discovered of the nature and properties of bodies, their source, effects, affections, operations, phenomena, and laws.

MECHANICS.

§ 18.—All phenomena of the physical world arise directly from the action of forces upon the various forms of bodies. That branch of science which treats of this action is called *Mechanics.* A careful study of a course of mechanics is, therefore, an indispensable preparation for that of any branch of physical science. Mechanics is the subject of the present volume. It will be treated under three heads, suggested by peculiarities of physical condition, viz.: *Mechanics of Solids*, *Mechanics of Fluids*, and *Mechanics of Molecules;* the first treating of the action of forces upon solid bodies; the second, upon fluid bodies; and the third, upon the molecules or elements of both solids and fluids.

Mechanics is founded in a single fact, viz.: that *all action is ever accompanied by an equal, contrary, and simultaneous reaction.*

INERTIA.

§ 19.—This reaction very often arises from a property, common to all bodies, by which they resist, of themselves, every change of their own state in regard to rest or motion, and with an effort equal to that which produces the change. This property, known from experience, is called *Inertia.* It is force, but *passive* and *conservative* force.

PART I.

MECHANICS OF SOLIDS.

SPACE, TIME, MOTION, AND FORCE.

§ 20.—*Space* is indefinite extension, without limit, and contains all bodies.

§ 21.—*Time* is any limited portion of duration. We may conceive of a time which is longer or shorter than a given time. Time has, therefore, magnitude, as well as lines, areas, &c.

To *measure a given time*, it is only necessary to assume a certain interval of time as unity, and to express, by a number, how often this unit is contained in the given time. When we give to this number the particular name of the unit, as *hour*, *minute*, *second*, &c., we have a complete expression for time.

The *Instruments* usually employed in measuring time are *clocks*, *chronometers*, and *common watches*, which are too well known to need a description in a work like this.

The smallest division of time indicated by these time-pieces is the *second*, of which there are 60 in a minute, 3600 in an *hour*, and 86400 in a day; and chronometers, which are nothing more than a species of watch, have been brought to such perfection as not to vary in their rate a half a second in 365 days, or 31536000 seconds.

Thus the number of hours, minutes, or seconds, between any two events or instants, may be estimated with as much precision and ease as the number of yards, feet, or inches between the extremities of any given distance.

Time may be represented by lines, by laying off upon a given right line AB, the equal distances from 0 to 1, 1 to 2, 2 to 3, &c., each one of these equal distances representing the unit of time.

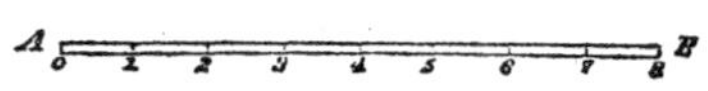

A second is usually taken as the unit of time, and a foot as the linear unit.

§ 22.—A body is in a state of *absolute rest* when it continues in the same place in space. There is perhaps no body absolutely at rest; our earth being in motion about the sun, nothing connected with it can be at rest. Rest must, therefore, be considered but as a *relative* term. A body is said to be at rest, when it preserves the same position in respect to other bodies which we may regard as fixed. A body, for example, which continues in the same place in a boat, is said to be at rest in relation to the boat, although the boat itself may be in motion in relation to the banks of a river on whose surface it is floating.

§ 23.—A body is in *motion* when it occupies successively different positions in space. Motion, like rest, is but relative. A body is in motion when it changes its place in reference to those which we may regard at rest.

Motion is essentially *continuous;* that is, a body cannot pass from one position to another without passing through a series of intermediate positions; a point, in motion, therefore describes a continuous line.

When we speak of the path described by a body, we are to understand that of a certain point connected with the body. Thus, the path of a ball, is that of its centre.

§ 24.—The motion of a body is said to be *curvilinear* or *rectilinear,* according as the path described is a *curve* or *right line.* Motion is either *uniform* or *varied.* A body is said to have *uniform motion* when it passes over equal spaces in equal successive portions of time: and it is said to have *varied motion* when it passes over unequal spaces in equal successive portions of time. The motion is said to be *accelerated* when the successive increments of space in equal times become greater and greater. It is *retarded* when these increments become smaller and smaller.

§ 25.—*Velocity* is the *rate* of a body's motion. Velocity is measured by the length of path described uniformly in a unit of time.

§ 26.—The spaces described in equal successive portions of time being equal in uniform motion, it is plain that the length of path described in any time will be equal to that described in a unit of time repeated as many times as there are units in the time. Let v denote the velocity, t the time, and s the length of path described, then will

$$s = v \,.\, t, \quad \ldots \ldots \ldots \ldots \quad (3)$$

If the position of the body be referred to any assumed origin whose distance from the point where the motion begins, estimated in the direction of the path described, be denoted by S, then will

$$s = S + v \,.\, t \quad \ldots \ldots \ldots \ldots \quad (4)$$

Equation (3) shows that in uniform motion, *the space described is always equal to the product of the time into the velocity;* that *the spaces described by different bodies moving with different velocities during the same time, are to each other as the velocities;* and that *when the velocities are the same, the spaces are to each other as the times.*

§ 27.—Differentiating Equation (3) or (4), we find

$$\frac{ds}{dt} = v \,; \quad \ldots \ldots \ldots \ldots \quad (5)$$

that is to say, *the velocity is equal to the first differential co-efficien of the space regarded as a function of the time.*

Dividing both members of Equation (3) by t, we have

$$\frac{s}{t} = v. \quad \ldots \ldots \ldots \ldots \quad (6)$$

which shows that, in *uniform motion, the velocity is equal to the whole space divided by the time in which it is described.*

§ 28.—Matter on the earth, in its unorganized state, is *inanimate* or *inert.* It cannot give itself motion, nor can it change of itself the motion which it may have received. A body at rest will forever remain so unless disturbed by something extraneous to itself; or if it be in motion in any direction, as from a to b, it will continue, after arriving at b, to move towards c in the prolongation of ab; for having arrived at b, there is no reason why it should deviate to one side more than another. Moreover, if the body have a certain velocity at b, it will retain this velocity unaltered, since no reason can be assigned why it should be increased rather than diminished in the absence of all extraneous causes.

If a billiard-ball, thrown upon the table, seem to diminish its rate of motion till it stops, it is because its motion is resisted by the cloth and the atmosphere. If a body thrown vertically downward seem to increase its velocity, it is because its weight is incessantly urging it onward. If the direction of the motion of a stone, thrown into the air, seem continually to change, it is because the weight of the stone urges it incessantly towards the surface of the earth. Experience proves that in proportion as the obstacles to a body's motion are removed, will the motion itself remain unchanged.

When a body is at rest, or moving with uniform motion, its *inertia* is not called into action.

§ 29.—A *force* has been defined to be that which changes or tends to change the state of a body in respect to rest or motion. *Weight* and *Elasticity* are examples. A body laid upon a table, or suspended from a fixed point by means of a thread, would move under the action of its weight, if the resistance of the table or that of the fixed point did not continually prevent by an equal, simultaneous, and contrary reaction. A body, subjected alone to the action of a spring, would change its state by moving faster or slower.

When we push or pull a body, be it free or fixed, we experience a sensation denominated *pressure*, *traction*, or, in general, *effort*. This effort is analogous to that which we exert in raising a weight. Forces are real pressures. Pressure may be strong or feeble; it therefore has magnitude, and may be expressed in numbers by assuming a certain pressure as *unity*. The unit of pressure will be taken to be tnat exerted by the weight of $\frac{1}{62,5}$ part of a cubic foot of distilled water, at 38°,75, and is called a *pound*.

§ 30.—The *intensity* of a force is its greater or less capacity to produce pressure. This intensity may be expressed in pounds, or in quantity of motion. Its value in pounds is called its *statical* measure; in quantity of motion, its *dynamical* measure.

§ 31.—The *point of application* of a force, is the material point to which the force may be regarded as directly applied.

§ 32.—The *line of direction* of a force is the right line which the point of application would describe, if it were perfectly free.

§ 33.—The effect of a force depends upon its intensity, point of application, and line of direction, and when these are given the force is known.

§ 34.—Two forces are equal when substituted, one for the other, in the same circumstances, they produce the same effect, or when directly opposed, they neutralize each other.

§ 35.—There can be no *action* of a force without an equal and contrary *reaction*. This is a law of nature, and our knowledge of it comes from experience. If a force act upon a body retained by a fixed obstacle, the latter will oppose an equal and contrary resistance. If it act upon a free body, the latter will change its state, and in the act of doing so, its inertia will oppose an equal and contrary resistance. *Action and reaction are ever equal, contrary and simultaneous.*

§ 36.—If a free body be drawn by a thread, the thread will stretch and even break if the action be too violent, and this will the more probably happen in proportion as the body is more massive. If a body be suspended by means of a vertical chain, and a weighing spring be interposed in the line of traction, the graduated scale of the spring will indicate the weight of the body when the latter is at rest; but if the upper end of the chain be suddenly elevated, the spring will immediately bend more in consequence of the resistance opposed by the inertia of the body while acquiring motion. When the motion acquired becomes uniform, the spring will resume and preserve the degree of flexure which it had at rest. If now, the motion be checked by relaxing the effort applied to the upper end of the chain, the spring will unbend and indicate a pressure less than the weight of the body, in consequence of the inertia acting in opposition to the retardation. The oscillations of the spring may therefore serve to indicate the variations in the motions of a body, and the energy of its force of inertia, which acts against or with a force, according as the velocity is increased or diminished.

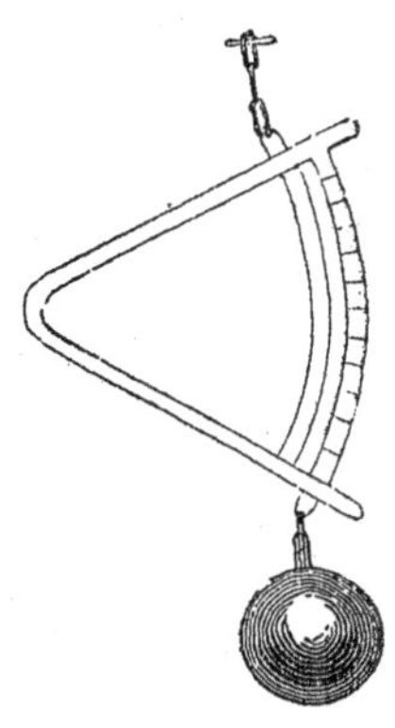

§ 37.—Forces produce various effects according to circumstances. They sometimes leave a body at rest, by balancing one another, through its intervention; sometimes they change its form or break it; sometimes they impress upon it motion, they accelerate or retard that which it has, or change its direction; sometimes these effects are produced gradually, sometimes abruptly, but however produced, they require some *definite time*, and are effected by *continuous degrees*. If a body is sometimes seen to change suddenly its state, either in respect to the direction or the rate of its motion, it is because the force is so great as to produce its effect in a time so short as to make its duration imperceptible to our senses, yet some definite portion of time is necessary for the change. A ball fired from a gun will break through a pane of glass, a piece of board, or a sheet of paper, when freely suspended, with a rapidity so great as to call into action a force of inertia in the parts which remain, greater than the molecular forces which connect the latter with those torn away.

In such cases the effects are obvious, while the times in which they are accomplished are so short as to elude the senses: and yet these times have had some definite duration, since the changes, corresponding to these effects, have passed in succession through their different degrees from the beginning to the ending.

§ 38.—Forces which give or tend to give motion to bodies, are called *motive forces*. The agent, by means of which the force is exerted, is called a *Motor*.

§ 39.—The statical measure of forces may be obtained by an instrument called the *Dynamometer*, which in principle does not differ from the spring balance. The dynamical measure will be explained further on.

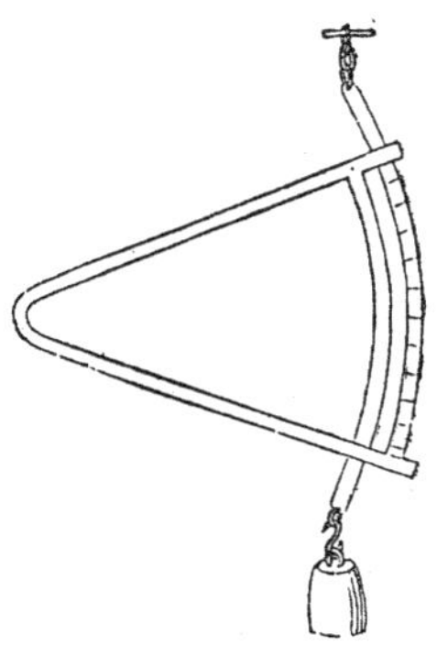

§ 40.—When a force acts against a point in the surface of a body, it exerts a pressure which crowds together the neighboring particles; the body yields, is compressed and its surface indented; the crowded particles make an effort, by their molecular forces, to regain their primitive places, and thus transmit this crowding action even to the remotest particles of the body. If these latter particles are fixed or prevented by obstacles from moving, the result will be a compression and change of figure throughout the body. If, on the contrary, these extreme particles are free, they will advance, and motion will be communicated by degrees to all the parts of the body. This internal motion, the result of a series of compressions, proves that a certain time is necessary for a force to produce its entire effect, and the error of supposing that a finite velocity may be generated instantaneously. The same kind of action will take place when the force is employed to destroy the motion which a body has already acquired; it will first destroy the motion of the molecules at and nearest the point of action, and then, by degrees, that of those which are more remote in the order of distance.

The molecular springs cannot be compressed without reacting in a contrary direction, and with an equal effort. The *agent* which presses a body will experience an equal pressure; *reaction* is equal and contrary to *action*. In pressing the finger against a body, in pulling it with a thread, or pushing it with a bar, we are pressed, drawn, or pushed in a contrary direction, and with an equal effort. Two weigh-

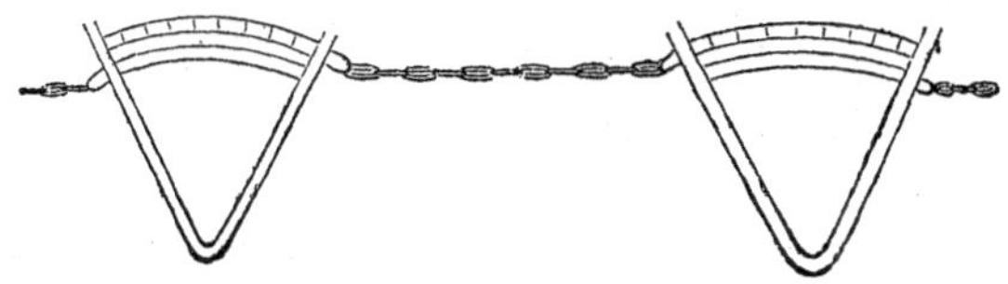

ing springs attached to the extremities of a chain or bar, will indicate the same degree of tension and in contrary directions when made to act upon each other through its intervention.

In every case, therefore, the action of a force is transmitted through a body to the ultimate point of resistance, by a series of equal and contrary actions and reactions which balance each other, and which the molecular springs of all bodies exert at every point of the right line, along which the force acts. It is in virtue of this property of bodies, that the action of a force may be assumed to be exerted at *any point in its line of direction within the boundary of the body.*

§ 41.—Bodies being more or less extensible and compressible, when interposed between the motor and resistance, will be stretched or compressed to a certain degree, depending upon the energy with which these forces act; but as long as the force and resistance remain the same, the body having attained its new dimensions, will cease to change. On this account, we may, in the investigations which follow, assume that the bodies employed to transmit the action of forces from one point to another, are inextensible and rigid.

WORK.

§ 42.—To *work* is to overcome a resistance continually recurring along some path. Thus, to raise a body through a vertical height, its weight must be overcome at every point of the vertical path. If a body fall through a vertical height, its weight develops its inertia at every point of the descent. To take a shaving from a board with a plane, the cohesion of the wood must be overcome at every point along the entire length of the path described by the edge of the chisel.

§ 43.—The resistance may be constant, or it may be variable. In the first case, the *quantity of work* performed is the constant resistance taken as many times as there are points at which it has acted, and is measured by the product of the resistance into the path described by its point of application, estimated in the direction of the resistance. When the resistance is variable, the quantity of work is obtained by estimating the elementary quantities of work and taking their sum. By the elementary quantity of work, is meant the intensity of the variable resistance taken as many times as there are points in the indefinitely small path over which the resistance may be regarded as constant; and is measured by the intensity of the resistance into the differential of the path, estimated in the direction of the resistance.

§ 44.—In general, let P denote any variable resistance, and s the path described by its point of application, estimated in the direction of the resistance; then will the quantity of work, denoted by Q, be given by

$$Q = \int P . ds \quad . \quad . \quad . \quad . \quad . \quad . \quad . \quad . \quad (7)$$

which integrated between certain limits, will give the value of Q.

§ 45.—The simplest kind of work is that performed in raising a weight through a vertical height. It is taken as a standard of comparison, and suggests at once an idea of the quantity of work expended in any particular case.

Let the weight be denoted by W, and the vertical height by H; then will

$$Q = W . H \quad . \quad . \quad . \quad . \quad . \quad . \quad . \quad . \quad (8).$$

If W become one pound, and H one foot, then will

$$Q = 1 ;$$

and the unit of work is, therefore, the unit of force, one pound, exerted over the unit of distance, one foot; and is measured by a square of which the adjacent sides are respectively one foot and one pound, taken from the same scale of equal parts.

§ 46.—To illustrate the use of Equation (7), let it be required to compute the quantity of work necessary to compress the spiral spring of the common spring balance to any given degree, say from the length AB to DB. Let the resistance vary directly as the degree of compression, and denote the distance AD' by x; then will

$$P = C \cdot x;$$

in which C denotes the resistance of the spring when the balance is compressed through the distance unity.

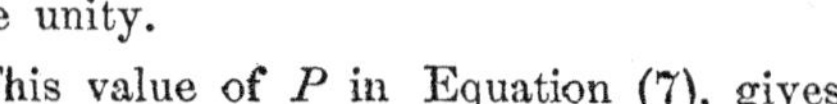

This value of P in Equation (7), gives

$$Q = \int P \cdot dx = \int C \cdot x dx = C \cdot \frac{x^2}{2} + C',$$

which integrated between the limits $x = 0$ and $x = AD = a$, gives

$$\overline{Q} = C \cdot \frac{a^2}{2}.$$

Let $C = 10$ pounds, $a = 3$ feet; then will

$$Q = 45 \quad \text{units of work,}$$

and the quantity of work will be equal to that required to raise 45 pounds through a vertical height of one foot, or one pound through a height of 45 feet, or 9 pounds through 5 feet, or 5 pounds through 9 feet, &c., all of which amounts to the same thing.

§ 47.—A *mean resistance* is that which, multiplied into the entire path described in the direction of the resistance, will give the entire quantity of work. Denote this by R, and the entire path by s, and from the definition, we have

$$R \cdot s = \int P \cdot ds;$$

whence,

$$R = \frac{\int P \cdot ds}{s} \quad . \quad . \quad . \quad . \quad . \quad . \quad . \quad . \quad (9).$$

That is, the mean resistance is equal to the entire work, divided by the entire path.

In the above example the path being 3 feet, the mean resistance would be 15 pounds.

§ 48.—Equation (7) shows that the quantity of work is equal to the area included between the path s, in the direction of the resistance, the curve whose ordinates are the different values of P, and the ordinates which denote the extreme resistances. Whenever, therefore, the curve which connects the resistance with the path is known, the process for finding the quantity of work is one of simple integration.

Sometimes this law cannot be found, and the intensity of the resistance is given only at certain points of the path. In this case we proceed as follows, viz.: At the several points of the path where the resistance is known, erect ordinates equal to the corresponding resistances, and connect their extremities by a curved line; then divide the path described into any *even* number of equal parts, and erect the ordinates at the points of division, and at the extremities; number the ordinates in the order of the natural numbers; *add together the extreme ordinates, increase this sum by four times that of the even ordinates and twice that of the uneven ordinates, and multiply by one-third of the distance between any two consecutive ordinates.*

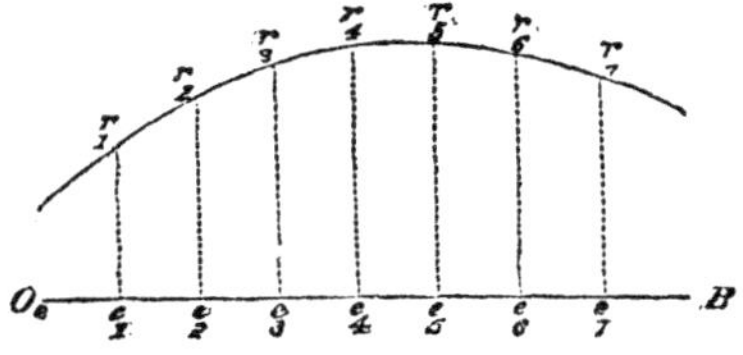

Demonstration: To compute the area comprised by a curve, any two of its ordinates and the axis of abscisses, by plane geometry, divide it into elementary areas, by drawing ordinates, as in the last figure, and regard each of the elementary figures, $e_1\ e_2\ r_2\ r_1$, $e_2\ e_3\ r_3\ r_2$, &c., as trapezoids; it is obvious that the error of this supposition will be less, in proportion as the number of trapezoids between given limits is greater. Take the first two trapezoids of the preceding figure, and divide the distance $e_1\,e_3$ into three equal parts, and at the points of division, erect the ordinates $m\,n$, $m_1\,n_1$; the area computed from the three trapezoids $e_1\,m\,n\,r_1$, $m\,m_1\,n_1\,n$, $m_1\,e_3\,r_3\,n_1$, will be more accurate than if computed from the two $e_1\,e_2\,r_2\,r_1$, $e_2\,e_3\,r_3\,r_2$.

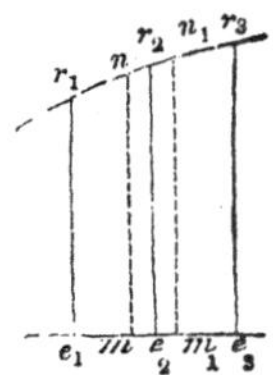

The area by the three trapezoids is

$$e_1\,m \times \frac{e_1\,r_1 + m\,n}{2} + m\,m_1\frac{m\,n + m_1\,n_1}{2} + m_1\,e_3\frac{m_1\,n_1 + e_3\,r}{2}$$

But by construction,

$$e_1\,m = m\,m_1 = m_1\,e_3 = \tfrac{1}{3}\,e_1\,e_3 = \tfrac{2}{3}\,e_1\,e_2,$$

and the above may be written,

$$\tfrac{1}{3}\,e_1\,e_2\,(e_1\,r_1 + 2\,m\,n + 2\,m_1\,n_1 + e_3\,r_3),$$

but in the trapezoid $m\,m_1\,n_1\,n$,

$$2\,m\,n + 2\,m_1\,n_1 = 4\,e_2\,r_2, \quad \text{very nearly};$$

whence the area becomes

$$\tfrac{1}{3}\,e_1\,e_2\,(e_1\,r_1 + 4\,e_2\,r_2 + e_3\,r_3);$$

the area of the next two trapezoids in order, of the preceding figure, will be

$$\tfrac{1}{3}\,e_1\,e_2\,(e_3\,r_3 + 4\,e_4\,r_4 + e_5\,r_5);$$

and similar expressions for each succeeding pair of trapezoids Taking the sum of these, and we have the whole area bounded by the curve, its extreme ordinates, and the axis of abscisses; or,

$$Q = \tfrac{1}{3}e_1e_2\left[e_1r_1 + 4\,e_2r_2 + 2\,e_3\,r_3 + 4\,e_4r_4 + 2e_5r_5 + 4\,e_6r_6 + e_7r_7\right]. \quad (10)$$

whence the rule.

§ 49.—By the processes now explained, it is easy to estimate the quantity of work of the weights of bodies, of the resistances due to the forces of affinity which hold their elements together, of their elasticity, &c. It remains to consider the rules by which the quantity of work of inertia may be computed. Inertia is exerted only during a change of state in respect to motion or rest, and this brings us to the subject of varied motion.

VARIED MOTION.

§ 50.—Varied motion has been defined to be that in which unequal spaces are described in equal successive portions of time. In this kind of motion the velocity is ever varying. It is measured at any given instant by the length of path it would enable a body to describe in the first subsequent unit of time, were it to remain unchanged. Denote the space described by s, and the time of its description by t.

However variable the motion, the velocity may be regarded as constant during the indefinitely small time, dt. In this time the body will describe the small space ds; and as this space is described uniformly, the space described in the unit of time would, were the velocity constant, be ds repeated as many times as the unit of time contains dt. Hence, denoting the value of the velocity at any instant by v, we have

$$v = ds \times \frac{1}{dt};$$

or,

$$v = \frac{ds}{dt} \quad . \quad . \quad . \quad . \quad . \quad . \quad . \quad . \quad . \quad . \quad (11)$$

§ 51.—Continual variation in a body's velocity can only be produced by the incessant action of some force. The body's inertia opposes an equal and contrary reaction. This reaction is directly proportional to the mass of the body and to the amount of change in its velocity; it is, therefore, directly proportional to the product of the mass into the increment or decrement of the velocity. The product of a mass into a velocity, represents *a quantity of motion.*

The intensity of a motive force, at any instant, is assumed to be measured by the quantity of motion which this intensity can generate in a unit of time.

The mass remaining the same, the velocities generated in equal successive portions of time, by a constant force, must be equal to each other. However a force may vary, it may be regarded as constant during the indefinitely short interval dt; in this time it will generate a velocity dv, and were it to remain constant, it would generate in a unit of time, a velocity equal to dv repeated as many times as dt is contained in this unit; that is, the velocity generated would be equal to

$$dv \cdot \frac{1}{dt} = \frac{dv}{dt};$$

and denoting the intensity of the force by P, and the mass by M, we shall have

$$P = M.\frac{dv}{dt}. \quad . \quad . \quad . \quad . \quad . \quad . \quad . \quad . \quad (12)$$

Again, differentiating Equation (11), regarding t as the independent variable, we get,

$$dv = \frac{d^2s}{dt};$$

and this, in Equation (12), gives

$$P = M.\frac{d^2s}{dt^2}. \quad . \quad . \quad . \quad . \quad . \quad . \quad . \quad . \quad (13)$$

From Equation (11), we conclude that in varied motion, *the velocity at any instant is equal to the first differential co-efficient of the space regarded as a function of the time.*

From Equation (12), that the intensity of any motive force, or of the inertia it develops, at any instant, is measured by the *product of the mass into the first differential co-efficient of the velocity regarded as a function of the time.*

And from Equation (13), that the intensity of the motive force or of inertia, is measured by the *product of the mass into the second differential co-efficient of the space regarded as a function of the time.*

§ 52.—To illustrate. Let there be the relation

$$s = at^3 + bt^2 \quad . \quad . \quad . \quad . \quad . \quad . \quad . \quad . \quad (14)$$

required the space described in three seconds, the velocity at the end of the third second, and the intensity of the motive force at the same instant.

Differentiating Equation (14) twice, dividing each result by dt, and multiplying the last by M, we find

$$\frac{ds}{dt} = v = 3at^2 + 2bt \quad . \quad . \quad . \quad . \quad (15)$$

$$M \cdot \frac{d^2s}{dt^2} = P = M[6at + 2b] \quad . \quad . \quad . \quad (16)$$

Make $a = 20$ feet, $b = 10$ feet, and $t = 3$ seconds, we have, from Equations (14), (15), and (16),

$$s = 20 . 3^3 + 10 . 3^2 = 630 \text{ feet};$$

$$v = 3 . 20 . 3^2 + 2 . 10 . 3 = 600 \text{ feet};$$

$$P = M(6 . 20 . 3 + 2 . 10) = 380 . M.$$

That is to say, the body will move over the distance 630 feet in three seconds, will have a velocity of 600 feet at the end of the third second, and the force will have at that instant an intensity capable of generating in the mass M, a velocity of 380 feet in one second, were it to retain that intensity unchanged.

§ 53.—Dividing Equations (12) and (13) by M, they give

$$\frac{P}{M} = \frac{dv}{dt} \quad . \quad . \quad . \quad . \quad . \quad . \quad . \quad . \quad (17)$$

$$\frac{P}{M} = \frac{d^2s}{dt^2} \quad . \quad . \quad . \quad . \quad . \quad . \quad . \quad . \quad (18)$$

The first member is the same in both, and it is obviously that portion of the force's intensity which is impressed upon the unit of mass. The second member in each is the velocity impressed in the unit of time, and is called the *acceleration* due to the motive force.

§ 54.—From Equation (11) we have,

$$ds = v \,.\, dt \quad . \quad . \quad . \quad . \quad . \quad . \quad . \quad (19)$$

multiplying this and Equation (12) together, there will result,

$$P \,.\, ds = M \,.\, v \,.\, dv \quad . \quad . \quad . \quad . \quad (20)$$

and integrating,

$$\int P \,.\, ds = \frac{M \,.\, v^2}{2} \quad . \quad . \quad . \quad . \quad . \quad . \quad (21)$$

The first member is the quantity of work of the motive force, which is equal to that of inertia; the product $M . v^2$, is called the *living force* of the body whose mass is M. Whence, we see that the *work of inertia is equal to half the living force;* and the living force of a body is *double the quantity of work expended by its inertia while it is acquiring its velocity.*

§ 55.—If the force become constant and equal to F, the motion will be *uniformly varied,* and we have, from Equation (18),

$$\frac{F}{M} = \frac{d^2s}{dt^2}.$$

Multiplying by dt and integrating, we get

$$\frac{F}{M} \cdot t = \frac{ds}{dt} + C = v + C \quad . \quad . \quad (22)$$

and if the body be moved from rest, the velocity will be equal to zero when t is zero; whence $C = 0$, and

$$\frac{F}{M} \cdot t = v \quad . \quad . \quad . \quad . \quad . \quad . \quad . \quad . \quad . \quad . \quad (23)$$

Multiplying Equation (22) by dt, after omitting C from it, and integrating again, we find

$$\frac{F}{M} \cdot \frac{t^2}{2} = s + C',$$

and if the body start from the origin of spaces, C' will be zero, and

$$\frac{F}{M} \cdot \frac{t^2}{2} = s \quad . \quad . \quad . \quad . \quad . \quad . \quad . \quad . \quad . \quad (24)$$

Making t equal to one second, in Equations (24) and (23), and dividing the last by the first, we have

$$\frac{1}{2} = \frac{s}{v},$$

or,

$$v = 2s \quad . \quad . \quad . \quad . \quad . \quad . \quad . \quad . \quad . \quad (25)$$

That is to say, the *velocity generated in the first unit of time is measured by double the space described in acquiring this velocity.* Equations (23), (24), and (25) express the laws of constant forces.

§ 56.—The dynamical measure for the intensity of a force, or the pressure it is capable of producing, is assumed to be the effect this pressure can produce in a unit of time, this effect being a quantity of motion, measured by the product of the mass into the velocity generated. This assumed measure must not be confounded with the quantity of work of the force while producing this effect. The former is the measure of a single pressure; the latter, this pressure repeated as many times as there are points in the path over which this pressure is exerted.

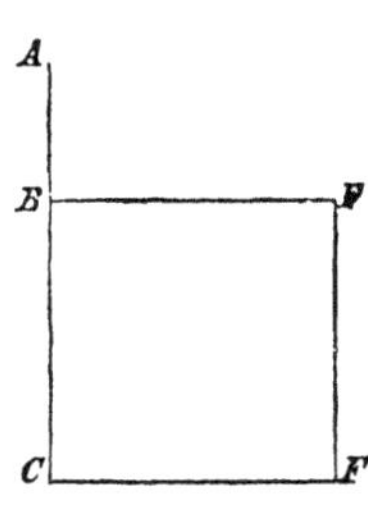

Thus, let the body be moved from A to B, under the action of a constant force, in one second; the velocity generated will, Equation (25), be $2AB$. Make $BC = 2AB$, and complete the square $BCFE$. BE will be equal to v; the intensity of the force will be $M.v$; and the quantity of work, the product of $M.v$ by AB, or by its equal $\frac{1}{2}v$; thus making the quantity of work $\frac{1}{2}Mv^2$, or the mass into one half the square BF; which agrees with the result obtained from Equation (21).

EQUILIBRIUM.

§ 57.—*Equilibrium* is a term employed to express the state of two or more forces which balance one another through the intervention of some body subjected to their simultaneous action. When applied to a body, it means that the body is at rest.

We must be careful to distinguish between the extraneous forces which act upon a body, and the forces of inertia which they may, or may not, develop.

If a body subjected to the simultaneous action of several extraneous forces, be at rest, or have uniform motion, the extraneous forces are in equilibrio, and the force of inertia is not developed. If the body have varied motion, the extraneous forces are not in equilibrio, but develop forces of inertia which, with the extraneous forces, are in equilibrio. Forces, therefore, including the force of inertia, are ever in equilibrio; and the indication of the presence or absence of the force of inertia, in any case, shows that the body is or is not changing its condition in respect to rest or motion. This is but a consequence of the universal law that every *action* is accompanied by an equal and contrary *reaction.*

THE CORD.

§ 58.—A *cord* is a collection of material points, so united as to form one continuous and flexible line. It will be considered, in what immediately follows, as perfectly *flexible*, *inextensible*, *and without thickness or weight.*

§ 59.—By the *tension* of a cord is meant, the effort by which any two of its adjacent particles are urged to separate from each other.

§ 60.—Two equal forces, P and P', applied at the extremities A, A' of a straight cord, and acting in opposite directions from its middle point, will maintain each other in equilibrio. For, all the points

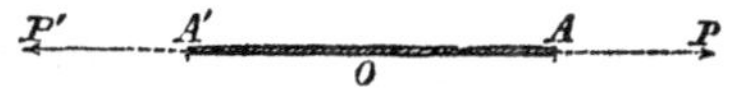

of the cord being situated on the line of direction of the forces, any one of them, as O, may be taken as the common point of application without altering their effects; but in this case, the forces being equal will, § 34, neutralize each other.

§ 61.—If two equal forces, P and P', solicit in opposite directions the extremities of the cord $A\,A'$, the tension of the cord will be measured by the intensity of one of the forces. For, the cord being in this case in equilibrio, if we suppose any one of its points as O, to become fixed, the equilibrium will not be disturbed, while all communication between the forces will be intercepted, and either force may be destroyed without affecting the other, or the part of the cord on which it acts. But if the part $A\,O$ of the cord be attached to a fixed point at O, and drawn by the force P alone, this force must measure the tension.

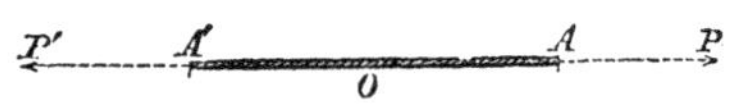

THE MUFFLE.

§ 62.—Suppose A, A', B, B', &c., to be several small wheels or pulleys *perfectly* free to move about their centres, which, conceive for the present to be fixed points. Let one end of a cord be fastened to a fixed point C, and be wound around the pulleys as represented in the figure; to the other extremity, attach a weight w. The weight w will be maintained in equilibrio by the resistance of the fixed point C, through the medium of the cord. The tension of the cord will be the same throughout its entire length, and equal to the weight w; for, the cord being perfectly flexible, and the wheels perfectly free to move about their centres, there is nothing to intercept the free transmission of tension from one end to the other.

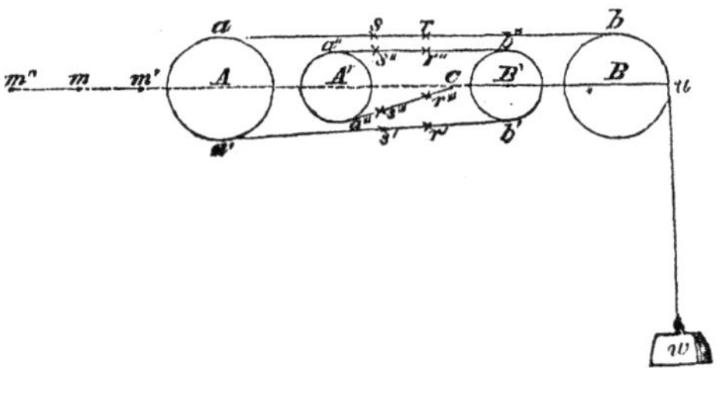

Let the points s and r of the cord be supposed for a moment, fixed; the intermediate portion $s\,r$ may be removed without affecting the tension of the cord, or the equilibrium of the weight w. At the point r, apply in the direction from r to a, a force whose intensity is equal to the tension of the cord, and at s an equal force acting in the direction from s to b; the points r and s may now be regarded as free. Do the same at the points s', r', s'', r'', s''' and r''', and the action of the weight w, upon the pulleys A and A' will be replaced by the four forces at s, s', s'' and s''', all of equal intensity and acting in the same direction.

Now, let the centres of the pulleys A and A' be firmly connected with each other, and with some other fixed point as m, in the direction of BA produced, and suppose the pulleys diminished indefinitely, or reduced to their centres. Each of the points A and A' will be solicited in the same direction, and along the same line, by a force equal to $2w$, and therefore the point m, by a force equal to $4w$.

Had there been six pulleys instead of four, the point m would have been solicited by a force equal to $6w$, and so of a greater number. That is to say, the point m would have been solicited by a force equal to w, repeated as many times as there are pulleys.

If the extremity C of the cord had been connected with the point m, after passing round a fifth pulley at C, the point m would have been subjected to the action of a force equal to $5w$; if seven pulleys had been employed, it would have been urged by a force $7w$; and it is therefore apparent, that the intensity of the force which solicits the point m, is found by *multiplying the tension of the cord, or weight w, by the number of pulleys.*

This combination of the cord with a number of wheels or pulleys, is called a *muffle.*

§ 63.—Conceive the point m to be transferred to the position m' or m'', on the line AB. The centres of the pulleys A, A', &c., being invariably connected with the point m, will describe equal paths, and each equal to $m m'$, or $m m''$, so that each of the parallel portions of the cord will be shortened in the first case, or lengthened in the second, by equal quantities; and if e denote the length of the path described by m, n the number of parallel portions of the cord, which is equal to the number of pulleys, and ξ, the change in length of the portion uw in consequence of the motion of m, we shall have, because the entire length of the cord remains the same,

$$n \cdot e = \xi \quad . \quad . \quad . \quad . \quad . \quad . \quad (26)$$

The first member of this equation we shall refer to as the *change in length of cord on the pulleys.*

§ 64.—The action of any force P, upon a material point, may be replaced by that of a muffle, *by making the tension of its cord equal to the intensity of the given force, divided by the number of parallel portions of the cord, or number of pullies.*

VIRTUAL VELOCITIES.

§ 65.—Let M represent a collection of material points, united in any manner whatever, forming a solid body, and subjected to the action of several forces, P, P', P'', P''', &c.; and suppose these forces in equilibrio.

Find the greatest force w, which will divide each of the given forces without a remainder; replace the force P by a *muffle*, having

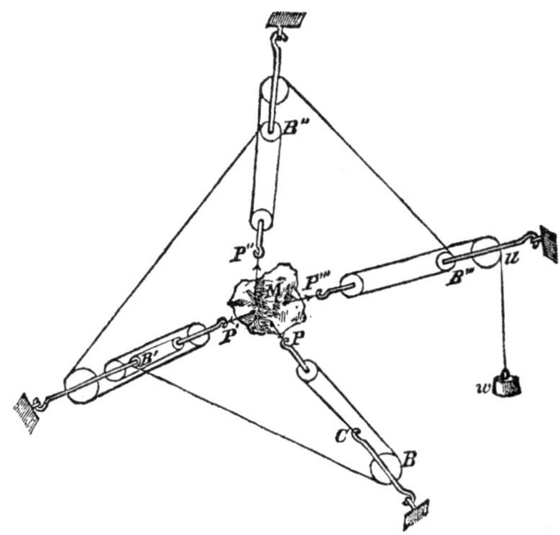

a number of pulleys denoted by $\frac{P}{w}$; the tension of the cord will be denoted by w. Do the same for each of the forces, and we shall have as many muffles as there are forces, and all the cords will have the same tension.

Let the several cords be united into one, as represented in the figure, one end being attached at C, the other acted upon by a weight equal to the force w. The action upon the body will remain unchanged; that is, the substituted forces, including w, will be in equilibrio.

In this state of the system, let a force Q be applied to put the body in motion, and at the instant motion begins, withdraw this force and stop the motion before the equilibrium of the forces is destroyed. The points of application of the original forces will each have described an indefinitely small path, as mn. Let mr be the projection of this path upon the original direction of the force, and denote the length of this projection by e. Join the point n with any point o, on the direction of the force and at some definite distance from m. From the triangle onr, we have

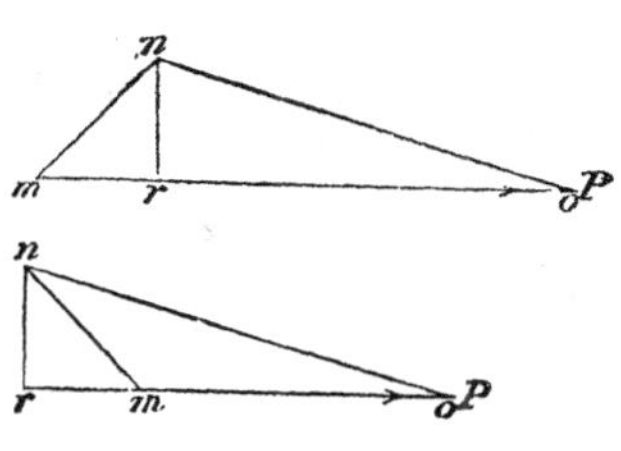

$$\overline{on}^2 = \overline{or}^2 + \overline{nr}^2;$$

the displacement being indefinitely small, $\overline{nr}^2$ may be neglected in comparison with $\overline{or}^2$, being an indefinitely small quantity of the second order; hence,

$$on = or,$$

and,

$$om - on = om - or = e.$$

But the number of pulleys in the muffle which acts along the direction of the force P is,

$$\frac{P}{w};$$

hence, the change in the length of the cord on the pulleys of this muffle, caused by the slight motion of the point of application of the force P, will, since the centre of the pulley B is fixed, be

$$\frac{P \cdot e}{w};$$

and denoting by e', e'', e''', &c., the projections of the paths described by the points to which the forces P', P'', P''', &c., are respectively applied, on the original directions of these forces, we shall have

$$\frac{P' \cdot e'}{w}, \quad \frac{P'' \cdot e''}{w}, \quad \frac{P''' \cdot e'''}{w}, \text{ \&c.,}$$

for the corresponding changes in the length of the cord on the other muffles.

In all these changes, the cord being inextensible, its entire length remains the same, and if the change in length which the portion uw undergoes be denoted by ξ, we shall have

$$\frac{1}{w}(P \cdot e + P' \cdot e' + P'' \cdot e'' + P''' \cdot e''' + \text{\&c.}) + \xi = 0 \quad . \quad . \quad (27)$$

This equation expresses the algebraic sum of all the changes in the lengths of the several parts of the cord, between the points of application, and the fixed points towards which the points of application are solicited; the effect of these changes being to shorten some and lengthen others, some of the terms of Equation (27) must be negative.

Now it is one of the essential properties of a system of forces in equilibrio, to leave a body subjected to their action as free to move as though these forces did not exist. The additional force Q, therefore, was wholly employed in developing the inertia of the body M; it was neither assisted nor opposed by the forces represented by the action of the muffles, because these forces balanced each other, and the motion was arrested before the points of application were sufficiently disturbed to break up the equilibrium; nor, reciprocally, § 35, was the action of the muffles, nor the tension of the cord which produced this action, affected by Q. Hence the tension of the cord was invariable during the disturbance. But an invariable tension must have kept the weight w at rest during the displacement, and we have

$$\xi = 0,$$

and Equation (27) will reduce to,

$$Pe + P'e' + P''e'' + P'''e''' + \&c. = 0 ; \quad . \quad . \quad . \quad . \quad (28)$$

§ 66.—It may be objected, that the given forces are incommensurable, and that therefore, a force cannot be found which will divide each without a remainder; to which it is answered, that Equation (28), being perfectly independent of the value of the weight w, or tension of the cord, this weight may be taken so small as to render the remainder after division in any particular case, perfectly inappreciable.

§ 67.—The indefinitely small paths $m\,n$, $m'n'$, described by the points of application of the forces, P and P', during the slight motion we have supposed, are called *virtual velocities;* and they are so called, because, being the actual distances passed over by the points to which the forces are applied, in the same time, they measure the relative rates of motion of these points. The distances $r\,m$ and $r'm'$, represented by e and e', are therefore, the projections of the virtual velocities upon the directions of the forces. These projections may fall on the side towards which the forces tend to urge these points, or the reverse, depending upon the direction of the motion imparted to the system. In the first case, the projections are regarded as *positive*, and in the second, as *negative*. Thus, in the case taken for illustration, $m\,r$ is positive, and $m'r'$ negative. The products $P\,e$ and $P'e'$, are called *virtual moments*. They are the elementary quantities of work of the forces P and P'. The forces are always regarded as positive; the sign of a virtual moment will, therefore, depend upon that of the projection of the virtual velocity.

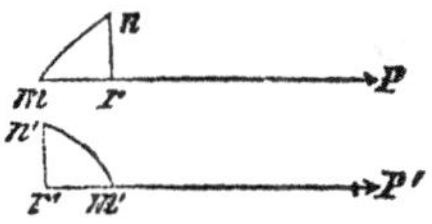

§ 68.—Referring to Equation (28), we conclude, therefore, that *whenever several forces are in equilibrio, the algebraic sum of their virtual*

moments is equal to zero; and in this consists what is called the principle of virtual velocities.

§ 69.—Conversely, if in any system of forces, the algebraic sum of the virtual moments be equal to zero, the forces will be in equilibrio. For, if they be not in equilibrio, some, if not all the points of application will have a motion. Let q, q', q'', &c., be the projections of the paths which these points describe in the first instant of time, and Q, Q', Q'', &c., the intensities of such forces as will, when applied to these points in a direction opposite to the actual motions, produce an equilibrium. Then, by the principle of virtual velocities, we shall have

$$Pe + P'e' + P''e'' + \&c. + Qq + Q'q' + Q''q'' + \&c. = 0$$

But by hypothesis,

$$Pe + P'e' + P''e'' + \&c. = 0,$$

and hence,

$$Qq + Q'q' + Q''q'' + \&c. = 0 \quad . \quad . \quad . \quad (28)'$$

Now, the forces Q, Q', Q'', &c., have each been applied in a direction contrary to the actual motion; hence, all the virtual moments in Equation (28)′ will have the negative sign; each term must, therefore, be equal to zero, which can only be the case by making Q, Q', Q'', &c., separately equal to zero, since by supposition the quantities denoted by q, q', q'', are not so. We therefore conclude, that when the algebraic sum of the virtual moments of a system of forces is equal to zero, the forces will be in equilibrio.

Whatever be its nature, the effect of a force will be the same if we attribute its effort to attraction between its point of application and some remote point assumed arbitrarily and as fixed upon its line of direction, the intensity of the attraction being equal to that of the force. Denote the distance from the point of application of P, to that towards which it is attracted, by p, and the corresponding distances in the case of the forces P', P'', &c., by p', p'', &c., respectively; also, let δp, $\delta p'$, $\delta p''$, &c., represent the augmentation or diminution of these distances caused by the displacement, supposed indefinitely small, then § 65, will

$$e = \delta p, \; e' = \delta p', \; e'' = \delta p'', \; \&c.,$$

and Equation (28) may be written

$$P\delta p + P'\delta p' + P''\delta p'' + \&c. = 0 \quad . \quad . \quad . \quad . \quad (29)$$

in which the Greek letter δ simply denotes change in the value of the letter written immediately after it, this change arising from the small displacement.

§ 70.—If the extraneous forces applied to a body be not in equilibrio, they will communicate motion to it, and will develop forces of inertia in its various elementary masses with which they will be in equilibrio; and if extraneous forces equal in all respects to these forces of inertia were introduced into the system, the algebraic sum of the virtual moments would be equal to zero.

But if m denote the mass of any element of the body, s the path it describes, its force of inertia will, Eq. (13), be

$$m \, . \, \frac{d^2s}{dt^2};$$

and denoting the projection of its virtual velocity on s by δs, its virtual moment will be

$$m \, . \, \frac{d^2s}{dt^2} \, . \, \delta s \, .$$

and because the forces of inertia act in opposition to the extraneous forces, their virtual moments must have signs contrary to those of the latter, and Equation (29) may be written

$$\Sigma P . \delta p - \Sigma m \, . \, \frac{d^2s}{dt^2} \, . \, \delta s = 0 \, ; \quad . \quad . \quad . \quad . \quad (30),$$

in which Σ denotes the algebraic sum of the terms similar to that written immediately after it.

PRINCIPLE OF D'ALEMBERT.

§ 71.—This simple equation involves the whole doctrine of Mechanics. The extraneous forces P, P', P'', &c., are called impressed forces. The forces of inertia which they develop may or may not be equal to them, depending upon the manner of their application. If the impressed forces be in equilibrio, for instance, they will develop no force of inertia;

but in all cases, the forces of inertia developed will be equal and contrary to so much of the impressed forces as determines the change of motion. The portions of the impressed forces which determine a change of motion are called *effective forces;* and from Equation (30), we infer that the impressed and effective forces are always in equilibrio when the directions of the latter are reversed. This is usually known as *D'Alembert's Principle*, and is nothing more than a plain consequence of the law that action and reaction are ever equal and contrary.

This same principle is also enunciated in another way. Since the effective forces reversed would maintain the impressed forces in equilibrio, and prevent them from producing a change of motion, it follows *that whatever forces may be lost and gained must be in equilibrio;* else a motion different from that which actually takes place must occur.

REFERENCE TO CO-ORDINATE AXES.

§ 72.—*First Transformation.* Equation (30) is of a form too general for easy discussion, and may be simplified by referring the forces and motions to rectangular axes.

Denote by α, β, γ, the angles which the direction of the force P makes with the axes x, y, z, respectively; by a, b, c, the angles which its virtual velocity makes with the same axes; and by φ, the angle which the virtual velocity and direction of the force make with each other, then will

$$\cos \varphi = \cos a \,.\, \cos \alpha + \cos b \,.\, \cos \beta + \cos c \,.\, \cos \gamma.$$

Denote by k, the virtual velocity, and multiply the above equation by Pk, and we have

$$Pk \cos \varphi = Pk \cos a \,.\, \cos \alpha + Pk \cos b \,.\, \cos \beta + Pk \cos c \,.\, \cos \gamma;$$

But denoting the co-ordinates of the point of application of P by x, y, z, we have

$$k \cos \varphi = \delta p \; ; \quad k \cos a = \delta x \; ; \quad k \cos b = \delta y \; ; \quad k \cos c = \delta z \; ;$$

and these values substituted above, give

$$P \,.\, \delta p = P \cos \alpha \,.\, \delta x + P \cos \beta \,.\, \delta y + P \cos \gamma \,.\, \delta z. \quad . \quad . \; (31).$$

Similar values may be found for the virtual moments of other forces.

§ 73.—If P be replaced by the force of inertia, then will α, β, and γ denote the inclinations of the direction of this force to the axes $x y z$; k its virtual velocity; a, b, and c the inclinations of the latter to the axes, and ϕ its inclination to the direction of the force of inertia, and we may, Eq. (13), write

$$m \cdot \frac{d^2 s}{d t^2} \cdot k \cos \phi = m \frac{d^2 s}{d t^2} \cdot \cos \alpha . k \cos a + m \frac{d^2 s}{d t^2} \cos \beta . k \cos b + m \frac{d^2 s}{d t^2} \cos \gamma . k \cos c.$$

But

$$k \cos \phi = \delta s ; \quad k \cos a = \delta x ; \quad k \cos b = \delta y ; \quad k \cos c = \delta z ;$$

$$d^2 s . \cos \alpha = d^2 x ; \quad d^2 s \cos \beta = d^2 y ; \quad d^2 s \cos \gamma = d^2 z ;$$

whence,

$$m \cdot \frac{d^2 s}{dt^2} \cdot \delta s = m \cdot \frac{d^2 x}{dt^2} \cdot \delta x + m \cdot \frac{d^2 y}{dt^2} \cdot \delta y + m \cdot \frac{d^2 z}{dt^2} \cdot \delta z \, ; \quad \cdot \quad (32),$$

and similar expressions may be found for the virtual moments of the forces of inertia of the other elementary masses.

§ 74.—If the intensity of the force P, be represented by a portion of its line of direction, which is the practice in all geometrical illustrations of Mechanics, the factors $P \cos \alpha$, $P \cos \beta$, and $P \cos \gamma$, in Equation (31), would represent the intensities of forces equal to the projections of the intensity P, on the axes; and regarding these as acting in the directions of the axes, the factors δx, δy, and δz, will represent the projections of their virtual velocities, which virtual velocities will coincide with that of the force P.

Again, Equation (32),

$$m \cdot \frac{d^2 x}{dt^2}, \quad m \cdot \frac{d^2 y}{dt^2}, \quad m \cdot \frac{d^2 z}{dt^2},$$

are forces of inertia in the directions of the axes, and δx, δy, δz, are the projections of their virtual velocities; these virtual velocities coincide with that of the inertia of m.

The values of these virtual velocities depend upon the nature of the displacement.

FREE MOTION OF A RIGID SYSTEM.

§ 75.—*Second Transformation.* By the substitution, in Equation (30), for $P\,\delta p$ and $m\,.\,\frac{d^2 s}{d\,t^2}\,.\,\delta\,s$, their values in Equations (31) and (32), there would result an equation containing, in general, three times as many variations of $x\,y\,z$ as there are extraneous forces and elementary masses, m. Where the forces are applied to a body whose elementary masses are invariably connected—that is, to a rigid solid—the number of these variations is greatly reduced, in consequence of the relations determined by this connection.

The most general motion we can attribute to a body is one of translation and of rotation combined. A motion of translation carries a body from place to place through space, and its position, at any instant, is determined by that of some one of its elements. A motion of rotation carries the elements of a body around some assumed point. In this investigation, let this point be that which determines the body's place.

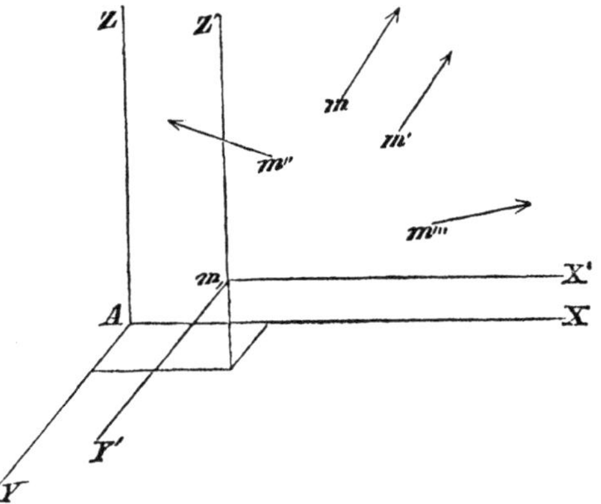

Denote its co-ordinates by $x_{\prime}\ y_{\prime}\ z_{\prime}$ and those of the element m, referred to this point as an origin by x', y', z'; there will thus be two sets of axes, and supposing them parallel, we have

$$\left.\begin{aligned} x &= x_{\prime} + x', \\ y &= y_{\prime} + y', \\ z &= z_{\prime} + z'; \end{aligned}\right\} \quad \dots\dots \quad (33),$$

and differentiating,

$$\left.\begin{aligned} dx &= dx_{\prime} + dx', \\ dy &= dy_{\prime} + dy', \\ dz &= dz_{\prime} + dz'. \end{aligned}\right\} \quad \dots\dots \quad (34).$$

Demit from m, the perpendiculars mX', mY', mZ', upon the movable axes. Denote the first by r', the second by r'', and the third by r'''. Let O', O'', O''', be the projections of m, on the planes $x\,y$, $x\,z$, $y\,z$, respectively. Join the several points by right lines as indicated in the figure.

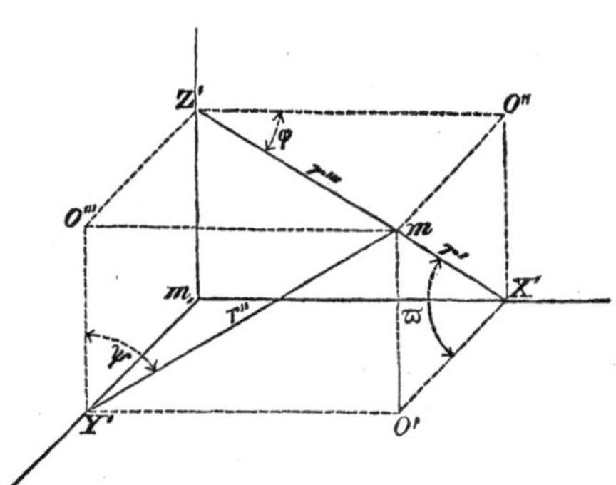

Denote the angle

$$m\,Z'\,O'' \text{ by } \varphi,$$
$$m\,X'\,O' \text{ by } \varpi,$$
$$m\,Y'\,O''' \text{ by } \psi.$$

Then will

the triangle $m\,Z'\,O''$ give $\left\{\begin{array}{l} x' = r''' \cos\varphi, \\ y' = r''' \sin\varphi, \end{array}\right\}$ . . . . . (35),

the triangle $m\,Y'\,O'''$, $\left\{\begin{array}{l} x' = r'' \sin\psi, \\ z' = r'' \cos\psi, \end{array}\right\}$ . . . . . (36),

the triangle $m\,X'\,O'$, $\left\{\begin{array}{l} y' = r' \cos\varpi, \\ z' = r' \sin\varpi, \end{array}\right\}$ . . . . . (37).

We here have two values of x', one dependent upon φ, and the other upon ψ. If the body be turned through an indefinitely small angle about the axis z', the corresponding increment of x' is obtained by differentiating the first of Equations (35); and we have

$$d\,x' = -r''' \sin\varphi\,.\,d\,\varphi;$$

if it be turned through a like angle about the axis y', the corresponding increment of x' is found by differentiating the first of Equations (36), and

$$d\,x' = r'' \cos\psi\,.\,d\,\psi.$$

If these motions take place simultaneously about both axes, the above become partial differentials of x', and we have for its total differential,

$$d\,x' = r'' \cos\psi\,.\,d\,\psi - r''' \sin\varphi\,.\,d\,\varphi,$$

replacing $r'' \cos \psi$ and $r''' \sin \varphi$, by their values in the above Equations, and we get

$$\left.\begin{array}{l} d x' = z' . d \psi - y' . d \varphi ; \\ \text{and in the same way,} \\ d y' = x' . d \varphi - z' . d \varpi, \\ d z' = y' . d \varpi - x' . d \psi, \end{array}\right\} \quad . \; . \; . \; (38)$$

which substituted in Equations (34), give

$$\left.\begin{array}{l} d x = d x_{,} + z' . d \psi - y' . d \varphi, \\ d y = d y_{,} + x' . d \varphi - z' . d \varpi, \\ d z = d z_{,} + y' . d \varpi - x' . d \psi. \end{array}\right\} \quad . \; . \; . \; (39)$$

and because the displacement is indefinitely small, we may write

$$\left.\begin{array}{l} \delta x = \delta x_{,} + z' . \delta \psi - y' . \delta \varphi, \\ \delta y = \delta y_{,} + x' . \delta \varphi - z' . \delta \varpi, \\ \delta z = \delta z_{,} + y' . \delta \varpi - x' . \delta \psi ; \end{array}\right\} \quad . \; . \; . \; (39)'$$

and these in Equations (31) and (32), give

$$P . \delta p = \left\{\begin{array}{l} P \cos \alpha . \delta x_{,} + P \cos \beta . \delta y_{,} + P \cos \gamma . \delta z_{,} \\ + P . (x' . \cos \beta - y' . \cos \alpha) . \delta \varphi \\ + P . (z' . \cos \alpha - x' . \cos \gamma) . \delta \psi \\ + P . (y' . \cos \gamma - z' . \cos \beta) . \delta \varpi. \end{array}\right\}$$

$$m \cdot \frac{d^2 s}{dt^2} \cdot \delta s = \left\{\begin{array}{l} m \cdot \frac{d^2 x}{dt^2} \cdot \delta x_{,} + m \cdot \frac{d^2 y}{dt^2} \cdot \delta y_{,} + m \cdot \frac{d^2 z}{dt^2} \cdot \delta z_{,} \\ + m \cdot \frac{x' . d^2 y - y' . d^2 x}{dt^2} \cdot \delta \varphi \\ + m \cdot \frac{z' . d^2 x - x' . d^2 z}{dt^2} \cdot \delta \psi \\ + m \cdot \frac{y' . d^2 z - z' . d^2 y}{dt^2} \cdot \delta \varpi. \end{array}\right\}$$

Similar values may be found for $P' . \delta p'$ and $m' . \frac{d^2 s'}{dt^2} \cdot \delta s'$, &c. In these values $\delta x_{,}$, $\delta y_{,}$ and $\delta z_{,}$, will be the same, as also $\delta \varphi$, $\delta \psi$, and $\delta \varpi$, for the first relate to the movable origin, and the latter to the angular rotation which, since the body is a solid, must be of equal values for all the elements; so that to find the values of the virtual moments of the other forces, it will be only necessary suitably to accent P, α, β, γ, x, y, z, x', y', z'.

These values being found and substituted in Equation (30), we shall find,

$$\left.\begin{aligned}
&\left(\Sigma P . \cos\alpha - \Sigma m \cdot \frac{d^2x}{dt^2}\right) \delta x_{,} \\
&+ \left(\Sigma P . \cos\beta - \Sigma m \cdot \frac{d^2y}{dt^2}\right) \delta y_{,} \\
&+ \left(\Sigma P . \cos\gamma - \Sigma m \cdot \frac{d^2z}{dt^2}\right) \delta z_{,} \\
&+ \left[\Sigma P . (x' . \cos\beta - y' . \cos\alpha) - \Sigma m \cdot \frac{x' . d^2y - y' . d^2x}{dt^2}\right] \delta\varphi \\
&+ \left[\Sigma P . (z' . \cos\alpha - x' . \cos\gamma) - \Sigma m \cdot \frac{z' . d^2x - x' . d^2z}{dt^2}\right] \delta\psi \\
&+ \left[\Sigma P . (y' . \cos\gamma - z' . \cos\beta) - \Sigma m \cdot \frac{y' . d^2z - z' . d^2y}{dt^2}\right] \delta\varpi
\end{aligned}\right\} = 0 . \quad (40)$$

Now the displacement was wholly arbitrary; the values of $\delta x_{,}$, $\delta y_{,}$, $\delta z_{,}$, $\delta\varphi$, $\delta\psi$, $\delta\varpi$, determined by this displacement, are also arbitrary; whence, by the principle of indeterminate co-efficients,

$$\left.\begin{aligned}
\Sigma P . \cos\alpha - \Sigma m \cdot \frac{d^2x}{dt^2} &= 0, \\
\Sigma P . \cos\beta - \Sigma m \cdot \frac{d^2y}{dt^2} &= 0, \\
\Sigma P . \cos\gamma - \Sigma m . \frac{d^2z}{dt^2} &= 0;
\end{aligned}\right\} \quad . \; . \; . \; . \; . \; (A)$$

$$\left.\begin{aligned}
\Sigma P . (x' . \cos\beta - y' . \cos\alpha) - \Sigma m \cdot \frac{x' . d^2y - y' . d^2x}{dt^2} &= 0, \\
\Sigma P . (z' . \cos\alpha - x' . \cos\gamma) - \Sigma m \cdot \frac{z' . d^2x - x' . d^2z}{dt^2} &= 0, \\
\Sigma P . (y' . \cos\gamma - z' . \cos\beta) - \Sigma m \cdot \frac{y' . d^2z - z' . d^2y}{dt^2} &= 0.
\end{aligned}\right\} \quad . \; . \; (B)$$

§ 76.—These six equations express either all the circumstances of motion attending the action of forces, or all the circumstances of equilibrium of the forces, according as inertia is or is not brought into action; and the study of the principles of Mechanics is little else than an attentive consideration of the conclusions which follow from their discussion.

Equations (A) relate to a motion of translation, and Equations (B) to a motion of rotation. They are perfectly symmetrical and may be memorized with great ease.

COMPOSITION AND RESOLUTION OF FORCES.

§ 77.—When a free body is subjected to the simultaneous action of several extraneous forces which are not in equilibrio, its state will be changed; and if this change may be produced by the action of a single force, this force is called the *resultant*, and the several forces are termed *components*.

The *resultant* of several forces is *a single force which, acting alone, will produce the same effect as the several forces acting simultaneously;* and the *components* of a single force, are *several forces whose simultaneous action produces the same effect as the single force.*

If, then, several extraneous forces applied to a body, be not in equilibrio, but have a resultant, a single force, equal in intensity to this resultant, and applied so as to be immediately opposed to it, will produce an equilibrium; or, what amounts to the same thing, if in any system of extraneous forces in equilibrio, the resultant of all the forces but one be found, this resultant will be equal in intensity and immediately opposed to the remaining force; otherwise the system could not be in equilibrio.

Conceive a system of extraneous forces, not in equilibrio, and applied to a solid body, and suppose that the equilibrium may be produced by the introduction of an additional extraneous force. Denote the intensity of this force by R, the angles which its direction makes with the axes x, y and z, by a, b and c, respectively, and the co-ordinates of its point of application by x, y, z. Then, because the inertia cannot act, d^2x, d^2y, d^2z will be zero, and taking the two origins to coincide, Equations (A) and (B), will give

$$R\cos a + P'\cos\alpha' + P''\cos\alpha'' + P'''\cos\alpha''' + \&c. = 0,$$
$$R\cos b + P'\cos\beta' + P''\cos\beta'' + P'''\cos\beta''' + \&c. = 0,$$
$$R\cos c + P'\cos\gamma' + P''\cos\gamma'' + P'''\cos\gamma''' + \&c. = 0;$$

$$\left.\begin{array}{l} R(x\cos b - y\cos a) + P'(x'\cos\beta' - y'\cos\alpha') \\ \quad + P''(x''\cos\beta'' - y''\cos\alpha'') + \&c. \end{array}\right\} = 0,$$

$$\left.\begin{array}{l} R(z\cos a - x\cos c) + P'(z'\cos\alpha' - x'\cos\gamma') \\ \quad + P''(z''\cos\alpha'' - x''\cos\gamma'') + \&c. \end{array}\right\} = 0,$$

$$\left.\begin{array}{l} R(y\cos c - z\cos b) + P'(y'\cos\gamma' - z'\cos\beta') \\ \quad + P''(y''\cos\gamma'' - z''\cos\beta'') + \&c. \end{array}\right\} = 0.$$

Now R is equal in intensity to the resultant of all the other forces of the system, or in other words, to the resultant of all the original forces; and if we give it a direction directly opposite to that in which it is supposed to act in the above equations, it becomes in all respects the same as that resultant, being equal to it in intensity and having the same point of application and line of direction. Adding, therefore, 180° to each of the angles a, b, and c, the first terms of the foregoing equations become negative, and transposing the other terms to the second member and changing all the signs, we have,

$$\left.\begin{array}{l} R\cos a = P'\cos\alpha' + P''\cos\alpha'' + P'''\cos\alpha''' + \&c. = X; \\ R\cos b = P'\cos\beta' + P''\cos\beta'' + P'''\cos\beta''' + \&c. = Y; \\ R\cos c = P'\cos\gamma' + P''\cos\gamma'' + P'''\cos\gamma''' + \&c. = Z. \end{array}\right\} \quad . . (41)$$

$$\left.\begin{array}{l} R(x\cos b - y\cos a) = \left\{\begin{array}{l} P'(x'\cos\beta' - y'\cos\alpha') \\ +P''(x''\cos\beta'' - y''\cos\alpha'') \\ + \&c. \end{array}\right\} = L; \\ R(z\cos a - x\cos c) = \left\{\begin{array}{l} P'(z'\cos\alpha' - x'\cos\gamma') \\ +P''(z''\cos\alpha'' - x''\cos\gamma'') \\ + \&c. \end{array}\right\} = M; \\ R(y\cos c - z\cos b) = \left\{\begin{array}{l} P'(y'\cos\gamma' - z'\cos\beta') \\ +P''(y''\cos\gamma'' - z''\cos\beta'') \\ + \&c. \end{array}\right\} = N. \end{array}\right\} \quad . \ (42)$$

or,

$$\left.\begin{aligned} R\cos a &= X, \\ R\cos b &= Y, \\ R\cos c &= Z. \end{aligned}\right\} \quad \cdots\cdots\cdots\cdots \quad (43)$$

$$\left.\begin{aligned} R\,(x\cos b - y\cos a) &= L, \\ R\,(z\cos a - x\cos c) &= M, \\ R\,(y\cos c - z\cos b) &= N. \end{aligned}\right\} \quad \cdots\cdots\cdots \quad (44)$$

Eliminating $R\cos a$, $R\cos b$ and $R\cos c$, from Equations (44), by means of Equations (43), we get, by transposing all the terms to the first member,

$$\left.\begin{aligned} Xy - Yx + L &= 0, \\ Zx - Xz + M &= 0, \\ Yz - Zy + N &= 0. \end{aligned}\right\} \quad \cdots\cdots\cdots \quad (45)$$

Either one of these equations is but a consequence of the other two. They are, therefore, the equations of a right line—the locus of the points of application; and from which it is apparent, that the point of application of a force may be taken anywhere on its line of direction, within the limits of the body, without altering the effects of the force. The condition expressive of the existence of the dependence of one of these equations on the others, will, also, express the existence of a single resultant.

§ 78.—To find this condition, multiply the first of these Equations by Z, the second by Y, the third by X, and add the products; we obtain,

$$ZL + YM + XN = 0 \quad \cdots\cdots \quad (46).$$

§ 79.—Having ascertained, by the verification of this Equation, that the forces have a single resultant, its intensity, direction, and the equations of its direction may be readily found from Equations (43) and (44).

Squaring each of the group (43), and adding, we obtain,

$$R^2\,(\cos^2 a + \cos^2 b + \cos^2 c) = X^2 + Y^2 + Z^2.$$

Extracting the square root and reducing by the relation,

$$\cos^2 a + \cos^2 b + \cos^2 c = 1,$$

there will result,

$$R = \sqrt{X^2 + Y^2 + Z^2} \quad . \quad . \quad . \quad . \quad . \quad . \quad (47)$$

which gives the *intensity* of the resultant, since X, Y and Z are known.

Again, from the same Equations,

$$\left.\begin{aligned} \cos a &= \frac{X}{R}, \\ \cos b &= \frac{Y}{R}, \\ \cos c &= \frac{Z}{R}. \end{aligned}\right\} \quad . \quad . \quad . \quad . \quad . \quad . \quad . \quad (48)$$

which make known the direction of the resultant.

The group of Equations (45) give,

$$\left.\begin{aligned} Xy - Yx + \Sigma P' (\cos \beta' x' - \cos a' y') &= 0, \\ Zx - Xz + \Sigma P' (\cos a' z' - \cos \gamma' x') &= 0, \\ Yz - Zy + \Sigma P' (\cos \gamma' y' - \cos \beta' z') &= 0. \end{aligned}\right\} \quad . \quad . \quad . \quad (49)$$

which are the equations of the line of the resultant.

PARALLELOGRAM OF FORCES.

§ 80.—If all the forces be applied to the same point, this point may be taken as the origin of co-ordinates, in which case,

$$\begin{aligned} x' = x'' &= x''' \text{ \&c.} = 0, \\ y' = y'' &= y''' \text{ \&c.} = 0, \\ z' = z'' &= z''' \text{ \&c.} = 0, \end{aligned}$$

and the last term in each of Equations (49), will reduce to zero. Hence, to determine the intensity, direction and equations of the line of direction of the resultant, we have, Equations (47), (48) and (49),

$$R = \sqrt{X^2 + Y^2 + Z^2} \quad . \quad . \quad . \quad . \qquad (50)$$

$$\left.\begin{array}{l} \cos a = \dfrac{X}{R}, \\ \cos b = \dfrac{Y}{R}, \\ \cos c = \dfrac{Z}{R}; \end{array}\right\} \quad . \quad . \quad . \quad . \quad . \quad . \quad . \qquad (51)$$

$$\left.\begin{array}{l} Xy - Yx = 0, \\ Zx - Xz = 0, \\ Yz - Zy = 0. \end{array}\right\} \quad . \quad . \quad . \quad . \quad . \quad . \quad . \quad . \qquad (52)$$

The last three equations show that the direction of the resultant passes through the common point of application of all the forces, which might have been anticipated.

§ 81.—Let the forces be now reduced to two, and take the plane of these forces as that of xy; then will

$$\gamma' = \gamma'' = \gamma''' = \&c. = 90^\circ; \; z = 0,$$

the last Equation of group (41) reduces to,

$$Z = 0;$$

and the above Equations become,

$$R = \sqrt{X^2 + Y^2} \quad . \quad . \quad . \quad . \quad . \quad . \quad . \qquad (53)$$

$$\left.\begin{array}{l} \cos a = \dfrac{X}{R}, \\ \cos b = \dfrac{Y}{R}, \end{array}\right\} \quad . \quad . \quad . \quad . \quad . \quad . \quad . \qquad (54)$$

$$\cos c = 0,$$

$$Xy - Yx = 0 \quad . \quad . \quad . \quad . \quad . \quad . \quad . \quad . \quad . \qquad (55)$$

The last is an equation of a right line passing through the origin. The *direction of the resultant will*, therefore, *pass through the point of application of the forces.* The cos c being zero, c is 90°, and *the direction of the resultant is therefore in the plane of the forces.*

Substituting in Equation (53), for X and Y, their values from Equations (41), we obtain,

$$R = \sqrt{(P' \cos \alpha' + P'' \cos \alpha'')^2 + (P' \cos \beta' + P'' \cos \beta'')^2};$$

and since

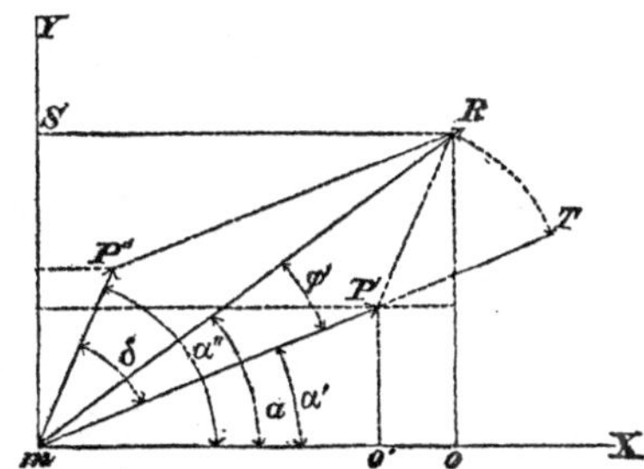

$$\cos^2 \alpha' + \cos^2 \beta' = 1,$$
$$\cos^2 \alpha'' + \cos^2 \beta'' = 1,$$

this reduces to

$$R = \sqrt{P'^2 + P''^2 + 2\,P'\,P''\,(\cos \alpha' \cos \alpha'' + \cos \beta' \cos \beta'')};$$

denoting the angle made by the directions of the forces by δ, we have,

$$\cos \alpha' \cos \alpha'' + \cos \beta' \cos \beta'' = \cos \delta;$$

and therefore,

$$R = \sqrt{P'^2 + P''^2 + 2\,P'\,P'' \cos \delta} \quad . \quad . \quad . \quad . \quad (56)$$

from which we conclude that the *intensity of the resultant is equal to that diagonal of a parallelogram whose adjacent sides represent the directions and intensities of the components, which passes through the point of application.*

§ 82.—Substituting in Equations (54), the values of X and Y, from Equations (41), we have,

$$R \cos a = P' \cos \alpha' + P'' \cos \alpha'',$$
$$R \cos b = P' \cos \beta' + P'' \cos \beta'',$$

and because

$$\alpha' = 90° - \beta',$$
$$\alpha'' = 90° - \beta'',$$
$$a = 90° - b,$$

these Equations reduce to,

$$R \cos a = P' \cos \alpha' + P'' \cos \alpha'',$$
$$R \sin a = P' \sin \alpha' + P'' \sin \alpha'';$$

and, by division,

$$\frac{\sin a}{\cos a} = \frac{P' \sin \alpha' + P'' \sin \alpha''}{P' \cos \alpha' + P'' \cos \alpha''};$$

clearing fractions and transposing, we find,

$$P'' (\sin \alpha'' \cos a - \cos \alpha'' \sin a) = P' (\sin a \cos \alpha' - \cos a \sin \alpha');$$

whence,

$$\frac{P'}{P''} = \frac{\sin \alpha'' \cos a - \cos \alpha'' \sin a}{\sin a \cos \alpha' - \cos a \sin \alpha'} = \frac{\sin (\alpha'' - a)}{\sin (a - \alpha')} \quad . \quad . \quad (56)'$$

That is to say, the intensities of the components are inversely proportional to the sines of the angles which their directions make with that of their resultant; but this is the relation that subsists between the two adjacent sides of a parallelogram and the sines of the angles which they make with the diagonal through their point of meeting. Whence, Eqs. (56) and (56)',

The resultant of any two forces, applied to the same point, is represented, in intensity and direction, by that diagonal of a parallelogram of which the adjacent sides represent the components.

Making

$$a - \alpha' = \text{the angle } R\, m\, P' = \varphi',$$

and

$$\frac{R + P' + P''}{2} = S,$$

we have, from the usual trigonometrical formula,

$$\sin \tfrac{1}{2}\varphi' = \sqrt{\frac{(S - P')\,(S - R)}{R \,.\, P'}} \quad . \quad . \quad . \quad . \quad . \quad (57)$$

§ 83.—In the triangle $R\, m\, P'$, since $P'\, R$ is equal and parallel to the line which represents the force P'', the angle $m\, P'\, R = \varphi$, is the supplement of the angle δ, made by the directions of the components, and there will result the following equation:

$$\cos \tfrac{1}{2}\delta = \sin \tfrac{1}{2}\varphi = \sqrt{\frac{(S - P')\,(S - P'')}{P'\, P''}} \quad . \quad . \quad . \quad (58)$$

Equation (57), will make known the angle made by the direction of the resultant with that of either of two oblique components, provided, the intensities of the components and resultant be known.

§ 84.—Also, from the two triangles $R\,m\,P'$ and $R\,m\,P''$, we find,

$$\left.\begin{aligned} \sin\varphi' &= \frac{P''\,.\,\sin\delta}{R}; \\ \sin\varphi'' &= \frac{P'\,.\,\sin\delta}{R}. \end{aligned}\right\} \cdots (59);$$

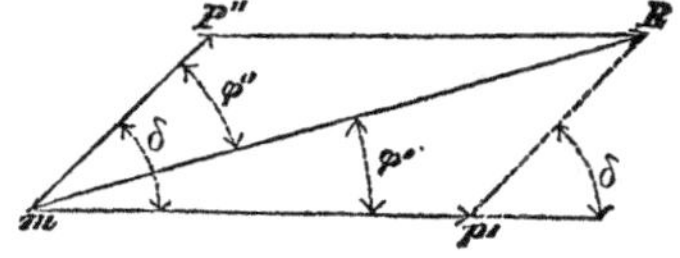

from which the angles made by the direction of the resultant with its two components may be found.

§ 85.—Let there now be the three forces P, P', P'', applied to the material point m, in the directions $m\,P$, $m\,P'$, $m\,P''$, not in the same plane; the resultant will be represented in intensity and direction by the diagonal of a parallelopipedon, constructed upon the lines representing the directions and intensities of these components. For, lay off the distances $m\,A$, $m\,C$, and $m\,E$, proportional to the intensities of the components which act in the direction of these lines, and construct the parallelopipedon $E\,B$; the resultant of the components P' and P will, § 82, be represented by the diagonal $m\,B$, of the parallelogram $m\,A\,B\,C$; and the resultant of this resultant and the remaining component P'', will be represented by the diagonal $m\,D$ of the parallelogram $E\,m\,B\,D$, which is that of the parallelopipedon.

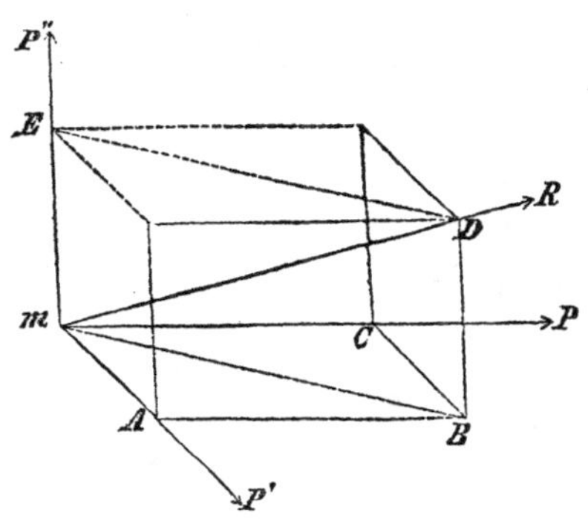

§ 86.—If the forces act at right angles to each other, the parallelopipedon will become rectangular, and the intensity of the resultant, denoted by R, will become known from the formula

$$R = \sqrt{P^2 + P'^2 + P''^2};$$

and if the angles which the direction of the resultant makes with those of the forces P, P' and P'', be represented by a, b, and c, respectively, then will

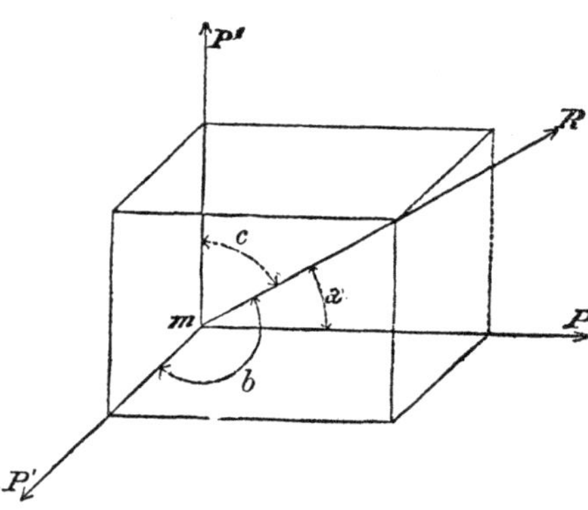

$$R \cos a = P,$$
$$R \cos b = P',$$
$$R \cos c = P''.$$

Let three lines be drawn through the point of application m', of the force P', parallel to any three rectangular axes x, y, z; and denote by α', β', γ', the angles which the direction of this force makes with these axes respectively; then will

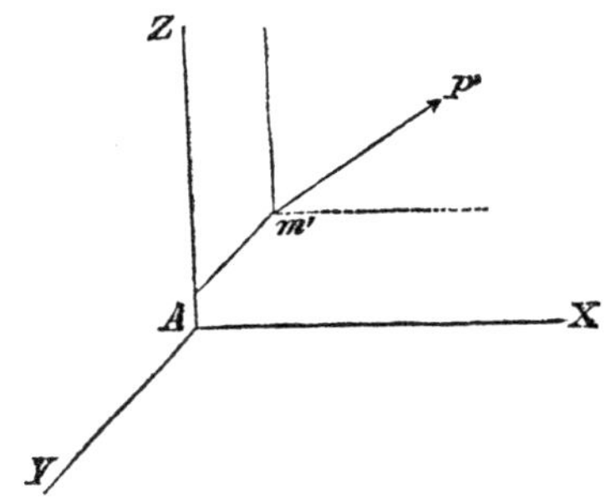

$$P' \cos \alpha',$$
$$P' \cos \beta',$$
$$P' \cos \gamma',$$

be the components of the force P', in the direction of the axes, and they will act along the lines drawn through the point m'. These are the same as the terms composing in part Equations (A), and as the effect of the components is identical with that of the resultant, these components may always be substituted for the force P'. The same for the forces of *inertia*, and $m \cdot \frac{d^2x}{dt^2}$, $m \cdot \frac{d^2y}{dt^2}$, and $m \cdot \frac{d^2z}{dt^2}$, denote the components of this force in the directions of the axes.

§ 87.—*Examples.*—1. Let the point m, be solicited by two forces whose intensities are 9 and 5, and whose directions make an angle with each other of 57° 30′. Required the intensity of the force by which the point is urged, and the direction in which it is compelled to move.

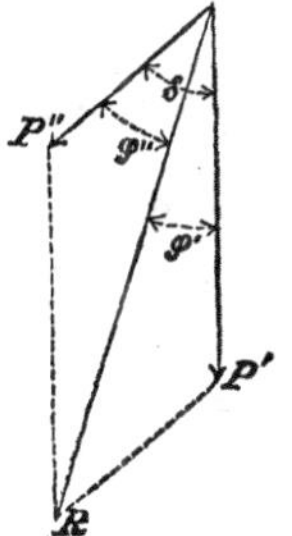

First, the intensity; make in Equation (56),

$$P' = 9,$$
$$P'' = 5,$$
$$\delta = 57° \; 30' ;$$

and there will result,

$$R = \sqrt{81 + 25 + 90 \times 0,537} = 12{,}422.$$

Again, substituting the values of δ, P' P'' and R in the first of Equations (59), we have,

$$\sin \varphi' = \frac{5 \times \sin 57° \; 30'}{12{,}422},$$

or,

$$\varphi' = 19° \; 50' \; 35'' \text{ nearly,}$$

which is the angle made by the direction of the force 9 with that of the resultant.

2.—Required the angle under which two equal components should act, in order that their resultant shall be the nth part of either of them separately.

By condition, we have

$$P' = P'' = nR;$$

hence,

$$\frac{P' + P'' + R}{2} = S = \frac{nR + nR + R}{2} = \frac{(2n + 1) R}{2};$$

and, Equation (58),

$$\sin \tfrac{1}{2} \varphi = \sqrt{\frac{(S - P') \; (S - P'')}{P' \; P''}},$$

which reduces to

$$\sin \tfrac{1}{2} \varphi = \pm \frac{1}{2n}.$$

If n be equal to unity, or the resultant be equal to either force,

$$\varphi = 60^\circ,$$

and, § 83, the angle of the components should be 120°.

3.—Required to resolve the force $18 = a$, into two components whose difference shall be $5 = b$, and whose directions make with each other an angle of $38^\circ = \delta$. Also, to find the angle which the direction of each component makes with that of the resultant.

Writing a for R in Equation (56), we have,

$$P'^2 + P''^2 + 2 P' P'' \cos \delta = a^2,$$

and by condition,

$$P' - P'' = b \quad . \quad . \quad . \quad . \quad . \quad . \quad . \quad (c).$$

Squaring the second and subtracting it from the first, we get

$$2P'P'' (1 + \cos \delta) = a^2 - b^2 \, ;$$

which, replacing $(1 + \cos \delta)$ by $2 \cos^2 \frac{1}{2} \delta$, reduces to

$$4P' P'' = \frac{a^2 - b^2}{\cos^2 \frac{1}{2} \delta}.$$

This added to the square of the Equation (c), gives

$$P' + P'' = \pm \sqrt{\frac{a^2 - b^2 (1 - \cos^2 \frac{1}{2} \delta)}{\cos^2 \frac{1}{2} \delta}} \, ;$$

from which and Equation (c) we finally obtain,

$$P' = \tfrac{1}{2} \left(\pm \sqrt{\frac{a^2 - b^2 (1 - \cos^2 \frac{1}{2} \delta)}{\cos^2 \frac{1}{2} \delta}} + b \right) = 12{,}049,$$

$$P'' = \tfrac{1}{2} \left(\pm \sqrt{\frac{a^2 - b^2 (1 - \cos^2 \frac{1}{2} \delta)}{\cos^2 \frac{1}{2} \delta}} - b \right) = 7{,}049,$$

which are the required components.

To find the angles which their directions make with the resultant, we have from Equations (59),

$\varphi'' = 24^\circ =$ the angle which P'' makes with the resultant.

and,

$$\varphi' = 14° = \text{angle which } P' \text{ makes with the resultant.}$$

4.—Required the angle under which two components whose intensities are denoted by 5 and 7 should act, to give a resultant whose intensity is represented by 9.

Ans. 84° 15′ 39″

5.—From Equation (56) it appears that the resultant of two components applied to the same point, is greatest when the angle made by their directions is 0°, and least when 180°. Required the angle under which the components should act, in order that the resultant may be a mean proportional between these values; and also the angle which the resultant makes with the greater component. Call P', the greater component.

$$Ans.\ \delta = \cos^{-1} - \frac{P''}{P'}.$$

$$\varphi' = \sin^{-1} \frac{P''}{P'}.$$

6.—Given a force whose intensity is denoted by 17. Required the two components which make with it angles of 27° and 43°.

§ 88.—The theorem of the parallelogram of forces, just explained, enables us to determine by an easy graphical construction the intensity and direction of the resultant of several forces applied to the same point.

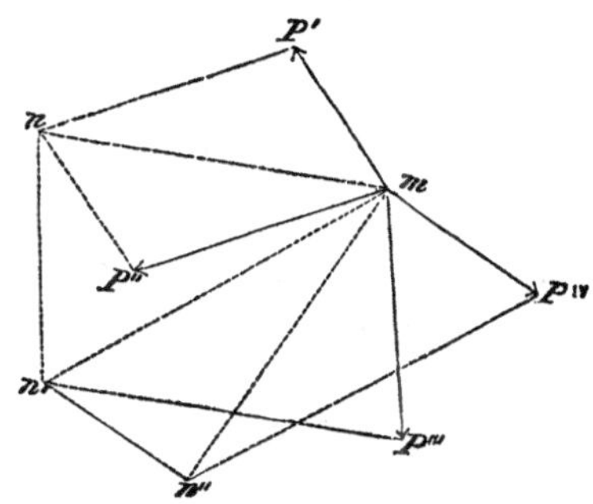

Let P', P'', P''', &c., be several forces applied to the same point m. Upon the directions of the forces, lay off from the point of application distances proportional to the intensities of the forces, and let these distances represent the forces. From the extremity P' of the line mP', which represents the first force, draw the line $P'n$ equal and parallel to $m\,P''$ which represents the second, then will the line joining the extremity of this line and the point of application, represent the resultant of these two forces. From the extremity n, draw the line $n\,n'$ equal and parallel to $m\,P'''$ which represents the third force; $m\,n'$ will represent the resultant of the first three forces. The construction being thus continued till a line be drawn equal and parallel to every line representing a force of the system, the resultant of the whole will be represented by the line, (in this instance $m\,n''$), joining the point of application with the last extremity of the last line drawn. Should the line which is drawn equal and parallel to that which represents the last force, terminate in the point of application, the resultant will be equal to zero.

The reason for this construction is too obvious to need explanation.

§ 89.—If the forces still be supposed to act in the same plane, but upon different points of the plane, the first of Equations (49) takes the form,

$$Yx - Xy = \Sigma\,[P'\,(\cos\beta'\,x' - \cos\alpha'\,y')\,],$$

thus, differing from Equation (55), in giving the equation of the line of direction of the resultant an independent term, and showing that this line no longer passes through the origin. It may be constructed from the above equation.

§ 90.—To find the resultant in this case, by a graphical construction, let the forces P', P'', P''' &c., be applied to the points m', m'', m''', &c., respectively. Produce the directions of the forces P' and P'' till they meet at O, and take this as their common point of application;

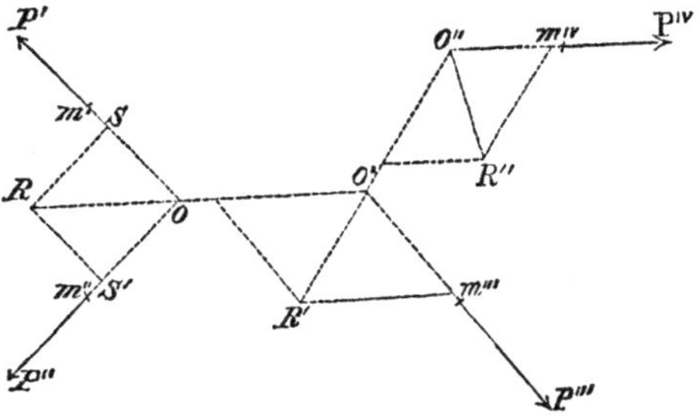

lay off from O, on the ines of direction, distances $O\,S$ and $O\,S'$, proportional to the intensities of the forces P' and P'', and construct the parallelogram $O\,S\,R\,S'$, then will $O\,R$ represent the resultant of these forces. The direction of this resultant being produced till it meet the direction of the force P''', produced, a similar construction will give the resultant of the first resultant and the force P''', which will be the resultant of the three forces P', P'' and P'''; and the same for the other forces.

OF PARALLEL FORCES.

§ 91.—If the forces act in parallel directions,

$$\cos\alpha' = \cos\alpha'' = \cos\alpha''' = \text{\&c.},$$
$$\cos\beta' = \cos\beta'' = \cos\beta''' = \text{\&c.},$$
$$\cos\gamma' = \cos\gamma'' = \cos\gamma''' = \text{\&c.},$$

and Equations (41) become,

$$X = (P' + P'' + P''' + \text{\&c.}) \cos\alpha',$$
$$Y = (P' + P'' + P''' + \text{\&c.}) \cos\beta',$$
$$Z = (P' + P'' + P''' + \text{\&c.}) \cos\gamma';$$

these values in Equation (47) give,

$$R = \pm \sqrt{(P' + P'' + P''' + \text{\&c.})^2 (\cos^2\alpha' + \cos^2\beta' + \cos^2\gamma')},$$

but,

$$\cos^2\alpha' + \cos^2\beta' + \cos^2\gamma' = 1;$$

hence,

$$R = P' + P'' + P''' + \text{\&c.} \quad \ldots \ldots \quad (60)$$

If some of the forces as P'', P''', act in directions opposite to the others, the cosines of α'' and α''' will be negative while they have the same numerical value; and the last equation will become

$$R = P' - P'' - P''' + \text{\&c.}$$

Whence we conclude, that the *resultant of a number of parallel forces is equal in intensity to the excess of the sum of the intensities of those which act in one direction over the sum of the intensities of those which act in the opposite direction.*

§ 92.—The values of R, X, Y and Z being substituted in Equations (48) give,

$$\cos a = \frac{(P' + P'' + P''' + \&c.) \cos \alpha'}{P' + P'' + P''' + \&c.} = \cos \alpha',$$

$$\cos b = \frac{(P' + P'' + P''' + \&c.) \cos \beta'}{P' + P'' + P''' + \&c.} = \cos \beta',$$

$$\cos c = \frac{(P' + P'' + P''' + \&c.) \cos \gamma'}{P' + P'' + P''' + \&c.} = \cos \gamma'.$$

The denominator of these expressions, being the resultant, is essentially positive; the signs of the cosines of the angles a, b and c, will, therefore, depend upon the numerators; these are sums of the components parallel to the three axes.

Hence, the resultant acts in the direction of those forces whose cosine coefficients are negative or positive according as the sum of the former or latter forces is the greater.

§ 93.—Equations (49), which are those of the resultant, become, after replacing X, Y, and Z, by their values in Equations (41),

$$Ry \,.\, \cos a - Rx \,.\, \cos b + \cos \beta' \,.\, \Sigma\, P'x' - \cos \alpha' \,.\, \Sigma\, P'y' = 0,$$

$$Rx \,.\, \cos c - Rz \,.\, \cos a + \cos \alpha' \,.\, \Sigma\, P'z' - \cos \gamma' \,.\, \Sigma\, P'x' = 0,$$

$$Rz \,.\, \cos b - Ry \,.\, \cos c + \cos \gamma' \,.\, \Sigma\, P'y' - \cos \beta' \,.\, \Sigma\, P'z' = 0 ;$$

and because,

$$\cos a = \cos \alpha',$$

$$\cos b = \cos \beta',$$

$$\cos c = \cos \gamma' ;$$

we have,

$$(Ry - \Sigma P'y') \,.\, \cos \alpha' - (Rx - \Sigma P'x') \,.\, \cos \beta' = 0,$$

$$(Rx - \Sigma P'x') \,.\, \cos \gamma' - (Rz - \Sigma P'z') \,.\, \cos \alpha' = 0,$$

$$(Rz - \Sigma P'z') \,.\, \cos \beta' - (Ry - \Sigma P'y') \,.\, \cos \gamma' = 0 ;$$

and because α', β' and γ', are connected only by the relation

$$\cos^2\alpha' + \cos^2\beta' + \cos^2\gamma' = 1\ ;$$

either two of the cosines of these angles are wholly arbitrary, and from the principle of indeterminate co-efficients, we have, by dispensing with the sign Σ and writing out the terms,

$$\left.\begin{array}{l} Rx = P'x' + P''x'' + P'''x''' + \&c. \\ Ry = P'y' + P''y'' + P'''y''' + \&c. \\ Rz = P'z' + P''z'' + P'''z''' + \&c. \end{array}\right\} \quad \cdots \quad (61)$$

The forces being given, the value of R, § 91, becomes known, and the co-ordinates x, y, z, are determined from the above equations; these co-ordinates will obviously remain the same whatever direction be given to the forces, provided, they remain parallel and retain the same intensity and points of application, these latter elements being the only ones upon which the values of x, y, z, depend.

The point whose co-ordinates are x, y, z, which is a point of application of the resultant, is called the *centre of parallel forces*, and may be defined to be, *that point in a system of parallel forces through which the resultant of the system will always pass, whatever be the direction of the forces, provided, their intensities and points of application remain the same.*

§ 94.—Dividing each of the above Equations by R, we shall have

$$\left.\begin{array}{l} x = \dfrac{P'x' + P''x'' + P'''x''' + \&c.}{P' + P'' + P''' + \&c.}, \\ y = \dfrac{P'y' + P''y'' + P'''y''' + \&c.}{P' + P'' + P''' + \&c.}, \\ z = \dfrac{P'z' + P''z'' + P'''z''' + \&c.}{P' + P'' + P''' + \&c.}, \end{array}\right\} \quad \cdots \quad (62)$$

Hence, *either co-ordinate of the centre of a system of parallel forces is equal to the algebraic sum of the products which result from multiplying the intensity of each force by the corresponding co-ordinate of its point of application, divided by the algebraic sum of the forces.*

If the points of application of the forces be in the same plane.

the co-ordinate plane xy, may be taken parallel to this plane, in which case

$$z' = z'' = z''' = z'''' \text{ \&c.};$$

and,

$$z = \frac{(P' + P'' + P''' + \text{\&c.})\, z'}{P' + P'' + P''' + \text{\&c.}} = z';$$

from which it follows that the centre of parallel forces is also in this plane.

If the points of application be upon the same straight line, take the axis of x parallel to this line; then in addition to the above results we have

$$y' = y'' = y''' = \text{\&c.};$$

and,

$$y = \frac{(P' + P'' + P''' + \text{\&c.})\, y'}{P' + P'' + P''' + \text{\&c.}} = y';$$

whence, the centre of parallel forces is also upon this line.

§ 95.—If we suppose the parallel forces to be reduced to two, viz. P' and P'', we may assume the axis x to pass through their points of application, and the plane xy to contain their directions, in which case, Equations (60) and (61) become,

$$R = P' + P''$$
$$Rx = P'x' + P''x''$$
$$z = 0 \text{ and } y = 0.$$

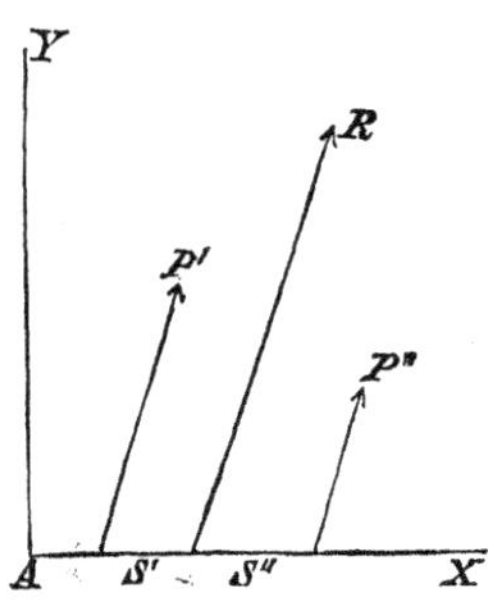

Multiplying the first by x', and subtracting the product from the second, we obtain

$$R\,(x - x') = P''\,(x'' - x') \quad . \ . \ (a)$$

Multiplying the first by x'' and subtracting the second from the product, we get

$$R\,(x'' - x) = P'\,(x'' - x') \quad . \ . \ . \ . \ (b)$$

Denoting by S' and S'', the distances from the points of application of P' and P'' to that of the resultant, which are $x - x'$ and $x'' - x$ respectively, we have

$$x'' - x' = S' + S'';$$

and from Equations (a) and (b), there will result

$$P' : P'' : R :: S'' : S' : S'' + S' \quad . \quad . \quad . \quad . \quad (63)$$

If the forces act in opposite directions, then, on the supposition that P' is the greater, will

$$R = P' - P''$$
$$Rx = P'x' - P''x''$$
$$z = 0, \ y = 0.$$

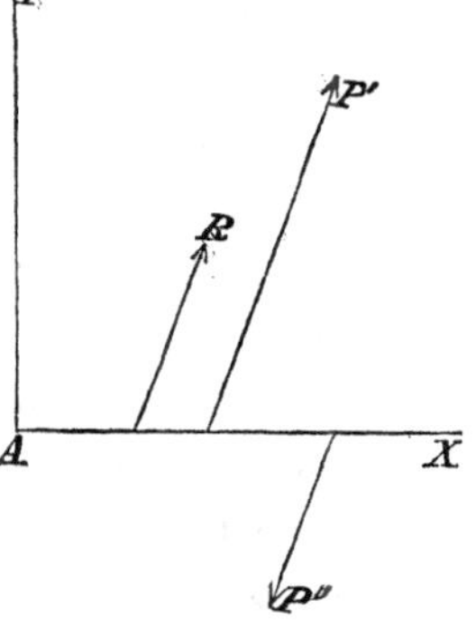

and by a process plainly indicated by what precedes,

$$P' : P'' : R :: S'' : S' : S'' - S'. \ . \quad (64).$$

From this and Proportion (63), it is obvious that the point of application of the resultant is always nearer that of the greater component; and that when the components act in the same direction, the distance between the point of application of the smaller component and that of the resultant, is less than the distance between the points of application of the components, while the reverse is the case when the components act in opposite directions. In the first case, then, the resultant is between the components, and in the second, the larger component is always between the smaller component and the resultant.

And we conclude, generally, *that the resultant of two forces which solicit two points of a right line in parallel directions, is equal in intensity to the sum or difference of the intensities of the components, according as they act in the same or opposite directions, that it always acts in the direction of the greater component, that its line of direction is contained in the plane of the components, and that the intensity of either component is to that of the resultant, as the distance between the point of application of the other component and that of the resultant, is to the distance between the points of application of the components.*

§ 96.—*Examples.*—1. The length of the line $m'm''$ joining the points of application of two parallel forces acting in the same direction, is 30 feet; the forces are represented by the numbers 15 and 5. Required the intensity of the resultant, and its point of application.

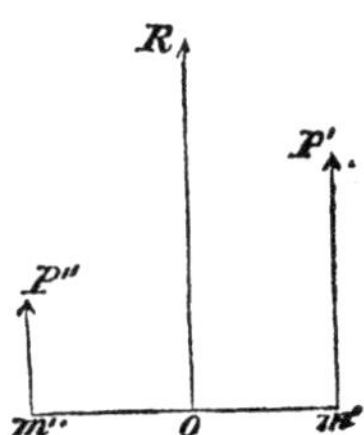

$$R = P' + P'' = 15 + 5 = 20;$$

$$R : P' :: m''m' : m''o,$$

$$20 : 15 :: 30 : m''o = 22{,}5 \text{ feet.}$$

A single force, therefore, whose intensity is represented by 20, applied at a distance from the point of application of the smaller force equal to 22,5 feet, will produce the same effect as the given forces applied at m'' and m'.

2.—Required the intensity and point of application of the resultant of two parallel forces, whose intensities are denoted by the numbers 11 and 3, and which solicit the extremities of a right line whose length is 16 feet in opposite directions.

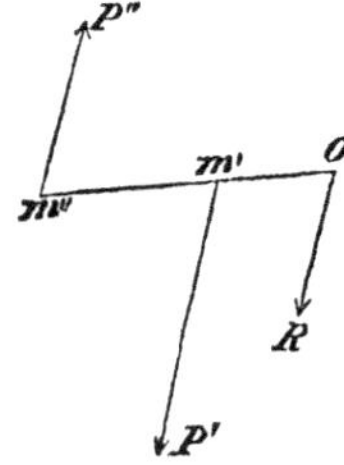

$$R = P' - P'' = 11 - 3 = 8,$$

$$P' - P'' : P' :: m''m' : m''o = \frac{P' \cdot m''m'}{P' - P''} = 22 \text{ feet.}$$

3.—Given the length of a line whose extremities are solicited in the same direction by two forces, the intensities of which differ by the nth part of that of the smaller. Required the distance of the point of application of the resultant from the middle of the line. Let $2\,l$, denote the length of the line. Then, by the conditions,

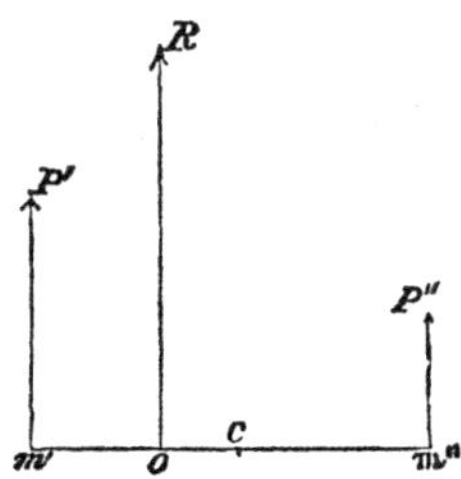

$$P' = P'' + \frac{1}{n} P'' = \left(\frac{n+1}{n}\right) P''$$

$$R = \left(\frac{n+1}{n}\right) P'' + P'' = \frac{2n+1}{n} P''$$

$$\left(\frac{2n+1}{n}\right) P'' : P'' :: 2l : m'o = \frac{2nl}{2n+1}$$

$$co = l - \frac{2nl}{2n+1} = \frac{1}{2n+1}\, l.$$

§ 97.—The rule at the close of § 95, enables us to determine by a very easy graphical construction, the position and point of application of the resultant of a number of parallel forces, whose directions, intensities, and points of application are given.

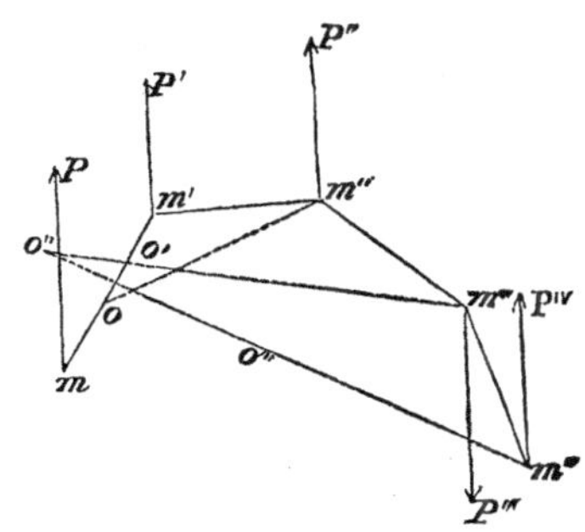

Let P, P', P'', P''', and P^{iv}, be several forces applied to the material points m, m', m'', m''', and m^{iv}, in parallel directions. Join the points m and m' by a straight line, and divide this line at the point o, in the inverse ratio of the intensities of the forces P and P'; join the points o and m'' by the straight line $o\,m''$, and divide this line at o', in the inverse ratio of the sum of the first two forces and the force P''; and continue this construction till the last point m^{iv} is included, then will the last point of division be the point of application of the resultant, through which its direction may be drawn parallel to that of the forces. The intensity of the resultant will be equal to the algebraic sum of the intensities of the forces.

The position of the point o will result from the proportion

$$P + P' : P' :: m\,m' : m\,o = \frac{P' \, . \, \overline{m\,m'}}{P + P'};$$

that of o' from

$$P + P' + P'' : P'' :: o\,m'' : o\,o' = \frac{P'' \, . \, \overline{o\,m''}}{P + P' + P''};$$

that of o'' from

$$P + P' + P'' - P''' : - P''' :: o'\,m''' : o'\,o'' = \frac{- P''' \, . \, \overline{o'\,m'''}}{P + P' + P'' - P'''};$$

and finally, that of o''' from

$$P+P'+P''-P'''+P^{iv} : P^{iv} :: o''\,m^{iv} : o''\,o''' = \frac{P^{iv} \, . \, \overline{o''\,m^{iv}}}{P+P'+P''-P'''+P^{iv}}.$$

OF COUPLES.

§ 98.—When two forces P' and P'' act in opposite directions, the distance of the point o, at which the resultant is applied, from the point m', at which the component P' is applied, is found from the formula

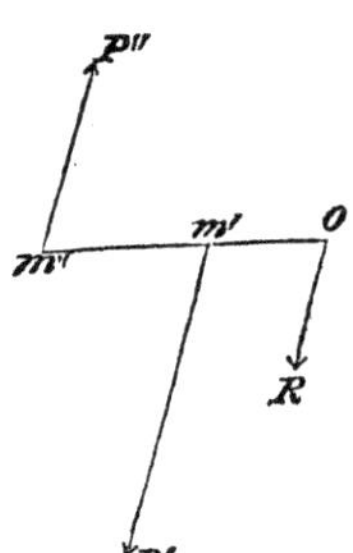

$$m'o = \frac{\overline{m''m'} \cdot P''}{P' - P''};$$

and if the components P' and P'' become equal, the distance $m'o$ will be infinite, and the resultant, zero. In other words, the forces will have no resultant, and their joint effect will be to turn the line $m''m'$, about some point between the points of application.

The forces in this case act in opposite directions, are equal, but not immediately opposed. To such forces the term *couple* is applied. A couple having no single resultant, their action cannot be compared to that of a single force.

§ 99.—The analytical condition, Equation (46), expressive of the existence of a single resultant in any system of forces, will obviously be fulfilled, when

$$X = 0, \quad Y = 0, \quad \text{and } Z = 0.$$

But this may arise from the parallel groups of forces whose sums are denoted by X, Y, and Z, reducing each to a couple. These three couples may easily be reduced by composition to a single couple, beyond which, no further reduction can be made. It is, therefore, a failing case of the general analytical condition referred to.

WORK OF THE RESULTANT AND OF ITS COMPONENTS.

§ 100.—We have seen that when the resultant of several forces is introduced as an additional force with its direction reversed, it will hold its components in equilibrio. Denoting the intensity of the resultant by R, and the projection of its virtual velocity by δr, we have from Equation (29),

$$- R\delta r + P.\delta p + P'.\delta p' + P''.\delta p'' + \&c. = 0,$$

or,

$$R\delta r = P.\delta p + P'\delta p' + P''\delta p'' + \&c., \quad \cdot \quad \cdot \quad \cdot \quad \cdot \quad (65)$$

in which P, P' P'', &c. are the components, and δp, $\delta p'$ $\delta p''$, &c. the projections of their virtual velocities.

§ 101.—Now, the displacement by which Equation (29) was deduced, was entirely arbitrary; it may, therefore, be made to conform in all respects to that which would be produced by the components P, P', &c., acting without the opposition of the force equal and contrary to their resultant; and writing dr for δr, dp for δp, &c., Equation (65) will become

$$R dr = P dp + P' dp' + P'' dp'' + \&c., \quad \cdot \quad \cdot \quad \cdot \quad (66)$$

and integrating,

$$\int R dr = \int P dp + \int P' dp' + \int P'' dp'' + \&c., \quad \cdot \quad \cdot \quad (67)$$

in which R, P, P', &c. may be constant or functions of r, p, p', &c., respectively.

From Equations (66) and (67), it appears that the quantity of work of the resultant of several forces is equal to the algebraic sum of the quantities of work of its components.

Again, replacing $P\delta p$, $P'\delta p'$, &c. in Equation (65), by their values in Equation (31), and writing dr for δr, dp for δp, &c., we find,

$$\int R dr = \int \Sigma P.\cos\alpha.dx + \int \Sigma P.\cos\beta.dy + \int \Sigma P.\cos\gamma.dz, \quad \cdot \quad \cdot \quad (68)$$

in which R may be constant or a function of r; P, constant or a function of x, y, z, &c.

If the forces be in equilibrio, then will $R = 0$, and,

$$\Sigma P.\cos\alpha.dx + \Sigma P.\cos\beta.dy + \Sigma P.\cos\gamma.dz = 0. \quad \cdot \quad \cdot \quad (69)$$

MOMENTS.

§ 102.—It is now apparent that in the transformation of Equation (30) to Equation (40), each force of the original system was replaced by its three components in directions of three rectangular axes, arbitrarily assumed.

The components parallel to either axis will, § 43, work during any motion which will carry their points of application in the direction of that axis, and will cease to work when the motion becomes perpendicular to the same line.

Let the points of application of the forces move in lines parallel to the axis z; the components parallel to z alone can work, for the paths being perpendicular to the directions of the other components, the work of the latter will be nothing, because the projections of the paths upon their lines of direction will be zero. The elementary work of the extraneous forces will, in this case, be found in the third term of Equation (40), and equal to

$$(\Sigma P \cos \gamma) \, . \, \delta z_{,}.$$

Again, let the points of application turn around the axis z, parallel to the plane $x y$; the components parallel to the axes x and y alone can work, since the paths will be perpendicular to the components in the direction of z, and their projections, therefore, zero. The elementary work in this case will be found in the fourth term of Equation (40), and equal to

$$[\Sigma P (x' \cos \beta - y' \cos \alpha)] \, \delta \varphi.$$

Now let both of these motions take place simultaneously; that is, let the points of application move in the direction of the axis z, and also turn about that line; all the components will work, because the paths will be oblique to their directions, and, therefore, have projections of measurable values. The amount of elementary work of the extraneous forces will, in this case be found in the third and fourth terms of Equation (40), and equal to

$$[(\Sigma P \cos \gamma)] \, . \, \delta z_{,} + [\Sigma P (x' \cos \beta - y' \cos \alpha)] \, . \, \delta \varphi.$$

The same remarks apply to motion in the direction of and about each of the other axes.

§ 103.—The rule for estimating the quantity of work when the motion is parallel to either axis or to a right line oblique to the three axes, is simple; that for getting the work during motion about an axis, is not so obvious. Let the motion take place around the axis z; and consider, first, the work of the force P. The two components of this force, viz., $P \cos \beta$ and $P \cos \alpha$, which enter the fourth term of Equation (40), have for their resultant $P \sin \gamma$. This resultant, § 81, acts in a plane parallel to that of $x\,y$, and, therefore, at right angles to the axis z. Denote by $\alpha_{,}$ the angle which this resultant makes with the axis x; then will

$$\left.\begin{aligned} P \cos \alpha &= P \sin \gamma \,.\, \cos \alpha_{,}, \\ P \cos \beta &= P \sin \gamma \,.\, \sin \alpha_{,} \end{aligned}\right\} \quad \ldots\ldots \quad (70)$$

and these values in the term $P\,(x' \cos \beta - y' \cos \alpha)$, give

$$P\,(x' \cos \beta - y' \cos \alpha) = P \,.\, \sin \gamma\,(x' \sin \alpha_{,} - y' \cos \alpha_{,}) \quad . \quad (71)$$

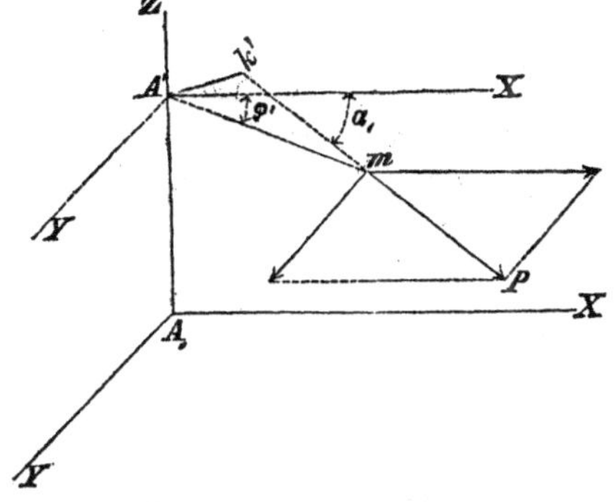

From the point of application m of P, draw the line $m\,A'$ perpendicular to the axis z; denote its length by h', and its inclination to the axis x by φ'. Multiply and divide Equation (71) by h' and reduce by the relations

$$\frac{x'}{h'} = \cos \varphi'; \quad \frac{y'}{h'} = \sin \varphi';$$

then will result

$$P\,(x' \cos \beta - y' \cos \alpha) = P \sin \gamma\, h'\,(\sin \alpha_{,} \,.\, \cos \varphi' - \cos \alpha_{,} \,.\, \sin \varphi') = P \sin \gamma\, h' \sin (\alpha_{,} - \varphi').$$

Draw from A' the line $A'\,k'$ perpendicular to the direction of the line $P\,m$ (produced), and denote its length by k'; then will

$$h' \sin (\alpha_{,} - \varphi') = k',$$

and then will result

$$P\,(x' \cos\beta - y' \cos\alpha) = P \sin\gamma \,.\, k', \quad . \quad . \quad . \quad . \quad (72)$$

and the same for the forces P', P'', &c.; so that we may write, omitting the accent from k,

$$\Sigma\, P\,(x' \cos\beta - y' \cos\alpha) = \Sigma\, P \,.\, \sin\gamma \,.\, k; \quad . \quad . \quad . \quad (73)$$

and the measure of the elementary work due to rotation about the axis z, will be given by either member of the Equation

$$[\Sigma\, P\,(x' \cos\beta - y' \cos\alpha)]\; \delta\varphi = [\Sigma\, P \sin\gamma \,.\, k]\; \delta\varphi \quad . \quad . \quad (74)$$

§ 104.—So that in estimating the work due to rotation alone about the axis z, each force is, in effect, replaced by its two components, the one parallel, the other perpendicular to that line, and the former is neglected because, in this motion, it cannot work.

§ 105.—The quantity of work obtained by multiplying that one of the two components of a force which is perpendicular, while the other is parallel, to a given line, into the perpendicular distance between this line and that of the force, is called the *component moment of the force in reference to the line.*

§ 106.—The line in reference to which the moment is taken, is called, in general, a component axis; the perpendicular distance from the axis to the line of direction of the force, is called the *lever arm of the force;* and the extremity of the lever arm on the axis is called a *centre of the moment.*

When the direction of the force is perpendicular to the axis, the latter is called the *moment axis of the force.* In this case the component parallel to the axis becomes zero, and the normal component the force itself.

The moment of the resultant of several component forces, taken in reference to its moment axis, is called the *resultant moment.* The moments of the component forces are called *component moments.*

§ 107.—Changing $\delta\varphi$ into $d\varphi$ in Equation (74), we may write

$$[\Sigma\, P\,(x' \cos\beta - y' \cos\alpha)]\, d\varphi = [\Sigma\, P \;\; \sin\gamma \,.\, k]\; d\varphi \quad . \quad . \quad (74)$$

or

$$\int[\Sigma\, P\,(x'\cos\beta - y'\cos\alpha)]\,d\varphi = \int[\Sigma\, P\,.\sin\gamma\,.\,k]\,d\varphi\quad . \quad (74)'$$

Whence it appears, that the elementary quantity of work a force will perform during the motion of its point of application about an axis, is equal to the *product of the moment of the force into the differential of the path described at the unit's distance from the axis.*

§ 108.—The whole quantity of work will result from the integration of Equation (74)′ between limits. In this integration two cases may arise, viz.; either the moment may be constant, or it may be variable. In the first case, the quantity of work is obtained by multiplying the constant moment into the path described by a point at the unit's distance from the axis. In the second, the force may be constant and the lever arm variable; the force variable and the lever arm constant; or both may be variable, and in such way as not to make their product constant. In all such cases, relations between the intensity of the force, its lever arm, and the path described at the unit's distance, must be known in order to reduce, by elimination, the second member of Equation (74)′ to a function of a single variable.

These remarks are equally true of the forces of inertia. The intensities of these depend upon the masses of the material elements and their degree of acceleration or retardation; their points of application are on the elements themselves; the elementary arc described at the unit's distance is the same for both sets of moments, and its value depends upon the distribution of the material with reference to the axis of motion.

The moments of the forces which urge a body to turn in opposite directions about any assumed axis must have contrary signs.

The sign of $P\sin\gamma\,k'$, or its equal $P\cos\beta\,.\,x' - P\cos\alpha\,.\,y'$, depends upon the angles which the direction of the force makes with the axes, and upon the signs and relative values of the co-ordinates of the point of application.

Let the angles which the direction of any force makes with the co-ordinate axes be estimated from the positive side of the origin; then, if the angles which this direction makes with both axes be acute, and the point of application lie in the first angle, $P\cos\beta\,.\,x'$

and $P \cos \alpha \,.\, y'$, will be positive, and if the first of these products exceed the second, the moment will be positive; but if the latter be the greater, the moment will be negative. The same remarks apply to the other axes.

COMPOSITION AND RESOLUTION OF MOMENTS.

§ 109.—The forces being supposed to act in any directions whatever, join the point of application of the resultant R and the origin by a right line, and denote its length by H. Multiply and divide each of the Equations (44) by H, and reduce by the relations,

$$\frac{x}{H} = \cos \zeta,$$

$$\frac{y}{H} = \cos \xi,$$

$$\frac{z}{H} = \cos \varepsilon,$$

in which ζ, ξ and ε, denote the angles which the line H makes with the axes x, y and z, respectively; then will

$$\left.\begin{array}{l} R \,.\, H \,.\, (\cos b \,.\, \cos \zeta - \cos a \,.\, \cos \xi) = L, \\ R \,.\, H \,.\, (\cos a \,.\, \cos \varepsilon - \cos c \,.\, \cos \zeta) = M, \\ R \,.\, H \,.\, (\cos c \,.\, \cos \xi - \cos b \,.\, \cos \varepsilon) = N. \end{array}\right\} \quad . \quad . \quad . \quad (75)$$

Squaring each of these Equations and adding, we find

$$R^2 \,.\, H^2 \left\{\begin{array}{l} \cos^2 b \,.\, \cos^2 \zeta - 2 \cos b \,.\, \cos a \,.\, \cos \zeta \,.\, \cos \xi + \cos^2 a \,.\, \cos^2 \xi \\ + \cos^2 a \,.\, \cos^2 \varepsilon - 2 \cos a \,.\, \cos c \,.\, \cos \varepsilon \,.\, \cos \zeta + \cos^2 c \,.\, \cos^2 \zeta \\ + \cos^2 c \,.\, \cos^2 \xi - 2 \cos b \,.\, \cos c \,.\, \cos \xi \,.\, \cos \varepsilon + \cos^2 b \,.\, \cos^2 \varepsilon \end{array}\right\}$$

$$= L^2 + M^2 + N^2 \quad . \quad . \quad . \quad . \quad . \quad . \quad . \quad (76)$$

But

$$\cos^2 a + \cos^2 b + \cos^2 c = 1, \quad . \quad . \quad . \quad . \quad . \quad . \quad (77)$$

$$\cos^2 \zeta + \cos^2 \xi + \cos^2 \varepsilon = 1, \quad . \quad . \quad . \quad . \quad . \quad . \quad (78)$$

$$\cos a \,.\, \cos \zeta + \cos b \,.\, \cos \xi + \cos c \,.\, \cos \varepsilon = \cos \varphi, \quad . \quad (79)$$

the angle φ, being that made by the line H, with the direction of the resultant.

Collecting the co-efficients of $\cos^2 a$, $\cos^2 b$, $\cos^2 c$, and reducing by the following relations, deduced from Equation (78); viz.:

$$\cos^2 \varepsilon + \cos^2 \xi = 1 - \cos^2 \zeta,$$
$$\cos^2 \zeta + \cos^2 \varepsilon = 1 - \cos^2 \xi,$$
$$\cos^2 \xi + \cos^2 \zeta = 1 - \cos^2 \varepsilon,$$

we find,

$$R^2 . H^2 . [1-(\cos a . \cos \zeta + \cos b . \cos \xi + \cos c . \cos \varepsilon)^2] = L^2 + M^2 + N^2;$$

from Equation (79),

$$1 - (\cos a . \cos \zeta + \cos b . \cos \xi + \cos c . \cos \varepsilon)^2 = 1 - \cos^2 \varphi = \sin^2 \varphi;$$

which reduces the above to

$$R^2 . H^2 . \sin^2 \varphi = L^2 + M^2 + N^2.$$

But $H^2 . \sin^2 \varphi$ is the square of the perpendicular drawn from the origin to the direction of the resultant; it is, therefore, the square of the lever arm of the resultant referred to the origin as a centre of moments. Denoting this lever arm by K, we have, after taking the square root,

$$R . K = \sqrt{L^2 + M^2 + N^2} \quad . \quad . \quad . \quad . \quad . \quad (80)$$

That is to say, *the resultant moment of any system of forces is equal to the square root of the sum of the squares of the sums of the component moments, taken in reference to any three rectangular axes through the point assumed as the centre of moments.*

§ 110.—Dividing the first of Equations (75), by Equation (80), we find,

$$\frac{H (\cos b . \cos \zeta - \cos a . \cos \xi)}{K} = \frac{L}{\sqrt{L^2 + M^2 + N^2}}.$$

The effect of a force is, § 77, independent of the position of its point of application, provided it be taken on the line of direction. Let the point of application of R, be taken at the extremity of its lever arm, then will H coincide with and be equal in length to K ζ and ξ will become the angles which the lever arm makes with the axes x and y, respectively, and the well known relation obtained from the formulas for the transformation of co-ordinates from one set of rectangular axes to another, will give

$$\cos \Theta_z = \cos b . \cos \zeta - \cos a . \cos \xi ;$$

in which Θ_z is the angle the resultant axis makes with the axis z; whence,*

$$\cos \Theta_z = \frac{L}{\sqrt{L^2 + M^2 + N^2}} . \quad . \quad . \quad . \quad . \quad . \quad . \quad (81)$$

In the same way, denoting by Θ_y and Θ_x the angles which the moment axis of R makes with the co-ordinate axes y and x respectively, will

$$\cos \Theta_y = \frac{M}{\sqrt{L^2 + M^2 + N^2}} . \quad . \quad . \quad . \quad . \quad . \quad . \quad (82)$$

$$\cos \Theta_x = \frac{N}{\sqrt{L^2 + M^2 + N^2}} . \quad . \quad . \quad . \quad . \quad . \quad . \quad (83)$$

whence we conclude that, *the cosine of the angle which the resultant axis makes with any assumed line is equal to the sum of the moments of the forces in reference to this line taken as a component axis divided by the resultant moment.*

§ 111.—Multiplying Equation (81) by Equation (80), there will result,

$$R . K . \cos \Theta_z = L \quad . \quad . \quad . \quad . \quad . \quad . \quad . \quad (84)$$

which shows that *the component moment of any system of forces in reference to any oblique axis is equal to the product of the resultant moment of the system into the cosine of the angle between the resultant and component axes.*

For the same system of forces and the same centre of moments, it is obvious that R and K will be constant; whence, Equation (80), the *sum of the squares of the sums of the moments in reference*

* See Appendix, No. I.

to any three rectangular axes through the centre of moments, taken as component axes, is a constant quantity. Also, since the axis z_1 may have an infinite number of positions and still satisfy the condition of making equal angles with the resultant axis, we see, Equation (84), that *the sum of the moments of the forces in reference to all component axes which make equal angles with the resultant axis will be constant.*

§ 112.—Denote by $\theta_z, \theta_y, \theta_x$, the angles which any component axis makes with the co-ordinate axes z, y and x, respectively, and by δ the angle which the component and resultant axes make with each other, then will

$$\cos\delta = \cos\Theta_z \cdot \cos\theta_z + \cos\Theta_y \cdot \cos\theta_y + \cos\Theta_x \cdot \cos\theta_x;$$

multiplying both members by $R.K$, we have

$$R.K.\cos\delta = R.K.\cos\Theta_z.\cos\theta_z + R.K.\cos\Theta_y\cos\theta_y + R.K.\cos\Theta_x.\cos\theta_x.$$

But, Equation (84),

$$R.K.\cos\Theta_z = L,$$
$$R.K.\cos\Theta_y = M,$$
$$R.K.\cos\Theta_x = N;$$

which substituted above, give

$$R.K.\cos\delta = L.\cos\theta_z + M.\cos\theta_y + N.\cos\theta_x \quad . \quad . \quad (85)$$

That is to say, *the component moment in reference to any assumed component axis, is equal to the sum of the products arising from multiplying the sum of the moments in reference to the co-ordinate axes, by the cosines of the angles which the direction of the component axis makes with these co-ordinate axes, respectively.*

TRANSLATION OF EQUATIONS (A) AND (B).

§ 113.—Equations (A) and (B) may now be translated. They express the conditions of equilibrium of a system of forces acting in various directions and upon different points of a solid body. These conditions are six in number; viz.:

1.—*The algebraic sum of the components of the forces in each of any three rectangular directions must be separately equal to zero ;*

2.—*The algebraic sum of the moments of the forces taken in reference to each of three rectangular axes drawn through any assumed centre of moments, must be separately equal to zero.*

If the extraneous forces be in equilibrio, the terms which measure the forces of inertia will disappear, and these conditions of equilibrium will be expressed by

$$\left.\begin{array}{l} \Sigma P . \cos \alpha = 0, \\ \Sigma P \cos \beta = 0, \\ \Sigma P . \cos \gamma = 0 ; \end{array}\right\} \quad . \quad . \quad . \quad . \quad . \quad . \quad (A)'$$

$$\left.\begin{array}{l} \Sigma P . (x' \cos \beta - y' \cos \alpha) = 0, \\ \Sigma P . (z' . \cos \alpha - x' \cos \gamma) = 0, \\ \Sigma P . (y' \cos \gamma - z' \cos \beta) = 0. \end{array}\right\} \quad . \quad . \quad . \quad (B)'$$

The above conditions, which relate to the action of a system of forces on a free body, are qualified by conditions of constraint that determine the possible motion.

§ 114.—If the body contain a *fixed point*, the origin of the movable co-ordinates, in Equation (40), may be taken at this point; in which case we shall have,

$$\delta x_{,} = 0,$$
$$\delta y_{,} = 0,$$
$$\delta z_{,} = 0;$$

and it will only be necessary that the forces satisfy Equations (B), these being the co-efficients of the indeterminate quantities that do not reduce to zero. Hence, in the case of a fixed point, *the sum of the moments of the forces, taken in reference to each of three rectangular axes, passing through the point, must separately reduce to zero.*

Should the system contain *two fixed points*, one of the axes, as that of x, may be assumed to coincide with the line joining these points, in which case, there will result in Equation (40),

$$\delta x_, = 0, \quad \delta \varphi = 0,$$
$$\delta y_, = 0, \quad \delta \psi = 0.$$
$$\delta z_, = 0,$$

and it will only be necessary that the forces satisfy the last Equation in group (B); or that *the sum of the moments of the forces in reference to the line joining the fixed points, reduce to zero.*

If the system be free to *slide along this line*, $\delta x_,$ will not reduce to zero, and it will be necessary that its co-efficient, in Equation (40), reduce to zero; or that *the algebraic sum of the components of the given forces parallel to the line joining the fixed points, also reduce to zero.*

If three points of the system be constrained to remain in a fixed plane, one of the co-ordinate planes, as that of $x y$, may be assumed parallel to this plane; in which case,

$$\delta z_, = 0,$$
$$\delta \varpi = 0,$$
$$\delta \psi = 0;$$

and the forces must satisfy the first and second of Equations (A) and the first of $(B)'$; that is, *the algebraic sum of the components of the given forces parallel to each of two rectangular axes parallel to the given plane, must separately reduce to zero, and the sum of the moments in reference to an axis perpendicular to this plane must reduce to zero.*

CENTRE OF GRAVITY.

§ 115.—Gravity is the name given to that force which urges all bodies towards the centre of the earth. This force acts upon every particle of matter. Every body may, therefore, be regarded as subjected to the action of a system of forces whose number is equal to the number of its particles, and whose points of application have, with respect to any system of axes, the same co-ordinates as these particles.

The *weight* of a body is the resultant of this system, or *the resultant of all the forces of gravity which act upon it*, and is equal, in intensity, but directly opposed to the force which is just sufficient to support the body.

The direction of the force of gravity is perpendicular to the earth's surface. The earth is an oblate spheroid, of small eccentricity, whose mean radius is nearly four thousand miles; hence, as the directions of the force of gravity converge towards the centre, it is obvious that these directions, when they appertain to particles of the same body of ordinary magnitude, are sensibly parallel, since the linear dimensions of such bodies may be neglected, in comparison with any radius of curvature of the earth.

The centre of such a system of forces is determined by Equations (62), § 94, which are

$$\left.\begin{aligned} x_{,} &= \frac{P'x' + P''x'' + P'''x''' + \&c.}{P' + P'' + P''' + \&c.}, \\ y_{,} &= \frac{P'y' + P''y'' + P'''y''' + \&c.}{P' + P'' + P''' + \&c.}, \\ z_{,} &= \frac{P'z' + P''z'' + P'''z''' + \&c.}{P' + P'' + P''' + \&c.}, \end{aligned}\right\} \quad . \quad . \quad . \quad . \quad (86)$$

in which $x_{,}$ $y_{,}$ $z_{,}$, are the co-ordinates of the centre; P', P'', &c., the forces arising from the action of the force of gravity, that is, the weights of the elementary masses m', m'', &c., of which the co-ordinates are respectively x' y' z', x'' y'' z'', &c.

This centre is called the *centre of gravity*. From the values of its co-ordinates, Equations (86), it is apparent that the position of this point is independent of the direction of the force of gravity in reference to any assumed line of the body; and the centre of gravity of a body may be defined to be *that point through which its weight always passes in whatever way the body may be turned in regard to the direction of the force of gravity.*

The values of P', P'', &c., being regarded as the weights w', w'', &c., of the elementary masses m', m'', &c., we have, Equation (1),

$$P' = w' = m'g'; \quad P'' = w'' = m''g''; \quad P''' = w''' = m'''g'''; \quad \&c.,$$

and, Equations (86),

$$\left.\begin{aligned} x_{\prime} &= \frac{m' g' x' + m'' g'' x'' + m''' g''' x''' + \&c.}{m' g' + m'' g'' + m''' g''' + \&c.}, \\ y_{\prime} &= \frac{m' g' y' + m'' g'' y'' + m''' g''' y''' + \&c.}{m' g' + m'' g'' + m''' g''' + \&c.}, \\ z_{\prime} &= \frac{m' g' z' + m'' g'' z'' + m''' g''' z''' + \&c.}{m' g' + m'' g'' + m''' g''' + \&c.}. \end{aligned}\right\} \quad \ldots \quad (87)$$

§ 116.—It will be shown by a process to be given in the proper place, that the intensity of the force of gravity varies inversely as the square of the distance from the centre of the earth. The distance from the surface to the centre of the earth is nearly four thousand miles; a change of half a mile in the distance at the surface would therefore, only cause a change of one four-thousandth part of its entire amount in the force of gravity; and hence, within the limits of bodies whose centres of gravity it may be desirable in practice to determine, the change would be inappreciable. Assuming, then, the force of gravity at the same place as constant, Equations (87), become

$$\left.\begin{aligned} x_{\prime} &= \frac{m' x' + m'' x'' + m''' x''' + \&c.}{m' + m'' + m''' + \&c.}, \\ y_{\prime} &= \frac{m' y' + m'' y'' + m''' y''' + \&c.}{m' + m'' + m''' + \&c.}, \\ z_{\prime} &= \frac{m' z' + m'' z'' + m''' z''' + \&c.}{m' + m'' + m''' + \&c.}; \end{aligned}\right\} \quad \ldots\ldots \quad (88)$$

from which it appears, that when the action of the force of gravity is constant throughout any collection of particles, the position of the centre of gravity is independent of the intensity of the force.

§ 117.—Substituting the value of the masses, given in Equation (1)′, there will result,

$$\left.\begin{aligned} x_{\prime} &= \frac{v' d' x' + v'' d'' x'' + v''' d''' x''' + \&c.}{v' d' + v'' d'' + v''' d''' + \&c.}, \\ y_{\prime} &= \frac{v' d' y' + v'' d'' y'' + v''' d''' y''' + \&c.}{v' d' + v'' d'' + v''' d''' + \&c.}, \\ z_{\prime} &= \frac{v' d' z' + v'' d'' z'' + v''' d''' z''' + \&c.}{v' d' + v'' d'' + v''' d''' + \&c.}, \end{aligned}\right\} \quad \ldots \quad (89)$$

and if the elements be of homogenous density throughout, we shall have,

$$d' = d'' = d''' = \&c.;$$

and Equations (89) become,

$$\left.\begin{aligned} x_{\prime} &= \frac{v'x' + v''x'' + v'''x''' + \&c.}{v' + v'' + v''' + \&c.}, \\ y_{\prime} &= \frac{v'y' + v''y'' + v'''y''' + \&c.}{v' + v'' + v''' + \&c.}, \\ z_{\prime} &= \frac{v'z' + v''z'' + v'''z''' + \&c.}{v' + v'' + v''' + \&c.}; \end{aligned}\right\} \quad \ldots \quad (90)$$

whence it follows, that in all homogeneous bodies, the position of the centre of gravity is independent of the density, provided the intensity of gravity is the same throughout.

§ 118.—Employing the character Σ, in its usual signification, Equations (90), may be written,

$$\left.\begin{aligned} x_{\prime} &= \frac{\Sigma\,(v\,x)}{\Sigma\,(v)}, \\ y_{\prime} &= \frac{\Sigma\,(v\,y)}{\Sigma\,(v)}, \\ z_{\prime} &= \frac{\Sigma\,(v\,z)}{\Sigma\,(v)}; \end{aligned}\right\} \quad \ldots \quad (91)$$

and if the system be so united as to be continuous,

$$\left.\begin{aligned} x_{\prime} &= \frac{\int_{v''}^{v'} x \,.\, d\,V}{V}, \\ y_{\prime} &= \frac{\int_{v''}^{v'} y \,.\, d\,V}{V}, \\ z_{\prime} &= \frac{\int_{v''}^{v'} z \,.\, d\,V}{V}. \end{aligned}\right\} \quad \ldots \quad (92)$$

§ 119.—If the collection be divided symmetrically by the plane $x\,y$, then will

$$\Sigma\,(v\,z) = 0,$$

and, therefore,

$$z_{,} = 0;$$

hence, the centre of gravity will lie in this plane.

If, at the same time, the collection of elements be symmetrically divided by the plane $x z$, we shall have,

$$\Sigma(vy) = 0,$$
$$y_{,} = 0;$$

the collection of elements will be symmetrically disposed about the axis x, and the centre of gravity will be on that line.

Although it is always true, that the centre of gravity will lie in a plane or line that divides a homogeneous collection of particles symmetrically; yet, the converse, it is obvious, is not always true, viz.: that the collection will be symmetrically divided by a plane or line that may contain the centre of gravity.

Equations (92) are employed to determine the centres of gravity of all geometrical figures.

THE CENTRE OF GRAVITY OF LINES.

§ 120.—Let s represent the entire length of an arc of any curve, whose centre of gravity is to be found, and of which the co-ordinates of the extremities are x', y', z', and x'', y'', z''.

To be applicable to this general case of a curve, included within the given limits, Equations (92) become

$$\left.\begin{aligned} x_{,} &= \frac{\int_{x''}^{x'} x\,dx \cdot \sqrt{1 + \frac{dy^2}{dx^2} + \frac{dz^2}{dx^2}}}{s}, \\ y_{,} &= \frac{\int_{x''}^{x'} y\,dx \cdot \sqrt{1 + \frac{dy^2}{dx^2} + \frac{dz^2}{dx^2}}}{s}, \\ z_{,} &= \frac{\int_{x''}^{x'} z\,dx \cdot \sqrt{1 + \frac{dy^2}{dx^2} + \frac{dz^2}{dx^2}}}{s}, \end{aligned}\right\} \quad . \quad . \quad . \quad (93)$$

in which

$$s = \int_{x''}^{x'} dx \sqrt{1 + \frac{dy^2}{dx^2} + \frac{dz^2}{dx^2}} \quad . \quad . \quad . \quad . \quad (94)$$

Example 1.—*Find the position of the centre of gravity of a right line.* Let,

$$y = \alpha x + \beta,$$
$$z = \alpha' x + \beta',$$

be the equations of the line.

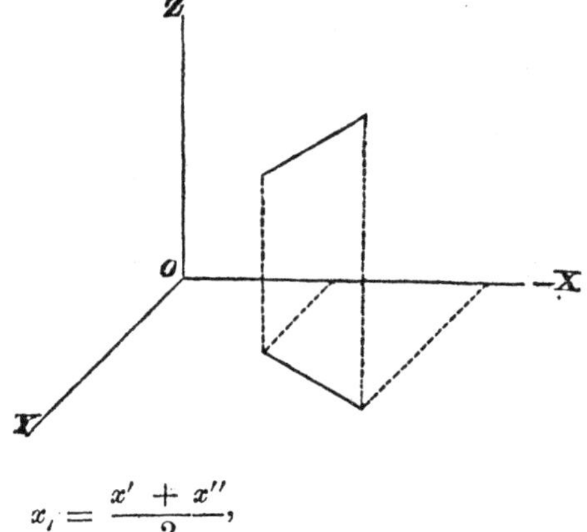

Differentiating, substituting in Equations (94) and (93), integrating between the proper limits, and reducing, there will result,

$$x_{,} = \frac{x' + x''}{2},$$

$$y_{,} = \frac{\alpha . (x' + x'')}{2} + \beta,$$

$$z_{,} = \frac{\alpha' . (x' + x'')}{2} + \beta',$$

which are the co-ordinates of the middle point of the line; x' y' z' and x'' y'' z'', being those of its extremities; whence we conclude *that the centre of gravity of a straight line is at its middle point.*

Example 2.—*Find the centre of gravity of the perimeter of a polygon.*

This may be done, according to Equations (90), by taking the sum of the products which result from multiplying the length of each side by the co-ordinate of its middle point, and dividing this sum by the length of the perimeter of the polygon. Or by construction, as follows:

The weights of the several sides of the polygon constitute a system of parallel forces, whose points of application are the centres of gravity of the sides. The sides being of homogeneous density, their weights are proportional to their lengths. Hence, to find the centre of gravity of the entire polygon, join the middle points of any two of the sides by a right line, and divide this line in the inverse ratio of the lengths of the adjacent sides, the point of division will, § 97, be the centre of gravity of these two sides; next, join this point with the middle of a third side by a straight line, and divide this line in the inverse ratio of the sum of first two sides, and this third side, the point of division will be the centre of gravity of the three sides. Continue this process till all the sides be taken, and the last point of division will be the centre of gravity of the polygon.

Find the position of the centre of gravity of a plane curve.

Assume the plane of $x\,y$ to coincide with the plane of the curve, in which case,

$$\frac{d\,z}{d\,x} = 0,$$

and Equations (93) and (94) become,

$$\left.\begin{aligned} x_{,} &= \frac{\int_{x''}^{x'} x\,d\,x\sqrt{1 + \frac{d\,y^2}{d\,x^2}}}{s}, \\ y_{,} &= \frac{\int_{x''}^{x'} y\,d\,x\sqrt{1 + \frac{d\,y^2}{d\,x^2}}}{s}, \end{aligned}\right\} \quad \cdots\cdots \quad (95)$$

$$s = \int_{x''}^{x'} d\,x\sqrt{1 + \frac{d\,y^2}{d\,x^2}}. \quad \cdots\cdots \quad (96)$$

Example 3.—*Find the centre of gravity of a circular arc.*

Take the origin at the centre of curvature, and the axis of y passing through the middle point of the arc. The equation of the curve is,

$$y^2 = a^2 - x^2,$$

whence,

$$\frac{dy}{dx} = -\frac{x}{y},$$

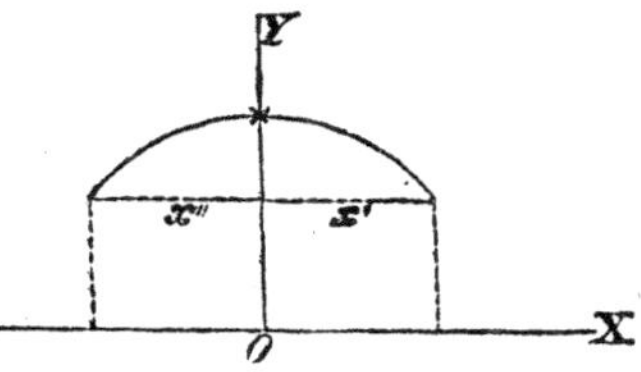

which substituted in Equations (95),

will give on reduction,

$$x_{\prime} = 0,$$

$$y_{\prime} = \frac{a\,(x' + x'')}{s};$$

and denoting the chord of the arc by $c = x' + x''$,

$$x_{\prime} = 0,$$

$$y_{\prime} = \frac{a\,c}{s};$$

whence we conclude that *the centre of gravity of a circular arc is on a line drawn through the centre of curvature and its middle point, and at a distance from the centre equal to a fourth proportional to the arc, radius and chord.*

Example 4.—Find the centre of gravity of the arc of a cycloid.

The radius of the generating circle being a, the differential equation of the curve is,

$$dx = \frac{y\,.\,d\,y}{\sqrt{2\,a\,y - y^2}}, \quad . \; . \; . \; (a)$$

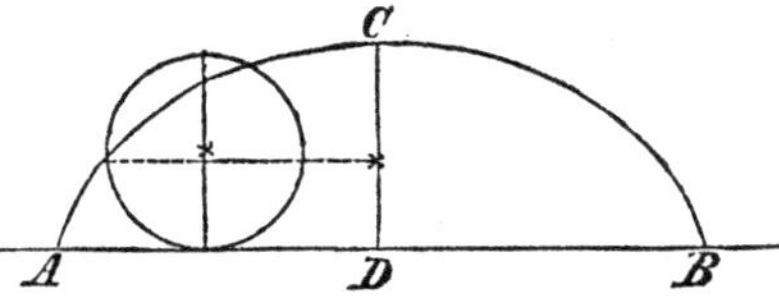

the origin being at A, and $A\,B$ being the axis of x.

Transfer the origin to C, and denote by x', y' the new co-ordinates, the former being estimated in the direction $C\,D$, and the latter in the direction $D\,A$. Then will

$$y = 2\,a - x',$$

$$x = a\,\pi - y';$$

and therefore,

$$\frac{d\,x}{d\,y} = \frac{d\,y'}{d\,x'} = \frac{2\,a - x'}{\sqrt{2\,a\,x' - x'^2}}; \quad . \; . \; . \; . \; . \; . \; (a)'$$

this, in Equations (96) and (95), gives, omitting the accent on the variables,

$$s = \int_{x''}^{x'} dx \sqrt{\frac{2a}{x}},$$

$$x_{,} = \frac{\int_{x''}^{x'} x\,dx \sqrt{\frac{2a}{x}}}{s},$$

$$y_{,} = \frac{\int_{x''}^{x'} y\,dx \sqrt{\frac{2a}{x}}}{s}.$$

Integrating the first two equations between the limits indicated, and substituting the value of s, deduced from the first, in the second, we have,

$$s = 2\sqrt{2a}\,(\sqrt{x''} - \sqrt{x'}),$$

$$x_{,} = \frac{1}{3} \cdot \frac{\sqrt{x''^3} - \sqrt{x'^3}}{\sqrt{x''} - \sqrt{x'}};$$

and from the third equation we have, after integrating by parts,

$$s\,y_{,} = 2\sqrt{2\,a}\,(y\sqrt{x} - \int\sqrt{x}\,dy);$$

substituting the value of dy, obtained from Equation $(a)'$, and reducing, there will result,

$$s\,y_{,} = 2\sqrt{2\,a}\,(y\sqrt{x} - \int\sqrt{2a - x}\,.\,dx),$$

and taking the integral between the indicated limits,

$$s\,y_{,} = 2\sqrt{2a}\,[y(\sqrt{x''} - \sqrt{x'}) + \tfrac{2}{3}(2a - x'')^{\frac{3}{2}} - \tfrac{2}{3}(2\,a - x')^{\frac{3}{2}}];$$

hence, replacing s by its value, and dividing,

$$y_{,} = y + \tfrac{2}{3} \cdot \frac{(2\,a - x'')^{\frac{3}{2}} - (2\,a - x')^{\frac{3}{2}}}{\sqrt{x''} - \sqrt{x'}}.$$

Supposing the arc to begin at C, we have,

$$x' = 0,$$

and,

$$x_{,} = \tfrac{1}{3}x'',$$

$$y_{,} = y + \frac{2}{3\sqrt{x''}} \cdot \left[(2a - x'')^{\frac{3}{2}} - 2\,a\sqrt{2a}\right].$$

If the entire semi-arc from C to A be taken, these values become,

$$x_{\prime} = \tfrac{2}{3} a,$$

$$y_{\prime} = a\left(\pi - \tfrac{4}{3}\right).$$

Taking the entire arc $A\,C\,B$, the curve will be symmetrical with respect to the axis of x'; and therefore,

$$y_{\prime} = 0;$$

hence, *the centre of gravity of the arc of the cycloid, generated by one entire revolution of the generating circle, is on the line which divides the curve symmetrically, and at a distance from the summit of the curve equal to one-third of its height.*

THE CENTRE OF GRAVITY OF SURFACES.

§ 121.—Let $L = 0$, be the equation of any surface; L being a function of $x\,y\,z$; then will $dx\,dy$, be the projection of an element of this surface, whose co-ordinates are $x\,y\,z$, upon the plane $x\,y$; and if θ'' denote the angle which a plane tangent to the surface at the same point makes with the plane $x\,y$, the value of the element itself will be

$$\frac{dx\,.\,dy}{\cos\theta''}.$$

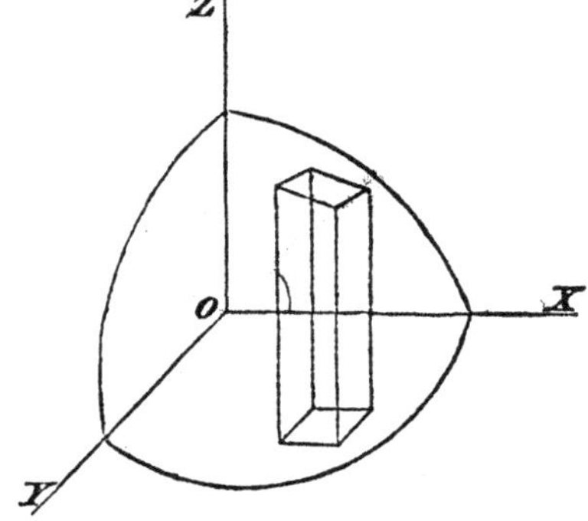

But the angle which a plane makes with the co-ordinate plane $x\,y$, is equal to the angle which the normal to the plane makes with the axis z, and, therefore,

$$\cos\theta'' = \pm\frac{\dfrac{dL}{dz}}{\sqrt{\left(\dfrac{dL}{dx}\right)^2 + \left(\dfrac{dL}{dy}\right)^2 + \left(\dfrac{dL}{dz}\right)^2}} = \pm\frac{1}{w} \quad . \quad . \quad (97)$$

and hence, in Equations (92), omitting the double sign,

$$d V = dx \cdot dy \cdot w, \quad \ldots \ldots \quad (98)$$

and those Equations become,

$$\left.\begin{aligned} x_{\prime} &= \frac{\int_{y''}^{y'}\int_{x''}^{x'} w \,.\, x \,.\, dx \,.\, dy}{s}, \\ y_{\prime} &= \frac{\int_{y''}^{y'}\int_{x''}^{x'} w \,.\, y\, dx \,.\, dy}{s}, \\ z_{\prime} &= \frac{\int_{y''}^{y'}\int_{x''}^{x'} w \,.\, z \,.\, dx \,.\, dy}{s}, \end{aligned}\right\} \quad \ldots \ldots \quad (99)$$

in which,

$$s = V = \int_{y''}^{y'}\int_{x''}^{x'} w \,.\, dx \,.\, dy \,; \quad \ldots \quad (100)$$

w being a function of x, y, z.

If the surface be plane, the plane of $x\,y$ may be taken in the surface, in which case,

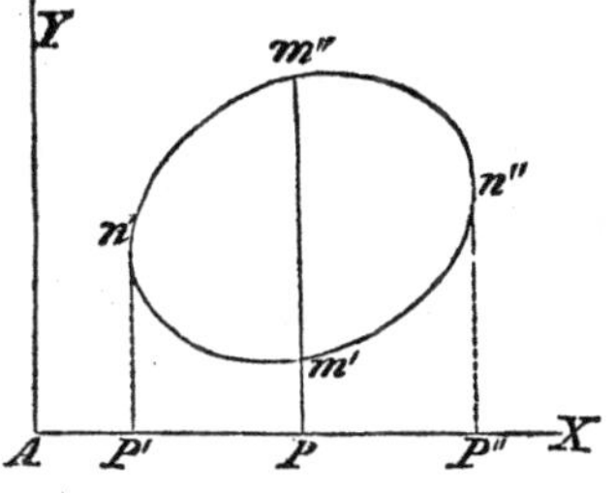

$$w = 1,$$

$$z = 0,$$

and Equations (99), and (100), become,

$$\left.\begin{aligned} x_{\prime} &= \frac{\int_{y''}^{y'}\int_{x''}^{x'} dy \,.\, x\, dx}{s}, \\ y_{\prime} &= \frac{\int_{y''}^{y'}\int_{x''}^{x'} dx \,.\, y\, dy}{s}, \end{aligned}\right\} \quad \ldots \ldots \quad (101)$$

$$s = \int_{y''}^{y'}\int_{x''}^{x'} dx \,.\, dy, \quad \ldots \ldots \quad (102)$$

in which the integral is to be taken first with respect to y, and between the limits $y'' = Pm''$ and $y' = Pm'$; then in respect to x, between the limits $x'' = AP''$, and $x' = AP'$. Hence

$$\left.\begin{aligned} x_{,} &= \frac{\int_{x''}^{x'} (y'' - y') \,.\, x\,dx}{s}, \\ y_{,} &= \frac{\frac{1}{2}\int_{x''}^{x'} (y''^2 - y'^2)\,dx}{s}, \end{aligned}\right\} \quad \ldots\ldots \quad (103)$$

$$s = \int_{x''}^{x'} (y'' - y')\,dx. \quad \ldots\ldots\ldots \quad (104)$$

y' and y'', denoting running co-ordinates, which may be either roots of the same equation, resulting from the same value of x, or they may belong to two distinct functions of x, the value of x being the same in each. For instance, if

$$F\ (x\,y) = 0,$$

be the equation of the curve $n'm''n''m'$, it is obvious that between the limits $x'' = AP''$ and $x' = AP'$, every value of x, as AP, must give two values for y, viz.: $y'' = Pm''$ and $y' = Pm'$. Or if

$$F\,(x\,y) = 0,$$
$$F'\,(x\,y) = 0,$$

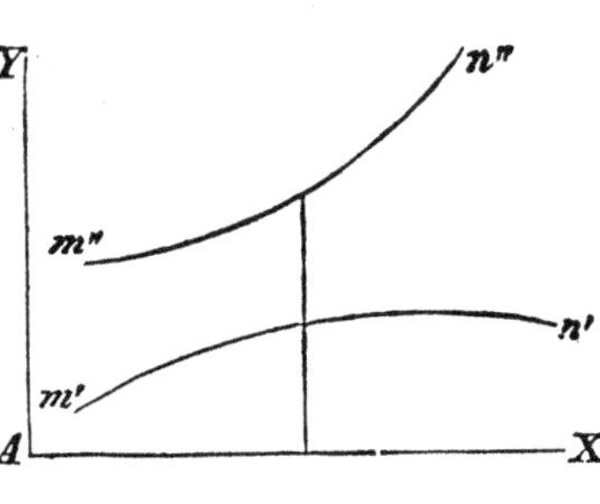

be the equations of two distinct curves $m''n''$ and $m'n'$, referred to the same origin A, then will y'' and y' result from these functions separately, when the same value is given to x in each.

Example 1.—*Required the position of the centre of gravity of the area of a triangle.*

Let $A\,B\,C$, be the triangle. Assume the origin of co-ordinates at one of the angles A, and draw the axis y parallel to the opposite side $B\,C$. Denote the distance $A\,P$ by x', and suppose,

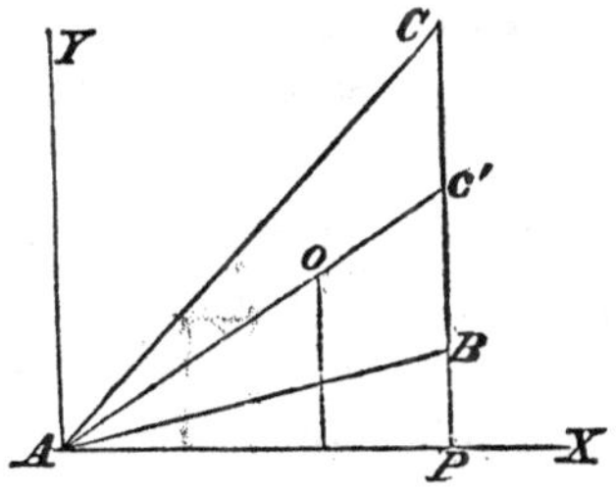

$$y'' = ax,$$
$$y' = bx,$$

to be the equations of the sides $A\,C$ and $A\,B$, respectively, then will

$$y'' - y' = (a - b)\,x,$$
$$y''^2 - y'^2 = (a^2 - b^2)\,x^2,$$

and,

$$x_{\prime} = \frac{\int_{x'}^{0} (a - b)\,x^2\,d\,x}{\int_{x'}^{0} (a - b)\,x\,d\,x} = \frac{2}{3}\,x',$$

$$y_{\prime} = \frac{\frac{1}{2}\int_{x'}^{0} (a^2 - b^2)\,x^2\,d\,x}{\int_{x'}^{0} (a - b)\,x\,d\,x} = \frac{2}{3}\,\frac{(a + b)\,x'}{2};$$

whence we conclude, *that the centre of gravity of a triangle is on a line drawn from any one of the angles to the middle of the opposite side, and at a distance from this angle equal to two-thirds of the line thus drawn.*

Example 2.—*Find the centre of gravity of the area of any polygon.*

From any one of the angles as A, of the polygon, draw lines to all the other angles except those which are adjacent on either side; the polygon will thus be divided into triangles. Find by the rule just given, the centre of gravity of each of the triangles;

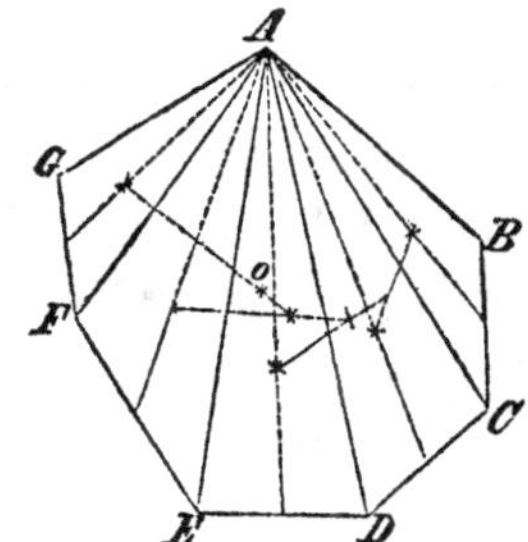

join any two of these centres by a right line, and divide this line in the inverse ratio of the areas of the triangles to which these centres belong; the point of division will be the centre of gravity of these two triangles. Join, by a straight line, this centre with the centre of gravity of a third triangle, and divide this line in the inverse ratio of the sum of the areas of the first two triangles and of the third, this point of division will be the centre of gravity of the three triangles. Continue this process till all the triangles be embraced by it, and the last point of division will be the centre of gravity of the polygon, the reasons for the rule being the same as those given for the determination of the centre of gravity of the perimeter of a polygon, it being only necessary to substitute the areas of the triangles for the lengths of the sides.

Example 3.—*Determine the position of the centre of gravity of a circular sector.*

The centre of gravity of the sector will be on the radius drawn to the middle point of the arc, since this radius divides the sector symmetrically. Conceive the sector CAB, to be divided into an indefinite number of elementary sectors; each one of these may be regarded as a triangle whose centre of gravity is at a distance from the centre C, equal to two-thirds of the radius. If, therefore, from this centre an arc be described with a radius equal to two-thirds the radius of the sector, this arc will be the locus of the centres of gravity of all the elementary sectors; and for reasons already explained, the centre of gravity of the entire sector will be the same as that of the portion of this arc which is included between the extreme radii of the sector. Hence, calling r the radius of the sector, a and c its arc and chord respectively, and $x_{,}$ the distance of the centre of gravity from the centre C, we have,

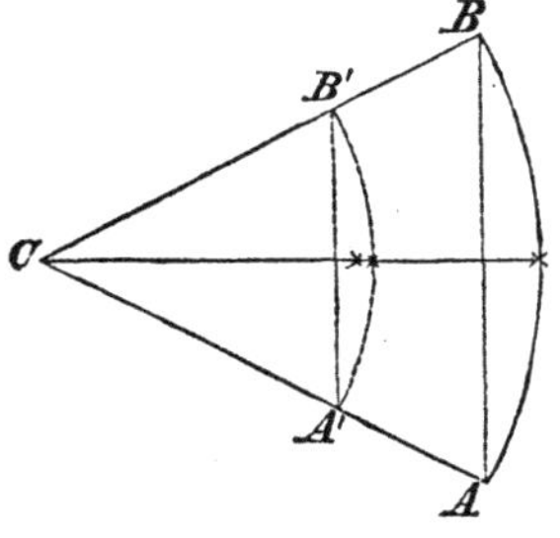

$$x_{,} = \frac{\frac{2}{3} r \cdot \frac{2}{3} c}{\frac{2}{3} a} = \frac{2}{3} \cdot \frac{r \cdot c}{a}.$$

The centre of gravity of a circular sector is therefore *on the radius drawn to the middle point of the arc of the sector, and at a distance from the centre of curvature equal to two-thirds of a fourth proportional to the arc, chord and radius of the sector.*

Example 4.—*Find the centre of gravity of a circular segment.*

Assume the origin at the centre C, and take the axis x passing through the middle point of the arc, the centre of gravity in question will be on this axis, and, therefore,

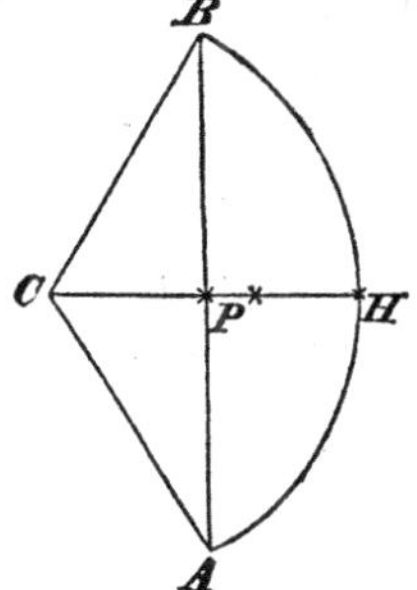

$$y_{\prime} = 0.$$

Let $ABHA$ be the segment, and

$$y = \pm \sqrt{a^2 - x^2},$$

the equation of the circle, the origin being at the centre C, then will

$$y'' = \sqrt{a^2 - x^2},$$
$$y' = -\sqrt{a^2 - x^2},$$

and, Equations (103) and (104),

$$x_{\prime} = \frac{2\int_a^{x'} \sqrt{a^2 - x^2} \, . \, x \, . \, dx}{s} = \frac{\frac{2}{3}(a^2 - x'^2)^{\frac{3}{2}}}{s},$$

$$s = 2\int_a^{x'} \sqrt{a^2 - x^2} \, . \, dx = a^2\left(\frac{\pi}{2} - \sin^{-1}\frac{x'}{a}\right) - x'\sqrt{a^2 - x'^2},$$

s being the area of the entire segment. Denoting the chord AB by c, we have,

$$\sqrt{a^2 - x'^2} = \tfrac{1}{2}c\,;$$

whence,

$$x_{\prime} = \frac{c^3}{12 \, . \, s}\,;$$

and we conclude, that *the centre of gravity of a circular segment is on the radius drawn to the middle of the arc, and at a distance from the centre equal to the cube of the chord, divided by twelve times the area of the segment.*

Replacing the value of s, and supposing x' to be zero, in which case the segment becomes a semicircle, we shall find,

$$c = 2a,$$

$$x_{,} = \frac{4a}{3\pi}.$$

§ 122.—If the surface be one of revolution, about the axis x for instance, it will be symmetrical with respect to this axis; hence,

$$y_{,} = 0; \quad z_{,} = 0;$$

and if $F(xy) = 0$, be the equation of a meridian section in the plane xy, then will the area of an elementary zone comprised between two planes perpendicular to the axis of revolution be,

$$2\pi . y . \sqrt{dx^2 + dy^2},$$

and therefore, Equations (92),

$$x_{,} = 2\pi \frac{\int_{x''}^{x'} yx\sqrt{1 + \frac{dy^2}{dx^2}} \cdot dx}{s} \quad . \quad . \quad . \quad . \quad (105)$$

$$s = 2\pi \int_{x''}^{x'} y . \sqrt{1 + \frac{dy^2}{dx^2}} . dx \quad . \quad . \quad . \quad . \quad (106)$$

Example 1.—*Find the position of the centre of gravity of a right conical surface.*

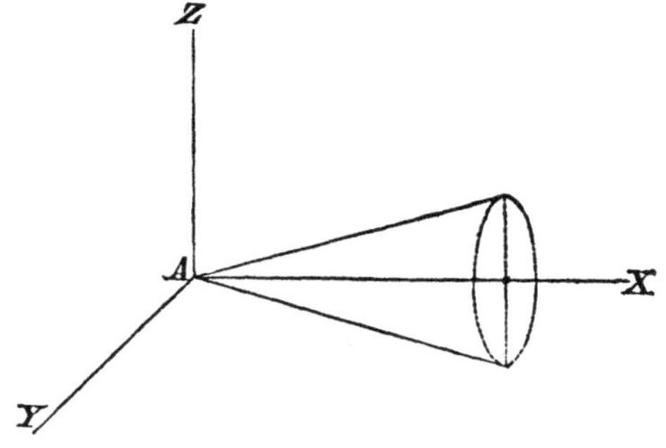

The equation of the element in the plane xy, is, assuming the origin at the vertex,

$$y = ax;$$

hence,

$$x_{,} = \frac{2\pi \int_{x''}^{0} ax^2\, dx \sqrt{1 + a^2}}{2\pi \int_{x''}^{0} ax\;\; dx \sqrt{1 + a^2}} = \frac{2}{3} x''.$$

Example 2.—Required the position of the centre of gravity of a spherical zone.

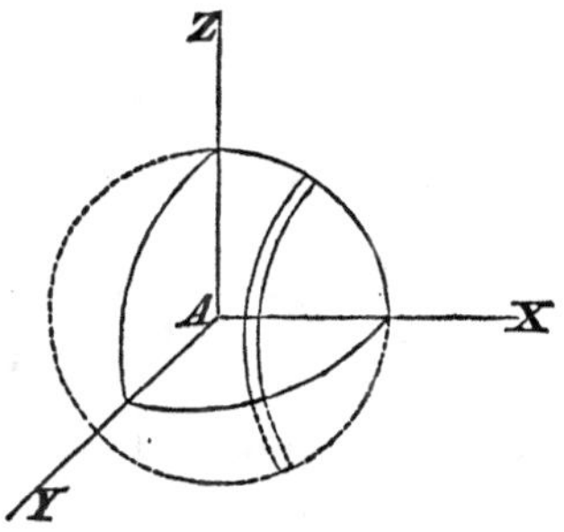

Assuming the origin at the centre, the equation of the meridian curve is,

$$y^2 = a^2 - x^2;$$

whence,

$$y\,dy = -\,x\,dx,$$

$$\frac{dy^2}{dx^2} = \frac{x^2}{y^2},$$

and,

$$x_{,} = \frac{\int_{x''}^{x'} a\,x\,dx}{\int_{x''}^{x'} a\,dx} = \frac{x''^2 - x'^2}{2\,(x'' - x')} = \frac{x'' + x'}{2}.$$

Hence, *the centre of gravity of a spherical zone, is at the middle point of a line joining the centres of its circular bases. And in the case of one base it is only necessary to make* $x'' = a$, *which gives,*

$$x_{,} = \frac{x' + a}{2}.$$

So that *the centre of gravity of a zone of one base is at the middle of the ver-sine of its meridian curve.*

THE CENTRES OF GRAVITY OF VOLUMES.

§ 123.—When it is the question to determine the centre of gravity of the *volume* of any body, we have

$$dV = dx\,.\,dy\,.\,dz,$$

and Equations (92) become,

$$x_{,} = \frac{\int_{x''}^{x'} \int_{y''}^{y'} \int_{z''}^{z'} x\,.\,dy\,.\,dz\,.\,dx}{V},$$

$$y_{\prime} = \frac{\int_{x''}^{x'} \int_{y''}^{y'} \int_{z''}^{z'} y \,.\, dy \,.\, dz \,.\, dx}{V},$$

$$z_{\prime} = \frac{\int_{x''}^{x'} \int_{y''}^{y'} \int_{z''}^{z'} z \,.\, dy \,.\, dz \,.\, dx}{V},$$

and,

$$V = \int_{x''}^{x'} \int_{y''}^{y'} \int_{z''}^{z'} dy \,.\, dz \,.\, dx.$$

In which the triple integral must be extended to include the entire space embraced by the surface of the body; this surface being given by its equation.

If the volume be symmetrical with respect to any line, this line may be assumed as one of the co-ordinate axes, as that of x; in which case, if X represent the area of a section perpendicular to this axis, and x, its distance from the plane yz, then will $X dx$, be an elementary volume symmetrically disposed in regard to the axis x, and Equations (92), become

$$x_{\prime} = \frac{\int_{x''}^{x'} X x\, dx}{V}, \quad \ldots \ldots \ldots \quad (107)$$

$$y_{\prime} = 0,$$

$$z_{\prime} = 0,$$

and,

$$V = \int_{x''}^{x'} X dx. \quad \ldots \ldots \ldots \quad (108)$$

Example 1.—Find the position of the centre of gravity of a semi-ellipsoid, the equation of whose surface is

$$\frac{x^2}{A^2} + \frac{y^2}{B^2} + \frac{z^2}{C^2} = 1.$$

The semi-axes of the elliptical section parallel to the plane yz, are,

$$y = B\sqrt{1 - \frac{x^2}{A^2}},$$

$$z = C\sqrt{1 - \frac{x^2}{A^2}};$$

whence,

$$X = \pi B C \left(1 - \frac{x^2}{A^2}\right),$$

and, Equations (107) and (108),

$$V = \int_A^0 \pi B C \left(1 - \frac{x^2}{A^2}\right) d x$$

$$x_, = \frac{\int_A^0 \pi B C \left(1 - \frac{x^2}{A^2}\right) x d x}{\int_A^0 \pi B C \left(1 - \frac{x^2}{A^2}\right) d x} = \frac{3}{8} A.$$

If the figure be one of revolution about the axis of x, then, denoting by

$$F(x y) = 0, \quad \ldots \ldots \ldots \quad (109)$$

the equation of the meridian section by the plane $x y$, will

$$X = \pi y^2,$$

and Equations (107) and (108), may be written,

$$x_, = \frac{\int_{x''}^{x'} \pi y^2 x d x.}{V}, \quad \ldots \ldots \quad (110)$$

$$V = \int_{x''}^{x'} \pi y^2 d x \quad \ldots \ldots \ldots \quad (111)$$

Example 1.—*Required the position of the centre of gravity of a paraboloid of revolution.*

In this case, Equaticn (109),

$$F(x y) = y^2 - 2 p x = 0,$$

whence,

$$V = 2 \pi p \int_a^0 x d x,$$

$$x_, = \frac{2 \pi p \int_a^0 x^2 d x}{2 \pi p \int_a^0 x d x} = \frac{2}{3} a.$$

Example 2.—*Required the position of the centre of gravity of the volume of a spherical segment.*

$$F(x\,y) = y^2 + x^2 - a^2 = 0,$$

whence,

$$V = \pi \int_{x''}^{x'} (a^2 - x^2)\, d\,x$$

$$x_{,} = \frac{\pi \int_{x''}^{x'} (a^2 - x^2)\,.\,x\,.\,d\,x}{\pi \int_{x''}^{x'} (a^2 - x^2)\, d\,x};$$

or,

$$x_{,} = \frac{3}{4} \cdot \left[\frac{x''^2(2a^2 - x''^2) - x'^2\,(2a^2 - x'^2)}{x''(3a^2 - x''^2) - x'\,(3a^2 - x'^2)}\right];$$

and for a segment of one base, $x'' = a$,

$$x_{,} = \frac{3}{4} \cdot \frac{a^4 - x'^2\,(2a^2 - x'^2)}{2a^3 - x'\,(3a^2 - x'^2)}.$$

If the volume have a plane face, and be of such figure that the areas of all sections parallel to this face, are connected by any law of their distances from it, the position of the centre of gravity, may also be found by the method of single integrals.

Example 1.—*Find the centre of gravity of any pyramid.*

Find by the method explained, the centre of gravity of the base of the pyramid, and join this point with the vertex by a straight line. All sections parallel to the base are similar to it, and will be pierced by this line in homologous points and therefore in their centres of gravity. Each section being supposed indefinitely thin, and its weight acting at its centre of gravity, the centre of gravity of the entire pyramid will, § 94, be found somewhere on the same line.

Take the origin at the vertex, draw the axis x perpendicular to the plane of the base, and the plane $x\,y$ through its centre of gravity; and let X represent any section parallel to the base, then will Equations (92) become,

$$x_{,} = \frac{\int_{x''}^{x'} X x\, d x}{V},$$

$$y_{,} = \frac{\int_{x''}^{x'} X y\, d x}{V},$$

$$z_{,} = 0,$$

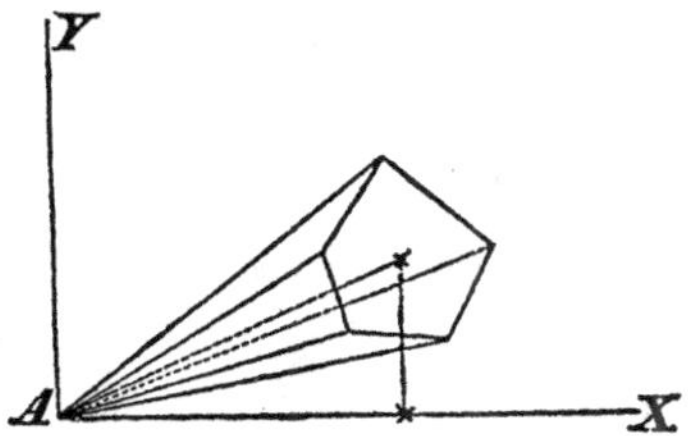

and,

$$V = \int_{x''}^{x'} X\, d x.$$

Represent by A the base of the pyramid, c its altitude, and let

$$y = a\, x,$$

be the equation of the line joining the vertex and centre of gravity of the base.

Then,

$$A : X : : c^2 : x^2,$$

$$X = \frac{A\, x^2}{c^2},$$

and for any frustum,

$$V = \int_{x''}^{x'} \frac{A\, x^2\, d x}{c^2},$$

$$x_{,} = \frac{\frac{A}{c^2}\int_{x''}^{x'} x^3\, d x}{\frac{A}{c^2}\int_{x''}^{x'} x^2\, d x} = \frac{3}{4}\left(\frac{x''^4 - x'^4}{x''^3 - x'^3}\right),$$

$$y_{,} = \frac{\frac{a\, A}{c^2}\int_{x''}^{x'} x^3\, d x}{\frac{A}{c^2}\int_{x''}^{x'} x^2\, d x} = \frac{3}{4}\, a \left(\frac{x''^4 - x'^4}{x''^3 - x'^3}\right);$$

and for the entire pyramid, make $x'' = c$, and $x' = 0$, which give

$$x_{,} = \tfrac{3}{4}\, c,$$

$$y_{,} = \tfrac{3}{4}\, a\, c;$$

whence we conclude that *the centre of gravity of a pyramid is on the line drawn from the vertex to the centre of gravity of the base, and at a distance from the vertex equal to three-fourths of the length of this line.*

The same rule obviously applies to a cone, since the result is independent of the figure of the base.

The weight of a body always acting at its centre of gravity, and in a vertical direction, it follows, that if the body be freely suspended in succession from any two of its points by a perfectly flexible thread, and the directions of this thread, when the body is in equilibrio, be produced, they will intersect at the centre of gravity; and hence it will only be necessary, in any particular case, to determine this point of intersection, to find, experimentally, the centre of gravity of a body.

THE CENTROBARYC METHOD.

§ 124.—Resuming the second of Equations (95) and (103), which are,

$$y_{,} = \frac{\int_{x''}^{x'} y\,dx\sqrt{1 + \frac{dy^2}{dx^2}}}{s},$$

in which

$$s = \int_{x''}^{x'} dx\sqrt{1 + \frac{dy^2}{dx^2}},$$

and

$$y_{,} = \frac{\frac{1}{2}\int_{x''}^{x'} (y''^2 - y'^2)\,dx}{s},$$

in which

$$s = \int_{x''}^{x'} (y'' - y')\,dx;$$

clearing the fractions and multiplying both members by 2π, we shall have,

$$2\pi . y_{,} s = \int_{x''}^{x'} 2\pi y\ \sqrt{dx^2 + dy^2}, \quad . \quad . \quad . \quad (112)$$

$$2\pi y_{,} s = \int_{x''}^{x'} \pi (y''^2 - y'^2)\,dx \quad . \quad . \quad . \quad . \quad (113)$$

The second member of Equation (112) is the area of a surface generated by the revolution of a plane curve, whose extremities are given by the ordinates answering to the abscisses x' and x'', about the axis x. In the first member, s is the entire length of this arc, and $2 \pi y_{,}$ is the circumference generated by its centre of gravity. Hence, we have this simple rule for finding the area of a figure of revolution, viz.:

Multiply the length of the generating curve by the circumference described by its centre of gravity about the axis of rotation; the product will be the required surface.

The second member of Equation (113) is the volume generated by a plane area, bounded by two branches of the same curve or by two different curves, and the ordinates answering to the abscisses x' and x'', about the axis x. s, in the first member, is the generating area, and $2 \pi y_{,}$ the circumference described by its centre of gravity. Hence, this rule for finding the volume of any figure of revolution, viz.:

Multiply the generating area by the circumference described by its centre of gravity about the axis of rotation; the product will be the volume sought.

Example 1.—*Required the measure of the surface of a right cone.*

Let the cone be generated by the rotation of the line $A B$ about the line $A C$. The centre of gravity of the generatrix is at its middle point G, and therefore, the radius of the circle described by it will be one-half of the radius $C B$, of the circular base of the cone. Hence,

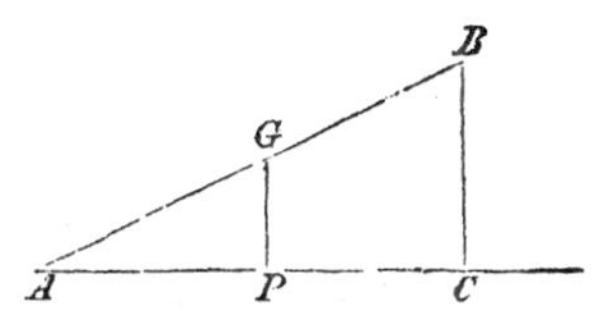

$$2 \pi y_{,} . s = 2 \pi . \frac{B C . A B}{2} = \pi B C . A B.$$

Example 2.—*Find the volume of the cone.*

The area of the generatrix $A B C$, is $\frac{1}{2} B C . A C$; and the radius of the circle described by its centre of gravity is $\frac{1}{3} B C$. Hence,

$$2 \pi y_{,} s = \tfrac{2}{3} \pi B C . \frac{B C . A C}{2} = \pi \frac{B C^2 . A C}{3}.$$

CENTRE OF INERTIA.

§ 125.—When the elementary masses of a body exert their forces of inertia simultaneously and in parallel directions, they must experience equal accelerations or retardations in the same time, and the factor

$$\frac{d^2s}{dt^2},$$

in the measures of these forces, as given in Equation (13), must be the same for all. Substituting these measures for P', P'', &c., in Equations (62), we find,

$$\left.\begin{aligned} x &= \frac{\frac{d^2s}{dt^2}\cdot\Sigma\, m\, x'}{\frac{d^2s}{dt^2}\cdot\Sigma\, m} = \frac{\Sigma\, m\, x'}{\Sigma\, m}; \\ y &= \frac{\frac{d^2s}{dt^2}\cdot\Sigma\, m\, y'}{\frac{d^2s}{dt^2}\cdot\Sigma\, m} = \frac{\Sigma\, m\, y'}{\Sigma\, m}; \\ z &= \frac{\frac{d^2s}{dt^2}\cdot\Sigma\, m\, z'}{\frac{d^2s}{dt^2}\cdot\Sigma\, m} = \frac{\Sigma\, m\, z'}{\Sigma\, m}. \end{aligned}\right\} \quad \ldots\ldots \quad (114)$$

Whence, Equations (88), the centre of inertia coincides with the centre of gravity when the force of gravity is constant, both being at the centre of mass. In strictness, however, the centre of gravity is always below the centre of inertia; for when the variation in the force of gravity, arising from change of distance, is taken into account, the lower of two equal masses will be found the heavier. And in bodies whose linear dimensions bear some appreciable proportion to their distances from the centre of attraction, the distance between these centres becomes sensible, and gives rise to some curious phenomena.

MOTION OF THE CENTRE OF INERTIA.

§ 126.—Substitute in Equations (A), the values of $d^2 x$, $d^2 y$, and $d^2 z$, given by Equations (34), and we have, because dt is constant, and $d^2 x_{\prime}$, $d^2 y_{\prime}$ and $d^2 z_{\prime}$, will each be a common factor for all the elementary masses,

$$\Sigma P \cos \alpha - M \cdot \frac{d^2 x_{\prime}}{dt^2} - \frac{1}{dt^2} \cdot \Sigma m . d^2 x' = 0,$$

$$\Sigma P \cos \beta - M \cdot \frac{d^2 y_{\prime}}{dt^2} - \frac{1}{dt^2} \cdot \Sigma m . d^2 y' = 0,$$

$$\Sigma P \cos \gamma - M \cdot \frac{d^2 z_{\prime}}{dt^2} - \frac{1}{dt^2} \cdot \Sigma m . d^2 z' = 0.$$

in which M, denotes the entire mass of the body, being equal to Σm.

Denote by $\bar{x}$, $\bar{y}$, $\bar{z}$, the co-ordinates of the centre of inertia referred to the movable origin, then, Equations (114),

$$M . \bar{x} = \Sigma m x',$$
$$M . \bar{y} = \Sigma m y',$$
$$M . \bar{z} = \Sigma m z',$$

and differentiating twice,

$$\left.\begin{array}{l} M . d^2\bar{x} = \Sigma m . d^2x', \\ M . d^2\bar{y} = \Sigma m . d^2y', \\ M . d^2\bar{z} = \Sigma m . d^2z', \end{array}\right\} \quad \ldots \ldots \quad (115)$$

which substituted in the preceding Equations, give,

$$\left.\begin{array}{l} \Sigma P \cdot \cos \alpha - M \cdot \dfrac{d^2x_{\prime}}{dt^2} - M \cdot \dfrac{d^2\bar{x}}{dt^2} = 0, \\ \Sigma P \cdot \cos \beta - M \cdot \dfrac{d^2y_{\prime}}{dt^2} - M \cdot \dfrac{d^2\bar{y}}{dt^2} = 0, \\ \Sigma P \cdot \cos \gamma - M \cdot \dfrac{d^2z_{\prime}}{dt^2} - M \cdot \dfrac{d^2\bar{z}}{dt^2} = 0, \end{array}\right\} \quad \ldots \quad (116)$$

and if the movable origin be taken at the centre of inertia, then will,

$$d^2\bar{x} = 0, \quad d^2\bar{y} = 0, \quad d^2\bar{z} = 0;$$

and $x_{,}$, $y_{,}$, $z_{,}$, will become the co-ordinates of the centre of inertia referred to the fixed origin, and we have,

$$\left.\begin{aligned} \Sigma P.\cos\alpha - M\cdot\frac{d^2x_{,}}{dt^2} &= 0, \\ \Sigma P.\cos\beta - M\cdot\frac{d^2y_{,}}{dt^2} &= 0, \\ \Sigma P.\cos\gamma - M\cdot\frac{d^2z_{,}}{dt^2} &= 0; \end{aligned}\right\} \quad \cdots \quad (117)$$

Equations which are wholly independent of the relative positions of the elementary masses m', m'' &c., since their co-ordinates x', y', z', &c., do not enter. It will also be observed that the resistance of inertia is the same as that of an equal mass concentrated at the body's centre of inertia.

Whence we conclude, that when a body is subjected to the action of any system of extraneous forces, the motion of its centre of inertia will be the same as though the entire mass were concentrated into that point, and the forces applied without change of intensity and direction, directly to it.

This is an important fact, and shows that in discussing the motion of translation of bodies, we may confine our attention to the motion of their centres of inertia regarded as material points.

ROTATION AROUND THE CENTRE OF INERTIA.

§ 127.—Now, retaining the movable origin at the centre of inertia, substitute in Equations (B), the values of d^2x, d^2y, and d^2z, as given by Equations (34), and reduce by the relations,

$$M.\bar{x} = \Sigma m.x' = 0,$$
$$M.\bar{y} = \Sigma m.y' = 0,$$
$$M.\bar{z} = \Sigma m.z' = 0;$$

and we have,

$$\left.\begin{aligned}
\Sigma P.(\cos\beta\,.\,x' - \cos\alpha\,.\,y') - \Sigma m\cdot\left(\frac{d^2y'}{dt^2}\cdot x' - \frac{d^2x'}{dt^2}\cdot y'\right) &= 0,\\
\Sigma P.(\cos\alpha\,.\,z' - \cos\gamma\,.\,x') - \Sigma m\cdot\left(\frac{d^2x'}{dt^2}\cdot z' - \frac{d^2z'}{dt^2}\cdot x'\right) &= 0,\\
\Sigma P.(\cos\gamma\,.\,y' - \cos\beta\,.\,z') - \Sigma m\cdot\left(\frac{d^2z'}{dt^2}\cdot y' - \frac{d^2y'}{dt^2}\cdot z'\right) &= 0;
\end{aligned}\right\}\quad(118)$$

from which all traces of the position of the centre of inertia have disappeared, and from which we infer that when a free body is acted upon by any system of forces, the body will rotate about its centre of inertia exactly the same whether that centre be at rest or in motion.

§ 128.—And we are to conclude, Equations (117) and (118), that when a body is subjected to the action of one or more forces, it will in general, take up two motions—one of translation, and one of rotation, each being perfectly independent of the other.

§ 129.—Multiply the first of Equations (117), by $y_{\prime}$, the second by $x_{\prime}$, and subtract the first product from the second; also, the first by $z_{\prime}$, the third by $x_{\prime}$, and subtract the second of these products from the first; also the third by $y_{\prime}$, and the second by $z_{\prime}$, and subtract the second of these products from the first, and we have,

$$\left.\begin{aligned}
\Sigma(P\cos\beta)\,.\,x_{\prime} - \Sigma(P\cos\alpha)\,.\,y_{\prime} - M\cdot\left(\frac{d^2y_{\prime}}{dt^2}\cdot x_{\prime} - \frac{d^2x_{\prime}}{dt^2}\cdot y_{\prime}\right) &= 0,\\
\Sigma(P\cos\alpha)\,.\,z_{\prime} - \Sigma(P\cos\gamma)\,.\,x_{\prime} - M\cdot\left(\frac{d^2x_{\prime}}{dt^2}\cdot z_{\prime} - \frac{d^2z_{\prime}}{dt^2}\cdot x_{\prime}\right) &= 0,\\
\Sigma(P\cos\gamma)\,.\,y_{\prime} - \Sigma(P\cos\beta)\,.\,z_{\prime} - M\cdot\left(\frac{d^2z_{\prime}}{dt^2}\cdot y_{\prime} - \frac{d^2y_{\prime}}{dt^2}\cdot z_{\prime}\right) &= 0;
\end{aligned}\right\}\quad(119)$$

Equations from which may be found the circumstances of motion of the centre of inertia about the fixed origin.

MOTION OF TRANSLATION.

§ 130.—Regarding the forces as applied directly to the centre of inertia, replace in Equations (117), the values $\Sigma P . \cos \alpha$, $\Sigma P . \cos \beta$, and $\Sigma P . \cos \gamma$, by X, Y, and Z, respectively, and we may write,

$$\left.\begin{aligned} X - M \cdot \frac{d^2x}{dt^2} &= 0, \\ Y - M . \frac{d^2y}{dt^2} &= 0, \\ Z - M . \frac{d^2z}{dt^2} &= 0; \end{aligned}\right\} \quad \dots \dots \quad (120)$$

from which the accents are omitted, and in which x, y, and z, must be understood as appertaining to the centre of inertia.

GENERAL THEOREM OF WORK, VELOCITY AND LIVING FORCE.

§ 131.—Multiply the first of Equations (120) by $2 d x$, the second by $2 d y$, the third by $2 d z$, add and integrate, we have

$$2 \int (X d x + Y d y + Z d z) - M . \frac{dx^2 + dy^2 + dz^2}{dt^2} + C = 0.$$

But,

$$\frac{dx^2 + dy^2 + dz^2}{dt^2} = \frac{ds^2}{dt^2} = V^2 ;$$

whence,

$$2 \int (X d x + Y d y + Z d z) - M . V^2 + C = 0 \quad . \quad . \quad (121)$$

The first term is, § 101, twice the quantity of work of the extraneous forces, the second is twice the quantity of work of the inertia, measured by the living force, and the third is the constant of integration.

If the forces X, Y, Z, be variable, they must be expressed in functions of x, y, z, before the integration can be performed

Supposing this latter condition fulfilled, and that the forms of the functions are such as make the integration possible, we may write,

$$F(x\,y\,z) - \tfrac{1}{2} M . V^2 + C' = 0, \quad . \quad . \quad . \quad . \quad (122)$$

and between the limits $x_{\prime}\ y_{\prime}\ z_{\prime}$ and $x_{\prime}'\ y_{\prime}'\ z_{\prime}'$,

$$F(x_{\prime}'\ y_{\prime}'\ z_{\prime}') - F(x_{\prime}\ y_{\prime}\ z_{\prime}) = \tfrac{1}{2} M (V'^2 - V_{\prime}^2) \quad . \quad . \quad (123)$$

whence we conclude, that the quantity of work expended by the extraneous forces impressed upon a body during its passage from one position to another, is equal to half the difference of the living forces of the body at these two positions.

We also see, from Equation (123), that whenever the body returns to any position it may have occupied before, its velocity will be the same as it was previously at that place. Also, that the velocity, at any point, is wholly independent of the path described.

If

$$X\,d\,x + Y\,d\,y + Z\,d\,z = 0,$$

the extraneous forces will, § 101, be in equilibrio, and

$$V = \sqrt{\frac{2 . C'}{M}};$$

that is, the velocity will be constant, and the motion, therefore, uniform.

§ 132.—Again, multiply the first of Equations (118) by $d\,\varphi$, the second by $d\,\psi$, the third by $d\,\varpi$; add and reduce by the relations given in Equations (38): we find

$$\Sigma P \cos\alpha\, d\,x' + \Sigma P \cos\beta\, d\,y' + \Sigma P \cos\gamma\, d\,z' = \Sigma m \left(\frac{d^2 x' . d\,x'}{d\,t^2} + \frac{d^2 y'\, d\,y'}{d\,t^2} + \frac{d^2 z' . dz'}{d\,t^2} \right);$$

integrating and replacing the first member by its equal in Equation (68), we have

$$\int R\,d\,r = \tfrac{1}{2} \Sigma m \left(\frac{d\,x'^2 + d\,y'^2 + d\,z'^2}{d\,t^2} \right) + C.$$

Denoting the lever arm of R by K, the velocity of the molecule m in reference to the centre of inertia by v, &c., and the arc described by a point in the plane of the resultant R and of its lever arm, at the unit's distance from the centre of inertia, by $s_{,}$, we have

$$\int R\,d\,r = \int R\,K\,.\,d\,s_{,}\,; \qquad \frac{d\,x'^2 + d\,y'^2 + d\,z'^2}{d\,t^2} = v^2, \text{ \&c.};$$

whence

$$\int R\,.\,K\,.\,d\,s_{,} = \tfrac{1}{2}\,\Sigma\,m\,v^2 + C$$

Adding this to Equation (121), there will result

$$2\int(X\,d\,x + Y\,d\,y + Z\,d\,z) + 2\int R\,.\,K\,.\,d\,s_{,} = M\,V^2 + \Sigma\,m\,v^2 + C \quad (121)'$$

From which it is apparent that the quantity of work impressed upon a body, or the living force with which it will move, is dependent not only upon the intensity of the force, but also upon the distance of its line of direction from the centre of inertia.

§ 133.—If Equation (121) be applied to each one of a collection of elements, of which the masses are m, m', &c., there will be as many equations as elements; and if the velocities of these elements be denoted by v, v', &c, we have, by addition,

$$2\,\Sigma\int(X\,d\,x + Y\,d\,y + Z\,d\,z) = \Sigma\,m\,v^2 - C \quad . \quad . \quad (121)''$$

Let the extraneous forces be only those arising from the mutual actions and reactions of the elements upon one another. If the elements m and m' be separated by the distance r, and their co-ordinates be $x\,y\,z$ and $x'\,y'\,z'$, respectively, then, the reciprocal action being along r, will

$$\cos\alpha = \frac{x - x'}{r}; \qquad \cos\beta = \frac{y - y'}{r}; \qquad \cos\gamma = \frac{z - z'}{r};$$

$$\cos\alpha' = -\frac{x - x'}{r}; \quad \cos\beta' = -\frac{y - y'}{r}; \quad \cos\gamma' = -\frac{z - z'}{r};$$

and for the element m we have

$$X\,d\,x + Y\,d\,y + Z\,d\,z = P\left(\frac{x - x'}{r}\,d\,x + \frac{y - y'}{r}\,d\,y + \frac{z - z'}{r}\,d\,z\right);$$

for the element m',

$$X'\,d\,x' + Y'\,d\,y' + Z'\,d\,z' = -P\left(\frac{x - x'}{r}\,d\,x' + \frac{y - y'}{r}\,d\,y' + \frac{z - z'}{r}\,d\,z'\right);$$

and by addition,

$$X\,d\,x + Y\,d\,y + Z\,d\,z + X'\,d\,x' + Y'\,d\,y' + Z'\,d\,z' = \frac{P}{r}[(x - x')\,d\,(x - x') + (y - y')\,d\,(y - y') + (z - z')\,d\,(z - z')].$$

But

$$r^2 = (x - x')^2 + (y - y')^2 + (z - z')^2,$$

and differentiating,

$$r\,d\,r = (x - x')\,d\,(x - x') + (y - y')\,d\,(y - y') + (z - z')\,d\,(z - z');$$

so that the second member above reduces to $P\,d\,r$; and Equation (121)'' to

$$2\,\Sigma \int P\,d\,r = \Sigma\,m\,v^2 - C \quad . \quad . \quad . \quad . \quad . \quad (121)'''$$

If the elements be invariably connected during the motion, the differentials of r will be zero, and

$$\Sigma\,m\,v^2 = C.$$

This is called the *conservation of living force.*

STABLE AND UNSTABLE EQUILIBRIUM.

§ 134.—Resuming Equation (123), omitting the subscript accents, and bearing in mind that the co-ordinates refer to the centre of inertia, into which we may suppose for simplification the body to be concentrated, we may write,

$$\tfrac{1}{2} M\,V'^2 - \tfrac{1}{2} M\,V^2 = F(x'\,y'\,z') - F(x\,y\,z),$$

in which

$$F(x\,y\,z) = \int (X\,d\,x + Y\,d\,y + Z\,d\,z),$$

and

$$d\,F(x\,y\,z) = X\,d\,x + Y\,d\,y + Z\,d\,z.$$

Now, if the limits $x'\,y'\,z'$ and $x\,y\,z$ be taken very near to each other, then will

$$x' = x + d\,x; \quad y' = y + d\,y; \quad z' = z + d\,z;$$

which substituted above, give

$$\tfrac{1}{2} M\,V'^2 - \tfrac{1}{2} M\,V^2 = F(x + d\,x,\ y + d\,y,\ z + d\,z) - F(x\,y\,z),$$

and developing by Taylor's theorem,

$$\tfrac{1}{2} M\,V'^2 - \tfrac{1}{2} M\,V^2 = \left\{ \begin{array}{l} A\,d\,x + B\,d\,y + C\,d\,z \\ + A'\,d\,x^2 + B'\,d\,y^2 + \&c. + D, \end{array} \right.$$

in which D denotes the sum of the terms involving the higher powers of $d\,x$, $d\,y$ and $d\,z$.

If $\frac{1}{2} M V^2$ be a maximum or minimum, then will

$$A dx + B dy + C dz = 0; \quad \cdot \quad \cdot \quad \cdot \quad \cdot \quad \cdot \quad (123)'$$

and since

$$A dx + B dy + C dz = d F(xyz) = X dx + Y dy + Z dz,$$

we have,

$$X dx + Y dy + Z dz = 0.$$

But when this condition is fulfilled, the forces will, Equation (69), be in equilibrio; and we therefore conclude that whenever a body whose centre of inertia is acted upon by forces not in equilibrio, reaches a position in which the living force or the quantity of work is a maximum or minimum, these forces will be in equilibrio.

And, reciprocally, it may be said, *in general*, that when the forces are in equilibrio, the body has a position such that the quantity of action will be a maximum or minimum, though this is not *always* true, since the function is not necessarily either a maximum or a minimum when its first differential co-efficient is zero.

§ 135.—Equation (123)′, being satisfied, we have

$$\tfrac{1}{2} M V'^2 - \tfrac{1}{2} M V^2 = \pm (A' d x^2 + B' d y^2 + \&c. + D) \cdot \cdot \cdot \quad (124)$$

The upper sign answers to the case of a minimum, and the lower to a maximum.

Now, if V be very small, and at the same time a maximum, V' must also be very small and less than V, in order that the second member may be negative; whence it appears that whenever the system arrives at a position in which the living force or quantity of work is a maximum and the system in a state bordering on rest, it cannot depart far from this position if subjected alone to the forces which brought it there. This position, which we have seen is one of equilibrium, is called a position of *stable equilibrium*. In fact, the quantity of work immediately succeeding the position in question becoming negative, shows that the projection of the virtual velocity is negative, and therefore that it is described in opposition to the resultant of the forces, which, as soon as it overcomes the living force already existing, will cause the body to retrace its course.

§ 136.—If, on the contrary, the body reach a position in which the quantity of work is a minimum, the upper sign in Equation (124) must be taken, the second member will always be positive and there will be no limit to the increase of V'. The body may therefore depart further and further from this position, however small V may be; and hence, this is called a position of *unstable equilibrium.*

§ 137.—If the entire second member of Equation (124), be zero, then will,

$$\tfrac{1}{2} M V'^2 - \tfrac{1}{2} M V^2 = 0,$$

and there will be neither increase nor diminution of quantity of work, and whatever position the body occupies the forces will be in equilibrio. This is called *equilibrium of indifference.*

§ 138.—If the system consist of the union of several bodies acted upon only by the force of gravity, the forces become the weights of the bodies which, being proportional to their masses, will be constant. Denoting these weights by W', W'', W''', &c., and assuming the axis of z vertical, we have from Equations (87),

$$R z_{,} = W' z' + W'' z'' + W''' z''' + \&c.,$$

in which R, is the weight of the entire system, and $z_{,}$ the co-ordinate of its centre of gravity; and differentiating,

$$R\, dz_{,} = W' dz' + W'' dz'' + W''' dz''' + \&c. \quad . \quad . \quad . \quad (125)$$

Now, if $z_{,}$ be a maximum or minimum, then will

$$W' dz' + W'' dz'' + W''' dz''' + \&c. = 0,$$

which is the condition of equilibrium of the weights. Whence, we conclude that when the centre of gravity of the system is at the highest or lowest point, the system will be in equilibrio.

In order that the virtual moment of a weight may be positive, vertical distances, when estimated downwards, must be regarded as positive. This will make the second differential of $z_{,}$, positive at the limit of the highest, and negative at the limit of the lowest point. The equilibrium will, therefore, be stable when the centre of gravity is at the lowest, and unstable when at the highest point.

Integrating Equation (125), between the limits $z_{\prime} = H_{\prime}$ and $z_{\prime} = H'$, $z' = h_{\prime}$ and $z' = h'$, &c., and we find,

$$R(H_{\prime} - H') = W'(h_{\prime} - h') + W''(h_{\prime\prime} - h'') + \&c.; \cdot \quad (126)$$

from which we see that the work of the entire weight of the system, acting at its centre of gravity, is equal to the sum of the quantities of work of the component weights, which descend diminished by the sum of the quantities of work of those which ascend.

INITIAL CONDITIONS, DIRECT AND INVERSE PROBLEM.

§ 139.—By integrating each of Equations (120) twice, we obtain three equations involving four variables, viz.: x, y, z and t. By eliminating t, there will result two equations between the variables x, y and z, which will be the equations of the path described by the centre of inertia of the body.

§ 140.—In the course of integration, six arbitrary constants will be introduced, whose values are determined by the *initial* circumstances of the motion. By the term *initial*, is meant the epoch from which t is estimated.

The initial elements are, 1st. The three co-ordinates which give the position of the centre of inertia at the epoch; and 2d. The component velocities in the direction of the three axes at the same instant.

The general integrals determine the nature only, and not the dimensions of the path.

§ 141.—Now two distinct propositions may arise. Either it may be required to find the path from given initial conditions, or to find the initial conditions necessary to describe a given path.

In the first case, by integrating Eqs. (120) twice, we obtain six equations in x, y, z, t, the component velocities, $\frac{dx}{dt}$, $\frac{dy}{dt}$, $\frac{dz}{dt}$, and six arbitrary constants of integration. Making in these equations $t = 0$, and substituting for the co-ordinates and component velocities their initial values, the constants become known. These, in the three equations obtained from last integration, give three equations in x, y, z and t, from which, if t be eliminated, two equations in x, y and z, will result. These will be the equations of the path, and the problem will be completely solved.

In the second case, the two equations of the path being differentiated twice and divided each time by dt, give only four equations involving *three* first, and *three* second differential co-efficients. The inverse problem is, therefore, indeterminate.

But Equation (121) being differentiated and divided by the differential of one of the variables, say dx, gives

$$\tfrac{1}{2} M . \frac{d V^2}{d x} = X + Y . \frac{d y}{d x} + Z \frac{d z}{d x} \quad . \quad . \quad . \quad . \quad (127)$$

which is a fifth equation involving X, Y, Z, and V. By assuming a value for any one of these four quantities, or any condition connecting them, the other five may be found in terms of x, y and z.

VERTICAL MOTION OF HEAVY BODIES.

§ 142.—When a body is abandoned to itself, it falls toward the earth's surface. To find the circumstances of motion, resume Equations (120), in which the only force acting, neglecting the resistance of the air, will be the weight $= Mg$; and we shall have, Equations (117),

$$\Sigma P \cos \alpha = X = Mg . \cos \alpha ;$$
$$\Sigma P \cos \beta = Y = Mg . \cos \beta ;$$
$$\Sigma P \cos \gamma = Z = Mg . \cos \gamma ;$$

in which M denotes the mass of the body. The force of gravity varies inversely as the square of the distance from the centre of the earth, but within moderate limits may be considered invariable. The weight will therefore be constant during the fall.

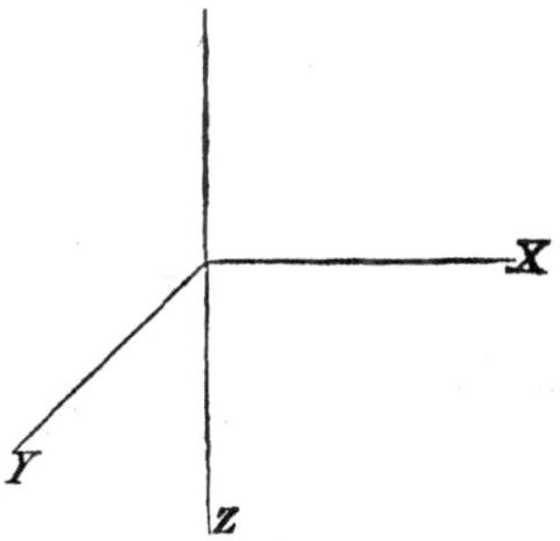

X
Y
Z

Take the co-ordinate z vertical, and positive when estimated downwards, then will

$$\cos \alpha = 0 ; \quad \cos \beta = 0 ; \quad \cos \gamma = 1,$$

and Equations (120) become, after omitting the common factor M,

$$\frac{d^2 x}{d t^2} = 0; \quad \frac{d^2 y}{d t^2} = 0; \quad \frac{d^2 z}{d t^2} = g,$$

and integrating,

$$\frac{d x}{d t} = u_x; \quad \frac{d y}{d t} = u_y;$$

$$\frac{d z}{d t} = v = g t + u_z \quad \ldots \ldots \quad (128)$$

in which v is the actual velocity in a vertical direction.

Making $t = 0$, we have

$$\frac{d z}{d t} = u_z.$$

The constants u_x, u_y and u_z, are the initial velocities in the directions of the axes x, y and z, respectively. Supposing the first two zero, and omitting the subscript z, from the third, we have,

$$\frac{d x}{d t} = 0; \quad \frac{d y}{d t} = 0;$$

$$v = \frac{d z}{d t} = g t + u \quad \ldots \ldots \quad (129)$$

Integrating again, we find

$$x = C; \quad y = C',$$

$$z = \tfrac{1}{2} g t^2 + u t + C'',$$

and if when $t = 0$, the body be on the axis z, and at a distance below the origin equal to a, then will

$$x = 0; \quad y = 0;$$

$$z = \tfrac{1}{2} g t^2 + u t + a \quad \ldots \ldots \quad (130)$$

If the body had been moving upwards at the epoch, then would u have been negative, and, Equations (129) and (130),

$$v = g t - u \quad \ldots \ldots \quad (131)$$

$$z = \tfrac{1}{2} g t^2 - u t + a \quad \ldots \ldots \quad (132)$$

If the body had moved from rest at the epoch and from the origin of co-ordinates, then would v be the actual velocity generated by the body's weight, and $z = h$, the actual space described in the time t; and Equations (129) and (130) would become,

$$v = g\,t \quad . \quad . \quad . \quad . \quad . \quad . \quad . \quad . \quad (133)$$

$$h = \tfrac{1}{2} g\,t^2 \quad . \quad . \quad . \quad . \quad . \quad . \quad . \quad . \quad (134)$$

and eliminating t,

$$v = \sqrt{2\,g\,h} \quad . \quad . \quad . \quad . \quad . \quad . \quad . \quad . \quad (135)$$

whence, we see that the velocity varies as the time in which it is generated; that the height fallen through varies as the square of the time of fall; and that the velocity varies directly as the square root of the height.

The value of h, is called the height due to the velocity v; and the value v, is called the velocity due to the height h.

If, in Equation (132), we suppose $a = 0$, we shall have the case of a body thrown vertically upwards with a velocity u, from the origin, and we may write,

$$v = g\,t - u, \quad . \quad . \quad . \quad . \quad . \quad . \quad . \quad . \quad (136)$$

$$z = \tfrac{1}{2} g\,t^2 - u\,t; \quad . \quad . \quad . \quad . \quad . \quad . \quad . \quad (137)$$

when the body has reached its highest point, v will be zero, and we find,

$$g\,t - u = 0;$$

or,

$$t = \frac{u}{g};$$

which is the time of ascent; and this value of t, in Equation (137), will give the greatest height, $h = z$, to which the body will attain,

$$h = -\frac{u^2}{2\,g}. \quad . \quad . \quad . \quad . \quad . \quad . \quad . \quad . \quad . \quad (138)$$

§ 143.—In the preceding discussion, no account is taken of the atmospheric resistance. For the same body, this resistance varies as the square of the velocity, so that if k, denote the velocity when the resistance becomes equal to the body's weight, then will

$$\frac{M \cdot g \cdot v^2}{k^2},$$

be the resistance when the velocity is v, and in Equations (117), we shall have,

$$\Sigma P \cos \alpha = X = M g \cos \alpha + M g \cdot \frac{v^2}{k^2} \cdot \cos \alpha',$$

$$\Sigma P \cos \beta = Y = M g \cos \beta + M g \cdot \frac{v^2}{k^2} \cdot \cos \beta',$$

$$\Sigma P \cos \gamma = Z = M g \cos \gamma + M g \cdot \frac{v^2}{k^2} \cdot \cos \gamma';$$

taking the co-ordinate z, vertical and positive downward, then will,

$$\cos \alpha = \cos \alpha' = 0,$$
$$\cos \beta = \cos \beta' = 0,$$
$$\cos \gamma = 1, \quad \cos \gamma' = -1;$$

and Equations (120) give,

$$M \cdot \frac{d^2 z}{d t^2} = M g - M g \cdot \frac{v^2}{k^2}$$

Omitting the common factor M, and replacing $\frac{d^2 z}{d t^2}$ by its value $\frac{d v}{d t}$.

$$\frac{d v}{d t} = g \left(1 - \frac{v^2}{k^2}\right)$$

whence,

$$g\, d t = \frac{k^2 \cdot d v}{k^2 - v^2} = \frac{k}{2} \left(\frac{d v}{k + v} + \frac{d v}{k - v}\right) \quad . \quad . \quad (139)$$

Integrating and supposing the initial velocity zero,

$$g t = \tfrac{1}{2} k \cdot \log \frac{k + v}{k - v} \quad . \quad . \quad . \quad . \quad . \quad (140)$$

which gives the time in terms of the velocity; or reciprocally,

$$\frac{k + v}{k - v} = e^{\frac{2gt}{k}} \quad . \quad . \quad . \quad . \quad . \quad . \quad . \quad . \quad (141)$$

in which e, is the base of the Naperian system of logarithms, and from which we find,

$$v = \frac{k\left(e^{\frac{gt}{k}} - e^{-\frac{gt}{k}}\right)}{e^{\frac{gt}{k}} + e^{-\frac{gt}{k}}}, \quad . \quad . \quad . \quad . \quad . \quad . \quad . \quad (142)$$

which gives the velocity in terms of the time. Substituting for v, its value $\frac{dz}{dt}$, integrating and supposing the initial space zero, we have

$$z = \frac{k^2}{g} \cdot \log \tfrac{1}{2}\left(e^{\frac{gt}{k}} + e^{-\frac{gt}{k}}\right) \quad . \quad . \quad . \quad . \quad . \quad (143)$$

Multiplying Equation (139) by

$$\frac{dz}{dt} = v,$$

we have,

$$g\,dz = \frac{k^2 . v . dv}{k^2 - v^2},$$

and integrating, observing the initial conditions as above,

$$z = \frac{k^2}{2g} \cdot \log \frac{k^2}{k^2 - v^2} \quad . \quad . \quad . \quad . \quad . \quad . \quad . \quad (144)$$

which gives the relation between the space and velocity.

As the time increases, the quantity $e^{-\frac{gt}{k}}$ becomes less and less, and the velocity, Equation (142), becomes more nearly uniform; for, if t be infinite, then will

$$e^{-\frac{gt}{k}} = 0,$$

and, Equation (142),

$$v = k;$$

making the resistance of the air equal to the body's weight.

§ 144.—If the body had been moving upwards with a velocity v, then, taking z positive upwards, would, Equations (120),

$$M.\frac{d^2 z}{d t^2} = - M g - M \frac{g v^2}{k^2};$$

substituting $\frac{dv}{dt}$ for $\frac{d^2 z}{d t^2}$, and omitting the common factor, we find,

$$\frac{k . d v}{k^2 + v^2} = - \frac{g d t}{k}; \quad \dots \dots \quad (145)$$

integrating,

$$\tan^{-1} \frac{v}{k} = - \frac{g t}{k} + C;$$

and supposing the initial velocity equal to a, we find

$$C = \tan^{-1} \frac{a}{k},$$

and,

$$\tan^{-1} \frac{v}{k} = \tan^{-1} \frac{a}{k} - \frac{g t}{k} \quad \dots \dots \quad (146)$$

Taking the tangent of both members and reducing, we find

$$v = k \cdot \frac{a - k . \tan \frac{g t}{k}}{k + a . \tan \frac{g t}{k}} \quad \dots \dots \quad (147)$$

which may be put under the form,

$$v = k \cdot \frac{a . \cos \frac{g t}{k} - k . \sin \frac{g t}{k}}{a . \sin \frac{g t}{k} + k . \cos \frac{g t}{k}} \quad \dots \quad (148)$$

Substituting for v its value $\frac{d z}{d t}$, integrating, and supposing the initial space zero, we have

$$z = \frac{k^2}{g} \cdot \log \left(\frac{a}{k} \cdot \sin \frac{gt}{k} + \cos \frac{gt}{k} \right) \quad \dots \quad (149)$$

Multiplying Equation (145), by

$$v = \frac{dz}{dt},$$

and we have,

$$g \, . \, dz = - \frac{k^2 \, . \, v \, . \, dv}{k^2 + v^2};$$

and integrating, with the same initial conditions of v being equal to a, when z is zero, there will result,

$$z = \frac{k^2}{2g} \cdot \log \frac{k^2 + a^2}{k^2 + v^2} \quad . \quad . \quad . \quad . \quad . \quad . \quad . \quad (150)$$

§ 145.—If we denote by h, the greatest height to which the body will ascend, we have $z = h$, when $v = 0$, and hence,

$$h = \frac{k^2}{2g} \cdot \log \frac{k^2 + a^2}{k^2} \quad . \quad . \quad . \quad . \quad . \quad . \quad . \quad (151)$$

Finding the value of t, from Equation (146), we have,

$$t = \frac{k}{g} \left(\tan^{-1} \frac{a}{k} - \tan^{-1} \frac{v}{k} \right) \quad . \quad . \quad . \quad . \quad (152)$$

from which, by making $v = 0$, we have,

$$t_a = \frac{k}{g} \cdot \tan^{-1} \frac{a}{k} \quad . \quad . \quad . \quad . \quad . \quad . \quad . \quad (153)$$

which is the time required for the body to attain the greatest elevation. Having attained the greatest height, the body will descend, and the circumstances of the fall will be given by the Equations of § 143. Denoting by a', the velocity when the body returns to the point of starting, Equation (144), gives,

$$h = \frac{k^2}{2g} \cdot \log \frac{k^2}{k^2 - a'^2}$$

and placing this value of h equal to that given by Equation (151), there will result,

$$\frac{k^2}{k^2 - a'^2} = \frac{k^2 + a^2}{k^2},$$

whence,

$$a'^2 = a^2 \cdot \frac{k^2}{a^2 + k^2};$$

that is, the velocity of the body when it returns to the point of departure is less than that with which it set out.

Making $v = a'$ in Equation (140), we have,

$$t_f = \frac{k}{2g} \cdot \log \frac{k + a'}{k - a'};$$

and, substituting for a', its value above,

$$t_f = \frac{k}{2g} \cdot \log \frac{\sqrt{a^2 + k^2} + a}{\sqrt{a^2 + k^2} - a}, \quad \ldots \quad (154)$$

a value very different from that of t_a, given by Equation (153), for the ascent.

Multiplying both numerator and denominator of the quantity whose logarithm is taken, by $\sqrt{a^2 + k^2} - a$, the above becomes,

$$t_f = \frac{k}{g} \cdot \log \frac{k}{\sqrt{k^2 + a^2} - a}. \quad \ldots \quad (155)$$

Adding Equations (153) and (155), we have,

$$t_a + t_f = \frac{k}{g}\left[\tan^{-1}\frac{a}{k} + \log \frac{k}{\sqrt{a^2 + k^2} - a}\right]$$

or, making $t = t_a + t_f$,

$$\frac{gt}{k} = \tan^{-1}\frac{a}{k} + \log \frac{k}{\sqrt{k^2 + a^2} - a} \quad \ldots \quad (156)$$

If a ball be thrown vertically upwards, and the time of its absence from the surface of the earth be carefully noted, t will be known, and the value of k may be found from this equation. This experiment being repeated with balls of different diameters, and the resulting values of k calculated, the resistance of the air, for any given velocity, will be known.

PROJECTILES.

§ **146.**—Any body projected or impelled forward, is called a *projectile*, and the curve described by its centre of inertia, is called a *trajectory*. The projectiles of artillery, which are usually thrown with great velocity, will be here discussed.

§ 147.—And first, let us consider what the trajectory would be in the absence of the atmosphere. In this case, the only force which acts upon the projectile after it leaves the cannon, is its own weight; and, Equations (117),

$$\Sigma P \cos \alpha = X = Mg \cos \alpha,$$
$$\Sigma P \cos \beta = Y = Mg \cos \beta,$$
$$\Sigma P \cos \gamma = Z = Mg \cos \gamma.$$

Assuming the origin at the point of departure, or the mouth of the piece, and taking the axis z vertical, and positive upwards, then will

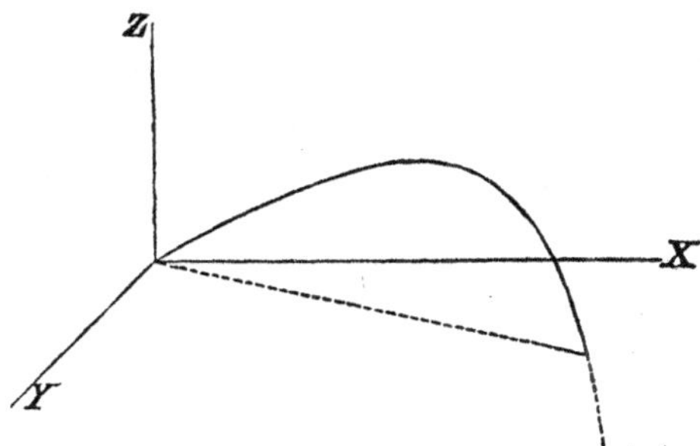

$$\cos \alpha = 0; \quad \cos \beta = 0; \quad \cos \gamma = -1; \text{ and, Equations (120),}$$

$$M \cdot \frac{d^2 x}{dt^2} = 0; \quad M \cdot \frac{d^2 y}{d t^2} = 0; \quad M \cdot \frac{d^2 z}{d t^2} = - Mg;$$

and integrating, omitting M,

$$\frac{d x}{d t} = u_x; \quad \frac{d y}{d t} = u_y; \quad \frac{d z}{d t} = -g t + u_z \quad . \quad . \quad (157)$$

Integrating again, and recollecting that the initial spaces are zero, we have,

$$x = u_x \cdot t; \quad y = u_y \cdot t; \quad z = -\tfrac{1}{2} g t^2 + u_z \cdot t \quad . \quad (158)$$

and eliminating t, from the first two, we obtain,

$$y = \frac{u_y}{u_x} \cdot x ;$$

which is the equation of a right line, and from which we see that the trajectory is a plane curve, and that its plane is vertical.

Assume the plane zx, in this plane, then will $y = 0$, and Equations (158), become,

$$x = u_x \cdot t ; \quad z = -\tfrac{1}{2} g t^2 + u_z \cdot t. \quad . \quad . \quad . \quad (159)$$

Denote by V, the velocity with which the ball leaves the piece, that is, the initial velocity, and by α, the angle which the axis of the piece makes with the axis x, then will,

$$V. \cos \alpha, \quad \text{and} \quad V . \sin \alpha,$$

be the lengths of the paths described in a unit of time, in the direction of the axes x and z, respectively, in virtue of the velocity V; they are, therefore, the initial velocities in the directions of these axes; and we have,

$$u_x = V \cos \alpha ; \quad u_z = V . \sin \alpha ;$$

which, in Equations (159), give

$$x = V . \cos \alpha . t ; \quad z = - \tfrac{1}{2} g t^2 + V . \sin \alpha . t \quad . \quad (160)$$

and eliminating t, we find

$$z = x \tan \alpha - \frac{g . x^2}{2 V^2 . \cos^2 \alpha} ;$$

or substituting for V its value in Equation (135),

$$z = x \tan \alpha - \frac{x^2}{4 h . \cos^2 \alpha} \quad . \quad . \quad . \quad . \quad (161)$$

which is the equation of a parabola.

§ 148.—The angle α is called *the angle of projection;* and the horizontal distance $A\ D$, from the place of departure A, to the point D, at which the projectile attains the same level, is called the *range.*

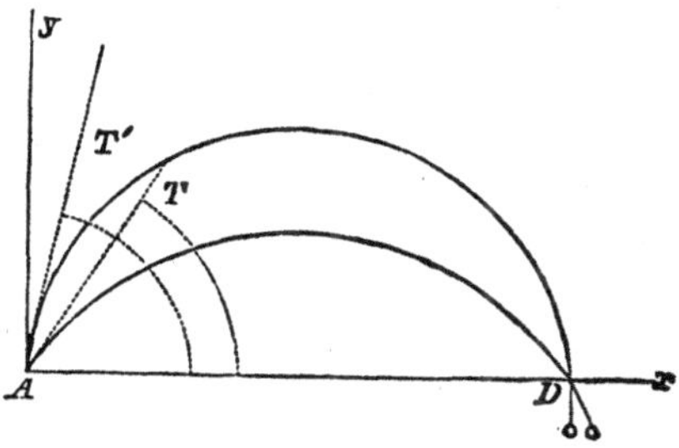

To find the range, make $z = 0$, and Equation (161) gives

$$x = 0, \text{ and } x = 4\,h \sin\alpha \cos\alpha = 2\,h \sin 2\,\alpha,$$

and denoting the range by R,

$$R = 2\,h\,.\sin 2\,\alpha \quad \ldots \quad (162)$$

the value of which becomes the greatest possible when the angle of projection is 45°. Making $\alpha = 45°$, we have

$$R = 2\,h \quad \ldots \quad (163)$$

that is, the maximum range is equal to twice the height due to the velocity of projection.

From the expression for its value, we also see that the same range will result from two different angles of projection, one of which is the complement of the other.

§ 149.—Denoting by v the velocity at the end of any time t, we have,

$$v^2 = \frac{d\,s^2}{d\,t^2} = \frac{d\,z^2 + d\,x^2}{d\,t^2}$$

or, replacing the values of $d\,z$ and $d\,x$, obtained from Equations (160),

$$v^2 = V^2 - 2\,V.g.t.\sin\alpha + g^2\,t^2 \quad \ldots \quad (164)$$

and eliminating t, by means of the first of Equations (160), and replacing V^2, in the last term by its value $2\,g\,h$,

$$v^2 = V^2 - 2\,g\,.\tan\alpha\,.\,x + g\cdot\frac{x^2}{2\,h\,.\cos^2\alpha} \quad \ldots \quad (165)$$

in which, if we make $x = 4h \,.\, \sin\alpha \cos\alpha$, we have the velocity at the point D,

$$v^2 = V^2,$$

which shows that the velocity at the furthest extremity of the range is equal to the initial velocity.

Differentiating Equation (161), we get

$$\frac{dz}{dx} = \tan\theta = \tan\alpha - \frac{x}{2h \,.\, \cos^2\alpha} \quad . \quad . \quad . \quad . \quad (166)$$

in which θ is the angle which the direction of the motion at any instant makes with the axis x.

Making $\tan\theta = 0$, we find

$$x = 2h \,.\, \cos\alpha \,.\, \sin\alpha,$$

which, in Equation (161), gives

$$z = h \,.\, \sin^2\alpha,$$

the elevation of the highest point.

Substituting for x, the range, $4h \cos\alpha \sin\alpha$, in Equation (166),

$$\tan\theta = -\tan\alpha,$$

which shows that the angle of fall is equal to minus the angle of projection.

§ 150.—The initial velocity V being given, let it be required to find the angle of projection which will cause the trajectory to pass through a given point whose co-ordinates are $x = a$ and $z = b$.

Substituting these in Equation (161), we have

$$b = a \tan\alpha - \frac{a^2}{4h \,.\, \cos^2\alpha},$$

from which to determine α.

Making $\tan\alpha = \varphi$, we find

$$\cos^2\alpha = \frac{1}{1 + \varphi^2},$$

which in the equation above, gives

$$4h.b + a^2 - 4h.a.\varphi + a^2\varphi^2 = 0;$$

whence,

$$\varphi = \tan\alpha = \frac{2h}{a} \pm \frac{1}{a}\sqrt{4h^2 - 4hb - a^2} \quad . \quad . \quad . \quad (167)$$

The double sign shows that the object is attained by two angles, and the radical shows that the solution of the problem will be possible as long as

$$4h^2 > 4hb + a^2.$$

Making,

$$4h^2 - 4h.b - a^2 = 0,$$

the question may be solved with only a single angle of projection. But the above equation is that of a parabola whose co-ordinates are a and b, and this curve being constructed and revolved about its vertical axis, will enclose the entire space within which the given point must be situated in order that it may be struck with the given initial velocity. This parabola will pass through the farthest extremity of the maximum range, and at a height above the piece equal to h.

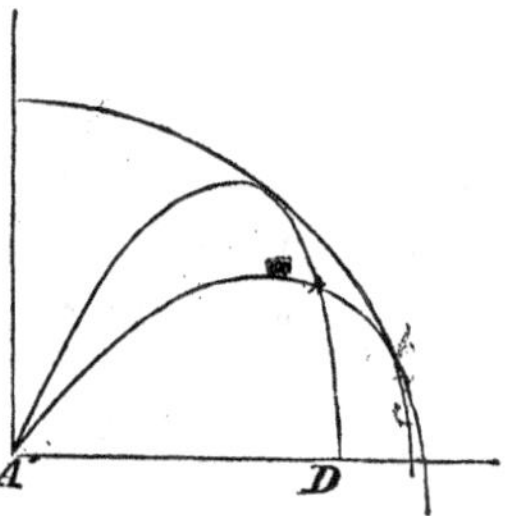

§ 151.—Thus we see that the theory of the motion of projectiles is a very simple matter as long as the motion takes place in vacuo. But in practice this is never the case, and where the velocity is considerable, the atmospheric resistance changes the nature of the trajectory, and gives to the subject no little complexity.

Denote, as before, the velocity of the projectile when the atmospheric resistance equals its weight, by k, and assuming that the resistance varies as the square of the velocity, the actual resistance at any instant when the velocity is v, will be,

$$\frac{M.g.v^2}{k^2} = Mcv^2,$$

by making,

$$\frac{g}{k^2} = c.$$

The forces acting upon the projectile after it leaves the piece being its weight and the atmospheric resistance, Equations (120), become,

$$M \cdot \frac{d^2 x}{d t^2} = Mg \,.\, \cos \alpha + Mc \,.\, v^2 \,.\, \cos \alpha',$$

$$M \cdot \frac{d^2 y}{d t^2} = Mg \,.\, \cos \beta + Mc \,.\, v^2 \,.\, \cos \beta',$$

$$M \cdot \frac{d^2 z}{d t^2} = Mg \,.\, \cos \gamma + Mc \,.\, v^2 \,.\, \cos \gamma'$$

Taking the co-ordinates z vertical, and positive when estimated upwards,

$$\cos \alpha = 0 ; \quad \cos \beta = 0 ; \quad \cos \gamma = -1,$$

and because the resistance takes place in the direction of the trajectory, and in opposition to the motion, if the projectile be thrown in the first angle, the angles α', β', and γ', will be obtuse,

$$\cos \alpha' = -\frac{d x}{d s} ; \quad \cos \beta' = -\frac{d y}{d s} ; \quad \cos \gamma' = -\frac{d z}{d s},$$

and the equations of motion become, after omitting the common factor M,

$$\frac{d^2 x}{d t^2} = -c \cdot v^2 \cdot \frac{d x}{d s} ;$$

$$\frac{d^2 y}{d t^2} = -c \cdot v^2 \cdot \frac{d y}{d s} ;$$

$$\frac{d^2 z}{d t^2} = -g - c \cdot v^2 \cdot \frac{d z}{d s}.$$

From the first two we have, by division,

$$\frac{d^2 y}{d y} = \frac{d^2 x}{d x} ;$$

and by integration,

$$\log dy = \log dx + \log C;$$

and, passing to the quantities,

$$dy = C dx.$$

Integrating again, we have,

$$y = Cx + C';$$

in which, if the projectile be thrown from the origin, $C' = 0$, thus giving an equation of a right line through the origin. Whence we see that the trajectory is a plane curve, and that its plane is vertical through the point of departure.

Assuming the plane zx, to coincide with that of the trajectory, and replacing v^2, by its value from the relation,

$$\frac{ds^2}{dt^2} = v^2,$$

we have,

$$\left.\begin{aligned} \frac{d^2 x}{dt^2} &= - c \cdot \frac{ds}{dt} \cdot \frac{dx}{dt}; \\ \frac{d^2 z}{dt^2} &= - g - c \cdot \frac{ds}{dt} \cdot \frac{dz}{dt}. \end{aligned}\right\} \quad \dots \dots \quad (168)$$

From the first we have,

$$\frac{\frac{d^2 x}{dt^2}}{\frac{dx}{dt}} = - c \cdot \frac{ds}{dt},$$

and by integration,

$$\log \frac{dx}{dt} = - c \,.\, s + C.$$

Denoting by e, the base of the Naperian system of logarithms, and making $C = \log A$, the above may be written,

$$\log \frac{dx}{dt} = - c \,.\, s \times \log e + \log A,$$

and passing from logarithms to the quantities,

$$\frac{dx}{dt} = A \cdot e^{-cs} \quad . \quad . \quad . \quad . \quad . \quad . \quad . \quad . \quad (169)$$

Denoting by V, the initial velocity, and by α, the angle of projection, we have, by making $s = 0$,

$$\frac{dx}{dt} = A = V \cos \alpha,$$

which substituted above, gives

$$\frac{dx}{dt} = V \cdot \cos \alpha \cdot e^{-cs} \quad . \quad . \quad . \quad . \quad . \quad . \quad . \quad (170)$$

To integrate the second of Equations (168), make

$$\frac{dz}{dt} = p \cdot \frac{dx}{dt}, \quad . \quad . \quad . \quad . \quad . \quad . \quad . \quad . \quad . \quad (171)$$

in which p is an additional unknown quantity.

Differentiating this equation, dividing by dt, and eliminating from the result, $\frac{d^2 x}{dt^2}$, by its value in the first of equations (168), we have,

$$\frac{d^2 z}{dt^2} = \frac{dp}{dt} \cdot \frac{dx}{dt} - p \cdot c \cdot \frac{ds}{dt} \cdot \frac{dx}{dt},$$

and substituting this value in the second of Equations (168), we have, after eliminating $\frac{dz}{dt}$ by its value, obtained from Equation (171),

$$\frac{dx}{dt} \cdot \frac{dp}{dt} = -g \quad . \quad . \quad . \quad . \quad . \quad . \quad . \quad . \quad (172)$$

and dividing this by the square of Equation (170),

$$\frac{\frac{dp}{dt}}{\frac{dx}{dt}} = -\frac{g}{V^2 \cos^2 \alpha} \cdot e^{2cs} \quad . \quad . \quad . \quad . \quad . \quad . \quad (173)$$

but regarding z and p as functions of x, we have, Equation (171),

$$p = \frac{\frac{dz}{dt}}{\frac{dx}{dt}} = \frac{dz}{dx}, \quad \ldots \ldots \ldots \quad (174)$$

and,

$$\frac{\frac{dp}{dt}}{\frac{dx}{dt}} = \frac{dp}{dx},$$

whence, making $V^2 = 2gh$, Equation (173) becomes

$$\frac{dp}{dx} = -\frac{e^{2cs}}{2h \cdot \cos^2 \alpha}, \quad \ldots \ldots \quad (175)$$

and multiplying this by the identical equation,

$$dx \cdot \sqrt{1 + p^2} = ds,$$

obtained from Equation (174), we find,

$$\sqrt{1 + p^2} \cdot dp = -\frac{e^{2cs} \cdot ds}{2h \cdot \cos^2 \alpha};$$

and integrating,

$$p \cdot \sqrt{1 + p^2} + \log\left(p + \sqrt{1 + p^2}\right) = C - \frac{e^{2cs}}{2ch \cdot \cos^2 \alpha}; \quad (176)$$

in which C is the constant of integration; to determine which, make $s = 0$; this gives $p = \tan \alpha$; and

$$C = \frac{1}{2ch \cos^2 \alpha} + \tan \alpha \cdot \sqrt{1 + \tan^2 \alpha} + \log\left(\tan \alpha + \sqrt{1 + \tan^2 \alpha}\right) \cdot \quad (177)$$

From Equation (175) we have,

$$dx = -2h \cdot \cos^2 \alpha \cdot e^{-2cs} \cdot dp$$

from Equation (171),

$$dz = p \cdot dx;$$

from Equation (172),

$$g\,d\,t^2 = -\,d\,x\,.\,dp\,;$$

and eliminating the exponential factor by means of Equation (176), we find,

$$c\,.\,d\,x = \frac{d\,p}{p\,\sqrt{1+p^2} + \log\,(p + \sqrt{1+p^2}) - C}\,; \quad (178)$$

$$c\,.\,d\,z = \frac{p\,d\,p}{p\,\sqrt{1+p^2} + \log\,(p + \sqrt{1+p^2}) - C}\,; \quad (179)$$

$$\sqrt{c\,g}\,.\,d\,t = \frac{-\,d\,p}{\sqrt{C - p\,\sqrt{1+p^2} - \log\,(p + \sqrt{1+p^2})}}\,; \quad (180)$$

Of the double sign due to the radical of the last equation, the negative is taken because p, which is the tangent of the angle made by any element of the curve with the axis of x, is a decreasing function of the time t.

These equations cannot be integrated under a finite form. But the trajectory may be constructed by means of auxiliary curves of which (178) and (179) are the differential equations. From the first, we have,

$$d\,x = T\,.\,d\,p\,; \quad \ldots\ldots \quad (181)$$

and from the second,

$$dz = T\,.\,p\,.\,d\,p\,; \quad \ldots\ldots \quad (182)$$

in which,

$$T = \frac{1}{c}\cdot\frac{1}{p\,\sqrt{1+p^2} + \log\,(p + \sqrt{1+p^2}) - C}\,; \quad (183)$$

and dividing Equations (181) and (182), by dp,

$$\frac{d\,x}{d\,p} = T\,; \quad \ldots\ldots \quad (184)$$

$$\frac{d\,z}{d\,p} = T\,.\,p\,; \quad \ldots\ldots \quad (185)$$

Now, regarding x, p, and z, p, as the variable co-ordinates of two auxiliary curves, T, and $T \cdot p$, will be the tangents of the angles which the elements of these curves make with the axis of p.

Any assumed value of p, being substituted in T, Equation (183), will give the tangent of this angle, and this, Equation (184), multiplied by dp, will give the difference of distances of the ends of the corresponding element of the curve from the axis of p. Beginning therefore, at the point in which the auxiliary curves cut the axis of p, and adding these successive differences together, a series of ordinates x and z, separated by intervals equal to dp, may be found, and the curves traced through their extremities.

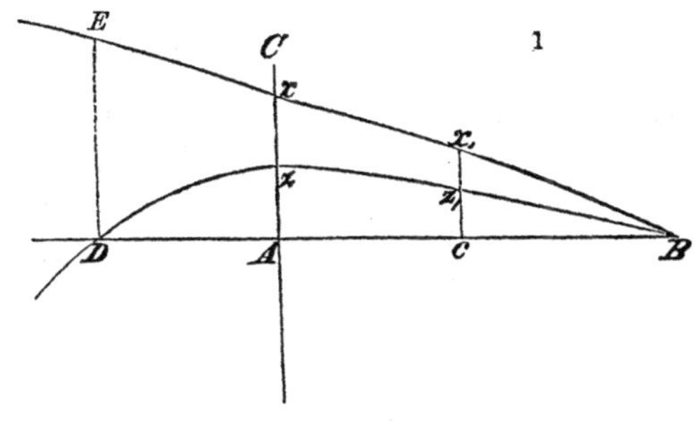

At the point from which the projectile is thrown, we have,

$$x=0\ ;\ z=0\ ;\ p=\tan\alpha,$$

and the auxiliary curves will cut the axis of p, in the same point, and at a distance from the origin equal to $\tan\alpha$. Let AB, be the axis of p, and AC, the axis of x and of z; take $AB = \tan\alpha$, and let BzD, and BxE, be constructed as above.

Draw the axes Ax and Az, through the point of departure A, Fig. (2); draw any ordinate $c\,z_{,}\,x_{,}$ to the auxiliary curves Fig. (1); lay off $Ax_{,}$ Fig. (2) equal to $Cx_{,}$ Fig. (1), and draw through $x_{,}$, the line $x_{,}\,z_{,}$ parallel to the axis Az, and equal to $c\,z_{,}$ Fig. (1); the point $z_{,}$ will be a point of the trajectory. The range AD, is equal to ED, Fig. (1).

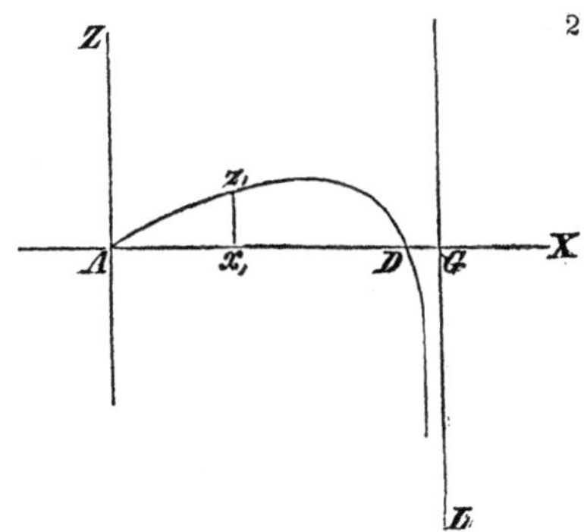

By reference to the value of C, Equation (177), it will be seen that the value of T, Equation (183), will always be negative, and that the auxiliary curve whose ordinates give the values of x, can, therefore, never approach the axis of p. As long as p is positive, the auxiliary curve, whose ordinates are z, will recede from the axis p; but when p becomes negative, as it will to the left of the axis $A\,C$, Fig. (1), the tangent of the angle which the element of the curve makes with the axis p, will, Equation (185), become positive, and this curve will approach the axis p, and intersect it at some point as D.

The value of p will continue to increase indefinitely to the left of the origin A, Fig. (1), and when it becomes exceedingly great, the logarithmic term as well as C, and unity may be neglected in comparison with p, which will reduce Equations (178) and (179) to

$$dx = \frac{dp}{c \cdot p^2}; \quad dz = \frac{dp}{c \cdot p};$$

and integrating,

$$x = C' - \frac{1}{cp}; \quad z = C'' + \frac{1}{c} \cdot \log p,$$

which will become, on making p very great,

$$x = C'; \quad z = C'' + \frac{1}{c} \log p,$$

which shows that the curve whose ordinates are the values of x, will ultimately become parallel to the axis p, while the other has no limit to its retrocession from this axis. Whence we conclude, that the descending branch of the trajectory approaches more and more to a vertical direction, which it ultimately attains; and that a line $G\,L$, Fig. (2), perpendicular to the axis x, and at a distance from the point of departure equal to C', will be an asymptote to the trajectory.

This curve is not, like the parabolic trajectory, symmetrical in reference to a vertical through the highest point of the curve; the angles of falling will exceed the corresponding angles of rising, the range will be less than double the abscissa of the highest point, and the angle which gives the greatest range will be less than 45°.

Denoting the velocity at any instant by v, we have

$$v^2 = \frac{d\,x^2 + d\,z^2}{d\,t^2} = (1 + p^2)\,\frac{d\,x^2}{d\,t^2},$$

and replacing $d\,x^2$ and $d\,t^2$ by their values in Equations (178) and (180), we find

$$v^2 = \frac{1}{c} \cdot \frac{g\,.\,(1 + p^2)}{C - p\,\sqrt{1 + p^2} - \log\,(p + \sqrt{1 + p^2})} \quad . \quad . \quad (186)$$

and supposing p to attain its greatest value, which supposes the projectile to be moving on the vertical portion of the trajectory, this equation reduces, for the reasons before stated, to

$$v = \sqrt{\frac{g}{c}} = k;$$

which shows that the final motion is uniform, and that the velocity will then be the same as that of a heavy body which has fallen in vacuo through a vertical distance equal to $\frac{1}{2\,c} = \frac{k^2}{2\,g}$.

§ 152.—When the angle of projection is very small, the projectile rises but a short distance above the line of the range, and the equation of so much of the trajectory as lies in the immediate neighborhood of this line may easily be found. For, the angle of projection being very small, p will be small, and its second power may be neglected in comparison with unity, and we may take,

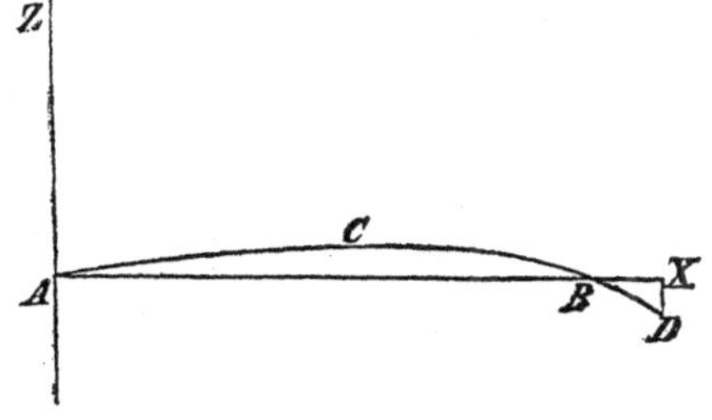

$$d\,s = d\,x; \text{ and } s = x;$$

which in Equation (175), gives,

$$\frac{d\,p}{d\,x} = \frac{d^2\,z}{d\,x^2} = -\,\frac{e^{2\,c\,x}}{2\,h\,.\cos^2\alpha} \quad . \quad . \quad . \quad . \quad . \quad (187)$$

Integrating,

$$\frac{dz}{dx} = -\frac{e^{2cx}}{4c.h.\cos^2\alpha} + C;$$

making $x = 0$, we have $\frac{dz}{dx} = \tan\alpha$,

whence,

$$C = \tan\alpha + \frac{1}{4c.h.\cos^2\alpha};$$

which substituted above, gives,

$$\frac{dz}{dx} = \tan\alpha - \frac{e^{2cx}}{4c.h.\cos^2\alpha} + \frac{1}{4c.h.\cos^2\alpha};$$

and integrating again

$$z = \tan\alpha.x - \frac{e^{2cx}}{8c^2.h.\cos^2\alpha} + \frac{x}{4c.h.\cos^2\alpha} + C',$$

making $x = 0$, then will $z = 0$, and

$$C' = \frac{1}{8c^2.h.\cos^2\alpha};$$

hence,

$$z = \tan\alpha\, x - \frac{1}{8c^2.h.\cos^2\alpha}\left(e^{2cx} - 2cx - 1\right) \quad . \quad . \quad (188)$$

From Equation (172), we have,

$$g.dt^2 = -dx.dp,$$

and substituting the value of dp, from Equation (187),

$$dt = \frac{e^{cx}.dx}{\sqrt{2gh}.\cos\alpha};$$

and integrating, making $x = 0$, when $t = 0$,

$$t = \frac{1}{c\sqrt{2gh}.\cos\alpha}.\left(e^{cx} - 1\right) \quad . \quad . \quad . \quad . \quad (189)$$

which will give the time of flight to any point whose horizontal distance from the piece is equal to x.

§ 153.—Let the projectile fall to the ground at the point D, and denote the co-ordinates of this point by $x = l$, and $z = \lambda$, and suppose the time of flight or $t = \tau$. These values in Equations (188) and (189), give

$$-8c^2 . h . \cos^2 \alpha (\lambda - l . \tan \alpha) = e^{2cl} - 2cl - 1 \quad . \quad (189)'$$

$$\cos \alpha . \tau . c . \sqrt{2gh} = e^{cl} - 1 \quad . \quad . \quad . \quad . \quad . \quad (189)''$$

When the two constants h and c, as well as α and λ, are known, these equations will give the horizontal distance l, and the time of flight. Conversely, when the quantities α, l, λ and τ are known, they give the co-efficient of resistance c, and the height h, due to the velocity of projection, and therefore, Equation (135), the initial velocity itself.

Eliminating the height h, we find

$$-4(\lambda - l . \tan \alpha)(e^{cl} - 1)^2 = g . \tau^2 . (e^{2cl} - 2cl - 1) ; \quad . \quad (189)'''$$

from which the value of c may be found, and one of the preceding equations will give h, or the initial velocity.

It may be worth while to remark that if the exponential term in Equation (188) be developed, and c be made equal to zero, which is equivalent to supposing the projectile in vacuo, we obtain Equation (161).

§ 154.—Assuming that the resistance of the air varies as the square of the velocity, some idea may be formed of its actual intensity from the fact that a twenty-four-pound ball projected with a velocity of 2,000 feet in vacuo, and under an angle of 45°, would have a range of 125,000 feet; whereas actual experiment in the air shows it to be but 7,300 feet—about one-seventeenth of the former.

Many circumstances qualify both the path and velocity of projectiles. The law of the resistance may be the same for all figures, but it is known, from actual trial, not to be that of the square of the velocity, except for very small rates of motion. For the same velocity, the intensity of the resistance varies with the size and figure of the ball. Much depends upon the facility with which the compressed air in front may escape latterly and make its way to the rear. The actual resistance at any instant is composed of two terms, the one due to the inertia of the displaced particles, the other to the difference of atmospheric pressure, as such, in front and rear. If during the motion the air could close in behind and exert the same pressure as in front, the resistance would be wholly due to inertia. If the ball were at rest, and all the air removed in rear of the plane of largest section perpendicular to the trajectory, the resistance would be due entirely to the barometric pressure on the extent of this section. Both terms of the resistance must be variable and a function of the velocity, till the latter is so great as to leave a vacuum behind, when the barometric term would become constant.

From a careful and elaborate investigation of the numerous experiments upon this subject, Col. Piobert has constructed this empirical formula for spherical projectiles, viz.:

$$\rho = A \,.\, \pi \,.\, r^2 \left(1 + \frac{v}{v_{\prime}}\right) . \, v^2 \quad . \quad . \quad . \quad . \quad . \quad . \quad (\quad)$$

in which ρ is the resistance in kilogrammes, v the velocity, π the ratio of the diameter to the circumference, r the radius of the ball, A the resistance on a square *mètre* when the velocity is one *mètre*, and $v_{\prime}$ the velocity which would make the resistance measured by the second term equal to that measured by the first.

§ 155.—If the ball be not perfectly homogeneous in density, the centre of inertia will, in general, be removed from that of figure; the resultant of the expansive action of the powder will pass through the latter centre and communicate to the ball a rotary motion about the former. The atmospheric resistance will be greater on the side of the greatest velocity, and deflect the projectile to the opposite side.

ROTARY MOTION.

§ 156.—Having discussed the motion of translation of a single body, we now come to its motion of rotation. To find the circumstances of a body's rotary motion, it will be convenient to transform Equations (118) from rectangular to polar co-ordinates. But before doing this, let us premise that the *angular velocity* of a body is *the rate of its rotation about a centre.* The angular velocity is measured by *the absolute velocity of a point at the unit's distance from the centre, and taken in such position as to make that velocity a maximum.*

§ 157.—Both members of Equations (38) being divided by $d\,t$, give

$$\left.\begin{aligned} \frac{d\,x'}{d\,t} &= z' \cdot \frac{d\,\psi}{d\,t} - y' \cdot \frac{d\,\varphi}{d\,t}, \\ \frac{d\,y'}{d\,t} &= x' \cdot \frac{d\,\varphi}{d\,t} - z' \cdot \frac{d\,\varpi}{d\,t}, \\ \frac{d\,z'}{d\,t} &= y' \cdot \frac{d\,\varpi}{d\,t} - x' \cdot \frac{d\,\psi}{d\,t}, \end{aligned}\right\} \quad \ldots\ldots \quad (190)$$

in which the first members taken in order, are the velocities of any element, as m, in the direction of the axes x, y, z, respectively, *in reference to the centre of inertia*, § 75, while

$$\frac{d\,\varpi}{d\,t}, \quad \frac{d\,\psi}{d\,t}, \quad \frac{d\,\varphi}{d\,t},$$

are the angular velocities about the same axes respectively.

Denoting the first of these by ν_x, the second by ν_y, and the third by ν_z, we have

$$\frac{d\,\varpi}{d\,t} = \nu_x; \quad \frac{d\,\psi}{d\,t} = \nu_y; \quad \frac{d\,\varphi}{d\,t} = \nu_z; \quad \ldots\ldots \quad (191)$$

and Equations (190) may be written

$$\left.\begin{aligned} \frac{dx'}{dt} &= z'.\nu_y - y'.\nu_z, \\ \frac{dy'}{dt} &= x'.\nu_z - z'.\nu_x, \\ \frac{dz'}{dt} &= y'.\nu_x - x'.\nu_y. \end{aligned}\right\} \quad \ldots\ldots \quad (192)$$

§ 158.—If an element m be so situated that its velocity shall be equal and parallel to that of the centre of inertia, then, for this element, will each of the first members of Equations (192) reduce to zero, and

$$\left.\begin{aligned} z'.\nu_y - y'.\nu_z &= 0, \\ x'.\nu_z - z'.\nu_x &= 0, \\ y'.\nu_x - x'.\nu_y &= 0; \end{aligned}\right\} \quad \ldots\ldots \quad (193)$$

the last being but a consequence of the two others, these equations are those of a right line passing through the centre of inertia, every point of which will have a simple motion of translation parallel and equal to that of the centre of inertia. The whole body must, for the instant, rotate about this line, and it is, therefore, called the *Axis of Instantaneous Rotation.*

§ 159.—Denote by $\alpha_{\prime}$, $\beta_{\prime}$, $\gamma_{\prime}$, the angles which this axis makes with the co ordinate axes x, y, z, respectively. Then, taking any point on the instantaneous axis, will,

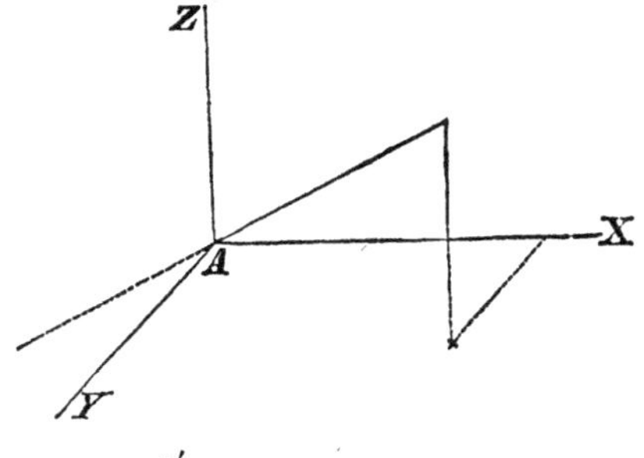

$$\cos\alpha_{\prime} = \frac{x'}{\sqrt{x'^2 + y'^2 + z'^2}},$$

$$\cos\beta_{\prime} = \frac{y'}{\sqrt{x'^2 + y'^2 + z'^2}},$$

$$\cos\gamma_{\prime} = \frac{z'}{\sqrt{x'^2 + y'^2 + z'^2}};$$

and eliminating x', y' and z', by Equations (193),

$$\left.\begin{aligned} \cos\alpha_{,} &= \frac{v_x}{\sqrt{v_x^2+v_y^2+v_z^2}}, \\ \cos\beta_{,} &= \frac{v_y}{\sqrt{v_x^2+v_y^2+v_z^2}}, \\ \cos\gamma_{,} &= \frac{v_z}{\sqrt{v_x^2+v_y^2+v_z^2}}, \end{aligned}\right\} \quad . \; . \; . \; . \; . \; . \quad (194)$$

which will give the position of the instantaneous axis as soon as the angular velocities about the axes are known.

§ 160.—Squaring each of Equations (192), taking their sum and extracting square root, we find

$$\sqrt{\frac{dx'^2+dy'^2+dz'^2}{dt^2}} = v = \sqrt{(z'.v_y - y'.v_z)^2 + (x'.v_z - z'.v_x)^2 + (y'.v_x - x'.v_y)^2};$$

Replacing v_x, v_y and v_z by their values obtained by simply clearing the fractions in Equations (194), this becomes

$$v = \sqrt{v_x^2+v_y^2+v_z^2} \times \sqrt{x'^2+y'^2+z'^2 - (x'\cos\alpha_{,} + y'\cos\beta_{,} + z'\cos\gamma_{,})^2},$$

which is the velocity of any element in reference to the centre of inertia.

Making

$$x'^2 + y'^2 + z'^2 = 1,$$

we have the element at a unit's distance from the centre of inertia; and making

$$x'\cos\alpha_{,} + y'\cos\beta_{,} + z'\cos\gamma_{,} = 0, \quad . \; . \; . \; . \quad (195)$$

the point takes the position; giving the maximum velocity. In this case v becomes the angular velocity, and we have, denoting the latter by v_i,

$$v_i = \sqrt{v_x^2 + v_y^2 + v_z^2} \quad . \; . \; . \; . \; . \; . \; . \quad (196)$$

Equation (195) is that of a plane passing through the centre of inertia, and perpendicular to the instantaneous axis. The position of the co-ordinate axes being arbitrary, Equation (196) shows that the sum of the squares of the angular velocities about the three co-ordinate axes is a constant quantity, and equal to the square of the angular velocity about the instantaneous axis.

§ 161.—Multiply Equation (196), by the first of Equations (194), and there will result

$$\nu_i \,.\, \cos \alpha_{,} = \nu_x \quad \ldots \ldots \ldots \quad (197)$$

whence the angular velocity about any axis oblique to the instantaneous axis, is equal to the angular velocity of the body multiplied by the cosine of the inclination of the two axes.

§ 162.—Equation (196) gives ν_i, when ν_x, ν_y, ν_z, are known. To find these, resume Equations (118), and write for the moments of the extraneous forces in reference to the axes $x,' y,' z,'$ through the centre of inertia, $N_{,}$, $M_{,}$, $L_{,}$, respectively, then will

$$\left.\begin{array}{l} \Sigma\, m \cdot \left(\dfrac{d^2 y'}{d t^2} \cdot x' - \dfrac{d^2 x'}{d t^2} \cdot y'\right) = L_{,} \,, \\[2ex] \Sigma\, m \cdot \left(\dfrac{d^2 x'}{d t^2} \cdot z' - \dfrac{d^2 z'}{d t^2} \cdot x'\right) = M_{,} \,, \\[2ex] \Sigma\, m \cdot \left(\dfrac{d^2 z'}{d t^2} \cdot y' - \dfrac{d^2 y'}{d t^2} \cdot z'\right) = N_{,} \,; \end{array}\right\} \ldots \quad (198)$$

differentiating the first of Equations (192), with respect to t, we find

$$\frac{d^2 x'}{d t^2} = \nu_y \cdot \frac{d z'}{d t} - \nu_z \cdot \frac{d y'}{d t} + \frac{d \nu_y}{d t} \cdot z' - \frac{d \nu_z}{d t} \cdot y' \,,$$

and replacing $\frac{d z'}{d t}$ and $\frac{d y'}{d t}$, by their values given in the second and third of Equations (192),

$$\frac{d^2 x'}{d t^2} = -\left(\nu_y^2 + \nu_z^2\right) \cdot x' + \nu_x \cdot \nu_y \cdot y' + \nu_x \cdot \nu_z \cdot z' + \frac{d \nu_y}{d t} \cdot z' - \frac{d \nu_z}{d t} \cdot y' \,;$$

in the same way

$$\frac{d^2 y'}{d t^2} = -\left(v_x^2 + v_z^2\right) \cdot y' + v_y \cdot v_x \cdot x' + v_y \cdot v_z \cdot z' + \frac{d v_z}{d t} \cdot x' - \frac{d v_x}{d t} \cdot z'$$

$$\frac{d^2 z'}{d t^2} = -\left(v_x^2 + v_y^2\right) \cdot z' + v_z \cdot v_x \cdot x' + v_z \cdot v_y \cdot y' + \frac{d v_x}{d t} \cdot y' - \frac{d v_y}{d t} \cdot x' .$$

and these values in the first of Equations, (198), give

$$\Sigma m \left(\frac{d^2 y'}{d t^2} \cdot x' - \frac{d^2 x'}{d t^2} \cdot y'\right) = \left\{ \begin{array}{l} \left(v_y^2 - v_x^2\right) \cdot \Sigma m . x' y' \\ + \left(v_y . v_z - \frac{d v_x}{d t}\right) \cdot \Sigma m . z' x' \\ - \left(v_x . v_z + \frac{d v_y}{d t}\right) \cdot \Sigma m . z' y' \\ + v_y . v_x . \Sigma m . \left(x'^2 - y'^2\right) \\ + \frac{d v_z}{d t} . \Sigma m \left(x'^2 + y'^2\right) \end{array} \right\} = L_{,} \cdot \quad (199)$$

Similar equations will result from the remaining two of Equations (198); then by elimination and integration, we might proceed to find the values of v_x, v_y and v_z, but the process would be long and tedious. It will be greatly simplified, however, if the co-ordinate axes be so chosen as to make at the instant corresponding to t,

$$\Sigma m x' y' = 0; \quad \Sigma m z' y' = 0; \quad \Sigma m z' x' = 0; \quad \cdot \quad \cdot \quad (200)$$

which is always possible, as will be shown presently. This will reduce Equation (199) to

$$\frac{d v_z}{d t} \cdot \Sigma m \left(y'^2 + x'^2\right) + v_x \cdot v_y . \Sigma m \left(x'^2 - y'^2\right) = L_{,} \cdot$$

The other two equations which refer to the motion about the axes y' and x', may be written from this one. They are,

$$\frac{d v_y}{d t} \cdot \Sigma m \left(x'^2 + z'^2\right) + v_x \cdot v_z . \Sigma m \left(z'^2 - x'^2\right) = M_{,} ,$$

$$\frac{d v_x}{d t} \cdot \Sigma m \left(y'^2 + z'^2\right) + v_y \cdot v_z . \Sigma m \left(y'^2 - z'^2\right) = N_{,} .$$

The axes x', y', z', which satisfy the conditions expressed in Equations (200), are called the *principal axes of figure* of the body. And if we make

$$\left.\begin{aligned} \Sigma\, m\,.\,(y'^2 + x'^2) &= C, \\ \Sigma\, m\,.\,(x'^2 + z'^2) &= B, \\ \Sigma\, m\,.\,(y'^2 + z'^2) &= A\,; \end{aligned}\right\} \quad . \quad . \quad . \quad . \quad . \qquad (201)$$

we find, by subtracting the third from the second,

$$\Sigma\, m\,.\,(x'^2 - y'^2) = B - A,$$

the first from the third,

$$\Sigma\, m\,.\,(z'^2 - x'^2) = A - C,$$

and the second from the first,

$$\Sigma\, m\,.\,(y'^2 - z'^2) = C - B;$$

which substituted above, give,

$$\left.\begin{aligned} C\,.\,\frac{d\,\mathrm{v}_z}{d\,t} + \mathrm{v}_x\,.\,\mathrm{v}_y\,.\,(B - A) &= L_{,}\,, \\ B\,.\,\frac{d\,\mathrm{v}_y}{d\,t} + \mathrm{v}_x\,.\,\mathrm{v}_z\,.\,(A - C) &= M_{,}\,, \\ A\,.\,\frac{d\,\mathrm{v}_x}{d\,t} + \mathrm{v}_y\,.\,\mathrm{v}_z\,.\,(C - B) &= N_{,}\,. \end{aligned}\right\} \quad . \quad . \quad . \quad . \quad . \qquad (202)$$

By means of these equations, the angular velocities v_x, v_y, v_z, must be found by the operations of elimination and integration.

§ 163.—It is plain that the quantities C, B and A, are constant for the same body; the first being the sum of the products arising from multiplying each elementary mass into the square of its distance from the principal axis z', the second the same for the principal axis y', and the third for the principal axis x'. The sum of the products of the elementary masses into the square of their distances from any axis, is called the *moment of inertia* of the body in reference to this axis. A, B and C are called principal moments of inertia.

§ 164.—Through any assumed point there may always be drawn one set of rectangular axes, and, in general, only one which will satisfy the conditions of Equations (200). To show this, assume the formulas for the transformation from one system of rectangular axes to another, also rectangular. These are

$$\left.\begin{aligned} x' &= x \,.\cos(x'x) + y\cos(x'y) + z\cos(x'z), \\ y' &= x\cos(y'x) + y \,.\cos(y'y) + z \,.\cos(y'z), \\ z' &= x\cos(z'x) + y\cos(z'y) + z \,.\cos(z'z), \end{aligned}\right\} \quad \ldots \quad (203)$$

in which $(x'x)$, $(y'x)$ and $(z'x)$, denote the angles which the new axes x', y', z', make with the primitive axis of x; $(x'y)$, $(y'y)$ and $(z'y)$, the angles which the same axes make with the primitive axis of y, and $(x'z)$, $(y'z)$ and $(z'z)$, the angles they make with the axis z.

Assume the common origin as the centre of a sphere of which the radius is unity; and conceive the points in which the two sets of axes pierce its surface to be joined by the arcs of great circles; also let these points be connected with the point N, in which the intersection of the planes xy and $x'y'$ pierces the spherical surface nearest to that in which the positive axis x pierces the same. Also, let

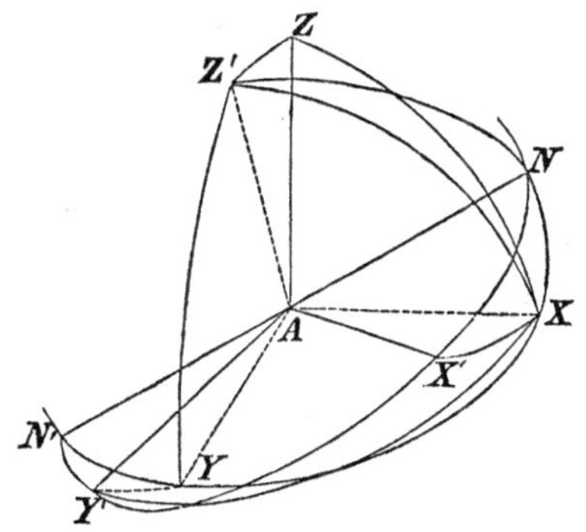

$\theta = Z'AZ = X'NX$, being the inclination of the plane $x'y'$ to that of xy.

$\psi = NAX$ being the angular distance of the intersection of the planes xy and $x'y'$, from the axis x.

$\varphi = NAX'$ being the angular distance of the same intersection from the axis x'.

Then, in the spherical triangle $X' N X$,

$$\cos(x'x) = \cos\psi \, . \cos\varphi + \sin\psi \, . \sin\varphi \, . \cos\theta ;$$

In the triangle $Y' N X$, the side $N Y' = \frac{\pi}{2} + \varphi$, and

$$\cos(y'x) = -\cos\psi \, . \sin\varphi + \sin\psi \, . \cos\varphi \, . \cos\theta ,$$

In the triangle $Z' N X$, the side $N Z' = \frac{\pi}{2}$, and

$$\cos(z'x) = \sin\psi \, . \sin\theta .$$

And in the same way it will be found that

$$\cos(x'y) = -\sin\psi \, . \cos\varphi + \cos\psi \, . \sin\varphi \, . \cos\theta ;$$
$$\cos(y'y) = \sin\psi \, . \sin\varphi + \cos\psi \, . \cos\varphi \, . \cos\theta ;$$
$$\cos(z'y) = \cos\psi \, . \sin\theta ;$$
$$\cos(x'z) = -\sin\varphi \, . \sin\theta ;$$
$$\cos(y'z) = -\cos\varphi \, . \sin\theta ;$$
$$\cos(z'z) = \cos\theta ;$$

and by substitution in Equations (203),

$$x' = x(\sin\psi \, . \sin\varphi \, . \cos\theta + \cos\psi \, . \cos\varphi)$$
$$+ y(\cos\psi \, . \sin\varphi \, . \cos\theta - \sin\psi \, . \cos\varphi) - z \sin\varphi \, . \sin\theta ,$$
$$y' = x(\sin\psi \, . \cos\varphi \, . \cos\theta - \cos\psi \, . \sin\varphi)$$
$$+ y(\cos\psi \, . \cos\varphi \, . \cos\theta + \sin\psi \, . \sin\varphi) - z \cos\varphi \, . \sin\theta ,$$
$$z' = x \sin\psi \, . \sin\theta + y \cos\psi \, . \sin\theta + z \cos\theta ;$$

or making, for sake of abbreviation,

$$D = x \cos\psi - y \sin\psi ,$$
$$E = x \sin\psi \, . \cos\theta + y \cos\psi \, . \cos\theta - z \sin\theta ,$$

the above reduce to

$$x' = E \, . \sin\varphi + D \, . \cos\varphi ,$$
$$y' = E \, . \cos\varphi - D \, . \sin\varphi ,$$
$$z' = x \, . \sin\psi \, . \sin\theta + y \, . \cos\psi \, . \sin\theta + z \, . \cos\theta .$$

Substituting these values in the equations

$$\Sigma m . x' . y' = 0 ; \quad \Sigma m . x' . z' = 0 ; \quad \Sigma m . y' . z' = 0 ;$$

we obtain from the first,

$$\sin \varphi . \cos \varphi . \Sigma m (E^2 - D^2) + (\cos^2 \varphi - \sin^2 \varphi) \Sigma m E . D = 0,$$

or, replacing $\sin \varphi . \cos \varphi$, and $\cos^2 \varphi - \sin^2 \varphi$, by their equals $\frac{1}{2} \sin 2 \varphi$, and $\cos 2 \varphi$, respectively,

$$\sin 2 \varphi . \Sigma m (E^2 - D^2) + 2 \cos 2 \varphi . \Sigma m D . E = 0 ; \cdot \cdot \cdot \quad (204)$$

and from the third and second, respectively,

$$\cos \varphi . \Sigma m . E . z' - \sin \varphi . \Sigma m D . z' = 0, \cdot \cdot \cdot \quad (205)$$

$$\sin \varphi . \Sigma m . E . z' + \cos \varphi . \Sigma m D . z' = 0. \cdot \cdot \cdot \quad (206)$$

Squaring the last two and adding, we find

$$(\Sigma m . E . z')^2 + (\Sigma m . D . z')^2 = 0.$$

which can only be satisfied by making

$$\left. \begin{array}{l} \Sigma m . E . z' = 0 ; \\ \Sigma m . D . z' = 0. \end{array} \right\} \cdot \cdot \cdot \cdot \cdot \cdot \cdot \quad (207)$$

These equations are independent of the angle φ, and will give the values of ψ and θ; and these being known, Equation (204) will give the angle φ.

Replacing E and D by their values, we have

$$\begin{aligned} E . z' = & \sin \theta . \cos \theta (x^2 \sin^2 \psi + 2 x y \sin \psi \cos \psi + y^2 \cos^2 \psi - z^2) \\ & + (\cos^2 \theta - \sin^2 \theta) (x z \sin \psi + y z . \cos \psi), \end{aligned}$$

$$\begin{aligned} D . z' = & \sin \theta \{ x y (\cos^2 \psi - \sin^2 \psi) + (x^2 - y^2) \sin \psi \cos \psi \} \\ & + \cos \theta (x z \cos \psi - y z \sin \psi). \end{aligned}$$

and assuming

$$\Sigma m x^2 = A' ; \ \Sigma m y^2 = B' ; \ \Sigma m z^2 = C' ;$$
$$\Sigma m x y = E' ; \ \Sigma m x z = F' ; \ \Sigma m y z = H',$$

and replacing $\sin \theta . \cos \theta$, and $\cos^2 \theta - \sin^2 \theta$, by their respective values, $\frac{1}{2} \sin 2 \theta$, and $\cos 2 \theta$, Equations (207) become

$$\left. \begin{array}{l} \sin 2 \theta (A' \sin^2 \psi + 2 E' \sin \psi \cos \psi + B' \cos^2 \psi - C') \\ + 2 \cos 2 \theta (F' \sin \psi + H' \cos \psi) \end{array} \right\} = 0 ;$$

$$\left.\begin{array}{l}\sin\theta\,\{E'.(\cos^2\psi - \sin^2\psi) + (A' - B').\sin\psi\cos\psi\} \\ \quad + \cos\theta\,(F'\cos\psi - H'\sin\psi)\end{array}\right\} = 0,$$

in which A', B', C'', E', F' and H', are constants, depending only upon the shape of the body and the position of the assumed axes x, y, z.

Dividing the first by $\cos 2\theta$, and the second by $\cos\theta$, they become

$$\left.\begin{array}{l}\tan 2\theta\,.(A'\sin^2\psi + 2E'\sin\psi\cos\psi + B'\cos^2\psi - C') \\ \quad + 2(F'\sin\psi + H'\cos\psi)\end{array}\right\} = 0\,;\quad(207)'$$

$$\left.\begin{array}{l}\tan\theta\,.\,\{E'(\cos^2\psi - \sin^2\psi) + (A' - B')\sin\psi\cos\psi\} \\ \quad + F'\cos\psi - H'\sin\psi\end{array}\right\} = 0.\quad(207)''$$

From the first of these we may find $\tan 2\theta$, and from the second, $\tan\theta$, in terms of $\sin\psi$, and $\cos\psi$; and these values in the equation

$$\tan 2\theta = \frac{2\tan\theta}{1 - \tan^2\theta} \quad . \quad . \quad . \quad . \quad . \quad . \quad . \quad (208)$$

will give an equation from which ψ may be found.

In order to effect this elimination more easily, make

$$\tan\psi = u,$$

whence

$$\sin\psi = \frac{u}{\sqrt{1 + u^2}}\,;\quad \cos\psi = \frac{1}{\sqrt{1 + u^2}}\,;$$

making these substitutions above, we find

$$\tan 2\theta = -\frac{2(F'u + H')\sqrt{1 + u^2}}{A'u^2 + 2E'u + B' - C'(1 + u^2)},$$

$$\tan\theta = -\frac{(F' - H'u)\sqrt{1 + u^2}}{E'(1 - u^2) + (A' - B')u},$$

which in Equation (208) give

$$\left.\begin{array}{l}\{E'(1 - u^2) + (A' - B')u\}\left\{\begin{array}{l}B'F' - F'C' - E'H' \\ + (C'H' - A'H' + E'F')u\end{array}\right\} \\ \qquad + (F'u + H').(F' - H'u)^2\end{array}\right\} = 0 \quad . \quad . \quad (209)$$

which is an equation of the third degree, and must have at least one real root, and, therefore, give one real value for ψ. This value being substituted in either of the preceding equations, must give a real value for θ, and this with ψ, in either of the Equations (205) or (206), a real value for φ; whence we conclude, that it is always possible to assume the axes so as to satisfy the required conditions, and that through every point there may be drawn at least one set of principal axes at right angles to each other.

The three roots of this cubic equation are necessarily real; and they represent the tangents of the angles which the axis x makes with the lines in which the three co-ordinate planes $x'\,y'$, $y'\,z'$, $x'\,z'$, cut that of $x\,y$; for there is no reason why we should consider one of these angles as given by the equation rather than the others, and the equations of condition are satisfied when we interchange the axes x' y' z'. Hence, in general, there exists only one set of principal axes. If there were more, the degree of the equation would be higher, and would, from what we have just said, give three times as many real roots as there are systems.

If $E' = H' = F' = 0$, Equation (209) will become identical; the problem will be indeterminate, have an infinite number of solutions, and the body consequently an infinite number of sets of principal axes. Such is obviously the case with the sphere, spheroid, &c.

MOMENT OF INERTIA, CENTRE AND RADIUS OF GYRATION.

§ 165.—The quantities A, B and C, in Equations (201) are the moments of inertia of the body in reference to the principal axes. To find these moments in reference to any other axes having the same origin as the principal axes, denote by

x', y', z', the co-ordinates of m referred to the principal axes; by

x, y, z, the co-ordinates of the same element referred to any other rectangular system having the same origin; and by

C', the moment of inertia referred to the axis z;

then from the definition,

$$C' = \Sigma\, m \,.\, (x^2 + y^2) = \Sigma\, m\, x^2 + \Sigma\, m\, y^2;$$

but by the usual formulas for transformation,

$$x = a\,x' + b\,y' + c\,z',$$
$$y = a'\,x' + b'\,y' + c'\,z',$$
$$z = a''\,x' + b''\,y' + c''\,z',$$

in which a, b, &c., denote the cosines of the angles which the axes of the same name as the co-ordinates into which they are respectively multiplied make with the axis corresponding to the variable in the first member.

Substituting the values of x and y in that of C', and reducing by the relations,

$$\Sigma\, m\, x'\, y' = 0\,; \quad \Sigma\, m\, x'\, z' = 0\,; \quad \Sigma\, m\, y'\, z' = 0\,;$$

and we have,

$$C' = a''^2 \,.\, \Sigma\, m\,(y'^2 + z'^2) + b''^2 \,.\, \Sigma\, m\,(x'^2 + z'^2) + c''^2 \,.\, \Sigma\, m\,(x'^2 + y'^2)\,;$$

and by substituting A, B and C for their values, this reduces to

$$C' = a''^2\, A + b''^2\, B + c''^2\, C \quad \cdot \quad \cdot \quad \cdot \quad \cdot \quad (210)$$

That is to say, the moment of inertia with reference to any axis passing through the common point of intersection of the principal axes, is equal to the sum of the products obtained by multiplying the moment of inertia with reference to each of the principal axes, by the square of the cosine of the angle which the axis in question makes with these axes.

§ 166.—Let A, be the greatest, and C, the least of the moments of inertia, with reference to the principal axes; then, substituting for a''^2, its value, $1 - b''^2 - c''^2$, in Equation (210), we have

$$C' = A - b''^2\,(A - B) - c''^2\,(A - C). \quad \cdot \quad \cdot \quad (211)$$

By hypothesis, $A - B$, and $A - C$, are positive; therefore, C' is always less than A, whatever be the value of b'', and c''.

Again, substituting for c''^2 its value $1 - a''^2 - b''^2$ in Equation (210), we get

$$C' = C + a''^2\,(A - C) + b''^2\,(B - C) \quad \cdot \quad \cdot \quad \cdot \quad (212)$$

and C' must always be greater than C.

Whence, we conclude that the principal axes give the greatest and least moments of inertia in reference to axes through the same point.

If A be equal to B, then will Equation (211) become

$$C'' = (1 - c''^2) A + c''^2 C, \quad . \quad . \quad . \quad . \quad . \quad (213)$$

and this only depending upon c'', we conclude that the moment of inertia will be the same for all axes making equal angles with the principal axis, z'. The moments of inertia, with reference to all axes in the plane $x' y'$, are, therefore, equal to one another. But all the axes in the plane $x' y'$, which are at right angles to one another, are, § 164, when taken with z', principal axes, and we, therefore, conclude that the body has an indefinite number of sets of principal axes.

If, at the same time, we have $A = B = C$, then will Equation (210) reduce to

$$C' = C = A = B.$$

that is, the moments of inertia are all equal to one another, and all axes are principal, the Equation (210) being satisfied independently of a'', b'', c''.

§ 167.—Resuming Equations, (33), and substituting the values of x, y, z, in the general expression,

$$\Sigma\, m\, (x^2 + y^2)$$

which is the moment of inertia with reference to any axis, z, parallel to the axis z', through the centre of inertia, we have

$$\begin{aligned} \Sigma\, m\, (x^2 + y^2) &= \Sigma\, m\, [\,(x_{,} + x')^2 + (y_{,} + y')^2] \\ &= \Sigma\, m\, (x'^2 + y'^2) + (x_{,}^2 + y_{,}^2) \,.\, \Sigma\, m \\ &\quad + 2\, x_{,} \,.\, \Sigma\, m\, x' + 2\, y_{,} \,.\, \Sigma\, m\, y'; \end{aligned}$$

but from the principle of the centre of inertia,

$$\Sigma\, m\, x' = 0, \quad \text{and} \quad \Sigma\, m\, y' = 0;$$

whence, denoting by d the distance between the axes z and z', and by M the whole mass,

$$\Sigma\, m \,.\, (x^2 + y^2) = \Sigma\, m\, (x'^2 + y'^2) + M d^2 \quad . \quad . \quad . \quad (214)$$

That is, the moment of inertia of any body in reference to a given axis, is equal to the moment of inertia with reference to a parallel axis through the centre of inertia, increased by the product of the whole mass into the square of the distance of the given axis from that centre.

And we conclude that the least of all the moments of inertia is that taken with reference to a principal axis through the centre of inertia.

§ 168.—Denote by r the distance of the elementary mass m from the axis z, then will

$$r^2 = x^2 + y^2,$$

and

$$\Sigma\, m\, (x^2 + y^2) = \Sigma\, m\, r^2.$$

Now, denoting the whole mass by M, and assuming

$$\Sigma\, m\, r^2 = Mk^2,$$

we have

$$k = \sqrt{\frac{\Sigma\, m\, r^2}{M}} \quad . \quad . \quad . \quad . \quad . \quad . \quad . \quad . \quad (215)$$

The distance k is called the *radius of gyration*, and it obviously measures the distance from the axis to that point into which if the whole mass were concentrated the moment of inertia would not be altered. The point into which this concentration might take place and satisfy the condition above, is called the *centre of gyration*. When the axis passes through the centre of inertia, the radius k and the point of concentration are called *principal radius* and *principal centre of gyration.*

The least radius of gyration is, Equation (215), that relating to the principal axis with reference to which the moment of inertia is the least.

If $k_{\prime}$ denote a principal radius of gyration, we may replace $\Sigma\, m\, (x'^2 + y'^2)$ in Equation (214) by $Mk_{\prime}^2$, and we shall have

$$\Sigma\, m\, r^2 = Mk^2 = M(k_{\prime}^2 + d^2) \quad . \quad . \quad . \quad . \quad (216)$$

If the linear dimensions of the body be very small as compared with d, we may write the moment of inertia equal to Md^2.

The letter k with the subscript accent, will denote a principal radius of gyration.

The determination of the moments of inertia and radii of gyration of geometrical figures, is purely an operation of the calculus. Such bodies are supposed to be continuous throughout, and of uniform density. Hence, we may write dM for m, and the sign of integration for Σ, and the formula becomes

$$\Sigma m r^2 = \int dM . r^2 \quad \ldots \ldots \quad (217)$$

Example 1.—*A physical line about an axis through its centre and perpendicular to its length.*

Denote the whole length by $2a$; then

$$2a : dr :: M : dM,$$

whence,

$$dM = M \cdot \frac{dr}{2a},$$

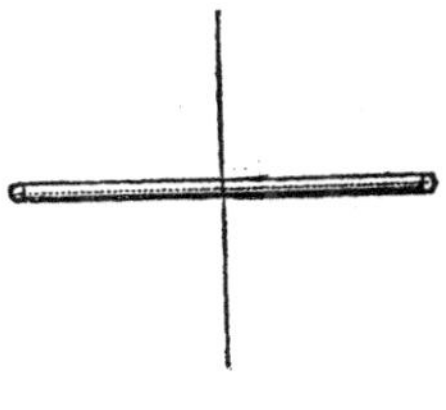

and

$$Mk_{\prime}^2 = \int_a^{-a} M \cdot \frac{r^2}{2a} \cdot dr = \frac{Ma^2}{3};$$

$$k_{\prime} = \frac{a}{\sqrt{3}}.$$

If the axis be at a distance d from the centre, and parallel to that above, then, Equation (216),

$$k = \sqrt{\tfrac{1}{3}a^2 + d^2}.$$

Example 2.—*A circular plate of uniform density and thickness, about an axis through its centre and perpendicular to its plane.*

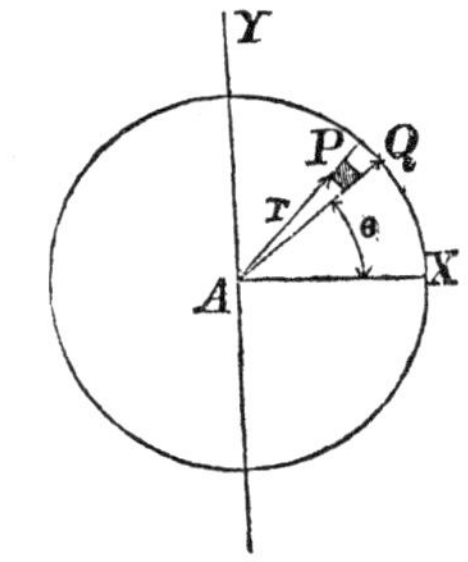

Denote the radius by a; the angle $X A Q$ by θ; the distance of dM from the centre by r; then,

$$\pi a^2 : r . d\theta . dr :: M : dM;$$

whence,

$$dM = M \cdot \frac{r . dr . d\theta}{\pi a^2},$$

and

$$M k_{,}^2 = \int_0^a \int_0^{2\pi} M \cdot \frac{r^3 . dr}{\pi a^2} \cdot d\theta = \int_0^a 2 M . \frac{r^3}{a^2} \cdot dr = \frac{M a^2}{2},$$

$$k_{,} = a \sqrt{\tfrac{1}{2}},$$

and for an axis parallel to the above at the distance d,

$$k = \sqrt{\tfrac{1}{2} a^2 + d^2}.$$

Example 3.—The same body about an axis through its centre and in its plane.

As before,

$$dM = M \cdot \frac{r . dr . d\theta}{\pi a^2},$$

in which r denotes the distance of dM from the centre; and taking the axis to be that from which θ is estimated, the distance of the elementary mass from the axis will be $r \sin \theta$, and

$$M k_{,}^2 = \int_0^a \int_0^{2\pi} M \frac{r^3 . \sin^2 \theta}{\pi a^2} \cdot dr . d\theta = \frac{M}{2 \pi a^2} \int_0^a \int_0^{2\pi} r^3 (1 - \cos 2\theta) \, dr . d\theta,$$

$$M k_{,}^2 = \frac{M}{a^2} \int_0^a r^3 . dr = M \frac{a^2}{4},$$

and

$$k_{,} = \tfrac{1}{2} a,$$

and about an axis parallel to the above and at the distance d.

$$k = \sqrt{\frac{1}{4} a^2 + d^2}.$$

It is obvious that both the axes first considered in Examples 2 and 3 are principal axes, as are also all others in the plane of the plate and through the centre, and if it were required to find the moment of inertia of the plate about an axis through the centre and inclined to its surface under an angle φ, the answer would be given by the Equation (210),

$$M k_{,}^2 = \tfrac{1}{2} M a^2 \sin^2 \varphi + \tfrac{1}{4} M a^2 \cos^2 \varphi$$
$$= \tfrac{1}{4} M a^2 (1 + \sin^2 \varphi),$$

and for a parallel axis whose distance is d,

$$M k^2 = M \left(\tfrac{1}{4} a^2 (1 + \sin^2 \varphi) + d^2 \right) \cdot$$

Example 4.—*A solid of revolution about any axis perpendicular to the axis of the solid.*

Let $D\,A'\,E$ be the given axis, cutting that of the solid in A'. Let A' be the origin of co-ordinates, $P\,M = y$; $A'\,P = x$; $A\,A' = m$; $A'\,B = n$; and $V =$ volume of the solid.

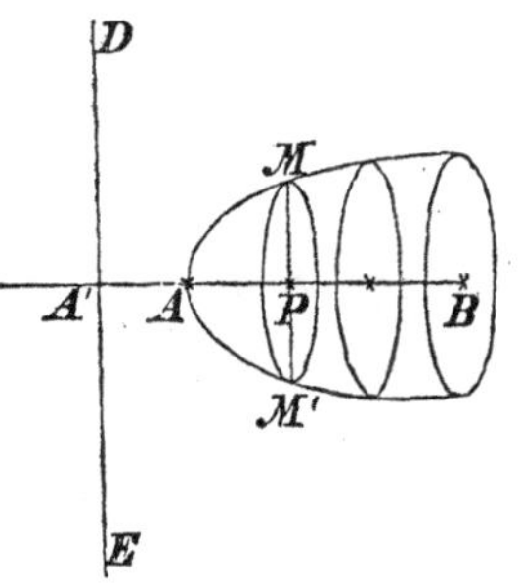

The volume of the elementary section at P will be

$$\pi\, y^2 \,.\, d\,x,$$

and

$$V : M :: \pi \,.\, y^2 \,.\, d\,x : d\,M;$$

whence,

$$d\,M = \frac{M}{V} \cdot \pi \cdot y^2 \cdot d\,x,$$

and its moment of inertia about $M\,M'$, is, Example 3,

$$\frac{M}{V} \cdot \pi \cdot y^2 \cdot d\,x \cdot \frac{y^2}{4} \cdot$$

and about the parallel axis, $D\,E$,

$$\frac{M}{V} \cdot \pi \,.\, y^2 \,.\, d\,x \left(\tfrac{1}{4}\, y^2 + x^2\right)$$

therefore,

$$M k^2 = \int_m^n \frac{M}{V} \pi \, . \, y^2 \left(\tfrac{1}{4} y^2 + x^2\right) d x.$$

But

$$V = \int_m^n \pi \, y^2 \, d x ;$$

whence,

$$k^2 = \frac{\int_m^n \left(\frac{1}{4} y^4 + x^2 y^2\right) . \, d x}{\int_m^n y^2 \, d x} .$$

The equation of the generating curve being given, y may be eliminated and the integration performed.

Example 5.—*A sphere about a line tangent to its surface.*
The equation of the generatrix is

$$y^2 = 2 a x - x^2 ;$$

in which a is the radius of the sphere. Substituting the value of y^2 in the last equation, recollecting that $m = 0$, and $n = 2 a$, we have

$$k^2 = \frac{\int_0^{2a} \left(a^2 x^2 + a x^3 - \frac{3}{4} x^4\right) d x}{\int_0^{2a} \left(2 a x - x^2\right) d x} = \frac{7}{5} a^2 .$$

Also Equation (216),

$$k_{,}^2 = k^2 - a^2 = \frac{2}{5} a^2 ,$$

and

$$k_{,} = a \sqrt{\frac{2}{5}} .$$

Thus, when the boundary of a rotating body and the law of its density may be defined by equations, its moment of inertia is readily found by the ordinary operations of the calculus; but when the figure is irregular and the density discontinuous, recourse is had to the properties of the compound pendulum, to be explained presently.

Example 6.—Find the points in reference to which the principal moments are equal.

Take the origin at the centre of inertia, and the principal axes through that point as the co-ordinate axes. Denote by $x_{,} \, y_{,} \, z_{,}$ the co-ordinates of one of the points sought; by $A_{,}$, $B_{,}$, and $C_{,}$ the principal moments with reference to this point, and by $x' \, y' \, z'$ the co-ordinates of the element m. Then, because the moments through the point $x_{,} \, y_{,} \, z_{,}$ are to be principal, will

$$\Sigma m (x' - x_{,}) (y' - y_{,}) = 0 \, ; \; \Sigma m (x' - x_{,}) (z' - z_{,}) = 0 \, ; \; \Sigma m (y' - y_{,}) (z' - z_{,}) = 0.$$

Performing the multiplication and reducing by the properties of the centre of inertia and principal axes, we have

$$M . x_{,} y_{,} = 0 \, ; \quad M x_{,} z_{,} = 0 \, ; \quad M y_{,} z_{,} = 0 :$$

which can only be satisfied by making two of the co-ordinates $x_{,} \, y_{,} \, z_{,}$ separately zero. Let $y_{,} = 0$, and $z_{,} = 0$; then, § 166 and Eq. (216),

$$A_{,} = A \, ; \quad B_{,} = B + M x_{,}^{2} \, ; \quad C_{,} = C + M x_{,}^{2} \, ;$$

but, by the conditions, the first members are equal. Whence

$$A = B + M x_{,}^{2} = C + M x_{,}^{2} \, ;$$

and, therefore,

$$B = C \, ; \text{ and } x_{,} = \pm \sqrt{\frac{A - B}{M}} \, ;$$

and from which it is apparent: 1st, that if all the principal moments in reference to the centre of inertia be unequal, there is no point in reference to which they can be equal; 2d, that if two of them be equal in reference to the centre of inertia and the third be the greatest, there are two points, equally distant from the centre of inertia and on the axis of the greatest moment, with reference to which they are equal; 3d, that if all three, with reference to the centre of inertia, be equal to one another, there is no other point with respect to which they can be equal.

IMPULSIVE FORCES.

§ 169.—We have thus far only been concerned with forces whose action may be likened to, and indeed represented by, the pressure arising from the weight of some definite body, as a cubic foot of distilled water at a standard temperature. Such forces are called *incessant*, because they extend their action through a definite and measurable portion of time. A single and instantaneous effort of such a force, called its intensity, is assumed to be measured by the whole effect which its incessant repetition for a unit of time can produce upon a free body. The effect here referred to is called the quantity of motion, being the product of the mass into the velocity generated. That is, Equations (12) and (13),

$$P = M . V_{,} = M \frac{d V}{d t} = M \frac{d^2 s}{d t^2} ; \quad . \quad . \quad . \quad . \quad (218)$$

in which $V_{,}$, denotes the velocity generated in a unit of time.

The force P, acting for one, two, or more units of time, or for any fractional portion of a unit of time, may communicate any other velocity V, and a quantity of motion measured by $M V$. And if the body which has thus received its motion gradually, impinge upon another which is free to move, experience tells us that it may suddenly transfer the whole of its motion to the latter by what seems to be a single blow, and although we know that this transfer can only take place by a series of successive actions and reactions between the molecular springs of the bodies, so to speak, and the inertia of their different elements, yet the whole effect is produced in a time so short as to elude the senses, and we are, therefore, apt to assume, though erroneously, that the effect is instantaneous. Such an assumption implies that a definite velocity can be generated in an indefinitely short time, and that the measure of the force's intensity is, Equation (218), infinite.

In all such cases, to avoid this difficulty, it is agreed to take the actual motion generated by these blows during the entire period of their action, as the measure of their intensity. Thus, denoting the mass impinged upon by M, and the actual velocity generated in it when perfectly free by V, we have

$$P = M V = M . \frac{d s}{d t} , \quad . \quad . \quad . \quad . \quad . \quad . \quad (219)$$

in which P, denotes the intensity of the force's action, and the second member of the equation the resistances of the body's inertia.

Forces which act in the manner just described, by a blow, are called *impulsive forces.*

MOTION OF A BODY UNDER THE ACTION OF IMPULSIVE FORCES.

§ 170.—The components of the inertia in the direction of the axes $x\,y\,z$, are respectively

$$M \cdot \frac{d\,s}{d\,t} \cdot \frac{d\,x}{d\,s} = M \cdot \frac{dx}{d\,t};$$

$$M \cdot \frac{d\,s}{d\,t} \cdot \frac{dy}{d\,s} = M \cdot \frac{dy}{d\,t};$$

$$M \cdot \frac{d\,s}{d\,t} \cdot \frac{d\,z}{d\,s} = M \cdot \frac{d\,z}{d\,t};$$

which, substituted for the corresponding components of inertia in Equations (A) and (B), give

$$\left.\begin{array}{l} \Sigma\, P \cos \alpha = \Sigma\, m \cdot \dfrac{dx}{d\,t}; \\ \Sigma\, P \cos \beta = \Sigma\, m \cdot \dfrac{dy}{d\,t}; \\ \Sigma\, P \cos \gamma = \Sigma\, m \cdot \dfrac{d\,z}{d\,t}; \end{array}\right\} \quad \cdots\cdots \quad (220)$$

$$\left.\begin{array}{l} \Sigma\, P\,(x' \cos \beta - y' \cos \alpha) = \Sigma\, m \left(x' \cdot \dfrac{dy}{d\,t} - y' \cdot \dfrac{d\,x}{d\,t}\right), \\ \Sigma\, P\,(z' \cos \alpha - x' \cos \gamma) = \Sigma\, m \left(z' \cdot \dfrac{dx}{d\,t} - x' \cdot \dfrac{d\,z}{d\,t}\right), \\ \Sigma\, P\,(y' \cos \gamma - z' \cos \beta) = \Sigma\, m \left(y' \cdot \dfrac{d\,z}{d\,t} - z' \cdot \dfrac{dy}{d\,t}\right). \end{array}\right\} \cdot \quad (221)$$

In which it will be recollected that $x\,y\,z$ are the co-ordinates of m, referred to the fixed origin, and $x'\,y'\,z'$, those of the same mass referred to the centre of inertia.

MOTION OF THE CENTRE OF INERTIA.

§ 171.—Substituting in Equations (220), for $d\,x$, $d\,y$, $d\,z$, their values obtained from Equations (34), and reducing by the relations

$$\Sigma\, m\, d\,x' = 0; \quad \Sigma\, m\, d\,y' = 0; \quad \Sigma\, m\, d\,z' = 0; \quad \cdot$$

given by the principle of the centre of inertia, we find

$$\left.\begin{aligned} \Sigma P \cos \alpha &= \frac{dx_{,}}{dt} . \Sigma m ; \\ \Sigma P \cos \beta &= \frac{dy_{,}}{dt} . \Sigma m ; \\ \Sigma P \cos \gamma &= \frac{dz_{,}}{dt} . \Sigma m ; \end{aligned}\right\} \quad . \quad . \quad . \quad . \quad . \quad . \quad (223)$$

and substituting M for Σm, we have

$$\Sigma P \cos \alpha = M . \frac{dx_{,}}{dt} ;$$

$$\Sigma P \cos \beta = M . \frac{dy_{,}}{dt} ;$$

$$\Sigma P \cos \gamma = M . \frac{dz_{,}}{dt} ;$$

which are wholly independent of the relative positions of the elements of the body, and from which we conclude that the motion of the centre of inertia will be the same as though the mass were concentrated in it, and the forces applied immediately to that point.

§ 172.—Replacing the first members of the above equations by their values given in Equations (41), and denoting **by** V the velocity which the resultant R can impress upon the whole mass, then will

$$\Sigma P \cos \alpha = M V \cos a ; \quad \Sigma P \cos \beta = M V \cos b ; \quad \Sigma P \cos \gamma = M V \cos c ;$$

substituting these above, we find

$$\left.\begin{aligned} V . \cos a &= \frac{dx_{,}}{dt} ; \\ V . \cos b &= \frac{dy_{,}}{dt} ; \\ V . \cos c &= \frac{dz_{,}}{dt} ; \end{aligned}\right\} \quad . \quad . \quad . \quad . \quad . \quad . \quad . \quad (224)$$

and integrating,

$$\left.\begin{aligned} x_{,} &= V \cdot \cos a \,.\, t + C', \\ y_{,} &= V \,.\, \cos b \,.\, t + C'', \\ z_{,} &= V \,.\, \cos c \,.\, t + C''', \end{aligned}\right\} \quad \ldots\ldots \quad (225)$$

and eliminating t from these equations, V will also disappear, and we find,

$$\left.\begin{aligned} z_{,} &= x_{,} \cdot \frac{\cos c}{\cos a} - \frac{C' \cos c - C''' \cos a}{\cos a}, \\ z_{,} &= y_{,} \cdot \frac{\cos c}{\cos b} - \frac{C'' \cos c - C''' \cos b}{\cos b}, \\ y_{,} &= x_{,} \cdot \frac{\cos b}{\cos a} - \frac{C' \cos b - C'' \cos a}{\cos a}, \end{aligned}\right\} \quad \ldots \quad (226)$$

which being of the first degree and either one but the consequence of the other two, are the equations of a straight line. This line makes with the axes x, y, z, the angles a, b, c, respectively, and is, therefore, parallel to the resultant of the impressed forces.

Whence we conclude, that the centre of inertia of a body acted upon simultaneously by any number of impulsive forces, will move uniformly in a straight line parallel to their common resultant.

MOTION ABOUT THE CENTRE OF INERTIA.

§ 173.—Substituting, in Equations (221), for dx, dy and dz, their values from Equations (34), reducing by

$$\Sigma\, m\, x' = 0,$$

$$\Sigma\, m\, y' = 0,$$

$$\Sigma\, m\, z' = 0,$$

and we find,

$$\left.\begin{aligned} \Sigma\, P\,(x' \cos \beta - y' \cos \alpha) &= \Sigma\, m \left(x' \cdot \frac{d\,y'}{d\,t} - y'\, \frac{d\,x'}{d\,t}\right); \\ \Sigma\, P\,(z' \cos \alpha - x' \cos \gamma) &= \Sigma\, m \left(z' \cdot \frac{d\,x'}{d\,t} - x' \cdot \frac{d\,z'}{d\,t}\right); \\ \Sigma\, P\,(y' \cos \gamma - z' \cos \beta) &= \Sigma\, m \left(y' \cdot \frac{d\,z'}{d\,t} - z' \cdot \frac{d\,y'}{d\,t}\right); \end{aligned}\right\} \quad \ldots \quad (227)$$

whence, the motion of the body about its centre of inertia will be the same whether that point be at rest or in motion, its co-ordinates having disappeared entirely from the equations.

ANGULAR VELOCITY.

§ 174.—Replacing the first members of Eqs. (227) by $L_{,}$, $M_{,}$, and $N_{,}$, respectively, § 162; and substituting in the second members for dx', dy' and dz', their values in Eqs. (190), we readily find

$$\left.\begin{aligned} \frac{d\varphi}{dt} &= \frac{L_{,} + \Sigma\, m\, x'z' \cdot \dfrac{d\varpi}{dt} + \Sigma\, m\, y'z' \cdot \dfrac{d\psi}{dt}}{\Sigma\, m\, (x'^2 + y'^2)} \\ \frac{d\psi}{dt} &= \frac{M_{,} + \Sigma\, m\, z'y' \cdot \dfrac{d\varphi}{dt} + \Sigma\, m\, x'y' \cdot \dfrac{d\varpi}{dt}}{\Sigma\, m\, (x'^2 + z'^2)} \\ \frac{d\varpi}{dt} &= \frac{N_{,} + \Sigma\, m\, x'y' \cdot \dfrac{d\psi}{dt} + \Sigma\, m\, x'z' \cdot \dfrac{d\varphi}{dt}}{\Sigma\, m\, (y'^2 + z'^2)} \end{aligned}\right\} \quad \dots \quad (228)$$

If the axes be principal, then will $\Sigma\, m\, x'y' = 0$, $\Sigma\, m\, y'z' = 0$, $\Sigma\, m\, x'z' = 0$; or if the axes be fixed in succession, then for the axis x' will $d\psi = 0$; $d\varphi = 0$; for the axis y, $d\varphi = 0$; $d\tilde{\omega} = 0$; and for the axis z, $d\varpi = 0$; $d\psi = 0$, and the above become

$$\left.\begin{aligned} \frac{d\varphi}{dt} &= \frac{L_{,}}{\Sigma\, m \,.\, (x'^2 + y'^2)}; \\ \frac{d\psi}{dt} &= \frac{M_{,}}{\Sigma\, m \,.\, (x'^2 + z'^2)}; \\ \frac{d\varpi}{dt} &= \frac{N_{,}}{\Sigma\, m \,.\, (y'^2 + z'^2)}. \end{aligned}\right\} \quad \dots \dots \quad (229)$$

That is, the component angular velocity about either a principal or fixed axis, is equal to the moment of the impressed forces divided by the moment of inertia with reference to that axis.

The resultant angular velocity being denoted by $\frac{ds_{,}}{dt}$, we also have, (Eq. 196),

$$\frac{ds_{,}}{dt} = \frac{1}{dt}\sqrt{d\varphi^2 + d\psi^2 + d\varpi^2}. \quad \dots \dots \quad (230)$$

AXIS OF INSTANTANEOUS ROTATION.

§ 175.—The axis of instantaneous rotation is found as in § 158, by making, in Equations (192), $d\,x' = 0$, $d\,y' = 0$, $d\,z' = 0$; and, therefore,

$$z' . \nu_y - y' . \nu_z = 0\,; \quad x' . \nu_z - z' . \nu_x = 0\,; \quad y' . \nu_x - x' . \nu_y = 0 \quad . \quad (231)$$

which, as the last is but a consequence of the others, are the equations of a right line through the centre of inertia.

The equations of the line of the resultant impact are, Eqs. (45),

$$z' = \frac{Z}{X} . x' + \frac{M_{\prime}}{X}, \qquad y' = \frac{Y}{X} . x' - \frac{L_{\prime}}{X}\,;$$

and the inclination θ of this line to the instantaneous axis, is given by

$$\cos\theta = \frac{\nu_z . Z + \nu_y . Y + \nu_x . X}{\sqrt{\nu_z^2 + \nu_y^2 + \nu_x^2} . \sqrt{Z^2 + Y^2 + X^2}}\,;$$

or, substituting for ν_z, ν_y, and ν_x their values, Eqs. (229) and (191),

$$\cos\theta = \frac{\frac{L_{\prime} . Z}{C} + \frac{M_{\prime} . Y}{B} + \frac{N_{\prime} . X}{A}}{\sqrt{\left(\frac{L_{\prime}}{C}\right)^2 + \left(\frac{M_{\prime}}{B}\right)^2 + \left(\frac{N_{\prime}}{A}\right)^2} . \sqrt{Z^2 + Y^2 + X^2}} \quad . \quad (232)$$

The point in which the line of the impact pierces the plane $y\,z$ is given by

$$z' = \frac{M_{\prime}}{X}\,; \qquad y' = -\frac{L_{\prime}}{X}\,;$$

dividing one by the other, we have, for the equation of the line through this point and the centre of inertia,

$$y' = -\frac{L_{\prime}}{M_{\prime}} z'.$$

Denote the angle which this line makes with the instantaneous axis by θ'; then from the equations of these lines will

$$\cos\theta' = \frac{-\frac{\nu_y}{\nu_z} . \frac{L_{\prime}}{M_{\prime}} + 1}{\sqrt{\left(\frac{\nu_y}{\nu_z}\right)^2 + \left(\frac{\nu_x}{\nu_z}\right)^2 + 1} . \sqrt{\left(-\frac{L_{\prime}}{M_{\prime}}\right)^2 + 1}}\,;$$

or, Eqs. (229) and (191),

$$\cos\theta' = \frac{-\frac{C}{B} + 1}{\sqrt{\left(\frac{M_{\prime}}{L_{\prime}} . \frac{C}{B}\right)^2 + \left(\frac{N_{\prime}}{L_{\prime}} . \frac{C}{A}\right)^2 + 1} . \sqrt{\frac{L_{\prime}^2}{M_{\prime}^2} + 1}} \quad . \quad (233)$$

AXIS OF SPONTANEOUS ROTATION.

§ 176.—If both members of Eqs. (34) be divided by $d\,t$, we have

$$\frac{d\,x}{d\,t} = \frac{d\,x_{\prime}}{d\,t} + \frac{d\,x'}{d\,t};$$

$$\frac{d\,y}{d\,t} = \frac{d\,y_{\prime}}{d\,t} + \frac{d\,y'}{d\,t};$$

$$\frac{d\,z}{d\,t} = \frac{d\,z_{\prime}}{d\,t} + \frac{d\,z'}{d\,t};$$

and if for any element

$$\frac{d\,x}{d\,t} = 0; \quad \frac{d\,y}{d\,t} = 0; \quad \frac{d\,z}{d\,t} = 0 \quad . \quad . \quad . \quad . \quad . \quad (234)$$

then will

$$\frac{d\,x_{\prime}}{d\,t} = -\frac{d\,x'}{d\,t}; \quad \frac{d\,y_{\prime}}{d\,t} = -\frac{d\,y'}{d\,t}; \quad \frac{d\,z_{\prime}}{d\,t} = -\frac{d\,z'}{d\,t} \quad . \quad . \quad (235)$$

Substituting for the first members their values given in Equations (224), and for the second members their values given in Equations (192), we have

$$\left.\begin{array}{l} z' \,.\, \nu_y - y' \,.\, \nu_z + V \,.\, \cos a = 0 \\ x' \,.\, \nu_z - z' \,.\, \nu_x + V \,.\, \cos b = 0 \\ y' \,.\, \nu_x - x' \,.\, \nu_y + V \,.\, \cos c = 0 \end{array}\right\} \quad . \quad . \quad . \quad . \quad . \quad (236)$$

Now, if either of these equations be but a consequence of the other two, then will they be the equations of a right line parallel, Equations (231), to the instantaneous axis; and all points upon this line will be at rest during the body's motion. This line is called the *axis of spontaneous rotation*.

To find the conditions which shall express the dependence of either of the Equations (236) upon the other two, multiply each by the angular velocity it does not already contain, add the products, and divide the sum by the resultant angular velocity ν_i; there will result,

$$\cos a \,.\, \frac{\nu_x}{\nu_i} + \cos b \,.\, \frac{\nu_y}{\nu_i} + \cos c \,.\, \frac{\nu_z}{\nu_i} = 0 \quad . \quad . \quad . \quad (236)'$$

The first member is the cosine of the angle which the resultant impact makes with the instantaneous axis; this being zero, it follows that whenever a body is struck so as to make the instantaneous axis perpendicular to the direction of the impact, the spontaneous axis will exist.

Denote by l, the distance from the spontaneous axis to the line of the impact; by e_z and e_y, the absolute terms in the second and third of Eqs. (236), solved with respect to z and y, then will

$$l = \frac{\left(e_z - \frac{M_{,}}{X}\right)\left(\frac{\nu_y}{\nu_x} - \frac{Y}{X}\right) + \left(e_y + \frac{L_{,}}{X}\right)\left(\frac{\nu_z}{\nu_x} - \frac{Z}{X}\right)}{\sqrt{\left(\frac{\nu_z}{\nu_x}\right)^2 + \left(\frac{\nu_y}{\nu_x}\right)^2 + 1} \cdot \sqrt{\left(\frac{Z}{X}\right)^2 + \left(\frac{Y}{X}\right)^2 + 1}} \quad . \quad (237)$$

Equation (232) will make known the circumstances of the impact and shape of the body which will determine the existence of the spontaneous axis.

Make the impact in the plane of the principal axes $x' y'$, and parallel to the axis x'. Then, Equation (232), will $\theta = 90°$, and the spontaneous axis will exist. Also, Equation (233), $\theta' = 90°$. And, Equations (237) and (229), and because $X = M . V$, and $V = e_y . \nu_z$,

$$l = e_y + \frac{L_{,}}{X} = e_y + \frac{M . k_{,}^2 . \nu_z}{M . e_y . \nu_z} = e_y + \frac{k_{,}^2}{e_y};$$

whence,

$$(l - e_y) . e_y = k_{,}^2 \quad . \quad . \quad . \quad . \quad . \quad . \quad . \quad . \quad (238).$$

That is, when the line of the impact is in the plane of two of the principal axes and parallel to one of them, there will be a spontaneous axis, and the product of its distance from the centre of inertia by that of the line of the impact from the same point, is equal to the square of the principal radius of gyration in reference to the instantaneous axis.

§ 177.—The body being free, and the axis of spontaneous rotation at rest, while the other parts of the body are acquiring motion, the forces, both extraneous and of inertia, are so balanced about that line as to impress no action upon it. The line of the impact and the points of the body on this line are called, respectively, the *axis* and *centres of percussion*, in reference to the spontaneous axis. A centre of percussion in reference to an axis is, therefore, any point at which a body may be struck without communicating a shock to a physical line coincident in position with that axis.

STABLE AND UNSTABLE ROTATION.

§ 178.—Now suppose the rotation to have been impressed, the instantaneous axis nearly coincident with the principal axis z, and the body abandoned to itself. What will be the circumstances of the motion?

The first member of the third of Equations (194) will be sensibly equal to unity, v_y and v_x, therefore, indefinitely small, their product an indefinitely small quantity of the second order; $L_{,}$, $M_{,}$, and $N_{,}$ will be zero, and Equations (202) may be written,

$$C \cdot \frac{d v_z}{d t} = 0 ; \; B \cdot \frac{d v_y}{d t} + v_x \cdot v_z \cdot (A - C) = 0 ; \; A \cdot \frac{d v_x}{d t} + v_y \cdot v_z \cdot (C - B) = 0.$$

Integrating the first, we have

$$v_z = \frac{c}{C} = n ;$$

in which c is the constant of integration; and this in the other equations gives

$$B \cdot \frac{d v_y}{d t} + n \cdot (A - C) \cdot v_x = 0 ; \quad A \cdot \frac{d v_x}{d t} + n \, (C - B) \, v_y = 0 ;$$

differentiating and substituting in each of the derived equations the values of the first differential coefficients obtained from the primitive, we find

$$\frac{d^2 v_y}{d t^2} + n^2 \cdot \frac{(A - C) \cdot (B - C)}{A \cdot B} \cdot v_y = 0 ;$$

$$\frac{d^2 v_x}{d t^2} + n^2 \cdot \frac{(A - C) \cdot (B - C)}{A \cdot B} \cdot v_x = 0.$$

If $A - C$ and $B - C$ be both positive or both negative, their product will be positive, and the integrals are

$$v_y = a_y \cdot \sin \left\{ n \cdot \sqrt{\frac{(A - C) \cdot (B - C)}{A \cdot B}} \cdot t + c_y \right\} ;$$

$$v_x = a_x \cdot \sin \left\{ n \cdot \sqrt{\frac{(A - C) \cdot (B - C)}{A \cdot B}} \cdot t + c_x \right\} .$$

If one of the factors $A - C$ and $B - C$ be positive and the other negative, their product will be negative, and the integrals will be

$$v_y = a_y \cdot e^{\left\{ n \cdot \sqrt{\frac{(A - C) \cdot (C - B)}{A \cdot B}} \cdot t + c_y \right\}} ,$$

$$v_x = a_x \cdot e^{\left\{ n \cdot \sqrt{\frac{(A - C) \cdot (C - B)}{A \cdot B}} \cdot t + c_x \right\}} .$$

In these integrals, a_y, a_x, c_y, and c_x, are constants whose values result from the initial conditions of the rotation. They are small at the epoch, because v_y and v_z are small. In the first integration, v_y and v_z will continue small and resume periodically their initial values; in the second, they will increase with the time indefinitely. If the instantaneous axis coincide with the axis z, then will a_y and a_z be zero; v_y and v_z will be zero, and, hence, *a principal axis is always a permanent axis of rotation;* and the rotation will be *stable* about the axes of greatest and least moments of inertia, and *unstable* about all others.

MOTION OF A SYSTEM OF BODIES.

§ 179—We have seen that the Equations (117) and (119) give all the circumstances of motion of the centre of inertia of a single body in reference to any assumed point taken as an origin of co ordinates. For a second, third, and indeed any number of bodies, referred to the same origin, we would have similar equations, the only difference being in the values of the co-ordinates, of the intensities and directions of the forces, and of the magnitudes of the masses. This difference being indicated in the usual way by accents, we should obtain by addition,

$$\left.\begin{aligned} \Sigma\, M \cdot \frac{d^2 x}{d\, t^2} &= \Sigma\, X; \\ \Sigma\, M \cdot \frac{d^2 y}{d\, t^2} &= \Sigma\, Y; \\ \Sigma\, M \cdot \frac{d^2 z}{d\, t^2} &= \Sigma\, Z; \end{aligned}\right\} \quad \ldots \ldots \quad (239)$$

$$\left.\begin{aligned} \Sigma\, M \left(x \cdot \frac{d^2 y}{d\, t^2} - y \cdot \frac{d^2 x}{d\, t^2}\right) &= \Sigma\, (Y x - X y)\,; \\ \Sigma\, M \left(z \cdot \frac{d^2 x}{d\, t^2} - x \cdot \frac{d^2 z}{d\, t^2}\right) &= \Sigma\, (X z - Z x)\,; \\ \Sigma\, M \left(y \cdot \frac{d^2 z}{d\, t^2} - z \cdot \frac{d^2 y}{d\, t^2}\right) &= \Sigma\, (Z y - Y z)\,; \end{aligned}\right\} \quad \ldots \quad (240)$$

in which it must be recollected that x, y, z, &c., denote the co ordinates of the centres of inertia of the several masses M, &c., referred to a fixed origin.

MOTION OF THE CENTRE OF INERTIA OF THE SYSTEM.

§ 180.—Taking a movable origin at the centre of inertia of the entire system, denoting the co-ordinates of this point referred to the fixed origin by $x_{,}$, $y_{,}$, $z_{,}$, and the co-ordinates of the centres of inertia of the several masses referred to the movable origin by x', y', z', &c., we have, the axes of the same name in the two systems being parallel,

$$\left.\begin{array}{l} x = x_{,} + x', \\ y = y_{,} + y', \\ z = z_{,} + z', \\ \text{and,} \\ d^2 x = d^2 x_{,} + d^2 x', \\ d^2 y = d^2 y_{,} + d^2 y', \\ d^2 z = d^2 z_{,} + d^2 z', \end{array}\right\} \quad \ldots \ldots \quad (241)$$

which substituted in Equations (239), and reducing by the relations,

$$\Sigma M . d^2 x' = 0 ; \quad \Sigma M d^2 y' = 0 ; \quad \Sigma M d^2 z' = 0 ; \quad \ldots \quad (242)$$

obtained from the property of the centre of inertia, we find

$$\left.\begin{array}{l} \dfrac{d^2 x_{,}}{d t^2} . \Sigma M = \Sigma X ; \\ \dfrac{d^2 y_{,}}{d t^2} . \Sigma M = \Sigma Y ; \\ \dfrac{d^2 z_{,}}{d t^2} . \Sigma M = \Sigma Z ; \end{array}\right\} \quad \ldots \ldots \quad (243)$$

which being wholly independent of the relative positions of the several bodies, show that the motion of the centre of inertia of the system will be the same as though its entire mass were concentrated in that point, and the forces applied directly to it.

§ 181.—Multiplying the first of Equations, (243), by $y_{,}$, the second by $x_{,}$, and taking the difference; also, their first by $z_{,}$ the third by $x_{,}$, and taking the difference, and again the second by $z_{,}$, the third by $y_{,}$, and taking the difference, we find

$$\left.\begin{aligned} \left(x_{,}\cdot\frac{d^2 y_{,}}{d t^2} - y_{,}\cdot\frac{d^2 x_{,}}{d t^2}\right)\cdot\Sigma M &= x_{,}\cdot\Sigma Y - y_{,}\cdot\Sigma X; \\ \left(z_{,}\cdot\frac{d^2 x_{,}}{d t^2} - x_{,}\cdot\frac{d^2 z_{,}}{d t^2}\right)\cdot\Sigma M &= z_{,}\cdot\Sigma X - x_{,}\cdot\Sigma Z; \\ \left(y_{,}\cdot\frac{d^2 z_{,}}{d t^2} - z_{,}\cdot\frac{d^2 y_{,}}{d t^2}\right)\cdot\Sigma M &= y_{,}\cdot\Sigma Z - z_{,}.\Sigma Y; \end{aligned}\right\} \quad (244)$$

which will make known the circumstances of motion of the common centre of inertia about the fixed origin.

MOTION OF THE SYSTEM ABOUT ITS COMMON CENTRE OF INERTIA.

§ 182.—Substituting the values of x, y, z, $d^2 x$, &c., given by Equations (241), in Equations (240) and reducing by Equations (244) and (242), there will result

$$\left.\begin{aligned} \Sigma M\cdot\left(x'\cdot\frac{d^2 y'}{d t^2} - y'\cdot\frac{d^2 x'}{d t^2}\right) &= \Sigma\,(Y x' - X y') \\ \Sigma M.\left(z'.\frac{d^2 x'}{d t^2} - x'\cdot\frac{d^2 z'}{d t^2}\right) &= \Sigma\,(X z' - Z x') \\ \Sigma M\cdot\left(y'.\frac{d^2 z'}{d t^2} - z'\cdot\frac{d^2 y'}{d t^2}\right) &= \Sigma\,(Z y' - Y z') \end{aligned}\right\} \quad (245)$$

Equations from which all traces of the position of the centre of inertia have disappeared, and from which we conclude that the motion of the elements of the system about that point will be the same, whether it be at rest or in motion. These equations are identical in form with Equations (118); whence we conclude that the molecular forces disappear from the latter, and cannot, there fore, have any influence upon the motion due to the action of the extraneous forces.

CONSERVATION OF THE MOTION OF THE CENTRE OF INERTIA.

§ 183.—If the system be subjected only to the forces arising from the mutual attractions or repulsions of its several parts, then will

$$\Sigma X = 0;\ \Sigma Y = 0;\ \Sigma Z = 0.$$

For, the action of the mass M, upon a single element of M', will vary with the number of acting elements contained in M; and the effort necessary to prevent M' from moving under this action will be equal to the whole action of M upon a single element of M' repeated as many times as there are elements in M' acted upon; whence, the action of M upon M' will vary as the product $M M'$. In the same way it will appear that the force required to prevent M from moving under the action of M', will be proportional to the same product, and as these reciprocal actions are exerted at the same distance, they must be equal; and, acting in contrary directions, the cosines of the angles their directions make with the co-ordinate axes, will be equal, with contrary signs. Whence, for every set of components $P \cos \alpha$, $P \cos \beta$, $P \cos \gamma$, in the values of ΣX, ΣY, ΣZ, there will be the numerically equal components, $- P' \cos \alpha'$, $- P' \cos \beta'$, $- P' \cos \gamma'$, and, Equations (243), reduce, after dividing by ΣM, to

$$\frac{d^2 x_{,}}{d t^2} = 0; \quad \frac{d^2 y_{,}}{d t^2} = 0; \quad \frac{d^2 z_{,}}{d t^2} = 0 \quad . \quad . \quad . \quad . \quad (246)$$

and from which we obtain, after two integrations,

$$\left. \begin{aligned} x_{,} &= C' . t + D'; \\ y_{,} &= C'' . t + D''; \\ z_{,} &= C''' . t + D'''; \end{aligned} \right\} \quad . \quad . \quad . \quad . \quad . \quad . \quad . \quad (247)$$

in which C', C'', C''', D', D'' and D''' are the constants of integration; and from which, by eliminating t, we find two equations of the first degree between the variables $x_{,}$, $y_{,}$, $z_{,}$, whence the path of the centre of inertia, if it have any at all, is a right line.

Also multiplying Equations (246) by $2 d x_{,}$, $2 d y_{,}$, $2 d z_{,}$, respectively, adding and integrating, we have

$$\frac{d x_{,}^2 + d y_{,}^2 + d z_{,}^2}{d t^2} = V^2 = C \quad . \quad . \quad . \quad . \quad (248)$$

in which C is the constant of integration and V the velocity of the centre of inertia of the system. From all of which we conclude, that when a system of bodies is subjected only to forces arising from the action of its elements upon each other, its centre of inertia will either be at rest or move uniformly in a right line. This is called the conservation of the motion of the centre of inertia.

CONSERVATION OF AREAS.

§ 184.—The second member of the first of Equations (245) may be written,

$$Y x' - X y' + Y' x'' - X' y'' + \&c.;$$

and considering the bodies by pairs, we have

$$X = - X'; \quad Y = - Y';$$

and eliminating X' and Y' above by these values, we have

$$Y (x' - x'') - X (y' - y'') + \&c.$$

But,

$$X = P \cdot \frac{x' - x''}{p}; \quad Y = P \, \frac{y' - y''}{p};$$

in which p denotes the distance between the centres of inertia of the two bodies. And substituting these above, we get

$$P \cdot \frac{y' - y''}{p} (x' - x'') - P \cdot \frac{x' - x''}{p} (y' - y'') = 0;$$

and the same being true of every other pair, the second members of Equations (245), will be zero, and we have

$$\Sigma M \cdot \left(x' \cdot \frac{d^2 y'}{d t^2} - y' \cdot \frac{d^2 x'}{d t^2} \right) = 0;$$

$$\Sigma M \cdot \left(z' \cdot \frac{d^2 x'}{d t^2} - x' \cdot \frac{d^2 z'}{d t^2} \right) = 0;$$

$$\Sigma M \cdot \left(y' \cdot \frac{d^2 z'}{d t^2} - z' \cdot \frac{d^2 y'}{d t^2} \right) = 0;$$

and integrating

$$\left. \begin{aligned} \Sigma M \cdot \frac{x' d y' - y' d x'}{d t} &= C', \\ \Sigma M \cdot \frac{z' d x' - x' d z'}{d t} &= C'', \\ \Sigma M \cdot \frac{y' d z' - z' d y'}{d t} &= C'''. \end{aligned} \right\} \quad \ldots \ldots \quad (249)$$

But § 190, $x'\,dy' - y'\,dx'$, is twice the differential of the area swept over by the projection of the radius vector of the body M, on the co-ordinate plane $x'\ y'$, and the same of the similar expressions in the other equations, in reference to the other co-ordinate planes; whence, denoting by A_z, A_y, A_x, double the areas described in any interval of time, t, by the projections of the radius vector of the body M, on the co-ordinate planes, $x'\,y'$, $x'\,z'$, and $y'\,z'$, and adopting similar notations for the other bodies, we have

$$\Sigma M \cdot \frac{d\,A_z}{d\,t} = C';$$

$$\Sigma M \cdot \frac{d\,A_y}{d\,t} = C'';$$

$$\Sigma M \cdot \frac{d\,A_x}{d\,t} = C''';$$

in which C', C'', C''', denote the sums of the products obtained by multiplying each mass into twice the area swept over in a unit of time by the projection of its radius vector on the planes $x'\,y'$, $x'\,z'$, $y'\,z'$; and by integrating between the limits $t_{\prime}$ and t', giving an interval equal to t,

$$\Sigma M . A_z = C' . t;$$

$$\Sigma M . A_y = C''\,t;$$

$$\Sigma M . A_x = C'''\,t;$$

whence we find that when a system is in motion and is only subjected to the attractions or repulsions of its several elements upon each other, the sum of the products arising from multiplying the mass of each element by the projection, on any plane, of the area swept over by the radius vector of this element, measured from the centre of inertia of the entire system, varies as the time of the motion. This is called the principle of the *conservation of areas.*

§ 185.—It is important to remark that the same conclusions would be true if the bodies had been subjected to forces directed towards a fixed point. For, this point being assumed as the origin of co-ordinates, the equation of the direction of any one force, say that acting upon M, will be

$$Yx - Xy = 0;$$

and the second members of Equations (240) will reduce to zero; and the form of these equations being the same as Equations (245), they will give, by integration, the same consequences.

INVARIABLE PLANE.

§ 186.—If we examine Equations (249), we shall find that $M \cdot \frac{dy'}{dt}$ is the quantity of motion of the mass M, in the direction of the axis y', and is the measure of the component of the moving force in that direction; the same may be said of $M \cdot \frac{dx'}{dt}$, in the direction of the axis x'; whence the expression,

$$M \cdot \frac{x'dy' - y'dx'}{dt},$$

is the moment of the moving force of M, with respect to the axis z'. Designating, as before, the sum of the moments with respect to the axes z', y' and x', by $L_{,}$, $M_{,}$, $N_{,}$, respectively, Equations (249) become

$$L_{,} = C'; \quad M_{,} = C''; \quad N_{,} = C'''.$$

Denoting by θ_z, θ_y, and θ_x, the angles which the resultant axis makes with the axes z', y' and x', we have, § 110,

$$\left.\begin{aligned} \cos\theta_z &= \frac{L_{,}}{\sqrt{L_{,}^2 + M_{,}^2 + N_{,}^2}} = \frac{C'}{\sqrt{C'^2 + C''^2 + C'''^2}}; \\ \cos\theta_y &= \frac{M_{,}}{\sqrt{L_{,}^2 + M_{,}^2 + N_{,}^2}} = \frac{C''}{\sqrt{C'^2 + C''^2 + C'''^2}}; \\ \cos\theta_x &= \frac{N_{,}}{\sqrt{L_{,}^2 + M_{,}^2 + N_{,}^2}} = \frac{C'''}{\sqrt{C'^2 + C''^2 + C'''^2}}. \end{aligned}\right\} \quad \cdots \quad (250)$$

These determine the position of the resultant or *principal axis.* The plane at right angles to this axis is called the *principal plane.* The position of this plane is invariable, and it is therefore called the *invariable plane,* either when the only forces of the system are those arising from the mutual actions and reactions of the bodies upon each other, or when the forces are all directed towards a fixed centre.

PRINCIPLE OF LIVING FORCE.

§ 187.—If, during the motion, two or more bodies of the system impinge against each other so as to produce a sudden change in their velocities, the sum of the living forces will undergo a change. To estimate this change, let A, B, C be the velocities of the mass m, in the direction of the axes before the impact, and a, b, c what these velocities become at the instant of nearest approach of the centres of inertia of the impinging masses, then will

$$A - a, \quad B - b, \quad C - c,$$

be the components of the velocities lost or gained by m at the instant corresponding to this state of the impact, and

$$m(A - a), \quad m(B - b), \quad m(C - c),$$

the components of the forces lost or gained. The same expressions, with accents, will represent the components of the forces lost or gained by the other impinging bodies of the system. These, by the principle of D'Alembert, § 71, are in equilibrio, whence

$$\Sigma m(A - a)\,\delta x + \Sigma m(B - b)\,\delta y + \Sigma m(C - c)\,\delta z = 0.$$

The indefinitely small displacements δx, δy, δz, &c., must be made consistently with the connection by virtue of which the velocities are lost or gained; but as a, b, c denote the components of the actual velocities of the body whose mass is m, at the instant of its nearest approach to that with which it collides, this condition is fulfilled if we make

$$\delta x = a\,.\,\delta t; \quad \delta y = b\,.\,\delta t; \quad \delta z = c\,.\,\delta t.$$

These values being substituted in the above equation, we have, after dividing by δt,

$$\Sigma m(A - a)\,a + \Sigma m(B - b)\,b + \Sigma m(C - c)\,c = 0 \quad . \quad . \quad (251)$$

or,

$$\Sigma m(Aa + Bb + Cc) - \Sigma m(a^2 + b^2 + c^2) = 0 \quad . \quad . \quad (252)$$

But we have the identical equation,

$$(A-a)^2+(B-b)^2+(C-c)^2=\begin{cases} A^2+B^2+C^2+a^2+b^2 \\ +c^2-2(Aa+Bb+Cc), \end{cases}$$

or,

$$Aa+Bb+Cc=\begin{cases} \dfrac{A^2+B^2+C^2}{2}+\dfrac{a^2+b^2+c^2}{2} \\ -\dfrac{(A-a)^2+(B-b)^2+(C-c)^2}{2}, \end{cases}$$

which in Equation (252) gives,

$$\Sigma\, m(A^2+B^2+C^2)-\Sigma\, m(a^2+b^2+c^2)=\Sigma\, m\,[(A-a)^2+(B-b)^2+(C-c)^2],$$

and making

$$A^2+B^2+C^2=V^2,$$
$$a^2+b^2+c^2=u^2,$$
$$\Sigma\, mV^2-\Sigma\, mu^2=\Sigma\, m\,[(A-a)^2+(B-b)^2+(C-c)^2]\cdot\cdot \quad (253)$$

whence we conclude, *that the difference of the sums of the living forces before the collision, and at the instant of greatest compression, is equal to the sum of the living forces which the system would have, if the masses moved with the velocities lost and gained at this stage of the collision.*

Since all the terms of the preceding equation are essentially positive, it follows that at the instant of nearest approach of the impinging bodies, there is a loss of living force.

If the impinging masses now react upon each other in a way to cause them to be thrown asunder, and A', B', C', &c., denote the components of the actual velocities, in the direction of the axes, at the instant of separation, then will the components of the velocities lost and gained while the separation is taking place, be

$$a-A', \quad b-B', \quad c-C', \quad \&c., \&c.;$$

and Equation (251) will become

$$\Sigma\, m\,(a-A')\,a+\Sigma\, m\,(b-B')\,b+\Sigma\, m\,(c-C')\,c=0,$$

or,

$$\Sigma\, m\,(a^2+b^2+c^2)-\Sigma\, m\,(A'a+B'b+C'c)=0;$$

and eliminating $A'a + B'b + C'c$, by means of the identical equation,

$$(a - A')^2 + (b - B')^2 + (c - C')^2 = \left\{ \begin{array}{l} a^2 + b^2 + c^2 + A'^2 + B'^2 \\ + C'^2 - 2(A'a + B'b + C'c), \end{array} \right.$$

we obtain,

$$\Sigma m (a^2 + b^2 + c^2) - \Sigma m (A'^2 + B'^2 + C'^2) = - \Sigma m \left\{ \begin{array}{c} (a - A')^2 \\ + (b - B')^2 \\ + (c - C')^2 \end{array} \right\},$$

and making

$$A'^2 + B'^2 + C'^2 = V'^2,$$

$$\Sigma m u^2 - \Sigma m V'^2 = - \Sigma m \left[(a - A')^2 + (b - B')^2 + (c - C')^2 \right] \cdot\cdot (254)$$

All the terms of this equation being essentially positive, it follows, from the sign of the second member, that during the reaction of the bodies by which they are separated, there is a gain of living force.

If the loss and gain of velocities after, be the same as before the instant of greatest compression, then will there be no loss or gain of living force by the collision.

PLANETARY MOTIONS.

§ 188.—When the only forces are those arising from the mutual attractions of the several bodies of the system for one another, the second members of Equations (239) reduce, as we have seen, § 183, to zero, and those equations become

$$\left. \begin{array}{l} \Sigma M \cdot \dfrac{d^2 x}{d t^2} = 0, \\ \Sigma M \cdot \dfrac{d^2 y}{d t^2} = 0, \\ \Sigma M \cdot \dfrac{d^2 z}{d t^2} = 0, \end{array} \right\} \quad . \quad . \quad . \quad . \quad . \quad . \quad . \quad . \quad (255)$$

Let us now find the motion of any one body of the system in reference to any other, taken at pleasure. This latter body will be called the *central*, the former the *primary*, and the others, collectively, the *perturbating* bodies. Let the central and primary bodies be those whose masses are M and $M_{\prime}$ respectively; the perturbating bodies those whose masses are $M_{\prime\prime}$, $M_{\prime\prime\prime}$, &c. The first of the above equations may be written

$$M \cdot \frac{d^2 x}{d t^2} + M_{\prime} \cdot \frac{d^2 x_{\prime}}{d t^2} + \Sigma M_{\prime\prime} \cdot \frac{d^2 x_{\prime\prime}}{d t^2} = 0 \quad . \quad . \quad . \quad (256)$$

If the perturbating bodies alone acted upon one another, the last term would be zero; and when the action of the central and primary are included, the numerical value of this term will result from the action of these latter bodies. Denote the reciprocal action of any two bodies upon one another by writing their masses within the parenthetic sign, and use the subscript x to denote the component of this action parallel to the axis x. Then will

$$\Sigma (M M_{\prime\prime})_x + \Sigma (M_{\prime} M_{\prime\prime})_x - \Sigma M_{\prime\prime} \frac{d^2 x_{\prime\prime}}{d t^2} = 0\,;$$

adding this to the next equation above, we get

$$M \cdot \frac{d^2 x}{d t^2} + M_{\prime} \cdot \frac{d^2 x_{\prime}}{d t^2} + \Sigma (M M_{\prime\prime})_x + \Sigma (M_{\prime} M_{\prime\prime})_x = 0 \quad . \quad . \quad (257)$$

Taking the movable origin at the centre of the body M, we have

$$x_{\prime} = x - x', \text{ and } d^2 x_{\prime} = d^2 x - d^2 x',$$

which, substituted above, gives

$$(M + M_{\prime}) \frac{d^2 x}{d t^2} - M_{\prime} \cdot \frac{d^2 x'}{d t^2} + \Sigma (M M_{\prime\prime})_x + \Sigma (M_{\prime} M_{\prime\prime})_x = 0\,;$$

dividing by $M + M_{\prime}$ and multiplying by M, there will result

$$M \cdot \frac{d^2 x}{d t^2} - \frac{M \cdot M_{\prime}}{M + M_{\prime}} \cdot \frac{d^2 x'}{d t^2} + \frac{M}{M + M_{\prime}} \cdot \Sigma (M M_{\prime\prime})_x + \frac{M}{M + M_{\prime}} \cdot \Sigma (M_{\prime} M_{\prime\prime})_x = 0.$$

The value of the first term results from the component action of the primary and perturbating bodies upon M; whence

$$M \cdot \frac{d^2 x}{d t^2} - [(M M_{\prime})_x - \Sigma (M M_{\prime\prime})_x] = 0\,;$$

from which subtracting the equation above, there will result

$$\frac{M M_{\prime}}{M + M_{\prime}} \cdot \frac{d^2 x'}{d t^2} - (M M_{\prime})_x + \frac{M_{\prime}}{M + M_{\prime}} \cdot \Sigma (M M_{\prime\prime})_x - \frac{M}{M + M_{\prime}} \cdot \Sigma (M_{\prime} M_{\prime\prime})_x = 0.$$

Dividing by the coefficient of the first term, and treating the other two of Equations (255) in the same way, we finally get

$$\left.\begin{aligned}
&\frac{d^2 x'}{d t^2} - \frac{M+M_{\prime}}{M \cdot M_{\prime}} \cdot (M M_{\prime})_x + \frac{1}{M} \cdot \Sigma\,(M M_{\prime\prime})_x - \frac{1}{M_{\prime}} \cdot \Sigma\,(M_{\prime} M_{\prime\prime})_x = 0,\\
&\frac{d^2 y'}{d t^2} - \frac{M+M_{\prime}}{M \cdot M_{\prime}} \cdot (M M_{\prime})_y + \frac{1}{M} \cdot \Sigma\,(M M_{\prime\prime})_y - \frac{1}{M_{\prime}} \cdot \Sigma\,(M_{\prime} M_{\prime\prime})_y = 0,\\
&\frac{d^2 z'}{d t^2} - \frac{M+M_{\prime}}{M \cdot M_{\prime}} \cdot (M M_{\prime})_z + \frac{1}{M} \cdot \Sigma\,(M M_{\prime\prime})_z - \frac{1}{M_{\prime}} \cdot \Sigma\,(M_{\prime} M_{\prime\prime})_z = 0.
\end{aligned}\right\} \quad (258)$$

Which, by integration, will give all the circumstances of the primary's motion in reference to the central body.

LAWS OF CENTRAL FORCES.

§ 189.—A central force is one which is directed towards a centre, movable or fixed, and of which the intensity is a function of the distance from the centre. The forces of nature are of this description.

If the perturbating bodies did not exist, then would the action on the primary be directed to the central body as a centre, the Equations (258) would reduce to their first two terms, and, denoting the distance from the central to the primary by r', they would be written,

$$\left.\begin{aligned}
&\frac{d^2 x'}{d t^2} = \frac{M+M_{\prime}}{M \cdot M_{\prime}} \cdot (M M_{\prime})_x = \frac{M+M_{\prime}}{M \cdot M_{\prime}} \cdot (M M_{\prime}) \cdot \frac{x'}{r'};\\
&\frac{d^2 y'}{d t^2} = \frac{M+M_{\prime}}{M \cdot M_{\prime}} \cdot (M M_{\prime})_y = \frac{M+M_{\prime}}{M \cdot M_{\prime}} \cdot (M M_{\prime}) \cdot \frac{y'}{r'};\\
&\frac{d^2 z'}{d t^2} = \frac{M+M_{\prime}}{M \cdot M_{\prime}} \cdot (M \cdot M_{\prime})_z = \frac{M+M_{\prime}}{M \cdot M_{\prime}} \cdot (M M_{\prime}) \cdot \frac{z'}{r'}.
\end{aligned}\right\} \quad . \; . \; (259)$$

Multiply the first by y', the second by x', and take the difference of the products; also multiply the first by z', the third by x', and take the difference of the products; and again the second by z', the third by y', and take the difference of the products: there will result, omitting the accents,

$$\frac{d^2 y}{d t^2} \cdot x - \frac{d^2 x}{d t^2} \cdot y = 0,$$

$$\frac{d^2 x}{d t^2} \cdot z - \frac{d^2 z}{d t^2} \cdot x = 0,$$

$$\frac{d^2 z}{d t^2} \cdot y - \frac{d^2 y}{d t^2} \cdot z = 0;$$

which, being integrated, give

$$\left.\begin{array}{l}\frac{dy}{dt}.x-\frac{dx}{dt}.y=C',\\ \frac{dx}{dt}.z-\frac{dz}{dt}.x=C'',\\ \frac{dz}{dt}.y-\frac{dy}{dt}.z=C''';\end{array}\right\} \quad \ldots \ldots \quad (260)$$

in which C', C'', and C''' are the constants of integration.

Multiplying each by the first power of the variable which it does not contain, and adding, we have

$$C'z+C''y+C'''x=0\,;$$

which is the equation of an invariable plane passing through the centre, and of which the position depends upon the constants C', C'', C'''. Whence we conclude that the primary deflected by the central body alone, will describe a plane curve of which the plane will contain the centres of both.

§ 190.—Take the co-ordinate plane $x\,y$ to coincide with this plane, and the Equations (260) will reduce to

$$\frac{dy}{dt}.x-\frac{dx}{dt}.y=C' \quad \ldots \ldots \quad (261)$$

Transform to polar co-ordinates; for this purpose we have

$$x=r.\cos\alpha\,;\quad y=r.\sin\alpha\,;$$

differentiating,

$$dx=dr\cos\alpha-r\sin\alpha\,d\alpha,$$

$$dy=dr\sin\alpha+r\cos\alpha\,d\alpha.$$

Substituting in Equation (261), we find

$$\frac{dy}{dt}.x-\frac{dx}{dt}.y=r^2.\frac{d\alpha}{dt}=C' \quad \ldots \ldots \quad (262)$$

integrating again, we have

$$\int r^2.d\alpha=C't+C'',$$

and taking between the limits $r_{,}$, $\alpha_{,}$ and $r_{,,}$, $\alpha_{,,}$, corresponding to the time $t_{,}$ and $t_{,,}$,

$$\int_{r_{,,}\alpha_{,,}}^{r_{,}\alpha_{,}} r^2.d\alpha=C'(t_{,,}-t_{,}) \quad \ldots \ldots \quad (263)$$

But $\int r^2 \, d\alpha$ is double the area described by the motion of the radius vector; whence we see, Equation (263), that the areas described by the radius vector of a body revolving about a centre, are proportional to the intervals of time required to describe them.

Making, in Equation (263), $t_{,,} - t_{,}$ equal to unity, the first member becomes double the area described in a unit of time. Denoting this by $2c$, that equation gives

$$C' = 2c.$$

Placing this in Equation (263), we find

$$t_{,,} - t_{,} = \frac{\int_{r_{,,}\alpha_{,,}}^{r_{,}\alpha_{,}} r^2 . d\alpha}{2c} \quad . \quad . \quad . \quad . \quad . \quad . \quad . \quad (264)$$

That is to say, any interval of time is equal to the area described in that interval, divided by the area described in the unit of time.

§ 191.—The converse is also true; for, differentiating Equation (262), we find

$$\frac{d^2 y}{d t^2} x - \frac{d^2 x}{d t^2} y = 0;$$

Multiplying by M, and replacing $M . \frac{d^2 y}{d t^2}$ and $M . \frac{d^2 x}{d t^2}$ by their values in Equations (120), there will result

$$Y x - X y = 0 \quad . \quad . \quad . \quad . \quad . \quad . \quad . \quad . \quad (265)$$

which is the Equation of the line of direction of the force; and having no independent term, this line passes through the centre. Whence we conclude, that a body whose radius vector describes about any point areas proportional to the times, is acted upon by a force of which the line of direction passes through that point as a centre. The force will be attractive or repulsive according as the orbit turns its concave or convex side towards the centre.

§ 192.—Replacing C' by its value $2c$, in Equation (262), and dividing by r^2, we have

$$\frac{d\alpha}{dt} = \frac{2c}{r^2} \quad . \quad . \quad . \quad . \quad . \quad . \quad . \quad . \quad . \quad (266)$$

The first member being the actual velocity of a point on the radius vector at the distance unity from the centre, is called the *angular velocity* of the body. *The angular velocity therefore varies inversely as the square of the radius vector.*

§ 193.—Multiply Equation (266) by $d\,s$, and it may be put under the form,

$$\frac{d\,s}{d\,t} = V = \frac{2\,c}{r\,.\,\frac{r\,d\,\alpha}{d\,s}};$$

but $\frac{r\,.\,d\,\alpha}{d\,s}$, is equal to the sine of the angle which the element of the orbit makes with the radius vector, and denoting by p the length of the perpendicular from the centre on the tangent to the orbit at the place of the body, we have

$$p = r\,.\,\frac{r\,.\,d\,\alpha}{d\,s},$$

and

$$V = \frac{2\,c}{p} \quad . \quad . \quad . \quad . \quad . \quad . \quad . \quad . \quad . \quad . \quad (267)$$

whence, the actual velocity of the body varies inversely as the distance of the tangent to the orbit at the body's place, from the centre.

§ 194.—Denoting the intensity of the acceleration on $M_{\prime}$ by F; substituting $M_{\prime}\,.\,F\,.\,d\,r$ for $X\,d\,x + Y\,d\,y + Z\,d\,z$, writing $M_{\prime}$ for M in the coefficient of V^2 in Equation (121), and differentiating, we find

$$V\,d\,V = -\,F\,d\,r\,;$$

and taking the logarithms of both members of Equation (267),

$$\log\,V = \log\,2\,c - \log p\,;$$

differentiating,

$$\frac{d\,V}{V} = -\,\frac{d\,p}{p},$$

and dividing the equation above by this,

$$V^2 = F\,.\,p\,.\,\frac{d\,r}{d\,p} = 2\,F\,.\,\frac{1}{2}\,p\,.\,\frac{d\,r}{d\,p} \quad . \quad . \quad . \quad (268)$$

Whence we conclude that, the velocity of a body at any point of its orbit is the same as that which it would have acquired had it fallen freely from rest at that point over the distance ME, equal to one-fourth of the chord of curvature MG, through the fixed centre—the force retaining unchanged its intensity at M.

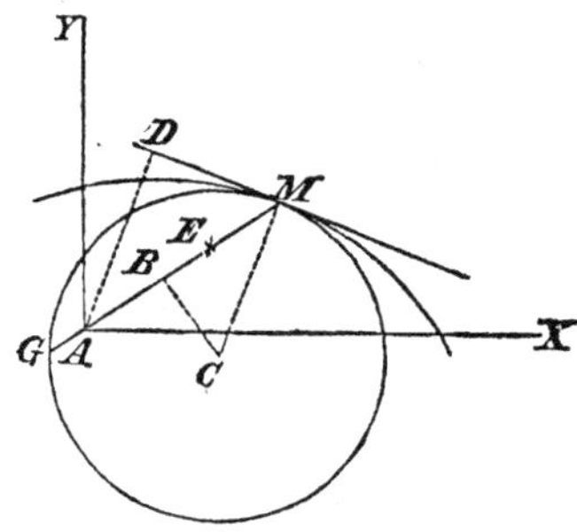

§ 195.—Resuming Equations (120), we have

$$X = M \cdot \frac{d^2 x}{d t^2} = M \cdot \frac{d \dfrac{d x}{d t}}{d t},$$

and performing the operation indicated, regarding the arc of the orbit as the independent variable, we have, after dividing both numerator and denominator by $d s^3$,

$$X = M \cdot \frac{\dfrac{d t}{d s} \cdot \dfrac{d^2 x}{d s^2} - \dfrac{d x}{d s} \cdot \dfrac{d^2 t}{d s^2}}{\dfrac{d t^3}{d s^3}} = M \cdot \left[\frac{d s^2}{d t^2} \cdot \frac{d^2 x}{d s^2} - \frac{d x}{d s} \cdot \frac{d s^3}{d t^3} \cdot \frac{d^2 t}{d s^2}\right];$$

but

$$\frac{d s^3}{d t^3} \cdot \frac{d^2 t}{d s^2} = - \frac{d^2 s}{d t^2}; \quad \frac{d s}{d t} = V;$$

whence,

$$X = M \cdot \left[V^2 \cdot \frac{d^2 x}{d s^2} + \frac{d x}{d s} \cdot \frac{d^2 s}{d t^2}\right].$$

In like manner,

$$Y = M \cdot \left[V^2 \cdot \frac{d^2 y}{d s^2} + \frac{d y}{d s} \cdot \frac{d^2 s}{d t^2}\right];$$

$$Z = M \cdot \left[V^2 \cdot \frac{d^2 z}{d s^2} + \frac{d z}{d s} \cdot \frac{d^2 s}{d t^2}\right].$$

Squaring and adding,

$$X^2+Y^2+Z^2=\begin{cases} V^4\left\{\left(\frac{d^2 x}{d s^2}\right)^2+\left(\frac{d^2 y}{d s^2}\right)^2+\left(\frac{d^2 z}{d s^2}\right)^2\right\}.M^2 \\ +2V^2.\frac{d^2 s}{d t^2}\left(\frac{d x}{d s}.\frac{d^2 x}{d s^2}+\frac{d y}{d s}.\frac{d^2 y}{d s^2}+\frac{d z}{d s}.\frac{d^2 z}{d s^2}\right).M^2 \\ +\left(\frac{d x^2}{d s^2}+\frac{d y^2}{d s^2}+\frac{d z^2}{d s^2}\right).\left(\frac{d^2 s}{d t^2}\right)^2.M^2; \end{cases}$$

but, denoting the radius of curvature by ρ, we have

$$\left(\frac{d^2 x}{d s^2}\right)^2+\left(\frac{d^2 y}{d s^2}\right)^2+\left(\frac{d^2 z}{d s^2}\right)^2=\frac{1}{\rho^2};^*$$

and multiplying the second term of the second member of the preceding equation by $\frac{\rho}{\rho}$, it may be put under the form,

$$2\frac{M V^2}{\rho}.\frac{M.d^2 s}{d t^2}\left(\frac{d x}{d s}.\rho\frac{d^2 x}{d s^2}+\frac{d y}{d s}.\rho\frac{d^2 y}{d s^2}+\frac{d z}{d s}.\rho\frac{d^2 z}{d s^2}\right);$$

or,

$$2\frac{M V^2}{\rho}.\frac{M.d^2 s}{d t^2}.\cos\delta;^*$$

in which δ denotes the angle made by the element of the curve and radius of curvature; also

$$\frac{d x^2}{d s^2}+\frac{d y^2}{d s^2}+\frac{d z^2}{d s^2}=1;$$

whence, substituting for $X^2+Y^2+Z^2$ its value R^2, we have

$$R^2=\frac{M^2 V^4}{\rho^2}+2.\frac{M V^2}{\rho}.\frac{M.d^2 s}{d t^2}.\cos\delta+M^2.\left(\frac{d^2 s}{d t^2}\right)^2;$$

and comparing this with Equation (56) we find that R is equal to the resultant of the two component forces

$$\frac{M V^2}{\rho} \text{ and } M.\frac{d^2 s}{d t^2},$$

which make with each other the angle δ. But δ is equal to 90°, and therefore

$$R^2=\frac{M^2 V^4}{\rho^2}+M^2.\left(\frac{d^2 s}{d t^2}\right)^2 \quad . \quad . \quad . \quad . \quad . \quad (269)$$

* See Appendix No. 2.

The second of these components is, Equation (13), the intensity of the reaction of inertia in the direction of the tangent, and the first is therefore its reaction in the direction of the radius of curvature.

This first component is called the *centrifugal force*, and may be defined to be *the resistance which the inertia of a body in motion opposes to whatever deflects it from its rectilinear path.* It is measured, Equation (269), by the living force of the body divided by the radius of curvature. The direction of its action is from the centre of curvature, and it thus differs from the force which acts towards a centre, and which is called *centripetal force.* The two are called *central forces.*

If the component in the direction of the orbit be zero, then will

$$M \cdot \frac{d^2 s}{d t^2} = 0;$$

and denoting the centrifugal force by $F_{,}$, we have

$$F_{,} = \frac{M V^2}{\rho} \quad . \quad . \quad . \quad . \quad . \quad . \quad . \quad . \quad (270)$$

and integrating the next to the last equation, we have

$$\frac{d s}{d t} = V = C;$$

in which C is the constant of integration. Whence, the velocity will be constant, and we conclude that a body in motion and acted upon by a force whose direction is always normal to the path described, will preserve its velocity unchanged.

These laws, except that expressed by Equation (268), are wholly independent of the intensity of the extraneous force and of the law of its variation. Not so, however, of

THE ORBIT.

§ 196.—To find the differential equation of the orbit, multiply the first of Equations (259) by $2\,d\,x$, the second by $2\,d\,y$, add and integrate; we find, omitting the accents,

$$\frac{d x^2 + d y^2}{d t^2} = \frac{M + M_{,}}{M \, . \, M_{,}} \cdot \int (M M_{,}) \cdot \frac{2\,x\,d\,x + 2\,y\,d\,y}{r};$$

but

$$r^2 = x^2 + y^2, \text{ and } r\,d\,r = x\,d\,x + y\,d\,y;$$

also

$$x = r \cos \alpha; \quad y = r \,.\, \sin \alpha;$$

$$d\,x = - r \sin \alpha\, d\, \alpha + \cos \alpha\, d\, r;$$

$$d\,y = r \cos \alpha\, d\, \alpha + \sin \alpha\, d\, r;$$

and, Equation (266),

$$\frac{1}{d\,t} = \frac{2\,c}{r^2\,d\,\alpha}.$$

These substituted above, give

$$4\,c^2 \left(\frac{1}{r^2} + \frac{d\,r^2}{r^4\,d\,\alpha^2}\right) = 2 \cdot \frac{M + M_{,}}{M\,M_{,}} \cdot \int (M\,M_{,})\,d\,r.$$

Make

$$\frac{1}{r} = u, \text{ and therefore } \frac{d\,r}{r^2} = - d\,u,$$

substitute above, differentiate and reduce, there will result

$$4\,c^2\,u^2 \left(\frac{d^2\,u}{d\,\alpha^2} + u\right) = \frac{M + M_{,}}{M \,.\, M_{,}} \cdot (M\,M_{,}) \quad = \quad \left[\frac{(M\,M_{,})}{M} + \frac{(M\,M_{,})}{M_{,}}\right];$$

and making

$$F \quad = \quad \left[\frac{(M\,M_{,})}{M} + \frac{(M\,M_{,})}{M_{,}}\right] = \text{relative acceleration on } M_{,} \quad . \quad (271)$$

$$F = 4\,c^2\,u^2 \cdot \left(\frac{d^2\,u}{d\,\alpha^2} + u\right) \quad . \; . \; . \; . \; . \; . \; . \quad (272)$$

From which the equation of the orbit may be found by integration, when the law of the force is known; or the law of the force deduced, when the equation of the orbit is given.

In the first case, the integral will contain three arbitrary constants —two introduced in the process of integration, and the third, c, existing in the differential equation. These are determined by the initial or other circumstances of the motion, viz.: the body's velocity, its dis tance from the centre, and direction of the motion at a given instant. The general integral only determines the nature of the orbit described: the circumstances of the motion at any given time determine the *species* and *dimensions* of the orbit.

In the second case, find the second differential coefficient of u in regard to α, from the polar equation of the curve; substitute this in the above equation, eliminating α, if it occur, by means of the relation between u and α, and the result will be F, in terms of u alone.

SYSTEM OF THE WORLD.

§ 197.—The most remarkable system of bodies of which we have any knowledge, and to which the preceding principles have a direct application, is that called the solar system. It consists of the *Sun*, the *Planets*, of which the earth we inhabit is one, the *Satellites* of the planets, and the *Comets*. These bodies are of great dimensions, are spheroidal in figure, are separated by distances compared to which their diameters are almost insignificant, and the mass of the sun is so much greater than that of the sum of all the others, as to bring the common centre of inertia of the whole within the boundary of its own volume.

These bodies revolve about their respective centres of inertia, are ever shifting their relative positions, and our knowledge of them is the result of computations based upon data derived from actual observation.

Kepler found;

I. *That the areas swept over by the radius vector of each planet about the sun, in the same orbit, are proportional to the times of describing them.*

II. *That the planets move in ellipses, each having one of its foci in the sun's centre.*

III. *That the squares of the periodic times of the planets about the sun, are proportional to the cubes of their mean distances from that body.*

These are called the *laws* of Kepler, and lead directly to a knowledge of the nature of the forces which uphold the solar system.

CONSEQUENCES OF KEPLER'S LAWS.

§ 198.—The first law shows, § 191, that the centripetal forces which keep the planets in their orbits, are all directed to the sun's centre; and that the sun is, therefore, the *centre of the system*.

§ 199.—What law of the force will cause a primary to describe about a central body an ellipse having one of its foci at the centre of the latter? The equation of the ellipse referred to its focus as a pole is

$$r = \frac{a\,(1 - e^2)}{1 + e \cos \alpha};$$

whence,

$$\frac{1}{r} = u = \frac{1 + e \cos \alpha}{a\,(1 - e^2)},$$

and,

$$\frac{d^2 u}{d\,\alpha^2} = \frac{- e \cos \alpha}{a\,(1 - e^2)},$$

which, substituted in Equation (272), give

$$F = 4\,c^2\,u^2 \left(\frac{- e \cos \alpha}{a\,(1 - e^2)} + \frac{1 + e \cos \alpha}{a\,(1 - e^2)}\right);$$

reducing and replacing u by its value $\frac{1}{r}$, we have

$$F = \frac{4\,c^2}{a\,(1 - e^2)} \cdot \frac{1}{r^2} \quad . \quad . \quad . \quad . \quad . \quad . \quad . \quad . \quad (273)$$

and from which we conclude, that the only law for the relative acceleration, is that of the inverse square of the distance.

§ 200.—Conversely, let the force vary inversely as the square of the distance; required the orbit.

Denote by $k_{\prime}$ the reciprocal attraction of one unit of mass upon another at the unit's distance; then will

$$(M\,M_{\prime}) = M \,.\, M_{\prime} \cdot \frac{k_{\prime}}{r^2};$$

and, Equation (271),

$$F = k_{\prime} \,.\, (M + M_{\prime}) \,.\, u^2 = k_{\prime} \,.\, m \,.\, u^2;$$

in which

$$m = M + M_{\prime} \quad . \quad . \quad . \quad . \quad . \quad . \quad (273)'$$

and, Equation (272),

$$\frac{d^2 u}{d\alpha^2} + u = \frac{k_{,} . m}{4 c^2};$$

multiplying by $2\,d\,u$ and integrating,

$$\frac{d u^2}{d \alpha^2} = \frac{2 k_{,} . m}{4 . c^2} \cdot u - u^2 + C \quad . \quad . \quad . \quad . \quad . \quad . \quad (274)$$

whence

$$d\alpha = \frac{-d u}{\sqrt{C + \frac{2 k_{,} . m}{4 c^2} \cdot u - u^2}}$$

the negative sign being taken, because

$$\frac{d u}{d \alpha} = \frac{d\left(\frac{1}{r}\right)}{d \alpha} = -\frac{d r}{r^2 d \alpha} \quad . \quad . \quad . \quad . \quad . \quad . \quad (275)$$

Place under the radical $\left(\frac{k_{,} . m}{4 c^2}\right)^2 - \left(\frac{k_{,} . m}{4 c^2}\right)^2$, and we may write,

$$a\,\alpha = \frac{1}{\sqrt{\left(\frac{k_{,} m}{4 c^2}\right)^2 + C}} \cdot \frac{-d u}{\sqrt{1 - \frac{\left(u - \frac{k_{,} m}{4 c^2}\right)^2}{\left(\frac{k_{,} . m}{4 c^2}\right)^2 + C}}};$$

and integrating,

$$\alpha + \varphi = \cos^{-1} \frac{u - \frac{k_{,} . m}{4 c^2}}{\sqrt{\left(\frac{k_{,} . m}{4 c^2}\right)^2 + C}},$$

in which φ is the constant of integration.

Replacing u by its value, taking cosine of both members and solving with respect to r, there will result

$$r = \frac{\frac{4 c^2}{k_{,} . m}}{1 + \sqrt{1 + \left(\frac{4 c^2}{k_{,} . m}\right)^2 . C} . \cos(\alpha + \varphi)},$$

which is the equation of a conic section, having its pole at the central body. To find the precise curve, we must find C. To do this, denote by $r_{\prime}$ the initial value of the radius vector, and by $\varepsilon_{\prime}$ the angle which the orbit makes with $r_{\prime}$ at the point of intersection therewith. Then, Equation (275),

$$\frac{du}{d\alpha} = -\frac{1}{r_{\prime} \tan \varepsilon_{\prime}};$$

and this in Equation (274) gives

$$C = \frac{1}{r_{\prime}^2 \sin^2 \varepsilon_{\prime}} - \frac{2 k_{\prime} . m}{4 c^2 r_{\prime}};$$

but, Equation (267),

$$\frac{1}{r_{\prime}^2 \sin^2 \varepsilon_{\prime}} = \frac{V_{\prime}^2}{4 c^2} = \frac{V_{\prime}^2 r_{\prime}}{4 c^2 . r_{\prime}} \quad . \quad . \quad . \quad . \quad (275)'$$

in which $V_{\prime}$ is the velocity corresponding to $r_{\prime}$; hence,

$$C = \frac{V_{\prime}^2 . r_{\prime} - 2 k_{\prime} . m}{4 c^2 r_{\prime}};$$

which, substituted in the equation of the curve, gives

$$r = \frac{\frac{4 c^2}{k_{\prime} . m}}{1 + \sqrt{1 + \frac{4 c^2}{k_{\prime}^2 . m^2} . \left(V_{\prime}^2 - \frac{2 k_{\prime} . m}{r_{\prime}}\right)} . \cos(\alpha + \varphi)} \quad . \quad (276)$$

and comparing this with the general polar equation of a conic section referred to the focus as a pole, viz.:

$$r = \frac{a_{\prime} (1 - e^2)}{1 + e \cos(\alpha + \varphi)},$$

we find

$$a_{\prime} (1 - e^2) = \frac{4 c^2}{k_{\prime} . m} \quad . \quad . \quad . \quad . \quad . \quad . \quad . \quad . \quad (277)$$

$$e^2 = 1 + \frac{4 c^2}{k_{\prime}^2 . m^2} \left(V_{\prime}^2 - \frac{2 k_{\prime} . m}{r_{\prime}}\right) \quad . \quad . \quad . \quad . \quad (278)$$

and this last value will be greater or less than unity, according as $V_{\prime}^2$ is greater or less than $\frac{2 k_{\prime} . m}{r_{\prime}}$.

Multiplying and dividing the last factor by $M_{\prime} r_{\prime}$, and replacing m by its value, the orbit will be an ellipse, parabola, or hyperbola, according as

$$M_{,} V_{,}^{2} < \frac{2 k_{,} . (M + M_{,})}{r_{,}^{2}} . M_{,} . r_{,} ;$$

$$M_{,} . V_{,}^{2} = \frac{2 k_{,} . (M + M_{,})}{r_{,}^{2}} . M_{,} . r_{,} ;$$

$$M_{,} . V_{,}^{2} > \frac{2 k_{,} . (M + M_{,})}{r_{,}^{2}} . M_{,} . r_{,} .$$

That is, according as the living force of the primary at any point of its orbit is less than, equal to, or greater than twice the work its relative weight, at that point, would perform over a distance equal to its radius vector. So that a primary may describe any of the conic sections as well as the ellipse, the only condition for this purpose being an adequate value for its velocity.

Substituting the value of e^2 in Equation (277), we find

$$a_{,} = \frac{k_{,} . m . r_{,}}{2 k_{,} . m - V_{,}^{2} . r_{,}} ; \quad . \quad . \quad . \quad . \quad . \quad . \quad (279)$$

and denoting the semi-parameter by p, the equation of the curve gives, by making $\alpha + \varphi = 90°$,

$$p = \frac{4 c^2}{k_{,} . m} = \frac{V_{,}^{2} . \sin^2 \varepsilon_{,} . r_{,}^{2}}{k_{,} . m},$$

and denoting the semi-conjugate axis by $b_{,}$,

$$b_{,} = \sqrt{a_{,} . p} = V_{,} . \sin \varepsilon_{,} . r_{,} \sqrt{\frac{a_{,}}{k_{,} . m}} \quad . \quad . \quad . \quad (279)'$$

Whence it appears that the nature of the orbit and its transverse axis are independent of the direction of the primary's motion, while the conjugate axis is dependent upon this element.

§ 201.—The consequence of Kepler's third law is not less important. Denote the periodic time of the primary by $T_{,}$; then, Equation (264),

$$T_{,} = \frac{\pi . a_{,} b_{,}}{c} ;$$

and substituting the values of $b_{,}$, m, and c, Equations (279)′, (273)′, and (275)′,

$$T_{,} = 2 \pi . a_{,}^{\frac{3}{2}} . \sqrt{\frac{1}{(M + M_{,}) k_{,}}}$$

and for another body whose mass is $M_{,,}$, about the same central body,

$$T_{,,} = 2\,\pi\,a_{,,}^{\frac{3}{2}} \sqrt{\frac{1}{(M + M_{,,})\,k_{,,}}};$$

and by division,

$$\frac{T_{,}^2}{T_{,,}^2} = \frac{a_{,}^3}{a_{,,}^3} \cdot \frac{M + M_{,,}}{M + M_{,}} \cdot \frac{k_{,,}}{k_{,}} \quad \ldots\ldots \quad (280)$$

If the difference of the masses $M_{,}$ and $M_{,,}$ be so small in comparison with M as to make its omission insensible to ordinary observation, which is the case in the solar system, the above may be written,

$$\frac{T_{,}^2}{T_{,,}^2} = \frac{a_{,}^3}{a_{,,}^3} \cdot \frac{k_{,,}}{k_{,}}.$$

But by Kepler's third law,

$$\frac{T_{,}^2}{T_{,,}^2} = \frac{a_{,}^3}{a_{,,}^3};$$

whence

$$k_{,} = k_{,,}.$$

That is, the central body M would act equally on the unit of mass of each of the primaries $M_{,}$ and $M_{,,}$, were they at the same distance; so that not only is the law of the central force the same, but the absolute force at the same distance is the same, and it is one and the same force that keeps the planets in their orbits about the sun.

§ 202.—The observations of Dr. Maskelyne on the fixed stars, show that a neighboring mountain, Schehallien, drew the plumb-line of his instrument sensibly from the vertical; and those of Cavendish and Baily upon leaden and other balls, demonstrate this power of attraction to reside in every particle of matter wherever found; and that it is exerted under all circumstances, without the possibility of being inter cepted. It is, therefore, concluded that matter is endowed with a general gravitating principle by which every particle attracts every other particle, and according to the law before given.

PERTURBATIONS.

§ 203.—Granting, for the present, that universal gravitation is a principle of nature, and denoting the distances of the several bodies of the system from the central by r with subscript accents corresponding to those of the bodies to which they belong, and employing the same notation in regard to the co-ordinates, we shall have

$$(M M_{,})_x = k \cdot \frac{M \,.\, M_{,}}{r_{,}^2} \cdot \frac{x'}{r_{,}} = k \,.\, M \,.\, M_{,} \cdot \frac{x'}{r_{,}^3};$$

$$\Sigma\,(M M_{,,})_x = k\, M \,.\, \Sigma\, \frac{M_{,,}}{r_{,,}^2} \cdot \frac{x''}{r_{,,}} = k \,.\, M \,.\, \Sigma\, \frac{M_{,,}\, x''}{r_{,,}^3};$$

$$\Sigma\,(M_{,} M_{,,})_x = k \,.\, \Sigma\, \frac{M_{,} M_{,,}}{(x''-x')^2+(y''-y')^2+(z''-z')^2} \cdot \frac{x''-x'}{\sqrt{(x''-x')^2+(y''-y')^2+(z''-z')^2}}$$

which, substituted in first of Equations (258), give

$$\frac{d^2 x'}{d t^2} - k \left[(M + M_{,}) \cdot \frac{x'}{r_{,}^3} - \Sigma\, \frac{M_{,,}\, x''}{r_{,,}^3} + \frac{1}{M_{,}} \cdot \Sigma\, \frac{M_{,} \,.\, M_{,,} \,.\, (x''-x')}{[(x''-x')^2+(y''-y')^2+(z''-z')^2]^{\frac{3}{2}}} \right] = 0;$$

but

$$\frac{x''-x'}{[(x''-x')^2+(y''-y')^2+(z''-z')^2]^{\frac{3}{2}}} = \frac{1}{d x'} \cdot d\, \frac{1}{\sqrt{(x''-x')^2+(y''-y')^2+(z''-z')^2}},$$

and making

$$\lambda = \Sigma\, \frac{M_{,} \,.\, M_{,,}}{\sqrt{(x''-x')^2+(y''-y')^2+(z''-z')^2}} \quad . \quad . \quad (281)$$

the last term of the equation above becomes

$$k \cdot \frac{1}{M_{,}} \cdot \frac{d \lambda}{d x'},$$

and

$$\frac{d^2 x'}{d t^2} - k \left[(M + M_{,}) \cdot \frac{x'}{r_{,}^3} - \Sigma\, \frac{M_{,,}\, x''}{r_{,,}^3} + \frac{1}{M_{,}} \cdot \frac{d \lambda}{d x'} \right] = 0.$$

Make

$$R = \frac{M_{,,}(x'x''+y'y''+z'z'')}{r_{,,}^3} + \frac{M_{,,,}(x'x'''+y'y'''+z'z'''}{r_{,,,}^3} + \&c. - \frac{\lambda}{M_{,}}; \ (282)$$

then will

$$\frac{d R}{d x'} = \frac{M_{,,}\, x''}{r_{,,}^3} + \frac{M_{,,,}\, x'''}{r_{,,,}^3} + \&c. - \frac{d \lambda}{M_{,} \,.\, d x'} = \Sigma\, M_{,,} \cdot \frac{x''}{r_{,,}^3} - \frac{d \lambda}{M_{,} \,.\, d x'};$$

which, substituted above, give, after treating the other two of Equations (258) in the same way,

$$\left.\begin{aligned}
\frac{d^2 x'}{d t^2} - k\left[(M + M_{,}) \cdot \frac{x'}{r_{,}^3} - \frac{d R}{d x'}\right] = 0 ; \\
\frac{d^2 y'}{d t^2} - k\left[(M + M_{,}) \cdot \frac{y'}{r_{,}^3} - \frac{d R}{d y'}\right] = 0 ; \\
\frac{d^2 z'}{d t^2} - k\left[(M + M_{,}) \cdot \frac{z'}{r_{,}^3} - \frac{d R}{d z'}\right] = 0.
\end{aligned}\right\} \quad . \quad . \quad . \quad (283)$$

The curve which would be described by the primary about the central, under the reciprocal action of these two bodies alone, and which we have seen is a conic section, is called the *undisturbed orbit* of the primary. That which it actually describes under the joint action of all the bodies of the system, is called the *disturbed orbit.* The undisturbed orbit is given by the first two terms of Equations (283); the disturbed by all three. The departures of the disturbed from the undisturbed orbit are called *perturbations*, and the last terms of Equations (283), which determine them, are called *perturbating functions.* The constructions of the perturbating functions are given in Equations (281) and (282), and the methods of computing their values are greatly facilitated by the principle of the

COEXISTENCE AND SUPERPOSITION OF SMALL MOTIONS.

§ 204.—Denote by $\theta_{,,}$, $\theta_{,,,}$, &c., numerical quantities which depend upon the perturbating actions of the bodies whose masses are $M_{,,}$, $M_{,,,}$, &c., and of which the values are so small as to justify the omission of all terms into which their products enter as factors, in comparison with such as contain them singly. The co-ordinates of $M_{,}$, at the time t, when undisturbed, being $x'\ y'\ z'$, become, when the body $M_{,}$ is disturbed by $M_{,,}$ at the same time,

$$x' + \theta_{,,} \cdot x' ; \quad y' + \theta_{,,} \cdot y' ; \quad z' + \theta_{,,} \cdot z' ;$$

and for the same reason, when also disturbed by $M_{,,,}$,

$$x' + \theta_{,,}x' + \theta_{,,,}(x' + \theta_{,,}x') ; \quad y' + \theta_{,,}y' + \theta_{,,,}(y' + \theta_{,,}y') ; \quad z' + \theta_{,,}z' + \theta_{,,,}(z' + \theta_{,,}z'),$$

or, performing the multiplication and omitting the terms containing $\theta_{,,} \cdot \theta_{,,,}$,

$$x' + x'(\theta_{,,} + \theta_{,,,}) ; \quad y' + y'(\theta_{,,} + \theta_{,,,}) ; \quad z' + z'(\theta_{,,} + \theta_{,,,}) ;$$

in the same way, when also disturbed by $M_{,,,,}$,

$$x' + x'(\theta_{,,} + \theta_{,,,} + \theta_{,,,,}); \quad y' + y'(\theta_{,,} + \theta_{,,,} + \theta_{,,,,}); \quad z' + z'(\theta_{,,} + \theta_{,,,} + \theta_{,,,,})$$

and for the simultaneous disturbance of all the bodies of the system,

$$x' + x'\,\Sigma\,\theta_{,,}; \quad y' + y'\,\Sigma\,\theta_{,,}; \quad z' + z'\,\Sigma\,\theta_{,,};$$

in which $x'.\Sigma\,\theta_{,,}$, $y'.\Sigma\,\theta_{,,}$, $z'.\Sigma\,\theta_{,,}$ are the increments of $x'\,y'\,z'$ respectively, due to the joint action of all the disturbing bodies. Now let

$$u = \varphi\,(x'\,y'\,z'),$$

in which φ denotes any function of $x'\,y'\,z'$. Differentiating, we have

$$d\,u = \frac{d\,u}{d\,x'} \cdot x'\,\Sigma\,\theta_{,,} + \frac{d\,u}{d\,y'} \cdot y'\,\Sigma\,\theta_{,,} + \frac{d\,u}{d\,z'} \cdot z'\,\Sigma\,\theta_{,,},$$

and performing the multiplications indicated, we have

$$d\,u = \begin{cases} \dfrac{d\,u}{d\,x'} \cdot x'\,\theta_{,,} + \dfrac{d\,u}{d\,y'}\; y'\,\theta_{,,} + \dfrac{d\,u}{d\,z'} \cdot z'\,\theta_{,,}, \\ + \dfrac{d\,u}{d\,x'} \cdot x'\,\theta_{,,,} + \dfrac{d\,u}{d\,y'} \cdot y'\,\theta_{,,,} + \dfrac{d\,u}{d\,z'} \cdot z'\,\theta_{,,,}, \\ + \dfrac{d\,u}{d\,x'} \cdot x'\,\theta_{,,,,} + \quad \&c. \quad + \quad \&c. \\ + \quad \&c. \quad + \quad \&c. \quad + \quad \&c. \end{cases}$$

Whence it appears that the perturbation in u or $\varphi\,(x'\,y'\,z')$, is equal to the sum of the separate perturbations due to each of the perturbating bodies, supposing the others not to exist. The practical effect of this principle is to reduce the problem of the perturbations from one of several to one of a single perturbating body, and to give rise to what is known as the problem of the *three* bodies, viz.: the central, primary, and perturbating.

UNIVERSAL GRAVITATION.

§ 205.—From all of which it is manifest that either Kepler's laws cannot be rigorously true, or universal gravitation is not a Principle of Nature. Now, in point of fact, observations of far greater nicety than those of Kepler prove that his laws are not *accurately* true, though they differ but slightly from the truth; a circumstance arising entirely from the fact of the great mass of the sun as compared with the sum of the masses of all the planets. Were there but a single body in existence besides the sun, it would describe accurately an elliptical, parabolic or hyperbolic orbit about the centre of the sun, depending upon its living force and the sun's attraction. A third body would derange this motion and cause a departure from this simple path, and the degree of the disturbance would depend upon the mass, distance, and direction of the disturbing body as compared with those of the sun. The same remark would apply to a fourth, fifth, and to any number of additional bodies. The disturbed orbits in the solar system have been computed by Equations (283), and the complete harmony which is found to subsist between the numerical results deduced from theory and observation, is the strongest possible evidence in support of the Law of Universal Gravitation.

If the principal plane of the solar system, as determined at different and remote periods, be found to have undergone no change, this will show that the system is uninfluenced by the action of the fixed stars and other distant bodies, and its centre of inertia will, § 193, either be at rest or be moving uniformly through space in a right line; but if the principal plane be found to have changed its place, it will be a sign that the system is in motion, and that its centre of inertia is describing a curvilinear path about some distant centre.

§ 206.—Thus much for the larger bodies of nature. But these are themselves built up of innumerable molecules which are ever on the move about their respective places of relative rest. The molecular forces within the range of their natural action vary directly as the distance from their respective centres. Let it be required to determine the nature of the orbits under this law. Then will

$$F = m k_{,} . r = \frac{m k_{,}}{u};$$

which, in Equation (272), gives

$$\frac{d^2 u}{d \alpha^2} + u = \frac{k_{,} . m}{4 c^2 u^3};$$

multiplying by $2\,d\,u$, and integrating, we find

$$\frac{d\,u^2}{d\,\alpha^2} + u^2 = C - \frac{k_{,}\,.\,m}{4\,c^2\,u^2} \quad . \quad . \quad . \quad . \quad . \quad (284)$$

from which we get

$$d\,\alpha = -\frac{u\,d\,u}{\sqrt{C\,u^2 - \frac{k_{,}\,.\,m}{4\,c^2} - u^4}};$$

the negative sign being taken, because

$$\frac{d\,u}{d\,\alpha} = \frac{d\left(\frac{1}{r}\right)}{d\,\alpha} = -\frac{d\,r}{r^2\,d\,\alpha} \quad . \quad . \quad . \quad . \quad . \quad . \quad (284)'$$

Placing $\frac{1}{4}\,C^2 - \frac{1}{4}\,C^2$ under the radical, we may write

$$d\,\alpha = \tfrac{1}{2}\,.\,\frac{1}{\sqrt{\frac{C^2}{4} - \frac{k_{,}\,m}{4\,c^2}}}\,.\,\frac{-2\,u\,d\,u}{\sqrt{1 - \frac{(u^2 - \frac{1}{2}\,C)^2}{\frac{C^2}{4} - \frac{k_{,}\,.\,m}{4\,c^2}}}},$$

and integration,

$$2\,(\alpha + \varphi) = \cos^{-1}\frac{u^2 - \frac{1}{2}\,C}{\sqrt{\frac{C^2}{4} - \frac{k_{,}\,.\,m}{4\,c^2}}};$$

in which φ is the constant of integrating.

Taking cosine of both members, replacing u by its value and solving with respect to r, we find

$$r = \frac{1}{\sqrt{\frac{1}{2}\,C + \frac{1}{2}\sqrt{C^2 - \frac{k_{,}\,.\,m}{c^2}}\,.\,\cos 2\,(\alpha + \varphi)}}.$$

Denote by $r_{,}$ the radius vector which is normal to the orbit; corresponding to this value we have

$$\frac{d\,u}{d\,\alpha} = 0,$$

and, by Equation (284),

$$C = \frac{1}{r_{,}^2} + \frac{k_{,}\,.\,m\,.\,r_{,}^2}{4\,c^2};$$

and because

$$\cos 2(\alpha + \varphi) = \cos^2(\alpha + \varphi) - \sin^2(\alpha + \varphi),$$

the above reduces to

$$r = \frac{1}{\sqrt{\frac{1}{r_{\prime}^2}\cos^2(\alpha + \varphi) + \frac{k_{\prime} . m . r_{\prime}^2}{4 c^2}\sin^2(\alpha + \varphi)}} \quad . \quad . \quad (285)$$

which is the equation of an ellipse referred to its centre as a pole, the semi-axes being

$$r_{\prime} \text{ and } \frac{2c}{r_{\prime}}\sqrt{\frac{1}{k_{\prime} . m}}.$$

§ 207.—The time required to describe the entire ellipse being denoted by T, we have, Equation (264),

$$T = \frac{\pi . r_{\prime} . 2c\sqrt{\frac{1}{k_{\prime} . m}}}{r_{\prime} . c} = 2\pi\sqrt{\frac{1}{k_{\prime} . m}};$$

and replacing m by its value, Equation (273)′,

$$T = 2\pi\sqrt{\frac{1}{(M + M_{\prime}) k_{\prime}}} \quad . \quad . \quad . \quad . \quad . \quad . \quad (286)$$

Thus the time is wholly independent of the dimensions of the orbit, and will be the same in all orbits, great and small. This result finds its application in the subject of acoustics, thermotics, optics, &c.

§ 208.—Let us conclude the planetary motions with the centrifugal force on its surface, arising from the rotation of one of these bodies, say the earth, about its axis.

If V_1 denote the angular velocity of a body about a centre, then will $V = \rho V_1$, and Equation (270) becomes

$$F_{\prime} = M V_1^2 \rho.$$

The earth revolves about its axis $A A'$ once in twenty-four hours, and the circumferences of the parallels of latitude have velocities which diminish from the equator to the poles. The law of this diminution, on the supposition that the planet is a sphere, is given by

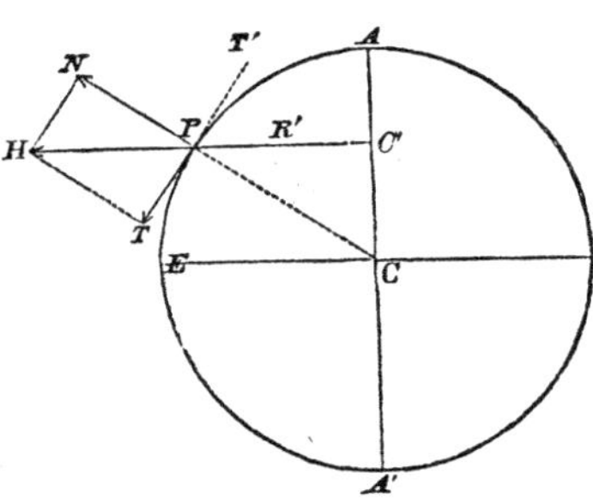

$$F_{,} = M V_1^2 R';$$

in which M is the body's mass, V_1 the earth's angular velocity, and R' the radius of one of its parallels of latitude.

Denoting the equatorial radius $C E = C P$, by R, and the angle $C P C' = P C E$, which is the latitude of the place, by φ, we have

$$R' = R \cos \varphi;$$

which substituted for R' above, gives

$$F_{,} = M V_1^2 R \cos \varphi \quad . \quad . \quad . \quad . \quad . \quad . \quad . \quad (286)'$$

The only variable quantity in this expression, when the same mass is taken from one latitude to another, is φ; *whence we conclude that the centrifugal force varies as the cosine of the latitude.*

The centrifugal force is exerted in the direction of the radius R' of the parallel of latitude, and therefore in a direction oblique to the horizon $T T'$. The normal and tangential components are, respectively,

$$F_{,} . \cos \varphi = M V_1^2 R \cos^2 \varphi,$$

$$F_{,} . \sin \varphi = M V_1^2 R . \sin \varphi \cos \varphi = \tfrac{1}{2} M V_1^2 R \sin 2 \varphi;$$

whence we conclude, *that the diminution of the weights of bodies arising from the centrifugal force at the earth's surface, varies as the square of the cosine of the latitude; and that all bodies are, in consequence of the centrifugal force, urged towards the equator by a force which varies as the sine of twice the latitude.*

At the equator the diminution of the force of gravity is a maximum, and equal to the entire centrifugal force; at the poles it is zero. The earth is not perfectly spherical, and all observations agree in demonstrating that it is protuberant at the equator and flattened at the poles, the difference between the equatorial and polar diameters being about twenty-six English miles. If we suppose the earth to have been at one time in a state of fluidity, or even approaching to it, its present figure is readily accounted for by the foregoing considerations.

To find the value of the centrifugal force at the equator, make, in Equation (286)', $M = 1$ and $\cos \varphi = 1$, which is equivalent to supposing a unit of mass on the equator, and we have

$$F_{,} = V_1^2 R,$$

in which, if the known radius of the equator and angular velocity be substituted, we shall find

$$F_{,} = V_1^2 . R = \overset{f}{0}, 1112.$$

To find the angular velocity with which the earth should rotate, to make the centrifugal force of a body at the equator equal to its weight, make

$$g = \overset{f}{32}, 1937 = V_1'^2 R;$$

in which $\overset{f}{32}, 1937$ is the force of gravity at the equator.

Dividing the second by the first, we find

$$\frac{32,1937}{0,1112} = \frac{V_1'^2}{V_1^2} = 289, \text{ nearly};$$

whence,

$$V_1' = 17 V_1;$$

that is to say, if the earth were to revolve seventeen times as fast as it does, bodies would possess no weight at the equator.

IMPACT OF BODIES.

§ 209.—When a body in motion comes into collision with another, either at rest or in motion, an *impact* is said to arise.

The action and reaction which take place between two bodies, when pressed together, are exerted along the same right line, perpendicular to the surfaces of both, at their common point of contact. This arises from the symmetrical disposition of the molecular springs about this line.

When the motions of the centres of inertia of the two bodies are *parallel* to this normal before collision, the impact is said to be *direct*.

When this normal passes through the centres of inertia of both bodies, and the motions of these centres *are along* that line, the impact is said to be *direct* and *central.*

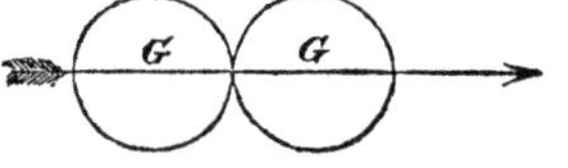

When the motion of the centre of inertia of one of the bodies is along the common normal, and the normal does not pass through the centre of inertia of the other, the impact is said to be *direct* and *eccentric.*

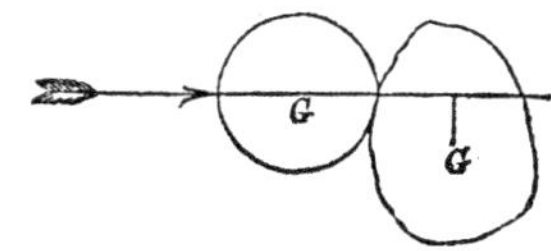

When the path described by the centre of inertia of one of the bodies, makes an angle with this normal, the impact is said to be *oblique.*

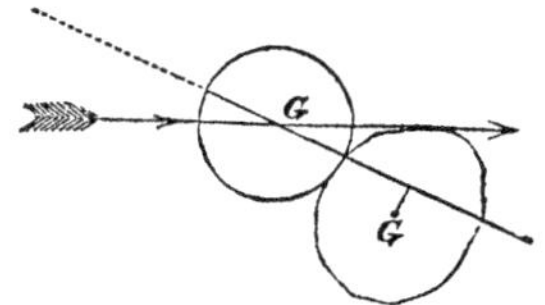

When two bodies come into collision, each will experience a pressure from the reaction of the other; and as all bodies are more or less compressible, this pressure will produce a change in the figure of both; the change of figure will increase till the instant the bodies cease to approach each other, when it will have attained its maximum. The molecular spring of each will now act to restore the former figures, the bodies will repel each other, and finally separate.

Three periods must, therefore, be distinguished, viz.: 1st., that occupied by the process of compression; 2d., that during which the greatest compression exists; 3d., that occupied by the process, as far as it extends, of restoring the figures. The *force of restitution* must also be distinguished from *the force of distortion;* the latter denoting the reciprocal action exerted between the bodies in the first, and the former in the third period.

The greater or less capacity of the molecular springs of a body to restore to it the figure of which it has been deprived by the application of some extraneous force when the latter ceases to act, is called its *elasticity.*

The ratio of the force of restitution to that of distortion, is the measure of a body's elasticity. This ratio is sometimes called the *co-efficient of elasticity.* When these two forces are equal, the ratio is unity, and the body is said to be *perfectly elastic;* when the ratio is zero, the body is said to be *non-elastic.* There are no bodies that satisfy these extreme conditions, all being more or less elastic, but none perfectly so.

Let the two bodies AB and $A'B'$, the former moving along the line HT, and the latter along $H'T'$, come into collision at the point O. Through O, draw the common normal NL. Denote the angle HGN by φ, and $H'EN$ by φ'—these being the angles which the directions of the two motions make with the normal. Also denote the velocity and mass of the body AB by V and M respectively, and the velocity and mass of $A'B'$ by V' and M'.

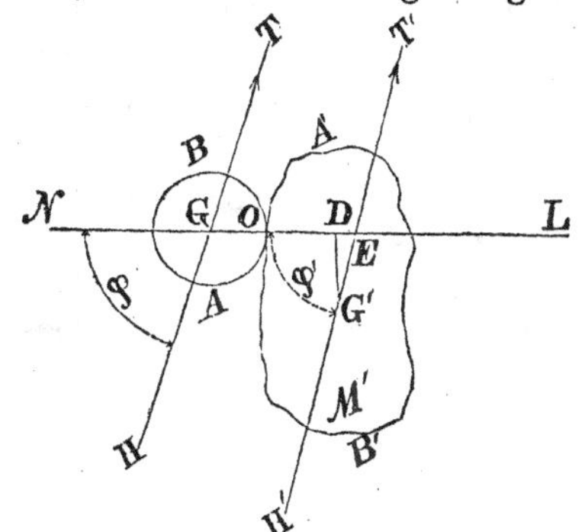

The components of the quantity of motion of the two bodies in the direction of the normal and of the perpendicular to the normal, will be

$$MV\cos\varphi, \quad M'V'\cos\varphi' \quad \text{and} \quad MV\sin\varphi, \quad M'V'\sin\varphi'.$$

The former of these components will alone be involved in the impact; for if the bodies were only animated by the latter, they would not collide, but would simply move the one by the other. For simplicity, let the body AB be spherical; the normal will pass through its centre of inertia.

Denote by u, the velocity of the body AB in the direction of the normal at the instant of greatest compression, and by u' the velocity of the body $A'B'$ at the same instant in the same direction. Then will

$$V\cos\varphi - u, \quad \text{and} \quad V'\cos\varphi' - u' \quad \cdot \quad \cdot \quad \cdot \quad (287)$$

be the velocities lost and gained in the direction of the normal, and

$$M(V\cos\varphi - u), \quad \text{and} \quad M'(V'\cos\varphi' - u') \quad \cdot \cdot \cdot \quad (288)$$

be the forces lost and gained at the instant of greatest compression; and hence,

$$M(V\cos\varphi - u) + M'(V'\cos\varphi' - u') = 0; \quad \cdot \quad \cdot \quad (289)$$

and denoting the angular velocity of the body $A'B'$ by $V_{\prime}'$, the distance $G'D$ from the centre of inertia of $A'B'$ to the normal by e, and the principal radius of gyration of $A'B'$, with reference to the instantaneous axis by $k_{\prime}$, then will

$$V_{\prime}' = \frac{M(V\cos\varphi - u).e}{M'k_{\prime}^2} \quad \cdot \quad \cdot \quad (290)$$

and since the velocity u must be equal to that of the point D at the end of the lever arm e, we have

$$u = u' + e.V_{\prime}' \quad \cdot \cdot \cdot \cdot \cdot \cdot \cdot \cdot \quad (291)$$

Substituting the values of u and u' from this equation successively in Equation (289), we find

$$u = \frac{MV\cos\varphi + M'V'\cos\varphi' + M'eV_{\prime}'}{M + M'} \cdot \cdot \cdot \quad (292)$$

$$u' = \frac{MV\cos\varphi + M'V'\cos\varphi' - MeV_{\prime}'}{M + M'} \cdot \cdot \cdot \quad (293)$$

After the instant of greatest compression, the molecular springs of the bodies will be exerted to restore the original figures, and if c denote the co-efficient of elasticity, then will the velocities lost by AB and gained by $A'B'$ during the process of restitution be, respectively,

$$c(V\cos\varphi - u) \quad \text{and} \quad c(V'\cos\varphi' - u');$$

and the entire loss of AB, and gain of $A'B'$, will be, respectively,

$$V\cos\varphi - u + c(V\cos\varphi - u), \quad \text{and} \quad V'\cos\varphi' - u' + c(V'\cos\varphi' - u').$$

Also the gain of angular velocity of the body $A'B'$, during the process of restitution, will be

$$cV_{\prime}' = c\frac{(V\cos\varphi - u).e}{k_{\prime}^2}\cdot\frac{M}{M'}$$

and the whole angular velocity produced by the impact and denoted by $V_{,}$, will be given by the equation,

$$V_{,} = (1 + c)\frac{(V\cos\varphi - u)\, e}{k_{,}^{2}}\cdot\frac{M}{M'} \quad . \quad . \quad . \quad (294)$$

Denoting the velocities of $A\,B$ and $A'\,B'$, after the collision by v and v', and the angles which the directions of these velocities make with the normal by θ and θ', respectively, then will

$$v\cos\theta = V\cos\varphi - V\cos\varphi + u - c\,(V\cos\varphi - u) = (1 + c)\,u - c\,V\cos\varphi,$$

$$v'\cos\theta' = V'\cos\varphi' - V'\cos\varphi' + u' - c(V'\cos\varphi' - u') = (1+c)u' - c\,V'\cos\varphi',$$

and replacing the values of u and u', as given by Equations (292) and (293),

$$v\cos\theta = (1+c)\,\frac{M\,V\cos\varphi + M'\,V'\cos\varphi' + M'e\,V_{,}'}{M + M'} - c\,V\cos\varphi, \quad (295)$$

$$v'\cos\theta' = (1+c)\,\frac{M\,V\cos\varphi + M'\,V'\cos\varphi' - M\,e\,V_{,}'}{M + M'} - c\,V'\cos\varphi' \quad (296)$$

Moreover, because the effects of the impact arising from the components of the quantities of motion in the direction of the normal will be wholly in that direction, the components of the quantities of motion before and after the impact at right angles to the normal will be the same, and hence

$$v\sin\theta = V\sin\varphi, \quad . \quad . \quad . \quad . \quad . \quad . \quad . \quad . \quad (297)$$

$$v'\sin\theta' = V'\sin\varphi'. \quad . \quad . \quad . \quad . \quad . \quad . \quad . \quad (298)$$

Squaring Equations (295) and (297) and adding; also Equations (296) and (298) and adding, we find after taking square root, and reducing by the relations

$$\cos^2\theta + \sin^2\theta = 1; \quad \cos^2\theta' + \sin^2\theta' = 1;$$

$$v = \sqrt{[(1+c)\frac{M\,V\cos\varphi + M'\,V'\cos\varphi' + M'e\,V_{,}'}{M + M'} - c\,V\cos\varphi]^2 + V^2\sin^2\varphi} \cdot (299)$$

$$v' = \sqrt{[(1+c)\frac{M\,V\cos\varphi + M'\,V'\cos\varphi' - MeV_{,}'}{M + M'} - c\,V'\cos\varphi']^2 + V'^2\sin^2\varphi'} \cdot (300)$$

Dividing Equation (297) by Equation (295), and Equation (298) by Equation (296), we have,

$$\tan\theta = \frac{V_{\prime}\,.\,\sin\varphi}{(1+c)\dfrac{M\,V\cos\varphi + M'\,V'\cos\varphi' + M'\,e\,V_{\prime}'}{M+M'} - c\,V\cos\varphi}, \quad (301)$$

$$\tan\theta' = \frac{V'\,.\,\sin\varphi'}{(1+c)\dfrac{M\,V\cos\varphi + M'\,V'\cos\varphi' - M\,e\,V_{\prime}'}{M+M'} - c\,V'\cos\varphi'} \cdot (302)$$

Equations (290) and (292), will give the values of u and $V_{\prime}'$, in known terms, and these in Equations (294), (295) and (296) will give the values of $V_{\prime}$, v, and v', and all the circumstances of the collision will be known.

§ 210.—If the bodies be both spherical, then will $e = 0$, and Equation (294) gives $V_{\prime} = 0$; and Equations (299) and (300), (301) and (302), become

$$v = \sqrt{[(1+c)\frac{M\,V\cos\varphi + M'\,V'\cos\varphi'}{M+M'} - c\,V\cos\varphi]^2 + V^2\sin^2\varphi} \quad \cdots \quad (303)$$

$$v' = \sqrt{[(1+c)\frac{M\,V\cos\varphi + M'\,V'\cos\varphi'}{M+M'} - c\,V'\cos\varphi']^2 + V'^2\sin^2\varphi'} \quad \cdots \quad (304)$$

$$\tan\theta = \frac{V\sin\varphi}{(1+c)\dfrac{M\,V\cos\varphi + M'\,V'\cos\varphi'}{M+M'} - c\,V\cos\varphi} \quad \cdots \quad (305)$$

$$\tan\theta' = \frac{V'\sin\varphi'}{(1+c)\dfrac{M\,V\cos\varphi + M'\,V'\cos\varphi'}{M+M'} - c\,V'\cos\varphi'} \quad \cdots \quad (306)$$

The Equations (303) and (304) will make known the velocities, and (305) and (306) the directions in which the bodies will move, after the impact.

Now, suppose the body $A'B'$ at rest, and its mass so great that the mass of $A\,B$ is insignificant in comparison, then will V' be zero, M' may be written for $M + M'$ and $\dfrac{M}{M'}$ will be a fraction so small that all the terms into which it enters as a factor may be neglected, and Equation (303) becomes

$$v = V\sqrt{c^2 \cos^2 \varphi + \sin^2 \varphi};$$

and Equation (305),

$$\tan \theta = -\frac{\tan \varphi}{c} \quad . \quad . \quad . \quad . \quad . \quad . \quad . \quad (307)$$

The tangent of θ being negative, shows that the angle NHK, which the direction of AB's motion makes with the normal NN' after the impact, is greater than 90 degrees; in other words, that the body AB is driven back or reflected from $A'B'$. This explains why it is that a cannon-ball, stone, or other body thrown obliquely against the surface of the earth, will rebound several times before it comes to rest.

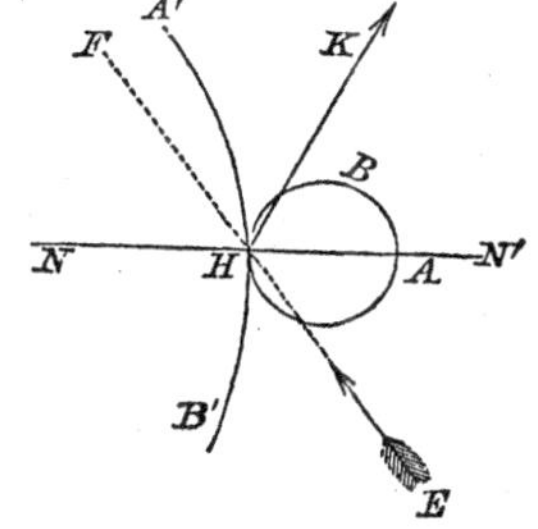

If the bodies be non-elastic, or, which is the same thing, if c be zero, the tangent of θ becomes infinite; that is to say, the body AB will move along the tangent plane, or if the body $A'B'$ were reduced at the place of impact to a smooth plane, the body AB would move along this plane.

If the body were perfectly elastic, or if c were equal to unity, which expresses this condition, then would Equation (307) become

$$\tan \theta = -\tan \varphi \quad . \quad . \quad . \quad . \quad . \quad . \quad . \quad . \quad (308)$$

which means that the angle $NHF = EHN'$ becomes equal to KHN'. The angle EHN' is called the angle of incidence, the angle KHN', commonly, the angle of reflection. Whence we see, that when a perfectly elastic body is thrown against a smooth, hard, and fixed plane, the angle of incidence will be equal to the angle of reflection.

If the angles φ and φ' be zero, then will $\cos \varphi = 1$, $\cos \varphi' = 1$,

$\sin\varphi = 0$, $\sin\varphi' = 0$; the impact will be direct and central, and Equations (303) and (304) become

$$v = (1 + c)\frac{MV + M'V'}{M + M'} - cV,$$

$$v' = (1 + c)\frac{MV + M'V'}{M + M'} - cV';$$

and passing to the limits, non-elasticity on the one hand and perfect elasticity on the other, we have in the first case, $c = 0$, and

$$v = \frac{MV + M'V'}{M + M'} \quad \dots\dots\dots \quad (309)$$

$$v' = \frac{MV + M'V'}{M + M'} \quad \dots\dots\dots \quad (310)$$

and in the second, $c = 1$, consequently,

$$v = 2\frac{MV + M'V'}{M + M'} - V \quad \dots\dots \quad (311)$$

$$v' = 2\frac{MV + M'V'}{M + M'} - V' \quad \dots\dots \quad (312)$$

CONSTRAINED MOTION.

§ 211.—Thus far we have only discussed the subject of *free motion*. We now come to *constrained motion*.

Motion is said to be constrained when by the interposition of some rigid surface or curve, or by connection with some one or more fixed points, a body is compelled to pursue a path different from that indicated by the forces which impart motion.

§ 212.—The centre of inertia of a body may be made to continue on a given surface, by causing it to slide or roll upon some other rigid surface.

§ 213.—We have seen, § 128, that the motion of translation of the centre of inertia, and of rotation about that point, are wholly independent of one another, and the generality of any discussion relating to the former will not, therefore, be affected by making, in Equation (40);

$$\delta \varphi = 0; \quad \delta \psi = 0; \quad \delta \varpi = 0;$$

which will reduce that equation to

$$\left. \begin{aligned} &(\Sigma P \cos \alpha - \frac{d^2 x}{d t^2} \cdot \Sigma m) \delta x_{,} \\ &+ (\Sigma P \cos \beta - \frac{d^2 y}{d t^2} \cdot \Sigma m) \delta y_{,} \\ &+ (\Sigma P \cos \gamma - \frac{d^2 z}{d t^2} \cdot \Sigma m) \delta z_{,} \end{aligned} \right\} = 0.$$

Making

$$\Sigma m = M; \quad \Sigma P \cos \alpha = X; \quad \Sigma P \cos \beta = Y; \quad \Sigma P \cos \gamma = Z;$$

and omitting the subscript accents, we may write

$$\left(X - M \cdot \frac{d^2 x}{d t^2}\right) \delta x + \left(Y - M \cdot \frac{d^2 y}{d t^2}\right) \delta y + \left(Z - M \cdot \frac{d^2 z}{d t^2}\right) \delta z = 0. \quad (313)$$

Now, assuming the movable origin at the centre of inertia, and supposing this latter point constrained to move on the surface of which the equation is

$$L = F(x y z) = 0, \quad \dots \dots \quad (314)$$

the virtual velocity must lie in this surface, and the generality of Equation (313), is restricted to the conditions imposed by this circumstance.

Supposing the variables $x\, y\, z$, in the above equations, to receive the increments or decrements δx, δy, δz, respectively; we have, from the principles of the calculus,

$$\frac{d L}{d x} \cdot \delta x + \frac{d L}{d y} \cdot \delta y + \frac{d L}{d z} \cdot \delta z = 0. \quad \dots \quad (315)$$

Multiplying by an indeterminate intensity λ, and adding the product to Equation (313), there will result

$$\left. \begin{aligned} &\left(X - M \cdot \frac{d^2 x}{d t^2} + \lambda \cdot \frac{d L}{d x}\right) \delta x \\ &+ \left(Y - M \cdot \frac{d^2 y}{d t^2} + \lambda \cdot \frac{d L}{d y}\right) \delta y \\ &+ \left(Z - M \cdot \frac{d^2 z}{d t^2} + \lambda \cdot \frac{d L}{d z}\right) \delta z \end{aligned} \right\} = 0.$$

The quantity λ, being entirely arbitrary, let its value be such as to reduce the coefficient of one of the variables δx, δy, δz, say that of δx, to zero; and there will result

$$X - M \cdot \frac{d^2 x}{d t^2} + \lambda \cdot \frac{d L}{d x} = 0, \quad \cdot \cdot \cdot \cdot \cdot \cdot \quad (316)$$

and

$$\left(Y - M \cdot \frac{d^2 y}{d t^2} + \lambda \cdot \frac{d L}{d y}\right) \delta y + \left(Z - M \cdot \frac{d^2 z}{d t^2} + \lambda \cdot \frac{d L}{d z}\right) \delta z = 0 \cdot \quad (317)$$

Now in Equation (315), δy and δz may be assumed arbitrarily, and δx will result; hence δy and δz in Equation (317) may be regarded as independent of each other, and by the principle of indeterminate coefficients,

$$\left.\begin{aligned} Y - M \cdot \frac{d^2 y}{d t^2} + \lambda \cdot \frac{d L}{d y} &= 0, \\ Z - M \cdot \frac{d^2 z}{d t^2} + \lambda \cdot \frac{d L}{d z} &= 0, \end{aligned}\right\} \quad \cdot \cdot \cdot \cdot \quad (318)$$

and eliminating λ by means of Equation (316), we find,

$$\left.\begin{aligned} \left(Y - M \cdot \frac{d^2 y}{d t^2}\right) \cdot \frac{d L}{d x} - \left(X - M \cdot \frac{d^2 x}{d t^2}\right) \cdot \frac{d L}{d y} &= 0, \\ \left(Z - M \cdot \frac{d^2 z}{d t^2}\right) \cdot \frac{d L}{d y} - \left(Y - M \cdot \frac{d^2 y}{d t^2}\right) \cdot \frac{d L}{d z} &= 0; \end{aligned}\right\} \cdot \cdot (319)$$

which, with the equation of the surface, will determine the place of the centre of inertia at the end of a given time.

MOTION ON A CURVE OF DOUBLE CURVATURE.

§ 214.—If the centre of inertia be constrained to move upon two surfaces at the same time, or, which is the same thing, upon a *curve of double curvature* resulting from their intersection, take

$$\left.\begin{aligned} L &= F(x\,y\,z) = 0, \\ H &= F'(x\,y\,z) = 0; \end{aligned}\right\} \quad \cdot \cdot \cdot \cdot \cdot \cdot \cdot \quad (320)$$

from which, by the process of differentiating and replacing $d x$, $d y$, $d z$, by the projections of the virtual velocity,

$$\frac{d L}{d x} \delta x + \frac{d L}{d y} \cdot \delta y + \frac{d L}{d z} \cdot \delta z = 0; \quad . \quad . \quad . \quad . \quad (321)$$

$$\frac{d H}{d x} \cdot \delta x + \frac{d H}{d y} \cdot \delta y + \frac{d H}{d z} \cdot \delta z = 0. \quad . \quad . \quad . \quad . \quad (322)$$

Multiplying the first of these by λ, and the second by λ', adding the products to Equation (313), and collecting the coefficients of δx, δy, and δz, we have

$$\left.\begin{array}{l} \left(X - M . \frac{d^2 x}{d t^2} + \lambda \cdot \frac{d L}{d x} + \lambda' \cdot \frac{d H}{d x}\right) \delta x \\ + \left(Y - M . \frac{d^2 y}{d t^2} + \lambda \cdot \frac{d L}{d y} + \lambda' \cdot \frac{d H}{d y}\right) \delta y \\ + \left(Z - M \cdot \frac{d^2 z}{d t^2} + \lambda . \frac{d L}{d z} + \lambda' \cdot \frac{d H}{d z}\right) \delta z \end{array}\right\} = 0 \quad \cdot \quad (323)$$

Now the coefficients of two of the three variables δx, δy and δz, say those of δx and δy, may be made equal to zero by assigning proper values for that purpose to the indeterminate intensities λ and λ', in which case, since δz is not equal to zero, its coefficient must also be equal to zero; whence

$$\left.\begin{array}{l} X - M \cdot \frac{d^2 x}{d t^2} + \lambda \cdot \frac{d L}{d x} + \lambda' \cdot \frac{d H}{d x} = 0, \\ Y - M \cdot \frac{d^2 y}{d t^2} + \lambda \cdot \frac{d L}{d y} + \lambda' \cdot \frac{d H}{d y} = 0, \\ Z - M \cdot \frac{d^2 z}{d t^2} + \lambda . \frac{d L}{d z} + \lambda' \cdot \frac{d H}{d z} = 0. \end{array}\right\} \cdot \cdot \cdot \cdot (324)$$

and eliminating λ and λ', there will result

$$\left.\begin{array}{l} \left(X - M \cdot \frac{d^2 x}{d t^2}\right) \cdot \left(\frac{d L}{d z} \cdot \frac{d H}{d y} - \frac{d L}{d y} \cdot \frac{d H}{d z}\right) \\ + \left(Y - M \cdot \frac{d^2 y}{d t^2}\right) \cdot \left(\frac{d L}{d x} \cdot \frac{d H}{d z} - \frac{d L}{d z} \cdot \frac{d H}{d x}\right) \\ + \left(Z - M \cdot \frac{d^2 z}{d t^2}\right) \cdot \left(\frac{d L}{d y} \cdot \frac{d H}{d x} - \frac{d L}{d x} \cdot \frac{d H}{d y}\right) \end{array}\right\} = 0. \quad (325)$$

which, with the equations of the surfaces, is sufficient to determine the co-ordinates of the centre of inertia when the time is given.

§ 215.—If the given surfaces be the projecting cylinders of a curve of double curvature, then will Equations (320) become

$$\left.\begin{array}{l} L = F(x\,z) = 0\,; \\ H = F'(y\,z) = 0\,. \end{array}\right\} \quad . \quad . \quad . \quad . \quad . \quad . \quad . \quad . \quad (326)$$

And because L is now independent of y, and H is independent of x, we have

$$\frac{d\,L}{d\,y} = 0\,; \quad \frac{d\,H}{d\,x} = 0\,;$$

which reduce Equations (324) to

$$\left.\begin{array}{l} X - M \cdot \dfrac{d^2\,x}{d\,t^2} + \lambda \cdot \dfrac{d\,L}{d\,x} = 0\,; \\[2ex] Y - M \cdot \dfrac{d^2\,y}{d\,t^2} + \lambda' \cdot \dfrac{d\,H}{d\,y} = 0\,; \\[2ex] Z - M \cdot \dfrac{d^2\,z}{d\,t^2} + \lambda \cdot \dfrac{d\,L}{d\,z} + \lambda' \cdot \dfrac{d\,H}{d\,z} = 0\,; \end{array}\right\} \quad . \quad . \quad . \quad (327)$$

and Equation (325) to

$$\left.\begin{array}{l} \left(X - M \cdot \dfrac{d^2\,x}{d\,t^2}\right) \cdot \dfrac{d\,L}{d\,z} \cdot \dfrac{d\,H}{d\,y} \\[2ex] + \left(Y - M \cdot \dfrac{d^2\,y}{d\,t^2}\right) \cdot \dfrac{d\,L}{d\,x} \cdot \dfrac{d\,H}{d\,z} \\[2ex] - \left(Z - M \cdot \dfrac{d^2\,z}{d\,t^2}\right) \cdot \dfrac{d\,L}{d\,x} \cdot \dfrac{d\,H}{d\,y} \end{array}\right\} = 0. \quad . \quad . \quad . \quad . \quad (328)$$

This, with the equations of the curve, will give the place of the centre of inertia at the end of a given time.

§ 216.—If the curve be plane, the co-ordinate plane $x\,z$, may be assumed to coincide with that of the curve; in which case the second of Equations (327), becomes independent of y, that variable reducing to zero, and

$$d^2\,y = 0, \quad \text{and} \quad \frac{d\,H}{d\,y} = 0\,;$$

hence Equations (327), bcome

$$\left.\begin{aligned} &X - M \cdot \frac{d^2 x}{d t^2} + \lambda \cdot \frac{d L}{d x} = 0; \\ &Y = 0; \\ &Z - M \cdot \frac{d^2 z}{d t^2} + \lambda \cdot \frac{d L}{d z} + \lambda' \cdot \frac{d H}{d z} = 0; \end{aligned}\right\} \quad . \quad . \quad . \quad (329)$$

and because the factor

$$Y - M \frac{d^2 y}{d t^2} = 0,$$

Equation (328) becomes, on dividing out the common factor $\frac{d H}{d y}$,

$$\left(X - M \cdot \frac{d^2 x}{d t^2}\right) \cdot \frac{d L}{d z} - \left(Z - M \cdot \frac{d^2 z}{d t^2}\right) \cdot \frac{d L}{d x} = 0 \cdot \quad . \quad (330)$$

§ 217.—By transposing the terms involving λ, in Equations (316) and (318) and squaring we have

$$\lambda^2 \left[\left(\frac{d L}{d x}\right)^2 + \left(\frac{d L}{d y}\right)^2 + \left(\frac{d L}{d z}\right)^2 \right] = \left\{ \begin{aligned} &\left(X - M \cdot \frac{d^2 x}{d t^2}\right)^2 \\ &+ \left(Y - M \cdot \frac{d^2 y}{d t^2}\right)^2 \\ &+ \left(Z - M \cdot \frac{d^2 z}{d t^2}\right)^2 \end{aligned} \right\}$$

The second member of this equation is, Equation (50), the square of the intensity of the resultant of the extraneous forces and the forces of inertia. Denoting this resultant by N, we may write

$$\lambda \sqrt{\left(\frac{d L}{d x}\right)^2 + \left(\frac{d L}{d y}\right)^2 + \left(\frac{d L}{d z}\right)^2} = N. \quad . \quad . \quad . \quad (331)$$

and dividing each of the equations

$$\lambda \cdot \frac{d L}{d x} = - \left(X - M \cdot \frac{d^2 x}{d t^2}\right),$$

$$\lambda \cdot \frac{d L}{d y} = - \left(Y - M \cdot \frac{d^2 y}{d t^2}\right),$$

$$\lambda \cdot \frac{d L}{d z} = - \left(Z - M \cdot \frac{d^2 z}{d t^2}\right),$$

obtained by the transposition just referred to, by Equation (331), we find,

$$\left.\begin{array}{l}\dfrac{\dfrac{dL}{dx}}{\sqrt{\left(\dfrac{dL}{dx}\right)^2+\left(\dfrac{dL}{dy}\right)^2+\left(\dfrac{dL}{dz}\right)^2}}=-\dfrac{X-M\cdot\dfrac{d^2x}{dt^2}}{N}\\[2em]\dfrac{\dfrac{dL}{dy}}{\sqrt{\left(\dfrac{dL}{dx}\right)^2+\left(\dfrac{dL}{dy}\right)^2+\left(\dfrac{dL}{dz}\right)^2}}=-\dfrac{Y-M\cdot\dfrac{d^2y}{dt^2}}{N}\\[2em]\dfrac{\dfrac{dL}{dz}}{\sqrt{\left(\dfrac{dL}{dx}\right)^2+\left(\dfrac{dL}{dy}\right)^2+\left(\dfrac{dL}{dz}\right)^2}}=-\dfrac{Z-M\cdot\dfrac{d^2z}{dt^2}}{N}\end{array}\right\}. \quad (332)$$

The second members are the cosines of the angles which the resultant of all the forces including those of inertia, makes with the axes; the first members are the cosines of the angles which the normal to the surface at the body's place makes with the same axes. These being equal, with contrary signs, it follows not only that the forces whose intensities are

$$\lambda\sqrt{\left(\frac{dL}{dx}\right)^2+\left(\frac{dL}{dy}\right)^2+\left(\frac{dL}{dz}\right)^2} \text{ and } N,$$

are equal, but that they are both normal to the surface, and act in opposite directions. The second is the direct action upon the surface; the first is the reaction of the surface.

Equation (331), will, therefore, give the value of a passive resistance sufficient to neutralize all action in the system which is inconsistent with the arbitrary condition imposed upon the body's path. If the body be constrained to move on a rigid surface or line, this resistance will arise from its reaction.

§ 218.—If Equations (332) be multiplied by

$$N,$$

and the angles which the normal resistance of the surface makes with the axes x, y, z, respectively, be denoted by θ_x, θ_y, and θ_z, those equations will take the form

$$\left.\begin{aligned} X - M \cdot \frac{d^2 x}{d t^2} + N \cdot \cos \theta_x &= 0; \\ Y - M \cdot \frac{d^2 y}{d t^2} + N \cdot \cos \theta_y &= 0; \\ Z - M \cdot \frac{d^2 z}{d t^2} + N \cdot \cos \theta_z &= 0. \end{aligned}\right\} \quad \ldots \ldots (333)$$

§ 219.—To impose the condition, therefore, that a body in motion shall remain on a rigid surface, is equivalent to introducing into the system an additional force, which shall be equal and directly opposed to the pressure upon the surface. The motion may then be regarded as perfectly free, and treated accordingly. The same might be shown from Equations (324) to be equally true of a rigid curve, but the principle is too obvious to require further elucidation.

Equations (333), may, therefore, be regarded as equally applicable to a rigid curve of any curvature, as to a surface; the normal reaction of the curve being denoted by N, and the angles which N makes with the axes x, y, z, by θ_x, θ_y, and θ_z.

§ 220.—To find the value of N, eliminate dt from Equations (333), by the relation

$$\frac{1}{d t} = \frac{V}{d s};$$

in which V and s are the velocity and the space; then by transposition these equations may be written

$$N \cdot \cos \theta_x = M \cdot V^2 \cdot \frac{d^2 x}{d s^2} - X;$$

$$N \cdot \cos \theta_y = M \cdot V^2 \cdot \frac{d^2 y}{d s^2} - Y;$$

$$N \cdot \cos \theta_z = M \cdot V^2 \cdot \frac{d^2 z}{d s^2} - Z.$$

15

Squaring, adding and reducing by the relations

$$R^2 = X^2 + Y^2 + Z^2,$$
$$\cos^2\theta_x + \cos^2\theta_y + \cos^2\theta_z = 1,$$

and we find

$$N^2 = \left\{\begin{array}{l} M^2 \cdot \frac{V^4}{d s^4}\left[(d^2 x)^2 + (d^2 y)^2 + (d^2 z)^2\right] + R^2 \\ -2 M \cdot V^2\left[X \cdot \frac{d^2 x}{d s^2} + Y \cdot \frac{d^2 y}{d s^2} + Z \cdot \frac{d^2 z}{d s^2}\right] \end{array}\right\}$$

Resolving R into two components, one parallel and the other perpendicular to the path, the former will be in equilibrio with the inertia it develops in the direction of the curve; and denoting by φ the inclination of R to the radius of curvature, we have

$$R \sin\varphi = M \cdot \frac{d^2 s}{d t^2} = M \cdot V^2 \cdot \frac{d^2 s}{d s^2};$$

or,

$$0 = R \cdot \sin\varphi - M \cdot V^2 \cdot \frac{d^2 s}{d s^2};$$

Squaring and subtracting from the equation above, there will result,

$$N^2 = \left\{\begin{array}{l} M^2 \cdot \frac{V^4}{d s^4} \cdot \left((d^2 x)^2 + (d^2 y)^2 + (d^2 z)^2 - (d^2 s)^2\right) + R^2 \cos^2\varphi \\ -2 M \cdot V^2 \cdot R\left(\frac{X}{R} \cdot \frac{d^2 x}{d s^2} + \frac{Y}{R} \cdot \frac{d^2 y}{d s^2} + \frac{Z}{R} \cdot \frac{d^2 z}{d s^2} - \sin\varphi \cdot \frac{d^2 s}{d s^2}\right) \end{array}\right\}$$

but

$$\sin\varphi = \frac{X}{R} \cdot \frac{d x}{d s} + \frac{Y}{R} \cdot \frac{d y}{d s} + \frac{Z}{R} \cdot \frac{d z}{d s};$$

multiplying the second member by $\rho \div \rho$, substituting above, and reducing by the relations,

$$\frac{d^2 x}{d s^2} - \frac{dx}{ds}\frac{d^2 s}{ds^2} = \frac{d\frac{dx}{ds}}{ds}; \quad \frac{d^2 y}{ds^2} - \frac{dy}{ds}\frac{d^2 s}{ds^2} = \frac{d\frac{dy}{ds}}{ds}; \quad \frac{d^2 z}{ds^2} - \frac{dz}{ds}\frac{d^2 s}{ds^2} = \frac{d\frac{dz}{ds}}{ds};$$

$$\cos\varphi = \frac{X}{R} \cdot \rho\frac{d\frac{dx}{ds}}{ds} + \frac{Y}{R} \cdot \rho\frac{d\frac{dy}{ds}}{ds} + \frac{Z}{R} \cdot \rho\frac{d\frac{dz}{ds}}{ds}; \; *$$

*See Appendix No. 2.

and

$$\rho = \frac{d s^2}{\sqrt{(d^2 x)^2 + (d^2 y)^2 + (d^2 z)^2 - (d^2 s)^2}},$$

in which ρ denotes the radius of curvature, we have,

$$N^2 = M^2 \cdot \frac{V^4}{\rho^2} - 2 \frac{M V^2}{\rho} \cdot R \cos \varphi + R^2 \cos^2 \varphi ;$$

and taking square root,

$$N = \frac{M V^2}{\rho} - R \cos \varphi. \quad \dots \dots \quad (334)$$

The first term of the second member is, § 195, the centrifugal force arising from the deflecting action of the curve, and the last term is the normal component of the resultant R. As the equation stands, its signs apply to the case in which the body is on the concave side of the curve, and the resultant acts from the curve. The angle φ, must be measured from the radius of curvature, or that radius produced, according as the body is on the concave or convex side of the curve. When the body is moving on the convex side of the curve, the first term of the second member must change its sign and become negative.

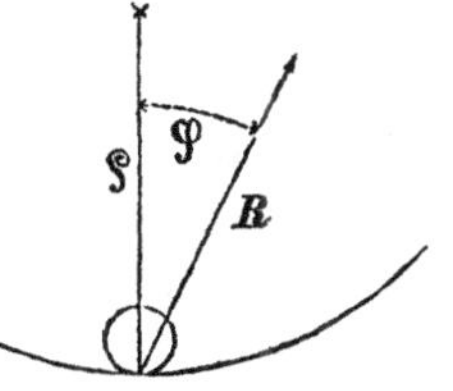

§ 221.—Writing Equations (333) under the form

$$M \cdot \frac{d^2 x}{d t^2} = X + N \cos \theta_x,$$

$$M \cdot \frac{d^2 y}{d t^2} = Y + N \cos \theta_y,$$

$$M \cdot \frac{d^2 z}{d t^2} = Z + N \cos \theta_z;$$

multiplying the first by $2 d x$, the second by $2 d y$, the third by $2 d z$, adding and reducing by the relation

$$d s \left(\frac{d x}{d s} \cdot \cos \theta_x + \frac{d y}{d s} \cdot \cos \theta_y + \frac{d z}{d s} \cdot \cos \theta_z \right) \cdot = 0,$$

the second factor being the cosine of the angle made by the normal and tangent to the curve, we have

$$M \cdot \left(\frac{2\,dx\,.\,d^2x + 2\,dy\,.\,d^2y + 2\,dz\,.\,d^2z}{dt^2}\right) = 2(Xdx + Ydy + Zdz);$$

integrating and reducing by

$$V^2 = \frac{dx^2 + dy^2 + dz^2}{dt^2},$$

we find

$$MV^2 = 2\int(Xdx + Ydy + Zdz) + C. \quad . \quad . \quad (335)$$

This being independent of the reaction of the curve, it can have no effect upon the velocity.

If the incessant forces be zero, then will

$$X = 0; \quad Y = 0; \quad \text{and} \quad Z = 0;$$

and

$$V^2 = \frac{C}{M};$$

that is, a body moving upon a rigid surface or curve, and not acted upon by incessant forces, will preserve its velocity constant, and the motion will be uniform.

We also recognize, in Equation (335), the general theorem of the living force and quantity of work; and from which, as before, it appears that the velocity is wholly independent of the path described.

Example 1.—Let the body be required to move upon the interior surface of a spherical bowl, under the action of its own weight. In this case,

$$L = x^2 + y^2 + z^2 - a^2 = 0; \quad . \quad . \quad . \quad . \quad (336)$$

$$\frac{dL}{dx} = 2x; \quad \frac{dL}{dy} = 2y; \quad \frac{dL}{dz} = 2z;$$

and the axis of z being vertical and positive downwards,

$$X = 0; \quad Y = 0; \quad Z = Mg;$$

which values in Equations (319), give

$$\left.\begin{array}{r} y \cdot \frac{d^2 x}{d t^2} - x \cdot \frac{d^2 y}{d t^2} = 0; \\ g y - y \cdot \frac{d^2 z}{d t^2} + z \cdot \frac{d^2 y}{d t^2} = 0; \end{array}\right\} \cdot (337)$$

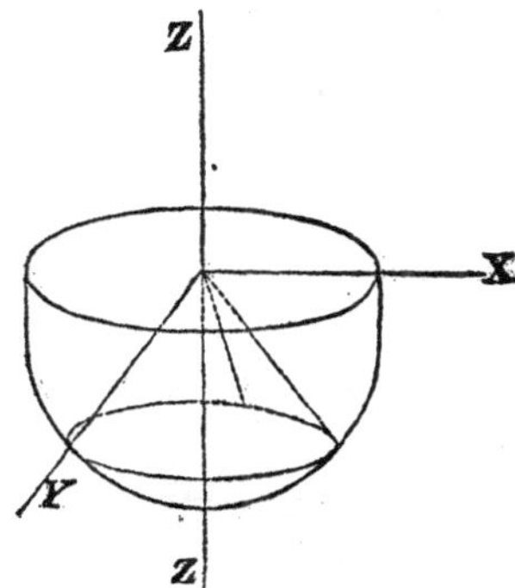

and differentiating the equation of the sphere twice, we have

$$x d^2 x + y d^2 y + z . d^2 z = - (d x^2 + d y^2 + d z^2);$$

dividing by $d t^2$, and replacing the second member by its value V^2, the velocity, we find,

$$x \cdot \frac{d^2 x}{d t^2} + y \cdot \frac{d^2 y}{d t^2} + z \cdot \frac{d^2 z}{d t^2} = - V^2.$$

But, Equation (335),

$$V^2 = 2 g z + C \quad . \quad . \quad . \quad . \quad . \quad . \quad . \qquad (338)$$

and denoting by V' and k, the initial values of V and z, respectively, we have

$$V^2 = V'^2 + 2 g (z - k),$$

which substituted above, gives

$$x \cdot \frac{d^2 x}{d t^2} + y \cdot \frac{d^2 y}{d t^2} + z \cdot \frac{d^2 z}{d t^2} = 2 g (k - z) - V'^2 \quad . \quad . \quad (339)$$

Eliminate x, y, $d^2 x$, $d^2 y$, from this equation by means of Equations (336) and (337).

From the latter we find,

$$\frac{d^2 y}{d t^2} = \frac{y}{z} \left(\frac{d^2 z}{d t^2} - g \right),$$

$$\frac{d^2 x}{d t^2} = \frac{x}{z} \left(\frac{d^2 z}{d t^2} - g \right),$$

which substituted in Equation (339), and reducing by means of Equation (336), we get

$$a^2 \cdot \frac{d^2 z}{d t^2} = g\,(a^2 - 3 z^2 + 2 k z) - V'^2 z\,;$$

multiplying by $2\,dz$, and integrating, we find

$$a^2 \cdot \frac{d z^2}{d t^2} = 2 g\,(a^2 z - z^3 + k z^2) - V'^2 z^2 + C\,;$$

in which C is the constant of integration, and to determine which, we denote the component of the velocity V', in the direction of the axis z, by V_z', and make $z = k$. This being done, we get

$$C = a^2 . V_z'^2 + V'^2 k^2 - 2 g a^2 k\,;$$

whence,

$$a^2 \cdot \frac{d z^2}{d t^2} = 2 g\,(a^2 z - z^3 + k z^2) - V'^2 z^2 + a^2 V_z'^2 + V'^2 k^2 - 2 g a^2 k,$$

adding and subtracting $a^2 V'^2$ in the second member, this reduces to

$$a^2 \cdot \frac{d z^2}{d t^2} = (a^2 - z^2)\,[V'^2 - 2 g\,(k - z)] - C_{,}\,,$$

in which

$$C_{,} = (a^2 - k^2)\,V'^2 - a^2\,V_z'^2.$$

Finding the value of $d\,t$, and integrating, we have

$$t = \int \frac{a\,d z}{\sqrt{(a^2 - z^2)\,[V'^2 - 2 g\,(k - z)] - C_{,}}} \quad \cdots \quad (340)$$

Could this equation be integrated in finite terms, then would z become known for a given value of t; and this value of z in Equation (336), and the first of Equations (337), after integration, would make known the values of x and y, and hence the position of the body; its velocity would be known from Equation (335). But this integration is not possible.

§ 222.—We may, however, approximate to the result when the initial impulse is small and in a horizontal direction, and the point of departure is near the bottom of the bowl. Let θ be the angle which the radius drawn to the variable position of the body makes with the axis of z; φ, the angle which the plane of the angle θ makes with the plane through the axis z and initial place of the body, supposed in the plane xz; $V' = \beta\sqrt{ga}$, the velocity of projection in a horizontal direction, β being a very small quantity; and α the initial value of θ. Then, because

$$z = a \cdot \cos\theta = a\,(1 - 2\sin^2 \tfrac{1}{2}\theta); \quad k = a\cos\alpha = a\,(1 - 2\sin^2\tfrac{1}{2}\alpha);$$

$$dz = -a \cdot \sin\theta \cdot d\theta; \quad V_z' = 0;$$

$$C_{,} = a^2 \cdot 4\sin^2\tfrac{1}{2}\alpha \cdot \cos^2\tfrac{1}{2}\alpha \cdot \beta^2 \cdot ag = a^3 \cdot \beta^2 \cdot g \cdot \sin^2\alpha;$$

Equation (340) becomes

$$t = -\sqrt{\frac{a}{g}} \cdot \int \frac{\sin\theta \cdot d\theta}{\sqrt{\beta^2(\sin^2\theta - \sin^2\alpha) - 4\sin^2\theta\,(\sin^2\tfrac{1}{2}\theta - \sin^2\tfrac{1}{2}\alpha)}};$$

and making θ and α very small, their arcs may be taken for their sines, and the above becomes, after differentiating,

$$\frac{dt}{d\theta} = -\sqrt{\frac{a}{g}} \cdot \frac{\theta}{\sqrt{(\alpha^2 - \theta^2)(\theta^2 - \beta^2)}} \quad . \quad . \quad (341)$$

which may be put under the form

$$2t = \sqrt{\frac{a}{g}} \cdot \int \frac{-4\theta \cdot d\theta}{\sqrt{(\alpha^2 - \beta^2)^2 - [2\theta^2 - (\alpha^2 + \beta^2)]^2}},$$

whence, by integration,

$$2t = \sqrt{\frac{a}{g}} \cdot \cos^{-1}\left[\frac{2\theta^2 - (\alpha^2 + \beta^2)}{\alpha^2 - \beta^2}\right] + C; \quad . \quad . \quad (342)$$

making $t = 0$, and $\theta = \alpha$, we have $C = -\overset{-1}{\cos}\,1 \cdot \sqrt{a} \div \sqrt{g}$, or $C = 0$; and solving the equation with reference to θ, we get

$$\theta^2 = \tfrac{1}{2}(\alpha^2 + \beta^2) + \tfrac{1}{2}(\alpha^2 - \beta^2) \cdot \cos 2\sqrt{\frac{g}{a}} \cdot t. \quad . \quad . \quad (343)$$

From which it appears that the greatest and least values of θ will occur periodically, and at equal intervals of time. The former of these values is found by making

$$\cos 2\sqrt{\frac{g}{a}} \cdot t = 1; \quad \text{whence } 2\sqrt{\frac{g}{a}} \cdot t = 0, \quad \text{or} = 2\pi, \quad \text{or} = 4\pi,$$

and so on; and for a single interval between two consecutive maxima, without respect to sign,

$$t = \pi\sqrt{\frac{a}{g}}; \quad \ldots \ldots \quad (344)$$

the maximum being α.

The least value occurs when

$$\cos 2\sqrt{\frac{g}{a}} \cdot t = -1, \quad \text{or } 2\sqrt{\frac{g}{a}} \cdot t = \pi, \text{ or} = 3\pi, \ \&c.$$

whence for a single interval between any maximum and the succeeding minimum,

$$t = \tfrac{1}{2}\pi\sqrt{\frac{a}{g}}; \quad \ldots \ldots \quad (345)$$

the minimum being β.

The movement by which these recurring values are brought about, is called *oscillatory motion;* that between any two equal values is called an *oscillation;* and when the oscillations are performed in equal times, they are said to be *Isochronous.*

Again,

$$\frac{d\varphi}{d\theta} = \frac{d\varphi}{dt} \cdot \frac{dt}{d\theta};$$

substituting for $\frac{d\varphi}{dt}$, its value obtained from the relation $y = x \tan\varphi$, we find

$$\frac{d\varphi}{d\theta} = \frac{1}{x^2 + y^2} \cdot \left(x \cdot \frac{dy}{dt} - y \cdot \frac{dx}{dt}\right) \cdot \frac{dt}{d\theta}.$$

Integrating the first of Equations (337), we get

$$x \cdot \frac{dy}{dt} - y \cdot \frac{dx}{dt} = 2c = V' a \alpha = \alpha \beta a \sqrt{g a};$$

substituting this above, and also the value of $\frac{dt}{d\theta}$, given by Equation (341), we find

$$\frac{d\varphi}{d\theta} = -\frac{\alpha \cdot \beta}{\theta\sqrt{(\alpha^2 - \theta^2)(\theta^2 - \beta^2)}}; \quad . \quad . \quad . \quad . \quad (346)$$

dividing this by Equation (341),

$$\frac{d\varphi}{dt} = \sqrt{\frac{g}{a}} \cdot \frac{\alpha \, . \, \beta}{\theta^2} = \sqrt{\frac{g}{a}} \cdot \frac{\alpha \, . \, \beta}{\frac{1}{2}(\alpha^2 + \beta^2) + \frac{1}{2}(\alpha^2 - \beta^2) \, . \cos 2\sqrt{\frac{g}{a}} \cdot t}.$$

but

$$\cos 2\sqrt{\frac{g}{a}} \cdot t = \cos^2 \cdot \sqrt{\frac{g}{a}} \cdot t - \sin^2 \sqrt{\frac{g}{a}} \cdot t;$$

whence

$$\frac{d\varphi}{dt} = \sqrt{\frac{g}{a}} \cdot \frac{\alpha \cdot \beta}{\alpha^2 \cdot \cos^2 \sqrt{\frac{g}{a}} \cdot t + \beta^2 \, . \sin^2 \sqrt{\frac{g}{a}} \cdot t}; \quad . \quad . \quad . \quad (347)$$

from which we find

$$d\varphi = \frac{\frac{\beta}{\alpha} \cdot \frac{\sqrt{\frac{g}{a}} \cdot dt}{\cos^2 \sqrt{\frac{g}{a}} \cdot t}}{1 + \frac{\beta^2}{\alpha^2} \cdot \tan^2 \sqrt{\frac{g}{a}} \cdot t};$$

integrating, and taking tangents of both members,

$$\tan \varphi = \frac{\beta}{\alpha} \cdot \tan \sqrt{\frac{g}{a}} \cdot t \quad . \quad . \quad . \quad . \quad . \quad . \quad (348)$$

from which the azimuth of the plane of oscillation may be found at the end of any time.

Making $\tan \varphi = \infty$, we have

$$\sqrt{\frac{g}{a}} \cdot t = \frac{1}{2}\pi; \quad \text{or} \quad = \frac{3}{2}\pi; \quad \text{or} \quad = \frac{5}{2}\pi, \text{ \&c.,}$$

and the interval from the epoch to the first azimuth of 90°, is

$$t_{\prime} = \frac{1}{2}\pi \cdot \sqrt{\frac{a}{g}},$$

and to the first azimuth of 270°,

$$t_{\prime\prime} = \frac{3}{2}\pi \cdot \sqrt{\frac{a}{g}},$$

and the interval from the azimuth of 90° to the next azimuth of 270°

$$t_{\prime\prime} - t_{\prime} = t = \pi \cdot \sqrt{\frac{a}{g}},$$

equal to the time of one entire oscillation.

From Equation (348) we have, after substituting for $\tan\varphi$ its value in the relation $y = x\tan\varphi$,

$$\frac{\alpha^2 y^2}{\beta^2 x^2} = \tan^2 \sqrt{\frac{g}{a}} \cdot t;$$

adding unity to both members,

$$\frac{\beta^2 x^2 + \alpha^2 y^2}{\beta^2 x^2} = 1 + \tan^2 \sqrt{\frac{g}{a}} \cdot t;$$

also from $y = x \,.\, \tan\varphi$,

$$\frac{x^2 + y^2}{x^2} = 1 + \tan^2\varphi;$$

dividing the last equation by this one, and replacing $x^2 + y^2$ by its value $a^2 - z^2$, from the equation of the surface, we get

$$\alpha^2 y^2 + \beta^2 x^2 = \beta^2 \cdot (a^2 - z^2) \cdot \frac{1 + \tan^2 \sqrt{\frac{g}{a}} \cdot t}{1 + \tan^2\varphi};$$

but, neglecting the term involving θ^4,

$$a^2 - z^2 = a^2 \theta^2;$$

substituting this above, replacing $\tan^2\varphi$ by its value in Equation (348), and θ^2 by its value in Equation (343), after making

$$\cos 2\sqrt{\frac{g}{a}} \cdot t = 1 - 2\sin^2 \sqrt{\frac{g}{a}} \cdot t$$

and reducing by the relation,

$$\sin^2 \sqrt{\frac{g}{a}} \cdot t = \frac{\tan^2 \sqrt{\frac{g}{a}} \cdot t}{1 + \tan^2 \sqrt{\frac{g}{a}} \cdot t}$$

we have

$$\frac{x^2}{\alpha^2} + \frac{y^2}{\beta^2} = a^2; \quad \ldots \ldots \quad (349)$$

which shows that the projection of the path of the body on the plane $x y$, is an ellipse whose centre is on the vertical radius of the sphere, and that the line connecting the body with the centre of the sphere, describes a conical surface.

If $\alpha = \beta$, then will, Equations (343) and (348),

$$\theta^2 = \alpha^2 = \beta^2; \quad \varphi = \sqrt{\frac{g}{a}} \cdot t;$$

and, Equation (349),

$$x^2 + y^2 = \alpha^2 a^2; \quad \ldots \ldots \quad (350)$$

hence, the body will describe a horizontal circle with a uniform motion.

The pressure upon the surface, at any point of the body's path, is given by the value of N in Equation (334).

§ 223.—*Example* 2.—Let the body, still reduced to its centre of inertia and acted upon by its own weight, be also repelled from the bottom point A of the bowl, by a force which varies inversely as the square of the distance; required the position of the body in which it would remain at rest.

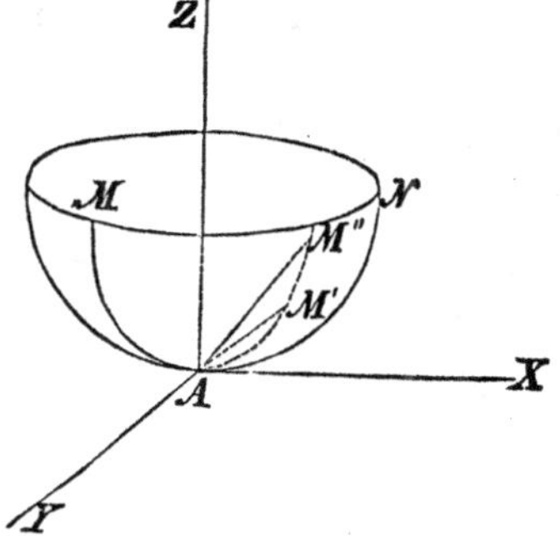

As the body is to be at rest, there will be no inertia exerted, and we have

$$\frac{d^2 x}{d t^2} = 0; \quad \frac{d^2 y}{d t^2} = 0; \quad \frac{d^2 z}{d t^2} = 0;$$

and assuming the axis z vertical, positive upwards, and the origin at the lowest point A,

$$L = x^2 + y^2 + z^2 - 2az = 0, \quad . \quad . \quad . \quad . \quad (351)$$

$$\frac{dL}{dx} = 2x; \quad \frac{dL}{dy} = 2y; \quad \frac{dL}{dz} = 2(z - a);$$

and denoting the distance of the body from the lowest point by r, the intensity of the repelling force at the unit's distance by F, and the force at any distance by P, then will

$$P = \frac{F}{r^2}; \quad r = \sqrt{x^2 + y^2 + z^2}; \quad . \quad . \quad . \quad . \quad (352)$$

for the force P, $\cos\alpha = \frac{x}{r}$; $\cos\beta = \frac{y}{r}$; $\cos\gamma = \frac{z}{r}$; for the weight Mg, $\cos\alpha' = 0$; $\cos\beta' = 0$; $\cos\gamma' = -1$; and

$$X = \frac{Fx}{r^3}; \quad Y = \frac{Fy}{r^3}; \quad Z = -Mg + \frac{Fz}{r^3}.$$

These several values being substituted in Equations (319), give

$$\frac{Fyx}{r^3} - \frac{Fyx}{r^3} = 0,$$

$$\left(\frac{Fz}{r^3} - Mg\right) \cdot y - \frac{Fy}{r^3} \cdot (z - a) = 0.$$

The first equation establishes no relation between x and y, since the equilibrium, which depends upon the distance of the particle from the source of repulsion, would obviously exist at any point of a horizontal circle whose circumference is at the proper height from the bottom.

From the second equation we deduce,

$$\frac{Fa}{r^3} = Mg,$$

$$r = \left(\frac{Fa}{Mg}\right)^{\frac{1}{3}},$$

$$\frac{F}{Mg} = \frac{r^3}{a}, \quad . \quad . \quad . \quad . \quad . \quad . \quad . \quad . \quad (353)$$

from which r becomes known; and to determine the position of the circle upon which the body must be placed, we have, by making $x = 0$ in Equations (352) and (351),

$$\sqrt{z^2 + y^2} = r,$$

$$y^2 + z^2 - 2az = 0.$$

Equation (353) makes known the relation between the weight of the body and the repulsive force at the unit's distance; the intensity of the force at any other distance may therefore be determined.

If there be substituted a repulsive force of different intensity, but whose law of variation is the same, we should have, in like manner,

$$\frac{F'}{Mg} = \frac{r'^3}{a};$$

hence,

$$F : F' :: r^3 : r'^3;$$

that is, the forces are as the cubes of the distances at which the body is brought to rest.

If, instead of being supported on the surface of a sphere, the body had been connected by a perfectly light and inflexible line with the centre of the sphere and the surface removed, the result would have been the same. In this form of the proposition, we have the common *Electroscope*.

The differential co-efficients of the second order, or the terms which measure the force of inertia, being equal to zero, Equations (332), show that the resultant of the extraneous forces, in this case the weight and repulsion, is normal to the surface, which should be the case; for then there is no reason why the body should move in one direction rather than another. The pressure upon the surface is given by the value of N, in Equation (334).

§ 224.—*Example* 3. Let it be required to find the circumstances of motion of a body acted upon by its own weight while on the arc of a cycloid, of which the plane is vertical, and directrix horizontal.

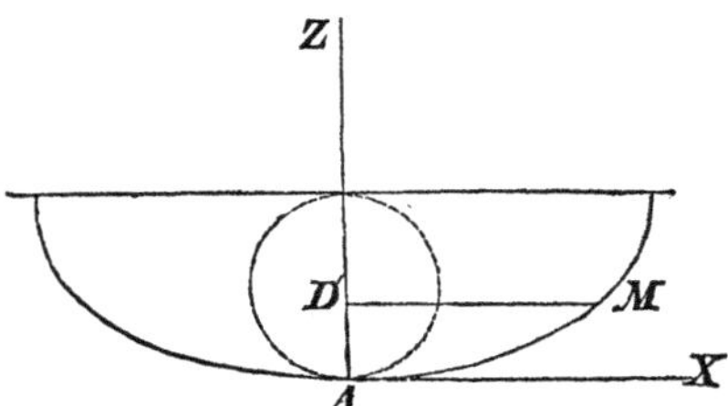

Taking the axis of z, vertical; the plane zx, in the plane of the curve; and the origin at the lowest point, then will

$$L = x - \sqrt{2az - z^2} - a \text{ versin}^{-1}\frac{z}{a} = 0; \quad . \quad . \quad (354)$$

in which z is taken positive upwards.

$$\frac{dL}{dx} = 1; \quad \frac{dL}{dz} = -\sqrt{\frac{2a - z}{z}}, \quad . \quad . \quad . \quad (355)$$

$$X = 0; \quad Z = -Mg,$$

and Equation (330) becomes

$$\frac{d^2x}{dt^2} \cdot \sqrt{\frac{2a - z}{z}} + g + \frac{d^2z}{dt^2} = 0. \quad . \quad . \quad . \quad . \quad (356)$$

and by transposition and division,

$$\frac{d^2x}{dt^2} = -\frac{g}{\sqrt{\frac{2a - z}{z}}} - \frac{d^2z}{dt^2} \cdot \frac{1}{\sqrt{\frac{2a - z}{z}}}. \quad . \quad . \quad . \quad (357)$$

From the equation of the curve we find,

$$2dx = 2dz \cdot \sqrt{\frac{2a - z}{z}}; \quad . \quad . \quad . \quad . \quad . \quad (358)$$

multiplying by Equation (357), there will result

$$\frac{2dx \cdot d^2x}{dt^2} = -2g\,dz - \frac{2dz \cdot d^2z}{dt^2};$$

and by integration,

$$\frac{dx^2}{dt^2} = C - 2gz - \frac{dz^2}{dt^2};$$

or,

$$\frac{dx^2 + dz^2}{dt^2} = V^2 = C - 2gz;$$

and supposing the velocity zero, when $z = h$;

$$0 = C - 2gh;$$

which subtracted from the above gives

$$\frac{dx^2 + dz^2}{dt^2} = 2g(h - z); \quad \ldots \ldots \quad (359)$$

and eliminating dx^2 by means of Equation (358),

$$\frac{dz^2}{dt^2} = \frac{g}{a} \cdot (hz - z^2)$$

whence,

$$dt = -\sqrt{\frac{a}{g}} \cdot \frac{dz}{\sqrt{hz - z^2}};$$

the negative sign being taken because z is a decreasing function of t.

By integration,

$$t = -\sqrt{\frac{a}{g}} \cdot \int \frac{dz}{\sqrt{hz - z^2}} = -\sqrt{\frac{a}{g}} \cdot \text{versin}^{-1} \cdot \frac{2z}{h} + C.$$

Making $z = h$, we have

$$0 = -\sqrt{\frac{a}{g}} \cdot \text{versin}^{-1} 2 + C;$$

whence,

$$C = \pi \sqrt{\frac{a}{g}},$$

and

$$t = \sqrt{\frac{a}{g}} \left(\pi - \text{versin}^{-1} \cdot \frac{2z}{h} \right) \quad . \quad . \quad . \quad . \quad . \quad (360)$$

When the body has reached the bottom, then will $z = 0$, and

$$t = \pi \sqrt{\frac{a}{g}},$$

which is wholly independent of h, or the point of departure, and we hence infer that the time of descent to the lowest point will be the same in the same cycloid, no matter from what point the body starts.

Whenever $z = h$, the body will, Equation (359), stop, and we shall have the times arranged in order before and after the epoch,

$$-4\pi \sqrt{\frac{a}{g}}; \quad -2\pi \sqrt{\frac{a}{g}}; \quad 0; \quad 2\pi \sqrt{\frac{a}{g}}; \quad 4\pi \sqrt{\frac{a}{g}}, \quad \&c.,$$

the difference between any two consecutive values being

$$2\pi \sqrt{\frac{a}{g}}.$$

The body will, therefore, oscillate back and forth, in equal times. The cycloid is a *Tautochrone.*

The pressure upon the curve is given by Equation (334).

The time being given and substituted in Equation (360), the value of z becomes known, and this, in Equations (359) and (354), will give the body's velocity and place.

§ 225.—*Example* 4.—Let a body reduced to its centre of inertia, and whose weight is denoted by W, be supported by the action of a constant force upon the branch EH of an hyperbola, of which the transverse axis is vertical, the force being directed to the centre of the curve. Required the position of equilibrium.

Denote the constant force by W', which may be a weight at the end of a cord passing over a small wheel at C, and attached to the body M. Denote the distance CM by r, and the axes of the curve by A and B. Take the axis z vertical, and the curve in the plane xz. Make

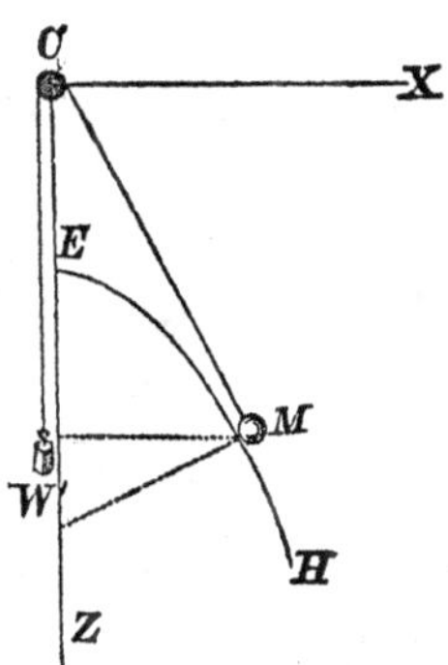

$$P' = W,$$

$$P'' = W'$$

then will

$$\cos\gamma' = 1, \quad \cos\alpha' = 0,$$

$$\cos\gamma'' = -\frac{z}{r}, \quad \cos\alpha'' = -\frac{x}{r},$$

$$X = P'\cos\alpha' + P''\cos\alpha'' = -W' \cdot \frac{x}{r},$$

$$Z = P'\cos\gamma' + P''\cos\gamma'' = W - W' \cdot \frac{z}{r},$$

and as the question relates to the state of rest,

$$\frac{d^2 x}{d t^2} = 0; \quad \frac{d^2 z}{d t^2} = 0.$$

The Equation of the curve is

$$L = A^2 x^2 - B^2 z^2 + A^2 B^2 = 0;$$

whence,

$$\frac{dL}{dx} = 2A^2 x,$$

$$\frac{dL}{dz} = -2B^2 z;$$

these values substituted in Equation (330), give

$$W' B^2 \frac{xz}{r} - W A^2 x + W' A^2 \frac{xz}{r} = 0:$$

whence,

$$(A^2 + B^2)\, W' \cdot z - W A^2 r = 0 \quad . \quad . \quad . \quad . \quad (361)$$

But

$$r^2 = x^2 + z^2 = z^2 + \frac{B^2}{A^2} z^2 - B^2 = z^2 \frac{A^2 + B^2}{A^2} - B^2;$$

whence, denoting the eccentricity by e,

$$r = \sqrt{e^2 z^2 - B^2}$$

and this, in Equation (361), gives after reduction,

$$z = \frac{B \cdot W}{e(W^2 - W'^2 e^2)^{\frac{1}{2}}};$$

which, with the equation of the curve, will give the position of equilibrium.

If $W' e$ be greater than W, the equilibrium will be impossible. If $W' e = W$, the body will be supported upon the asymptote.

The pressure upon the curve is given by Equation (334).

§ 226.—*Example* 5.—Required the circumstances of motion of a body moving from rest under the action of its own weight upon an inclined right line.

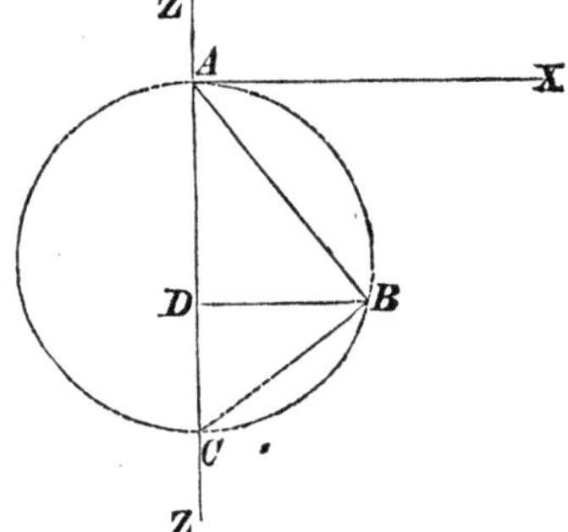

Take the axis of z vertical, the plane $z\,x$ to contain the line, and the origin at the point of departure, and let z be reckoned positive downwards. Then will

$$L = z - a\,x = 0,$$

$$\frac{dL}{dz} = 1; \quad \frac{dL}{dx} = -a;$$

$$X = 0; \quad Z = Mg;$$

which in Equation (330) give, after omitting the common factor M,

$$-\frac{d^2 x}{dt^2} + a g - a \frac{d^2 z}{dt^2} = 0. \quad \cdot \quad \cdot \quad \cdot \quad \cdot \quad \cdot \quad \cdot \qquad (362)$$

From the equation of the line we have

$$d^2 x = \frac{d^2 z}{a};$$

which in Equation (362), after slight reduction,

$$\frac{d^2 z}{d t^2} = \frac{a^2}{1 + a^2} \cdot g.$$

Multiplying by $2 d z$, and integrating,

$$\frac{d z^2}{d t^2} = 2 g \frac{a^2}{1 + a^2} \cdot z.$$

the constant of integration being zero.

Whence

$$d t = \sqrt{\frac{2 (1 + a^2)}{g \cdot a^2}} \cdot \frac{d z}{2 \sqrt{z}},$$

and

$$t = \sqrt{\frac{2 (1 + a^2)}{g a^2} \cdot z} = \sqrt{\frac{2 (1 + a^2)}{g a^2 z} \cdot z^2}; \quad . \quad . \quad . \quad (363)$$

the constant of integration being again zero.

The body being supposed at B, then will $z = A D$; and if we draw from B the perpendicular $B C$ to $A B$, we have

$$\frac{\overline{A B}^2}{z^2} = \frac{1 + a^2}{a^2};$$

which substituted above,

$$t = \sqrt{\frac{\overline{A B}^2}{z} \cdot \frac{2}{g}} = \sqrt{\frac{2 d}{g}}; \quad . \quad . \quad . \quad . \quad . \quad . \quad (364)$$

in which d denotes the distance $A C$.

But the second member is the time of falling freely through the vertical distance d; if, therefore, a circle be described upon $A C$ as a diameter, we see that the time down any one of its chords, terminating at the upper or lower point of this diameter, will be the same as that through the vertical diameter itself. This is called the mechanical property of the circle.

Example 6.—A spherical body placed on a plane inclined to the horizon, would, in the absence of friction, slide under the action of its own weight; but, owing to friction, it will roll. Required the circumstances of the motion.

If the sphere move from rest with no initial impulse, the centre will describe a straight line parallel to the element of steepest descent. Take the plane $x z$, to contain this element, the axis z vertical and positive upwards.

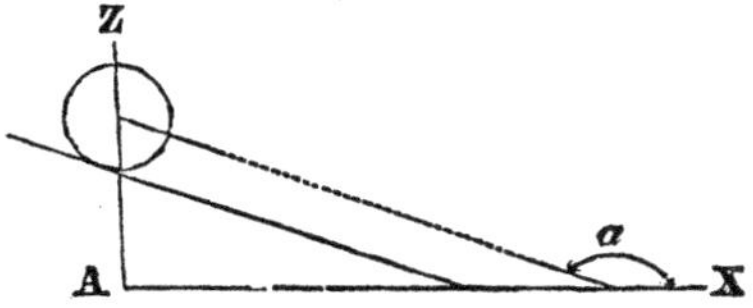

The equation of the path will be,

$$L = z + x \tan a - h = 0;$$

whence,

$$\frac{dL}{dz} = 1; \quad \frac{dL}{dx} = \tan a.$$

The extraneous forces are the weight of the sphere and the friction. Denote the first by W, and the second by F. The nature of friction and its mode of action will be explained in the proper place, § 354; it will be sufficient here to say that for the same weight of the sphere and inclination of the plane, it will be a constant force acting up the plane and opposed to the motion. We shall therefore have

$$Z = - Mg + F \sin a; \quad X = - F \cos a,$$

which values, and those above substituted in Equation (330), give

$$- F \cos a - M \cdot \frac{d^2 x}{d t^2} + \left(M g - F \sin a + M \cdot \frac{d^2 z}{d t^2}\right) \tan a = 0.$$

But from the equation of the path, we have

$$d^2 z = - d^2 x \cdot \tan a;$$

and eliminating $d^2 x$ by means of this relation, there will result

$$\frac{d^2 z}{d t^2} = \sin a \left(\frac{F}{M} - g \sin a\right) \cdot$$

Multiplying by $2\,dz$, integrating and making the velocity zero when $z = h$, we have

$$\frac{d\,z^2}{d\,t^2} = V_z^2 = 2 \sin a \left(\frac{F}{M} - g \sin a\right) \cdot (z - h).$$

This gives

$$d\,t = \frac{1}{\sqrt{2 \sin a \left(\frac{F}{M} - g \sin a\right)}} \cdot \frac{d\,z}{\sqrt{z - h}};$$

and by integration, the time being zero when $z = h$,

$$h - z = \tfrac{1}{2} \sin a \left(g \cdot \sin a - \frac{F}{M}\right) \cdot t^2. \quad . \quad . \quad . \quad (a).$$

Again, all axes in the sphere through its centre, are principal axes; the sphere will only rotate about the movable axis y, in which case v_x and v_z will each be zero, and Equations (202) will give

$$B \cdot \frac{d\,v_y}{d\,t} = M_{,};$$

wherein,

$$B = M k_{,}^2; \quad \frac{d\,v_y}{d\,t} = \frac{d^2 \psi}{d\,t^2}; \quad M_{,} = F r;$$

r being the radius of the sphere.

Whence,

$$\frac{d^2 \psi}{d\,t^2} = \frac{F r}{M k_{,}^2}.$$

Multiplying by $2\,d\,\psi$, integrating, and making the angular velocity and the arc ψ vanish together,

$$\frac{d\,\psi^2}{d\,t^2} = \frac{2\,F r}{M k_{,}^2} \cdot \psi;$$

whence,

$$d\,t = \sqrt{\frac{M k_{,}^2}{2\,F r}} \cdot \frac{d\,\psi}{\sqrt{\psi}};$$

and by integration, making t and ψ vanish together,

$$\psi = \frac{1}{2} \frac{F \cdot r}{M \cdot k_{,}^2} \cdot t^2.$$

Also, because the length of path described in the direction of the plane is $r \cdot \psi$, we have, in addition,

$$h - z = r \cdot \psi \cdot \sin a;$$

and eliminating ψ from this and the above equation, there will result

$$t = \sqrt{\frac{2 M k_{,}^2}{F \cdot r^2 \cdot \sin a} (h - z)}. \quad . \quad . \quad . \quad . \quad . \quad (b)$$

Dividing Equation (a) by Equation (b), and solving with respect to F,

$$F = g \cdot \sin a \frac{M \cdot k_{,}^2}{k_{,}^2 + r^2}; \quad . \quad . \quad . \quad . \quad . \quad . \quad . \quad . \quad . \quad (c)$$

and this in Equation (b), gives

$$t = \sqrt{\frac{2 (h - z)}{g \cdot \sin^2 a} \cdot \frac{k_{,}^2 + r^2}{r^2}}. \quad . \quad . \quad . \quad . \quad . \quad . \quad . \quad (d)$$

If the sphere be homogeneous, then will

$$k^2_{,} = \tfrac{2}{5} r^2, \qquad \text{and} \qquad t = \sqrt{\frac{2 (h - z)}{g \cdot \sin^2 a}} \cdot \sqrt{\frac{7}{5}};$$

if the matter be all concentrated into the surface, then will

$$k_{,}^2 = \tfrac{2}{3} r^2 \qquad \text{and} \qquad t = \sqrt{\frac{2 (h - z)}{g \cdot \sin^2 a}} \cdot \sqrt{\frac{5}{3}};$$

which times are to one another as $\sqrt{21}$ to $\sqrt{25}$.

CONSTRAINED MOTION ABOUT A FIXED POINT.

§ 227.—If a body be retained by a *fixed* point, the fixed and what has been thus far regarded as a movable origin may both be taken at this point; in which case, $\delta x_{,}$, $\delta y_{,}$, $\delta z_{,}$, in Equation (40), will be zero, the first three terms of that general equation of equilibrium will reduce to zero independently of the forces, and the equilibrium will be satisfied by simply making

$$\left.\begin{aligned}\Sigma P(x\cos\beta - y\cos\alpha) - \Sigma m\cdot\frac{x\,.\,d^2 y - y\,d^2 x}{d\,t^2} = 0;\\ \Sigma P(z\cos\alpha - x\cos\gamma) - \Sigma m\cdot\frac{z\,.\,d^2 x - x\,d^2 z}{d\,t^2} = 0;\\ \Sigma P(y\cos\gamma - z\cos\beta) - \Sigma m\cdot\frac{y\,.\,d^2 z - z\,.\,d^2 y}{d\,t^2} = 0;\end{aligned}\right\} \cdot\cdot (365)$$

the accents being omitted because the elements m, m', &c., being referred to the same origin, x', y', z' will become x, y, z.

The motion of the body about the fixed point might be discussed both for the cases of incessant and of impulsive forces, but the discussion being in all respects similar to that relating to the motion about the centre of inertia, § 127 and § 173, we pass to

CONSTRAINED MOTION ABOUT A FIXED AXIS.

§ 228.—If the body be constrained to turn about a fixed axis, both origins may be taken upon, and the co-ordinate axis y to coincide with this axis; in which case $\delta x_{,}$, $\delta y_{,}$, $\delta z_{,}$, $\delta\varphi$ and $\delta\varpi$, in Equation (40), will be zero, and to satisfy the conditions of equilibrium, it will only be necessary for the forces to fulfil the condition,

$$\Sigma P(z\cos\alpha - x\cos\gamma) - \Sigma m\frac{z\,d^2 x - x\cdot d^2 z}{d\,t^2} = 0 \quad \cdot\cdot (366)$$

the accents being omitted for reasons just stated.

§ 229.—The only possible motion being that of rotation, let us transform the above equation so as to contain angular co-ordinates.

For this purpose we have, Equations (36),

$$x' = r''\sin\psi; \quad z' = r''\cos\psi \quad \cdot\;\cdot\;\cdot\;\cdot\;\cdot \quad (367)$$

in which r'' denotes the distance of the element m from the axis y.

Omitting the accents, differentiating and dividing by $d\,t$, we have

$$\frac{dx}{d\,t} = r\cos\psi\frac{d\psi}{d\,t}; \quad \frac{d\,z}{d\,t} = -\,r\sin\psi\cdot\frac{d\psi}{d\,t} \quad \cdot\;\cdot\;\cdot \quad (368)$$

Now,

$$\frac{z \cdot d^2 x}{d t^2} - \frac{x \cdot d^2 z}{d t^2} = \frac{1}{d t} \cdot d \left(z \cdot \frac{d x}{d t} - x \cdot \frac{d z}{d t} \right);$$

whence by substitution, Equations (367) and (368),

$$z \cdot \frac{d^2 x}{d t^2} - x \cdot \frac{d^2 z}{d t^2} = \frac{1}{d t} \cdot d \left(r^2 \cdot \frac{d \psi}{d t} \right) = r^2 \cdot \frac{d^2 \psi}{d t^2};$$

and since $\frac{d^2 \psi}{d t^2}$ must be the same for every element, we have, Equation (366),

$$\Sigma \, m \, r^2 \cdot \frac{d^2 \psi}{d t^2} = \Sigma P \, (z \cos \alpha - x \cos \gamma),$$

and

$$\frac{d^2 \psi}{d t^2} = \frac{\Sigma P \cdot (z \cos \alpha - x \cos \gamma)}{\Sigma \, m \, r^2} \quad . \quad . \quad . \quad . \quad (369)$$

That is to say, the angular acceleration of a body retained by a fixed axis, and acted upon by incessant forces, is equal to the moment of the impressed forces divided by the moment of inertia with reference to this axis.

Denoting the angular velocity by V_1, and the moment of inertia by I, we find, by multiplying Equation (369) by $2 \, d \psi$ and integrating,

$$I \, V_1^2 = 2 \int \Sigma \, P (z \cos \alpha - x \cos \gamma) \, d \psi + C,$$

and supposing the initial angular velocity to be V_1', we have

$$I (V_1^2 - V_1'^2) = 2 \int \Sigma \, P \, (z \cos \alpha - x \cos \gamma) \, d \psi.$$

But the second member is, § 107, twice the quantity of work about the fixed axis; whence the quantity of work performed between the two instants at which the body has any two angular velocities, is equal to half the difference of the squares of these velocities into the moment of inertia, or to half the living force gained or lost in the interval.

Now, $I = M k_{,}^2 = M_{,} . (1)^2 = M_{,}$; so that, the moment of inertia measures that mass which would, if concentrated on the arc ψ, have a living force equal to that of the body which actually rotates.

COMPOUND PENDULUM.

§ 230.—Any body suspended from a horizontal axis $A B$, about which it may swing with freedom under the action of its own weight, is called a *compound pendulum.*

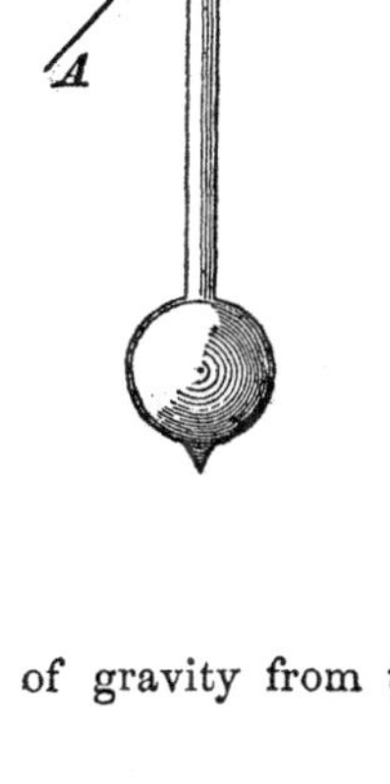

The elements of the pendulum being acted upon only by their own weights, we have

$$P = m g ; \quad P' = m' g, \text{ \&c. } ;$$

the axis of z being taken vertical and positive downwards,

$$\cos \alpha = \cos \alpha' = \text{\&c.} = 0 ;$$

$$\cos \gamma = \cos \gamma' = \text{\&c.} = 1,$$

and Equation (369) becomes

$$\frac{d^2 \psi}{d t^2} = - g \cdot \frac{\Sigma m x}{\Sigma m r^2} \cdot \cdot \cdot \cdot \quad (370)$$

Denote by e, the distance $A G$, of the centre of gravity from the axis; by ψ, the angle $H A G$, which $A G$ makes with the plane $y z$; by $x_{,}$, the distance of the centre of gravity from this plane; then will

$$x_{,} = e . \sin \psi ;$$

and from the principles of the centre of gravity,

$$\Sigma m x = M x_{,} = M . e . \sin \psi ;$$

which substituted above, gives

$$\frac{d^2 \psi}{d t^2} = - g . \frac{M . e . \sin \psi}{\Sigma m r^2} \cdot \cdot \cdot \cdot \cdot \cdot \quad (371)$$

Multiplying by $2\,d\psi$, and integrating,

$$\frac{d\psi^2}{dt^2} = 2g \cdot \frac{M.e.}{\Sigma m r^2} \cdot \cos\psi + C.$$

Denoting the initial value of ψ by α, we have

$$0 = 2g \cdot \frac{Me}{\Sigma m r^2} \cdot \cos\alpha + C;$$

whence,

$$\frac{d\psi^2}{dt^2} = 2g \cdot \frac{M.e}{\Sigma m r^2} (\cos\psi - \cos\alpha); \quad . \quad . \quad . \quad . \quad (372)$$

but

$$\cos\psi = 1 - \frac{\psi^2}{1.2} + \frac{\psi^4}{1.2.3.4} - \&c.$$

$$\cos\alpha = 1 - \frac{\alpha^2}{1.2} + \frac{\alpha^4}{1.2.3.4} - \&c.$$

and taking the value of ψ, so small that its fourth power may be neglected in comparison with radius, we have

$$\cos\psi - \cos\alpha = \frac{\alpha^2 - \psi^2}{2};$$

which substituted above, gives, after a slight reduction, and replacing $\Sigma m r^2$ by its value given in Equation (216),

$$dt = -\sqrt{\frac{k_{\prime}^2 + e^2}{e.g}} \cdot \frac{\frac{d\psi}{\alpha}}{\sqrt{1 - \frac{\psi^2}{\alpha^2}}};$$

the negative sign being taken because ψ is a decreasing function of the time.

Integrating, we have

$$t = \sqrt{\frac{k_{\prime}^2 + e^2}{e.g}} \cdot \cos^{-1}\frac{\psi}{\alpha} \quad . \quad . \quad . \quad . \quad . \quad . \quad . \quad . \quad (373)$$

The constant of integration is zero, because when $\psi = \alpha$, we have $t = 0$.

Making $\psi = -\alpha$, we have

$$t = \pi \sqrt{\frac{k_{,}^2 + e^2}{e \cdot g}}; \quad \ldots \ldots \quad (374)$$

which gives the time of one entire oscillation, and from which we conclude that the oscillations of the same pendulum will be isochronal, no matter what the lengths of the arcs of vibration, provided they be small.

If the number of oscillations performed in a given interval, say *ten* or *twenty minutes*, be counted, the duration of a single oscillation will be found by dividing the whole interval by this number.

Thus, let θ denote the time of observation, and N the number of oscillations, then will

$$t = \frac{\theta}{N} = \pi \cdot \sqrt{\frac{k_{,}^2 + e^2}{e \cdot g}};$$

and if the same pendulum be made to oscillate at some other location during the same interval θ, the force of gravity being different, the number N' of oscillations will be different; but we shall have, as before, g' being the new force of gravity,

$$\frac{\theta}{N'} = \pi \cdot \sqrt{\frac{k_{,}^2 + e^2}{e \cdot g'}}$$

Squaring and dividing the first by the second, we find

$$\frac{N'^2}{N^2} = \frac{g'}{g} \cdot \quad \ldots \ldots \quad (374)'$$

that is to say, the intensities of the force of gravity, at different places, are to each other as the squares of the number of oscillations performed in the same time, by the same pendulum. Hence, if the intensity of gravity at one station be known, it will be easy to find it at others.

§ 231.—From Equation (372), we have

$$\frac{d\psi^2}{dt^2} \cdot \Sigma m r^2 = 2 M \cdot g \cdot e (\cos \psi - \cos \alpha); \quad \ldots \quad (375)$$

and making

$$\frac{d\psi}{dt} = V_1; \quad \Sigma m r^2 = I; \quad e(\cos\psi - \cos\alpha) = H;$$

we have

$$I \, . \, V_1^2 = 2M \, . \, g \, . \, H; \quad . \quad . \quad . \quad . \quad . \quad . \quad . \quad (376)$$

in which H, denotes the vertical height passed over by the centre of gravity, and from which it appears that the pendulum will come to rest whenever ψ becomes equal to α, on either side of the vertical plane through the axis.

§ 232.—If the whole mass of the pendulum be conceived to be concentrated into a single point, the centre of gravity must go there also, and if this point be connected with the axis by a medium without weight and inertia, it becomes a *simple pendulum*. Denoting the distance of the point of concentration from the axis by l, we have

$$k_{\prime} = 0; \quad e = l,$$

which reduces Equation (374) to

$$t = \pi \cdot \sqrt{\frac{l}{g}}. \quad . \quad . \quad . \quad . \quad . \quad . \quad . \quad . \quad (377)$$

If the point be so chosen that

$$\sqrt{\frac{l}{g}} = \sqrt{\frac{k_{\prime}^2 + e^2}{e \, . \, g}};$$

or,

$$l = \frac{k_{\prime}^2 + e^2}{e}; \quad . \quad . \quad . \quad . \quad . \quad . \quad (378)$$

the simple and compound pendulum will perform their oscillations in the same time. The former is then called the *equivalent simple pendulum;* and the point of the compound pendulum into which the mass may be concentrated to satisfy this condition of equal duration, is called the *centre of oscillation*. A line through the centre of oscillation and parallel to the axis of suspension, is called an *axis of oscillation*.

§ 233.—The axes of oscillation and of suspension are reciprocal. Denote the length of the equivalent simple pendulum when the compound pendulum is inverted and suspended from its axis of oscillation, by l', and the distance of this latter axis from the centre of gravity by e', then will

$$l = e + e' \quad \text{or} \quad e' = l - e;$$

and, Equation (378),

$$l' = \frac{k_{,}^2 + e'^2}{e'} = \frac{k_{,}^2 + (l - e)^2}{l - e};$$

and replacing l, by its value in Equation (378), we find

$$l' = \frac{k_{,}^2 + e^2}{e} = l.$$

That is, if the old axis of oscillation be taken as a new axis of suspension, the old axis of suspension becomes the new axis of oscillation. This furnishes an easy method for finding the length of an equivalent simple pendulum.

Differentiating Equation (378), regarding l and e as variable, we have

$$\frac{d\,l}{d\,e} = \frac{e^2 - k_{,}^2}{e^2},$$

and if l be a minimum,

$$\frac{d\,l}{d\,e} = 0 = \frac{e^2 - k_{,}^2}{e^2};$$

whence,

$$e = k_{,}.$$

But when l is a minimum, then will t be a minimum, Equation (377). That is to say, the time of oscillation will be a minimum when the axis of suspension passes through the principal centre of *gyration*, and the time will be longer in proportion as the axis recedes from that centre.

Let A and A' be two acute parallel prismatic axes firmly connected with the pendulum, the acute edges being turned towards each other. The oscillation may be made to take place about either axis by simply inverting the pendulum. Also, let M be a sliding mass capable of being retained in any position by the clamp-screw H. For any assumed position of M, let the principal radius of gyration be $G\,C$; with G as a centre, $G\,C$ as radius, describe the circumference $C\,S\,S'$. From what has been explained, the time of oscillation about either axis will be shortened as it approaches, and lengthened as it recedes from this circumference, being a minimum, or least possible, when on it. By moving the mass M, the centre of gravity, and therefore the gyratory circle of which it is the centre, may be thrown towards either axis. The pendulum *b o b* being made heavy, the centre of gravity may be brought so near one of the axes, say A', as to place the latter within the gyratory circumference, keeping the centre of this circumference between the axes, as indicated in the figure. In this position, it is obvious that any motion in the mass M would at the same time either shorten or lengthen the duration of the oscillation about both axes, but unequally, in consequence of their unequal distances from the gyratory circumference.

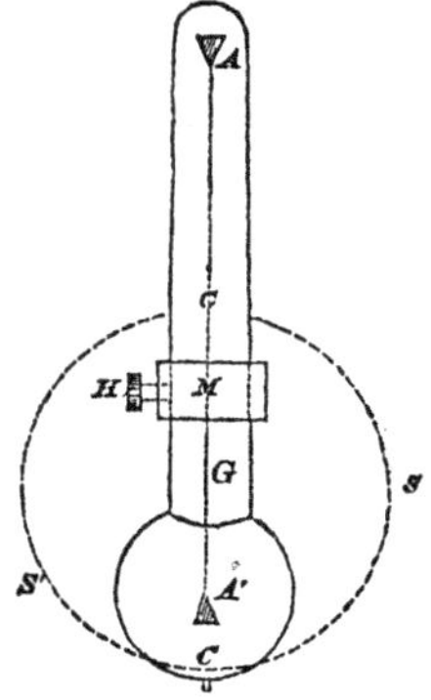

The pendulum thus arranged, is made to vibrate about each axis in succession during equal intervals, say an hour or a day, and the number of oscillations carefully noted; if these numbers be the same, the distance between the axes is the length l, of the equivalent simple pendulum; if not, then the weight M must be moved towards that axis whose number is the least, and the trial repeated till the numbers are made equal. The distance between the axes may be measured by a scale of equal parts.

§ 234.—From this value of l, we may easily find that of the *simple second's pendulum;* that is to say, the simple pendulum which will perform its vibration in one second. Let N, be the number of vibrations performed in one hour by the compound pendulum whose equivalent simple pendulum is l; the number performed in the same time by the second's pendulum, whose length we will denote by l', is of course 3600, being the number of seconds in 1 hour, and hence,

$$\frac{1^h}{N} = T = \pi \sqrt{\frac{l}{g}},$$

$$\frac{1^h}{3600^s} = T' = \pi \sqrt{\frac{l'}{g}};$$

and because the force of gravity at the same station is constant, we find, after squaring and dividing the second equation by the first,

$$l' = \frac{l \cdot N^2}{(3600^s)^2} \quad \ldots \ldots \quad (379)$$

Such is, in outline, the beautiful process by which KATER determined the length of the simple second's pendulum at the Tower of London to be 39,13908 inches, or 3,26159 feet.

As the force of gravity at the same place is not supposed to change its intensity, this length of the simple second's pendulum must remain forever invariable; and, on this account, the English have adopted it as the basis of their system of *weights* and *measures.* For this purpose, it was simply necessary to say that the $\frac{1}{3,26159}^{\text{th}}$ part of the *simple second's pendulum at the Tower of London* shall be *one English foot*, and all linear dimensions at once result from the relation they bear to the foot; that the *gallon* shall contain $\frac{231}{1728}^{\text{th}}$ of a cubic foot, and all measures of *volume* are fixed by the relations which other volumes bear to the gallon; and finally, that a *cubic* foot of distilled water at the temperature of sixty degrees Fahr. shall weigh *one thousand ounces*, and all weights are fixed by the relation they bear to the ounce.

§ 235.—It is now easy to find the apparent force of gravity at London; that is to say, the force of gravity as affected by the centrifugal force and the oblateness of the earth. The time of oscillation being one second, and the length of the simple pendulum 3,26159 feet, Equation (377) gives

$$1 = \pi \sqrt{\frac{\overset{ft.}{3{,}26159}}{g}};$$

whence,

$$g = \pi^2 (3{,}26159) = (3{,}1416)^2 . (3{,}26159) = 32{,}1908 \text{ feet.}$$

From Equation (377), we also find, by making t one second,

$$g = \pi^2 l,$$

and assuming

$$l = x + y \cos 2\psi,$$

we have

$$\frac{g}{\pi^2} = x + y \cos 2\psi \quad \cdot \cdot \cdot \cdot \cdot \cdot \cdot \quad (380)$$

Now starting with the value for g at London, and causing the same pendulum to vibrate at places whose latitudes are known, we obtain, from the relation given in Equation (374)′, the corresponding values of g, or the force of gravity at these places; and these values and the corresponding latitudes being substituted successively in Equation (380), give a series of Equations involving but two unknown quantities, which may easily be found by the method of least squares.

In this way it has been ascertained that

$$\pi^2 . x = 32{,}1808 \quad \text{and} \quad \pi^2 . y = -\ 0{,}0821;$$

whence, generally,

$$g = \overset{f}{32{,}1808} - \overset{f}{0{,}0821} \cos 2\psi; \quad \cdot \cdot \cdot \cdot \quad (381)$$

and substituting this value in Equation (377), and making $t = 1$, we find

$$l = \overset{f}{3{,}26058} - 0{,}008318 \cos 2\psi \quad \cdot \cdot \cdot \cdot \quad (382)$$

Such is the length of the simple second's pendulum at any place of which the latitude is ψ.

If we make $\psi = 40^\circ\ 42'\ 40''$, the latitude of the City Hall of New York, we shall find

$$l = \overset{ft.}{3{,}25938} = \overset{in.}{39{,}11256}.$$

§ 236.—The principles which have just been explained, enable us to find the moment of inertia of any body turning about a fixed axis, with great accuracy, no matter what its figure, density, or the distribution of its matter. If the axis do not pass through its centre of gravity, the body will, when deflected from its position of equilibrium, oscillate, and become, in fact, a compound pendulum; and denoting the length of its equivalent simple pendulum by l, we have, after multiplying Equation (378) by M,

$$M \cdot l \cdot e = M\,(k_{,}^{2} + e^{2}) = \Sigma\, m\, r^{2}; \quad . \quad . \quad . \quad . \quad (383)$$

or since

$$M = \frac{W}{g},$$

$$\frac{W}{g} \cdot l \cdot e = \Sigma\, m\, r^{2}, \quad . \quad . \quad . \quad . \quad . \quad . \quad . \quad . \quad (384)$$

in which W denotes the weight of the body.

Knowing the latitude of the place, the length l' of the simple second's pendulum is known from Equation (382); and counting the number N of oscillations performed by the body in one hour Equation (379) gives

$$l = \frac{l' \cdot (3600)^2}{N^2}.$$

To find the value of e, which is the distance of the centre of gravity from the axis, attach a spring or other balance to any point of the body, say its lower end, and bring the centre of gravity to a horizontal plane through the axis, which position will be indicated by the maximum reading of the balance. Denoting by a, the distance from the axis C to the point of support R,

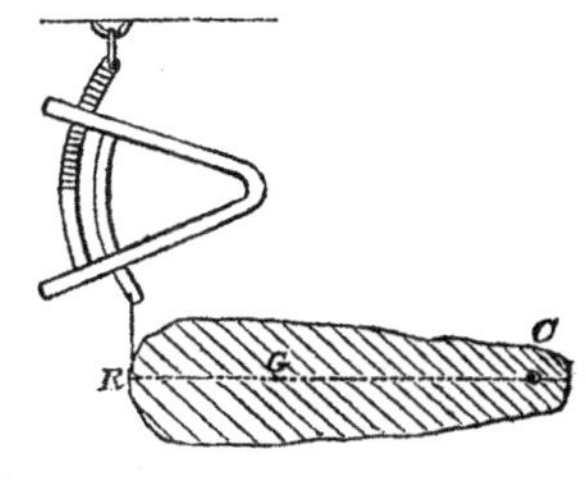

and by b, the maximum indication of the balance, we have, from the principle of moments,

$$b\,a = W e.$$

The distance a, may be measured by a scale of equal parts. Substituting the values of W, e and l in the expression for the moment of inertia, Equation (384), we get

$$\frac{b \,.\, a \,.\, l' \,.\, (3600)^2}{g \,.\, N^2} = I. \quad \cdot \quad \cdot \quad \cdot \quad \cdot \quad \cdot \quad \cdot \quad \cdot \quad (385)$$

If the axis pass through the centre of gravity, as, for example, in the *fly-wheel*, it will not oscillate; in which case, take Equation (383), from which we have

$$M k_{,}^2 = M \,.\, l \,.\, e - M e^2.$$

Mount the body upon a parallel axis A, not passing through the centre of gravity, and cause it to vibrate for an hour as before; from the number of these vibrations and the length of the simple second's pendulum, the value of l may be found; M is known, being the weight W divided by g; and e may be found by direct measurement, or by the aid of the spring balance, as already indicated; whence $k_{,}$ becomes known.

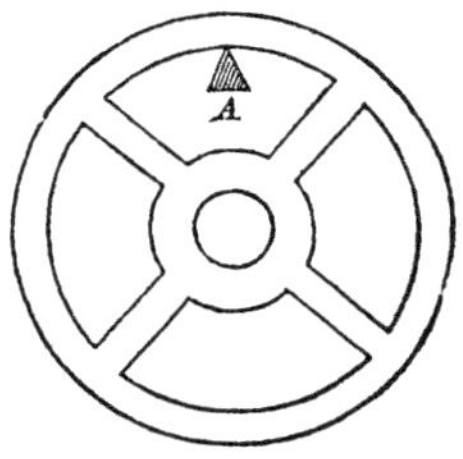

MOTION OF A BODY ABOUT AN AXIS UNDER THE ACTION OF IMPULSIVE FORCES.

§ 237.—If the forces be impulsive, we may, § 170, replace in Equation (366) the second differential co-efficients of x, y, z, by the first differential co-efficients of the same variables, which will reduce it to

$$\Sigma P \big(z \cos \alpha - x \cos \gamma \big) - \Sigma m \cdot \frac{z\,dx - x\,dz}{dt} = 0;$$

and replacing dx and dz, by their values in Equations (368), we find

$$\frac{d\psi}{dt} = \frac{\Sigma P(z\cos\alpha - x\cos\gamma)}{\Sigma m r^2} \quad \ldots \ldots \quad (386)$$

That is, *the angular velocity of a body retained by a fixed axis, and subjected to the simultaneous action of impulsive forces, is equal to the sum of the moments of the impressed forces divided by the moment of inertia with reference to this axis.*

BALLISTIC PENDULUM.

§ 238.—In artillery, the initial velocity of projectiles is ascertained by means of the *ballistic pendulum*, which consists of a mass of matter suspended from a horizontal axis in the shape of a knife-edge, after the manner of the compound pendulum. The bob is either made of some unelastic substance, as wood, or of metal provided with a large cavity filled with some soft matter, as dirt, which receives the projectile and retains the shape impressed upon it by the blow

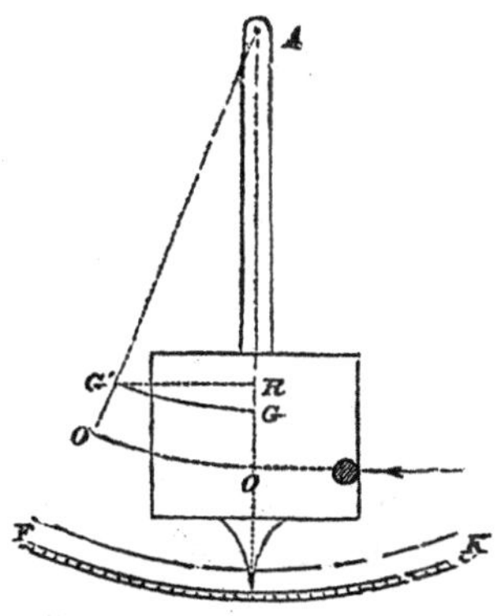

Denote by V and m, the initial velocity and mass of the ball; V_1 the angular velocity of the ballistic pendulum the instant after the blow, I and M its moment of inertia and mass. Also let l represent the distance of the centre of oscillation of the pendulum from the axis A. That no motion may be lost by the resistance of the axis arising from a shock, the ball must be received in the direction of a line passing through this centre and perpendicular to the plane of the axis and line AO. With this condition, Eq. (386) gives

$$\frac{d\psi}{dt} = V_1 = \frac{m \,.\, V \,.\, l}{\Sigma m r^2} = \frac{m \,.\, V \,.\, \frac{k_{,}^2 + e^2}{e}}{(M + m)(k_{,}^2 + e^2)} = \frac{m V}{(M + m) \,.\, e},$$

whence

$$V = \frac{M+m}{m} \cdot e \, . \, V_1 \, ;$$

and supposing the angular velocity communicated to the pendulum to be equal to that acquired by falling from rest through the initial arc α, in Equation (372), we have, from that equation and Equation (216), by writing e for d,

$$V_1 = \sqrt{2g \cdot \frac{(M+m)\,e}{(M+m)\,(k_{\prime}^2 + e^2)} \cdot (1 - \cos \alpha)} = 2\sqrt{\frac{g \, . \, e}{k_{\prime}^2 + e^2}} \cdot \sin \tfrac{1}{2}\alpha,$$

and Eq. (374),

$$\frac{\pi}{t} = \sqrt{\frac{g\,e}{k_{\prime}^2 + e^2}},$$

which substituted above gives

$$V_1 = 2\,\frac{\pi}{t} \cdot \sin \tfrac{1}{2}\alpha \, ;$$

and this in the value for V gives, after substituting for the ratio of the masses that of their weights,

$$V = 2\,\frac{W+w}{w} \cdot \frac{\pi}{t} \cdot e \cdot \sin \tfrac{1}{2}\alpha \quad . \quad . \quad . \quad . \quad . \quad . \quad . \quad . \quad (387).$$

From this equation we may find the initial velocity V; and for this purpose, it will only be necessary to have the duration of a single oscillation, and the amplitude of the arc described by the centre of gravity of the pendulum. The process for finding the time has been explained. To find the arc, it will be sufficient to attach to the lower extremity of the pendulum a pointer, and to fix, on a permanent stand below, a circular graduated groove, whose centre of curvature is at A; the groove being filled with some soft substance, as tallow, the pointer will mark on it the extent of the oscillation. Knowing thus the arc α, and the value of e, found as already described, § 236, we have V.

THE GUN PENDULUM.

This consists of a gun suspended from a horizontal axis. The shot is fired from the gun, and its velocity is inferred from the recoil, as in the Ballistic Pendulum. The forces measured by the quantities of motion developed by the expansive action of the exploded powder, must be in equilibrio. Make

$V =$ velocity of the ball on leaving the gun,
$nV =$ average velocity of the inflamed powder,
$V_{,} =$ angular velocity of pendulum on parting from shot,
$W_g =$ weight of gun pendulum,
$W_b =$ " ball and wad,
$W_c =$ " the charge of powder and bag,
$W_p =$ " " " of powder alone,
$\delta =$ diameter of bore,
$d =$ diameter of ball,
$\varepsilon =$ distance of axis of bore from axis of suspension.

The quantity of motion in ball and wad, on leaving the gun, will be $\frac{W_b}{g} V$; the corresponding pressure on the bottom of the gun is to that which generates this motion, as the area of a cross-section of the bore is to that of a great circle of the ball. Again, the blast of the powder will continue its action on the gun after the ball leaves it. Let this action be proportional to the charge of powder. The moment of the force impressed upon the pendulum, in reference to the axis of suspension, will be given by Eqs. (384) and (229); and taking the moments of the other forces in reference to the same axis, we have

$$\frac{W_g}{g} \cdot V_{,} \cdot l \cdot e - \frac{W_b}{g} \cdot V \cdot \varepsilon \cdot \frac{\delta^2}{d^2} - \frac{W_c}{g} \cdot n \cdot V \cdot \varepsilon - \frac{W_p}{g} \cdot n' \cdot V \cdot \varepsilon = 0;$$

in which n', like n, is a constant to be determined by experiment; and from which we find

$$V = \frac{W_g \cdot V_{,} \cdot l \cdot e}{W_b \cdot \varepsilon \cdot \frac{\delta^2}{d^2} + n W_c \cdot \varepsilon + n' W_p \cdot \varepsilon}.$$

The living force with which the pendulum separates from the ball must equal twice the work performed by the weight while the centre of gravity is moving to the highest point; whence

$$V_{\prime}^{2} \cdot \frac{W_g}{g} \cdot l \cdot e = 2 W_g \cdot e \cdot \text{versine } \alpha = 4 W_g \cdot e \cdot \sin^2 \tfrac{1}{2} \alpha,$$

in which α denotes the greatest inclination of e to the vertical. Whence

$$V_{\prime} = 2 \sqrt{\frac{g}{l}} \cdot \sin \tfrac{1}{2} \alpha ;$$

which substituted above gives,

$$V = \frac{2 \cdot W_g \cdot \frac{e}{\varepsilon} \cdot \sqrt{g \cdot l}}{W_b \cdot \frac{\delta^2}{d^2} + n W_c + n' W_p} \cdot \sin \tfrac{1}{2} \alpha \quad . \quad . \quad . \quad . \quad (388).$$

The methods for finding e and α are the same as in the ballistic pendulum. To find n and n', fire the ball from the gun into the ballistic pendulum; the effect upon the latter will give the initial velocity V. Repeat as often as may be thought desirable, and with different charges. The corresponding initial velocities substituted in Eq. (388), will give as many equations as trials. These equations will contain only n and n' as unknown quantities, which may be found by the method of least squares. For full and valuable information on this subject, consult Mordecai's "Experiments on Gunpowder."

PART II.

MECHANICS OF FLUIDS.

INTRODUCTORY REMARKS.

§ 239.—The physical condition of every body depends upon the relation subsisting among its molecular forces. When the attractions prevail greatly over the repulsions, the particles are held firmly together, and the body is *solid*. In proportion as the difference between these two sets of forces becomes less, the body is softer, and its figure yields more readily to external pressure. When these forces are equal, the particles will yield to the slightest force, the body will, under the action of its own weight, and the resistance of the sides of a vessel into which it is placed, readily take the figure of the latter, and is *liquid*. Finally, when the repulsive exceed the attractive forces, the elements of the body tend to separate from each other, and require either the application of some extraneous force or to be confined in a closed vessel to keep them together; the body is then a *gas*. In the vast range of relation among the molecular forces, from that which distinguishes a solid to that which determines a gas or vapor, bodies are found in all possible conditions—solids run imperceptibly into liquids, and liquids into gases. Hence all classification of bodies founded on their physical properties alone, must, of necessity, be arbitrary.

§ 240.—Any body whose elementary particles admit of motion

among each other, is called a *fluid*—such as water, wine, mercury, the air, and, in general, liquids and gases; all of which are distinguished from solids by the great mobility of their particles among themselves. This distinguishing property exists in different degrees in different liquids—it is greatest in the ethers and alcohol; it is less in water and wine; it is still less in the oils, the sirups, greases, and melted metals, that flow with difficulty, and rope when poured into the air. Such fluids are said to be *viscous*, or to possess *viscosity*. Finally, a body may approach so closely both a solid and liquid, as to make it difficult to assign it a place among either class, as *paste*, *putty*, and the like.

§ 241.—Fluids are divided in mechanics into two classes, viz.: *compressible* and *incompressible*. The term incompressible cannot, in strictness of propriety, be applied to any body in nature, all being more or less compressible; but the enormous power required to change, in any sensible degree, the volumes of liquids, seems to justify the term, when applied to them in a restricted sense. The *gases* are highly compressible. All *liquids* will, therefore, be regarded as incompressible; the *gases* as compressible.

§ 242.—The most important and remarkable of the gaseous bodies is the atmosphere. It envelops the entire earth, reaches far beyond the tops of our highest mountains, and pervades every depth from which it is not excluded by the presence of solids or liquids. It is even found in the pores of these latter bodies. It plays a most important part in all natural phenomena, and is ever at work to influence the motions within it. It is essentially composed of *oxygen* and *nitrogen*, in a state of mechanical mixture. The former is a supporter of combustion, and, with the various forms of carbon, is one of the principal agents employed in the development of mechanical power.

The existence of gases is proved by a multitude of facts. Contained in an inflexible and impermeable envelope, they resist pressure like solid bodies. Gas, in an inverted glass vessel plunged into water, will not yield its place to the liquid, unless some avenue of escape be provided for it. Tornadoes which uproot trees, overturn houses, and devastate entire districts, are but air in motion. Air opposes, by its inertia, the motion of other bodies through it, and this opposition is called its resistance. Finally, we know that wind is employed as a motor to turn mills and to give motion to ships of the largest kind.

§ 243.—In the discussions which are to follow, fluids will be considered as without viscosity; that is to say, the particles will be supposed to have the utmost freedom of motion among each other. Such fluids are said to be *perfect.* The results deduced upon the hypothesis of perfect fluidity will, of course, require modification when applied to fluids possessing sensible viscosity. The nature and extent of these modifications can be known only from experiments.

MARIOTTE'S LAW.

§ 244.—Gases readily contract into smaller volumes when pressed externally; they as readily expand and regain their former dimensions when the pressure is removed. They are therefore both *compressible* and *elastic.*

It is found by experiment, that the change in volume is, for a constant temperature, always directly proportional to the change of pressure. The density of the same body is inversely proportional to the volume it occupies. If, therefore, P denote the pressure upon a unit of surface which will produce, at a given temperature, say 0° Centr., a density equal to unity, and D any other density, and p the pressure upon a unit of surface which will, at the same temperature of the gas, produce this density, then, according to the experiments above referred to, will

$$p = P \, . \, D \quad \ldots \ldots \ldots \quad (389)$$

This law was investigated by Boyle and Mariotte, and is known as *Mariotte's Law.* By experiments made at Paris, it was found that this law obtains, when air, in its ordinary condition, is condensed 27 and rarefied 112 times.

LAW OF THE PRESSURE, DENSITY, AND TEMPERATURE.

§ 245.—Under a *constant pressure*, all bodies are expanded by heat; under a *constant volume*, their elastic force is increased by the same agent. Experiment has shown that the laws of these changes for gases are expressed by

$$p = P \,.\, D \,.\, (1 + \alpha\,\theta)\,; \quad \cdot \quad \cdot \quad \cdot \quad \cdot \quad \cdot \quad \cdot \quad \cdot \quad (390)$$

in which p denotes the pressure upon a unit of surface, D the density of the gas, θ the difference between the actual and some standard temperature, and α a constant which is equal to $\frac{1}{273} = 0{,}003665$ when the standard is $0°$ centr., and θ is expressed in units of that scale.

First supposing D and θ variable and p constant; then p and θ variable and D constant, Equation (390) gives

$$\frac{d\,D}{d\,\theta} = -\frac{\alpha\,.\,D}{1 + \alpha\,\theta}\,; \quad \frac{d\,p}{d\,\theta} = \frac{\alpha\,p}{1 + \alpha\,\theta} \quad \cdot \quad \cdot \quad \cdot \quad \cdot \quad (a)$$

The quantity of heat, denoted by q, necessary to change the temperature θ degrees from the assumed standard, will be a function of p, D, θ; but because of Equation (390,) we may write

$$q = f\,(D, p) \quad \cdot \quad \cdot \quad \cdot \quad \cdot \quad \cdot\,(b)$$

The increment of heat which will raise a body's temperature one degree, is called its *specific heat*. The specific heat being the increment of q for each unit of θ, if c denote the specific heat when the pressure is constant, and $c_{,}$ that when the density is constant, then will

$$c = \frac{d\,q}{d\,\theta} = \frac{d\,q}{d\,D} \cdot \frac{d\,D}{d\,\theta}\,; \quad c_{,} = \frac{d\,q}{d\,\theta} = \frac{d\,q}{d\,p} \cdot \frac{d\,p}{d\,\theta}\,;$$

or, Equations (a),

$$c = -\frac{d\,q}{d\,D} \cdot \frac{\alpha\,.\,D}{1 + \alpha\,\theta}\,; \quad c_{,} = \frac{d\,q}{d\,p} \cdot \frac{\alpha\,.\,p}{1 + \alpha\,\theta}\,;$$

and by division, making $c = \gamma\,.\,c_{,}$,

$$D \cdot \frac{d\,q}{d\,D} + \gamma \cdot p \cdot \frac{d\,q}{d\,p} = 0,$$

in which γ, denotes the ratio of the specific heat of the gas at a constant pressure to that at a constant density. This ratio is known from experiment to be constant for atmospheric air, and is probably so for all gases. The experiments of Desormes and

Clements make its value 1,3482; those of Gay-Lussac and Walter 1,3748; and those of Dulong on perfectly dry air 1,421. Regarding γ as constant, the integration of the foregoing equation gives

$$q = f\left(\frac{p^{\frac{1}{\gamma}}}{D}\right)$$ (See Appendix No. 3.)

in which f, denotes any arbitrary function of the quantity within the parenthesis, and from which, denoting the inverse functions by F, we may write

$$p = D^{\gamma} . F(q) \quad . \quad . \quad . \quad . \quad . \quad . \quad . \quad . \quad . \quad (c)$$

From Equation (390), we have

$$\theta = \frac{p}{\alpha . P . D} - \frac{1}{\alpha} = \frac{1}{\alpha . P} \cdot D^{\gamma - 1} . F(q) - \frac{1}{\alpha} \cdot \quad . \quad . \quad . \quad (d)$$

Sudden compression increases, and a sudden expansion decreases the temperature of bodies, and if q remain the same, while suddenly p, D, θ, become p', D', θ', we have

$$p' = D'^{\gamma} . F(q), \quad . \quad . \quad (e) \qquad \theta' = \frac{1}{\alpha . P} D'^{\gamma - 1} . F(q) - \frac{1}{\alpha} \cdot \quad . \quad . \quad (g)$$

Eliminating $F(q)$ first from Equations (c) and (e), and then from Equations (d) and (g), we have, replacing γ and α by their numerical values,

$$p' = p\left(\frac{D'}{D}\right)^{1,421} \quad . \quad . \quad . \quad . \quad (391)$$

$$\theta' = (273 + \theta)\left(\frac{D'}{D}\right)^{0,421} - 273 \quad . \quad . \quad . \quad . \quad (392)$$

These equations give the relation between the densities, elastic forces, and the temperatures of a gas suddenly compressed or dilated, and retaining the quantity of its heat unchanged.

The pressure being constant, make, in Eq. (390), $\theta = 0$, $D = D_{,}$, and divide same equation by the result; we find $D = D_{,} \div (1 + \alpha\theta)$. Make $p = D_{m} \cdot h_{,,} \cdot g'$ = weight of a column of mercury at standard temperature T, and resting on a base unity, in Lat. 45°, where gravity is g'. These in Eq. (389) give, after writing 0,00204 for α, and $t° - 32°$ for θ,

$$P = \frac{D_{m} . h_{,,} . g'}{D_{,}} \cdot [1 + (t° - 32°) . 0,00204] \cdot \quad . \quad . \quad (393)$$

If the temperature of the mercury vary from the sta

and become T' then will D_m also vary and become D'_m, and to exert the same pressure $h_{\prime\prime}$ must have a new value h, and such that

$$D_m . h_{\prime\prime} . g' = D'_m . h . g'.$$

Mercury expands or contracts 0,0001001th part of its entire volume for each degree of Fahr. by which it increases or diminishes its temperature. And as the density of the same body varies inversely as its volume, we have

$$D'_m = D_m [1 + (T - T') \cdot 0,0001001]$$

which substituted above gives

$$h_{\prime\prime} = h [1 + (T - T') . 0,0001001] \cdot \cdot \cdot \cdot \cdot (394)$$

EQUAL TRANSMISSION OF PRESSURE.

§ 246.—Let EHL, represent a closed vessel of any shape, with which two piston tubes AB' and DC' communicate, each tube being provided with a piston that fits it accurately and which may move within it with the utmost freedom. The vessel being filled with any fluid, let forces P and P', be applied, the former perpendicularly to the piston AB, and the latter in like direction to the piston CD, and suppose these forces in equilibrio, which they may be, since the fluid cannot escape. Now let the piston AB be moved to the position $A'B'$; the piston CD will take some new position, as $C'D'$. And denoting by s and s', the distances AA' and CC', respectively, we have, from the principle of virtual velocities,

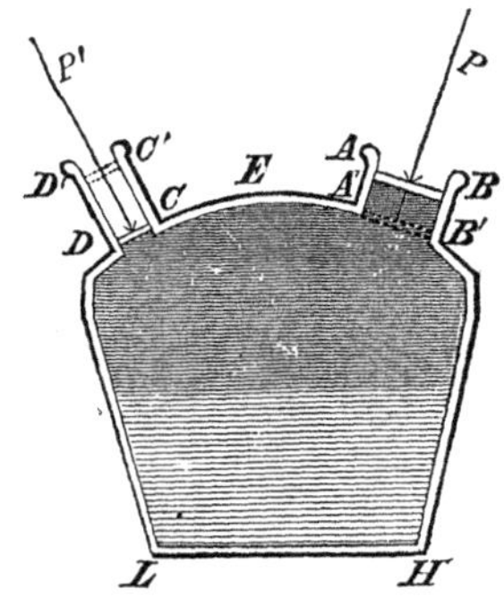

$$P s = P' s'.$$

Denote the area of the piston AB by a, and that of the piston CD by a', then will the volume of the fluid which was thrust from the tube AB', be measured by $a . s$, and that which entered the tube

$b\,C'$, will be measured by $a'\,s'$. But the pressure upon the pistons and the temperature remaining the same, the entire volume of the fluid in the vessel and tubes will be unchanged. Hence,

$$a\,s = a'\,s';$$

dividing the equation above by this one, we have

$$\frac{P}{a} = \frac{P'}{a'}. \qquad (396)$$

That is to say, *two forces applied to pistons which communicate freely with each other through the intervention of some confined fluid, will be in equilibrio when their intensities are directly proportional to the areas of the pistons upon which they act.*

This result is wholly independent of the relative dimensions and positions of the pistons; and hence we conclude that *any pressure communicated to one or more elements of a fluid mass in equilibrio, is equally transmitted throughout the whole fluid in every direction.* This law which is fully confirmed by experiment, is known as the principle of *equal transmission of pressure.*

§ 247.—Let a become the superficial unit, say a square inch or square foot, then will P be the pressure applied to a unit of surface, and, Equation (396),

$$P' = P\,a'. \qquad (397)$$

That is, the pressure transmitted to any portion of the surface of the containing vessel, will be equal to that applied to the unit of surface multiplied by the area of the surface to which the transmission is made.

§ 248.—Since the elements of the fluid are supposed in equilibrio, the pressure transmitted to the surface through the elements in contact with it, must, § 217 and Equations (332), be normal to the surface. That is, *the pressure of a fluid against any surface, acts always in the direction of the normal.*

MOTION OF THE FLUID PARTICLES.

§ 249.—The particles of a fluid having the utmost freedom of motion among one another, all the forces applied at each particle must be in equilibrio. Regarding the general Equation (40) as applicable to a single particle, whose co-ordinates are x, y, z, we shall have

$$x = x_{\prime}, \quad y = y_{\prime}, \quad z = z_{\prime},$$

and supposing the particle to have simply a motion of translation, we also have

$$\delta \varphi = 0; \quad \delta \psi = 0; \quad \delta \varpi = 0;$$

and that equation becomes

$$\left.\begin{array}{l} \left(\Sigma P \cos \alpha - m \cdot \dfrac{d^2 x}{d t^2}\right) \delta x \\ + \left(\Sigma P \cos \beta - m \cdot \dfrac{d^2 y}{d t^2}\right) \delta y \\ + \left(\Sigma P \cos \gamma - m \cdot \dfrac{d^2 z}{d t^2}\right) \delta z \end{array}\right\} = 0;$$

whence, upon the principle of indeterminate co-efficients,

$$\left.\begin{array}{l} \Sigma P \cos \alpha - m \cdot \dfrac{d^2 x}{d t^2} = 0; \\ \Sigma P \cos \beta - m \cdot \dfrac{d^2 y}{d t^2} = 0; \\ \Sigma P \cos \gamma - m \cdot \dfrac{d^2 z}{d t^2} = 0. \end{array}\right\} \quad \cdots \cdots \quad (398)$$

Now the terms $\Sigma P \cos \alpha$, $\Sigma P \cos \beta$ and $\Sigma P \cos \gamma$, are each composed of two distinct parts, viz.: 1st., the component of the resultant of the forces applied directly to the particle; and 2d., the component of the pressure transmitted to it from a distance, arising from the forces impressed upon other particles.

Denote by X, Y and Z, the accelerations, in the directions of the axes x, y, z, respectively, due to the forces applied directly to the particle; then m, being the mass of the particle, the components of the forces directly impressed will be

$$mX; \quad mY; \quad mZ.$$

The pressure transmitted will depend upon the particle's place, and will be a function of its co-ordinates of position. Denote by p, the pressure upon a unit of surface, on the supposition that every point of the unit sustains a pressure equal to that communicated to the particle from a distance; then, for a given time, will

$$p = F(x, y, z).$$

Conceive each particle of the fluid to consist of a small rectangular parallelopipedon whose faces are parallel to the co-ordinate planes, and whose contiguous edges at the time t, are dx, dy and dz; and let x, y, z, be the co-ordinates of the molecule in the solid angle nearest the origin of co-ordinates. Then would the difference of pressure on the opposite faces, which are parallel to the plane zy, were these faces equal to unity, be

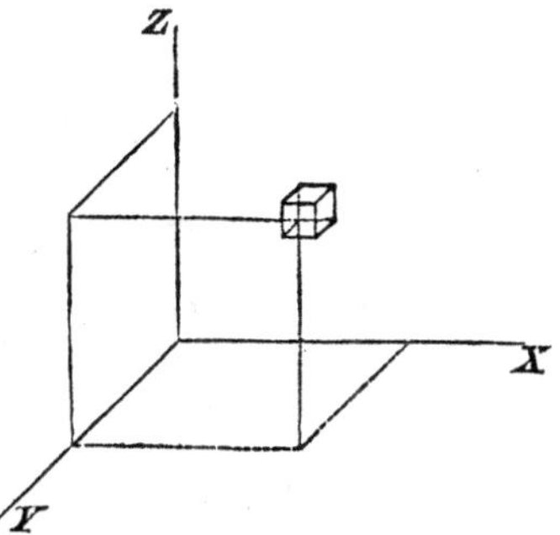

$$F(\overline{x + dx}, y, z,) - F(x, y, z,) = \frac{dp}{dx} \cdot dx;$$

and upon the actual faces whose dimensions are each $dz . dy$, this difference becomes, Equation (397),

$$\frac{dp}{dx} \cdot dx \cdot dy \cdot dz.$$

In like manner will the difference of the pressures transmitted to the opposite faces parallel to the planes zx and xy, be, respectively,

$$\frac{dp}{dy} \cdot dy \cdot dz \cdot dx, \quad \text{and} \quad \frac{dp}{dz} \cdot dz \cdot dx \cdot dy.$$

These pressures being normal to the surfaces to which they are respectively applied, they will act, the first in the direction of x, the second in the direction of y, and the third in the direction of z. And as these differences alone determine that portion of the motion due to the transmitted pressures, we have

$$\Sigma P \cos \alpha = m X - \frac{d p}{d x} \cdot d x \,.\, d y \,.\, d z \,;$$

$$\Sigma P \cos \beta = m Y - \frac{d p}{d y} \cdot d y \,.\, d x \,.\, d z \,;$$

$$\Sigma P \cos \gamma = m Z - \frac{d p}{d z} \cdot d z \,.\, d x \,.\, d y.$$

Denote by D the density of the mass m, then will, Equation (1)′,

$$m = D \,.\, d x \,.\, d y \,.\, d z,$$

and by substitution, Equations (398) become

$$\left.\begin{aligned} \frac{1}{D} \cdot \frac{d p}{d x} &= X - \frac{d^2 x}{d t^2} \,; \\ \frac{1}{D} \cdot \frac{d p}{d y} &= Y - \frac{d^2 y}{d t^2} \,; \\ \frac{1}{D} \cdot \frac{d p}{d z} &= Z - \frac{d^2 z}{d t^2} \,; \end{aligned}\right\} \quad \cdot \cdot \cdot \cdot \cdot \quad (399)$$

Denote by u, v and w, the velocities of the molecule whose co-ordinates are $x y z$, parallel to the axes x, y, z, respectively, at the time t. Each of these will be a function of the time and the co-ordinates of the molecule's place; and, reciprocally, each co-ordinate will be a function of t, u, v and w; whence, Equations (12) and (13),

$$\frac{d^2 x}{d t^2} = \frac{d u}{d t} = \left(\frac{d u}{d t}\right) \cdot \frac{d t}{d t} + \frac{d u}{d x} \cdot \frac{d x}{d t} + \frac{d u}{d y} \cdot \frac{d y}{d t} + \frac{d u}{d z} \cdot \frac{d z}{d t} \,;$$

and replacing $\frac{d x}{d t}$, $\frac{d y}{d t}$, $\frac{d z}{d t}$, by their values u, v, w, respectively, we have

$$\frac{d^2 x}{d t^2} = \left(\frac{d u}{d t}\right) + \frac{d u}{d x} \cdot u + \frac{d u}{d y} \cdot v + \frac{d u}{d z} \cdot w \,;$$

in the same way,

$$\frac{d^2 y}{d t^2} = \left(\frac{d v}{d t}\right) + \frac{d v}{d x} \cdot u + \frac{d v}{d y} \cdot v + \frac{d v}{d z} \cdot w,$$

$$\frac{d^2 z}{d t^2} = \left(\frac{d w}{d t}\right) + \frac{d w}{d x} \cdot u + \frac{d w}{d y} \cdot v + \frac{d w}{d z} \cdot w$$

which, substituted in Equations (399), give

$$\left.\begin{aligned} \frac{1}{D} \cdot \frac{dp}{dx} &= X - \left(\frac{du}{dt}\right) - \frac{du}{dx} \cdot u - \frac{du}{dy} \cdot v - \frac{du}{dz} \cdot w; \\ \frac{1}{D} \cdot \frac{dp}{dy} &= Y - \left(\frac{dv}{dt}\right) - \frac{dv}{dx} \cdot u - \frac{dv}{dy} \cdot v - \frac{dv}{dz} \cdot w; \\ \frac{1}{D} \cdot \frac{dp}{dz} &= Z - \left(\frac{dw}{dt}\right) - \frac{dw}{dx} \cdot u - \frac{dw}{dy} \cdot v - \frac{dw}{dz} \cdot w. \end{aligned}\right\} \quad (400)$$

Here are three equations involving five unknown quantities, viz.: u, v, w, p and D, which are to be found in terms of x, y, z and t.

Two other equations may be found from these considerations, viz: the velocity in the direction of x, of the molecule whose co-ordinates are $x\,y\,z$, is u; the velocity of the molecule in the angle of the parallelopipedon at the opposite end of the side $d\,x$, at the same time, is

$$u + \frac{d u}{d x} \cdot d x;$$

and hence the relative velocity of the two molecules is

$$u + \frac{d u}{d x} \cdot d x - u = \frac{d u}{d x} \cdot d x.$$

At the time t, the length of the edge joining these molecules is $d x$, and at the end of the time $t + d\,t$, this length will be

$$d x + \frac{d u}{d x} \cdot d x \,.\, d t = d x \left(1 + \frac{d u}{d x} \cdot d t\right);$$

the second term being the distance by which the molecules in question approach toward or recede from one another in the time $d\,t$.

In the same way the edges of the parallelopipedon which at the time t, were dy and dz, become respectively,

$$dy + \frac{dv}{dy} \cdot dy \,.\, dt = dy\,(1 + \frac{dv}{dy} \cdot dt);$$

$$dz + \frac{dw}{dz} \cdot dz \,.\, dt = dz\,(1 + \frac{dw}{dz} \cdot dt);$$

and the volume of the parallelopipedon, which at the time t, was $dx \,.\, dy \,.\, dz$, becomes at the time $t + dt$,

$$dx \,.\, dy \,.\, dz\,(1 + \frac{du}{dx} \cdot dt) \cdot (1 + \frac{dv}{dy} \cdot dt) \cdot (1 + \frac{dw}{dz} \cdot dt).$$

The density, which was D, at the time t, being a function of $x\,y\,z$ and t, becomes at the time $t + dt$,

$$D + \frac{dD}{dt} \cdot dt + \frac{dD}{dx} \cdot dx + \frac{dD}{dy} \cdot dy + \frac{dD}{dz} \cdot dz;$$

which may be put under the form,

$$D + \left(\frac{dD}{dt} + \frac{dD}{dx} \cdot \frac{dx}{dt} + \frac{dD}{dy} \cdot \frac{dy}{dt} + \frac{dD}{dz} \cdot \frac{dz}{dt}\right) dt;$$

and replacing

$$\frac{dx}{dt}, \quad \frac{dy}{dt}; \quad \frac{dz}{dt},$$

by their values u, v, w, respectively,

$$D + \left(\frac{dD}{dt} + \frac{dD}{dx} \cdot u + \frac{dD}{dy} \cdot v + \frac{dD}{dz} \cdot w\right) dt.$$

Multiplying this by the volume above, we have for the mass of the parallelopipedon, which was

$$D \,.\, dx \,.\, dy \,.\, dz,$$

at the time t, the value,

$$\left[D + \left(\frac{dD}{dt} + \frac{dD}{dx} \cdot u + \frac{dD}{dy} \cdot v + \frac{dD}{dz} \cdot w\right) dt\right]$$

$$\times\, dx \,.\, dy \,.\, dz \left(1 + \frac{du}{dx} \cdot dt\right) \cdot \left(1 + \frac{dv}{dy} \cdot dt\right) \cdot \left(1 + \frac{dw}{dz} \cdot dt\right)$$

at the time $t + dt$.

But these masses must be equal, since the quantity of matter is unchanged. Equating them, striking out the common factors, performing the multiplication, and neglecting the second powers of the differentials, we have

$$D\left(\frac{du}{dx}+\frac{dv}{dy}+\frac{dw}{dz}\right)+\frac{dD}{dt}+\frac{dD}{dx}\cdot u+\frac{dD}{dy}\cdot v+\frac{dD}{dz}\cdot w=0. \quad (401)$$

This is called the *Equation of continuity of the fluid*. It expresses the relation between the velocity of the molecules and the density of the fluid, which are necessarily dependent upon each other. This is a fourth equation.

§ 250.—If the fluid be compressible, then will the fifth equation be given by the relation,

$$F(D, p) = 0, \quad \cdots \cdots \quad (402)$$

as is illustrated in the particular instance of Mariotte's law, Equation (389). The form of the function designated by the letter F will depend upon the nature of the fluid.

§ 251.—If the fluid be incompressible, the total differential of D will be zero, and

$$\frac{dD}{dt}+\frac{dD}{dx}\cdot u+\frac{dD}{dy}\cdot v+\frac{dD}{dz}\cdot w=0; \quad \cdots \quad (403)$$

and consequently, the equation of continuity, Equation (401), becomes,

$$\frac{du}{dx}+\frac{dv}{dy}+\frac{dw}{dz}=0; \quad \cdots \cdots \quad (404)$$

and we have for the determination of u, v, w, D and p, the five Equations (400), (403), (404).

§ 252.—These equations admit of great simplification in the case of an *incompressible homogeneous fluid* when $u \cdot dx + v \cdot dy + w \cdot dz$, is a perfect differential. For if we make

$$u\,dx + v\,dy + w\,dz = d\varphi,$$

then from the partial differentials will

$$u = \frac{d\varphi}{dx}; \quad v = \frac{d\varphi}{dy}; \quad w = \frac{d\varphi}{dz}; \quad . \quad . \quad . \quad . \quad (405)$$

which, in Equation (404), gives for the equation of continuity,

$$\frac{d^2\varphi}{dx^2} + \frac{d^2\varphi}{dy^2} + \frac{d^2\varphi}{dz^2} = 0; \quad . \quad . \quad . \quad . \quad . \quad . \quad (406)$$

by the integration of which the function φ may be found.

Differentiating the values of u, v and w above, we have

$$du = \frac{d^2\varphi}{dx}; \quad dv = \frac{d^2\varphi}{dy}; \quad dw = \frac{d^2\varphi}{dz}.$$

Eliminating u, v, w, du, dv and dw, from Equation (400), by means of the values of these quantities above, we have

$$\frac{1}{D}\cdot\frac{dp}{dx} = X - \frac{d^2\varphi}{dx\cdot dt} - \frac{d\varphi}{dx}\cdot\frac{d^2\varphi}{dx^2} - \frac{d\varphi}{dy}\cdot\frac{d^2\varphi}{dx.dy} - \frac{d\varphi}{dz}\cdot\frac{d^2\varphi}{dx.dz},$$

$$\frac{1}{D}\cdot\frac{dp}{dy} = Y - \frac{d^2\varphi}{dy.dt} - \frac{d\varphi}{dx}\cdot\frac{d^2\varphi}{dy.dx} - \frac{d\varphi}{dy}\cdot\frac{d^2\varphi}{dy^2} - \frac{d\varphi}{dz}\cdot\frac{d^2\varphi}{dy.dz};$$

$$\frac{1}{D}\cdot\frac{dp}{dz} = Z - \frac{d^2\varphi}{dz.dt} - \frac{d\varphi}{dx}\cdot\frac{d^2\varphi}{dz.dx} - \frac{d\varphi}{dy}\cdot\frac{d^2\varphi}{dz.dy} - \frac{d\varphi}{dz}\cdot\frac{d^2\varphi}{dz^2}.$$

Multiplying the first by dx, the second by dy, the third by dz, and adding, we find,

$$\frac{1}{D}\cdot dp = Xdx + Ydy + Zdz - d\frac{d\varphi}{dt} - \tfrac{1}{2}d\left[\left(\frac{d\varphi}{dx}\right)^2 + \left(\frac{d\varphi}{dy}\right)^2 + \left(\frac{d\varphi}{dz}\right)^2\right] \quad (407)$$

From which, by integration, may be found the pressure at any point of an incompressible fluid mass in motion, when Equation (406) is the equation of continuity.

§ 253.—When the excursions of the molecules are small, the second powers of the velocities may be neglected, which will reduce Equation (407) to

$$\frac{1}{D}\cdot dp = Xdx + Ydy + Zdz - d\frac{d\varphi}{dt}. \quad . \quad . \quad (408)$$

§ 254.—If the condition expressed by Equation (406) be not fulfilled, then we must have recourse to Equation (404) to find the pressure.

§ 255.—Resuming Equation (401), which appertains to a compressible fluid, retaining the condition that

$$u\,dx + v\,dy + w\,dz = d\varphi$$

is a perfect differential, and from which, therefore,

$$u = \frac{d\varphi}{dx}; \quad v = \frac{d\varphi}{dy}; \quad w = \frac{d\varphi}{dz}; \quad . \quad . \quad . \quad (409)$$

we obtain by substitution,

$$D\left\{\frac{du}{dx} + \frac{dv}{dy} + \frac{dw}{dz}\right\} + \frac{dD}{dt} + \frac{dD}{dx}\frac{d\varphi}{dx} + \frac{dD}{dy}\frac{d\varphi}{dy} + \frac{dD}{dz}\frac{d\varphi}{dz} = 0.$$

If the excursions of the molecules from their places of rest be very small, both the change of density and velocity will be so small that the products which constitute the last three terms of this equation may be neglected, and the equation of *continuity* becomes

$$D\cdot\left(\frac{du}{dx} + \frac{dv}{dy} + \frac{dw}{dz}\right) + \frac{dD}{dt} = 0;$$

and replacing du, dv and dw, by their values from Equations (409), and dividing by D, we find

$$\frac{d\log D}{dt} + \frac{d^2\varphi}{dx^2} + \frac{d^2\varphi}{dy^2} + \frac{d^2\varphi}{dz^2} = 0. \quad . \quad . \quad . \quad (410)$$

from which, and Eq. (408), the equation connecting the extraneous forces with the co-ordinates $x\,y\,z$, and that expressive of Mariotte's law, the function φ may be found, then the value of D, and finally that of p.

The excursions being small, if we impose the additional condition that the molecules of the fluid are not acted upon by extraneous forces, in which case the motions can only arise from some arbitrary initial disturbance; then, Equation (408),

$$\frac{1}{D} \cdot dp = - d \frac{d\varphi}{dt} = - \frac{d^2\varphi}{dt},$$

and by Mariotte's law,

$$p = P.D = a^2.D \quad \dots \dots \quad (411)$$

from which

$$dp = a^2 dD \quad \dots \dots \quad (412)$$

and the above may be written, after dividing by dt,

$$\frac{a^2}{dt} \cdot \frac{dD}{D} = a^2 \cdot \frac{d \log D}{dt} = - \frac{d^2\varphi}{dt^2} \quad \dots \dots \quad (413)$$

which, in Equation (410), gives

$$\frac{d^2\varphi}{dt^2} = a^2 \left(\frac{d^2\varphi}{dx^2} + \frac{d^2\varphi}{dy^2} + \frac{d^2\varphi}{dz^2} \right) \quad \dots \quad (414)$$

From this Equation the function φ is to be determined, then the value of D, from Equation (410), and that of p, from either of the Equations (411) or (413).

§ 256.—Conceive a homogeneous elastic fluid to be disturbed at one of its points by the sudden expansion or contraction of the element there situated. This will break up the equilibrium of the surrounding molecular forces at that point, the particles adjacent will move to restore the uniformity of density, and an expanding disturbance will proceed outward from this as a centre. Take the origin at the point, and denote the distance of any particle involved in the disturbance, at any time t, subsequent to the disturbance by r, then will

$$x^2 + y^2 + z^2 = r^2.$$

Denote the velocity of the particle, supposed in the direction of r, by ζ; then will

$$u = \zeta \cdot \frac{x}{r}; \quad v = \zeta \cdot \frac{y}{r}; \quad w = \zeta \cdot \frac{z}{r}.$$

Differentiating the first of the above equations, we have

$$x\,d\,x + y\,d\,y + z\,d\,z = r\,.\,d\,r.$$

Substituting the values of x, y, and z from the second, third, and fourth, there will result

$$u\,d\,x + v\,d\,y + w\,d\,z = \zeta\,.\,d\,r;$$

so that this satisfies the condition of the first member being an exact differential; and, therefore, $d\,\varphi = \zeta\,.\,d\,r$; or

$$\zeta = \frac{d\,\varphi}{d\,r}.$$

And hence

$$u = \frac{d\,\varphi}{d\,x} = \frac{d\,\varphi}{d\,r}\cdot\frac{x}{r}; \quad v = \frac{d\,\varphi}{d\,y} = \frac{d\,\varphi}{d\,r}\cdot\frac{y}{r}; \quad w = \frac{d\,\varphi}{d\,z} = \frac{d\,\varphi}{d\,r}\cdot\frac{z}{r};$$

differentiating,

$$\frac{d^2\,\varphi}{d\,x^2} = \frac{d^2\,\varphi}{d\,r^2}\cdot\frac{x^2}{r^2} + \frac{d\,\varphi}{d\,r}\cdot\frac{y^2+z^2}{r^3};$$

$$\frac{d^2\,\varphi}{d\,y^2} = \frac{d^2\,\varphi}{d\,r^2}\cdot\frac{y^2}{r^2} + \frac{d\,\varphi}{d\,r}\cdot\frac{z^2+x^2}{r^3};$$

$$\frac{d^2\,\varphi}{d\,z^2} = \frac{d^2\,\varphi}{d\,r^2}\cdot\frac{z^2}{r^2} + \frac{d\,\varphi}{d\,r}\cdot\frac{x^2+y^2}{r^3};$$

and these values, substituted in Equation (414), give

$$\frac{d^2\,\varphi}{d\,t^2} = a^2\left(\frac{d^2\,\varphi}{d\,r^2} + \frac{2}{r}\cdot\frac{d\,\varphi}{d\,r}\right);$$

which may be written,

$$\frac{d^2\,r\,\varphi}{d\,t^2} = a^2\,.\,\frac{d^2\,r\,\varphi}{d\,r^2} \quad \ldots\ldots\ldots \quad (414)'$$

of which the integral is, Appendix No. IV.,

$$r\,\varphi = F(r + a\,t) + f(r - a\,t);$$

and in which F and f denote any arbitrary functions whatever. From this we have

$$\varphi = \frac{1}{r}\left[F(r + a\,t) + f(r - a\,t)\right] \quad . \quad . \quad . \quad . \quad (415)$$

Taking the first differential coefficient of φ with respect to r, and replacing its value by ζ,

$$\zeta = \frac{1}{r} \cdot \left[F'(r + a\,t) + f'(r - a\,t)\right] - \frac{1}{r^2}\left[F(r + a\,t) + f(r - a\,t)\right].$$

For any considerable distance from the origin, the second term may be omitted in comparison with the first; and there will result, after squaring and multiplying by m, the mass of the moving particle,

$$m\,\zeta^2 = \frac{m}{r^2} \cdot \left[F'(r + a\,t) + f'(r - a\,t)\right]^2 \quad . \quad . \quad . \quad (416)$$

The first member is the living force of the moving particle, or double the quantity of work it may impress upon the organs of sense exposed to its action. The effect it may produce will, therefore, all other things being equal, vary inversely as the square of its distance from the place of primitive disturbance. Equations (415) and (416) are employed in discussing the theory of sound.

EQUILIBRIUM OF FLUIDS.

§ 257.—If the fluid be at rest, then will

$$\frac{d^2 x}{d\,t^2} = 0; \quad \frac{d^2 y}{d\,t^2} = 0; \quad \frac{d^2 z}{d\,t^2} = 0;$$

and Equations (399) become

$$\left.\begin{aligned} \frac{dp}{dx} &= D.X; \\ \frac{dp}{dy} &= D.Y; \\ \frac{dp}{dz} &= D.Z. \end{aligned}\right\} \quad . \quad . \quad . \quad . \quad . \quad . \quad . \quad . \quad (417)$$

§ 258.—Multiplying the first by dx, the second by dy, the third by dz, and adding we find,

$$dp = D\,(X\,dx + Y\,dy + Z\,dz); \quad . \quad . \quad . \quad . \quad (418)$$

and by integration,

$$p = \int D \,.\, (X\,dx + Y\,dy + Z\,dz); \quad . \quad . \quad . \quad . (419)$$

whence, in order that the value of p may be possible for any point of the fluid mass, the product of the density by the function $X\,dx + Y\,dy + Z\,dz$, must be an exact differential of a function of the three independent variables x, y, z. Reciprocally, when this condition is fulfilled, not only will the pressure at any point become known by substituting its co-ordinates, but the Equations (417), will be satisfied, and the fluid will be in equilibrio.

§ 259.—Conceiving those points of the fluid which experience equal pressures to be connected by, indeed to form a surface, then in passing from one point to another of this surface, we shall have $dp = 0$, and

$$X\,dx + Y\,dy + Z\,dz = 0, \quad . \quad . \quad . \quad . \quad . \quad (420)$$

which is obviously the differential equation of the surface.

Dividing this by $R\,ds$, in which mR, denotes the resultant of the forces which act upon any particle, and ds, the element of any curve upon the surface passing through the particle, we have

$$\frac{X}{R} \cdot \frac{dx}{ds} + \frac{Y}{R} \cdot \frac{dy}{ds} + \frac{Z}{R} \cdot \frac{dz}{ds} = 0; \quad . \quad . \quad . \quad . \quad (421)$$

whence the resultant of the forces acting upon any one of the elements of a surface of equal pressure, is normal to that surface. This is the characteristic of what is called a *level surface*, which may be defined to be any surface which cuts at right angles the direction of the resultant of the forces which act upon its particles.

§ 260.—If Equation (420) be integrated, we have

$$\int (X\,dx + Y\,dy + Z\,dz) = C, \quad . \quad . \quad . \quad . \quad (422)$$

in which C is the constant of integration. The magnitudes of this constant must result from the dimensions of the surface, or from the volume of the fluid it envelops. By giving it different and suitable values, we may start from a single particle and proceed outwards to the boundary of the fluid, and if the successive values differ by a small quantity, we shall have a series of *level concentric* strata.

The last possible value for C will determine the exterior or bounding surface of the fluid; because this surface being free, the pressure upon it will be zero; the differential of the pressure from one point to another will, therefore, be zero, and the differential equation will be that numbered (420), or that of equal pressure. Every free surface of a fluid in equilibrio is, therefore, a level surface.

§ 261.—Putting Equation (418) under the form

$$\frac{dp}{D} = X\,dx + Y\,dy + Z\,dz, \quad \cdots \quad (423)$$

we see that whenever the second member is an exact differential, p must be a function of D, since the first member must also be an exact differential. Making, therefore,

$$p = F(D), \quad \cdots \quad (424)$$

in which F denotes any function whatever, the above equation becomes

$$\frac{d\,F(D)}{D} = X\,dx + Y\,dy + Z\,dz; \quad \cdots \quad (425)$$

but for a level surface or stratum, the second member reduces to zero; whence,

$$dF(D) = 0;$$

and by integration,

$$F(D) = C;$$

whence, not only will each level stratum be subjected to an equal pressure over its entire surface, but it will also have the same density throughout.

§ 262.—If the fluid be homogeneous and of the same temperature throughout, then will D be constant, and the condition of equilibrium simply requires that the function $X\,dx + Y\,dy + Z\,dz$, Equation (419), shall be an exact differential of the three independent variables x, y, z, and when this is not the case, the equilibrium will be impossible, no matter what the shape of the fluid mass, and though it were contained in a closed vessel.

But the function above referred to is, § 133, always an exact differential for the forces of nature, which are either attractions or repulsions, whose intensities are functions of the distances from the centres through which they are exerted. And to insure the equilibrium, it will only be necessary to give the exterior surface such shape as to cut perpendicularly the resultants of the forces which act upon the surface particles. This is illustrated in the simple example of a tumbler of water, or, on a larger scale, by ponds and lakes which only come to rest when their upper surfaces are normal to the resultant of the force of gravity and the centrifugal force arising from the earth's rotation on its axis.

In the case of a heterogeneous fluid subjected to the action of a central force, its equilibrium requires that it be arranged in concentric level strata, each stratum having the same density throughout. And the equilibrium will be stable when the centre of gravity of the whole is the lowest possible, § 138, and hence the denser strata should be the lowest.

When the fluid is incompressible, the density may be any function whatever of the co-ordinates of place. It may be continuous or discontinuous. When it is given, the value of the pressure is found from Equation (419).

§ 263.—In compressible fluids the density and pressure are connected by law, and the former is no longer arbitrary.

Dividing Equation (418) by Equation (389), we have

$$\frac{dp}{p} = \frac{X\,dx + Y\,dy + Z\,dz}{P} \quad . \quad . \quad . \quad . \quad (425)'$$

Integrating,

$$\log p = \int \frac{X\,dx + Y\,dy + Z\,dz}{P} + \log C; \quad . \quad . \quad . \quad (426)$$

denoting the base of the Naperian system by e, we have

$$p = C \, . \, e^{\int \frac{X dx + Y dy + Z dz}{P}}; \quad . \quad . \quad . \quad . \quad . \quad (427)$$

and this substituted in Equation (389), gives

$$D = \frac{C . e^{\int \frac{X dx + Y dy + Z dz}{P}}}{P}. \quad . \quad . \quad . \quad . \quad . \quad (428)$$

These equations determine the pressure and density.

For any surface of constant pressure, the exponent of e, in Equation (427), must be constant, its differential must, therefore, be zero, and all the consequences deduced from Equation (420) will follow; that is, when the fluid is at rest, it must be arranged in level strata, each stratum having the same density throughout, with the addition that the law of the varying density must be continuous by the requirements of Mariotte's law.

If the temperature vary, then will P vary, and in order that Equation (425)′ may be an exact differential, P must be a function of $x\,y\,z$, and hence, Equations (427) and (428), when p is constant, D will be constant; that is, each level stratum must be of uniform temperature throughout.

It is obvious that the atmosphere can never be in equilibrio; for the sun heating unequally its different portions as the earth turns upon its axis, the layers of equal pressure, density and temperature can never coincide. Hence, those perpetual currents of air known as the *trade winds*, and the periodical monsoons; also, the sea and land breezes, variable winds, &c., &c.

§ 264.—Rest is a relative term; when applied to a particle of a fluid mass, it means that that particle preserves unaltered its place in regard to the other particles; a condition consistent with a bodily movement of the entire mass.

If a liquid mass turn uniformly about an axis, the preceding equations will make known its permanent figure. For this purpose it will be sufficient to join to the forces X, Y, Z, the centrifugal force

Take the axis z as the axis of rotation; denote the angular velocity by φ, and the distance of the particle M from the axis z by r; then will

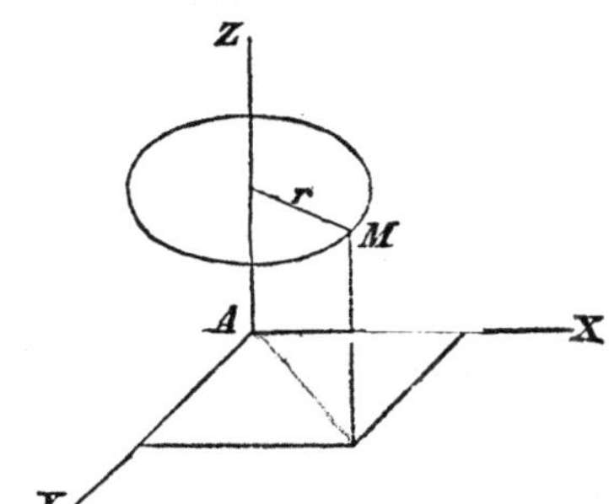

$$r^2 = x^2 + y^2;$$

the centrifugal force of M regarded as a unit of mass, will be

$$r\,\varphi^2,$$

and its components in the direction of x and y, respectively,

$$r \, . \, \varphi^2 \, . \, \frac{x}{r} = x\,\varphi^2;$$

$$r \, . \, \varphi^2 \, . \, \frac{y}{r} = y\,\varphi^2.$$

and these in Equation (418), give

$$dp = D \, . \, (X dx + Y dy + Z dz + \varphi^2 \, . \, x \, dx + \varphi^2 \, y \, . \, dy) \cdot \cdot (429)$$

When the second member is an exact differential, the permanent form will be possible.

For the free surface $dp = 0$, and we have

$$X dx + Y dy + Z dz + \varphi^2 \, . \, x \, . \, dx + \varphi^2 y \, dy = 0 \cdot \cdot \cdot (430)$$

Example 1.—Let it be required to find the figure assumed by the free surface of a heavy and homogeneous fluid contained in an open vessel and rotating about a vertical axis.

Here,

$$X = 0; \quad Y = 0; \quad Z = -g;$$

and Equation (430) becomes

$$g\,dz = \varphi^2 (x\,dx + y\,dy).$$

Integrating,

$$z = \frac{\varphi^2}{2g}(x^2 + y^2) + C; \quad . \quad . \quad . \quad . \quad . \quad (431)$$

which is the equation of a paraboloid whose axis is that of rotation.

To find the constant C, let the vessel be a right cylinder, with circular base, whose radius is a, and denote by h the height due to the velocity of the fluid at the circumference, then

$$a^2 \varphi^2 = 2 g h,$$

and

$$z = \frac{h \cdot r^2}{a^2} + C \quad \cdots \cdots \quad (432)$$

Denote by b the height of the liquid before the rotation; its volume will be $\pi a^2 . b$. Conceive the whole body of the liquid to be divided into concentric cylindrical layers, having for a common axis the axis of rotation. The base of any one of these layers will have for its area, neglecting $d r^2$, $2 \pi r . d r$, and for its volume, taking the origin of co-ordinates in the bottom of the vessel, $2 \pi r . d r . z$, which being integrated between the limits $r = 0$ and $r = a$, will give the whole volume of the fluid, and hence,

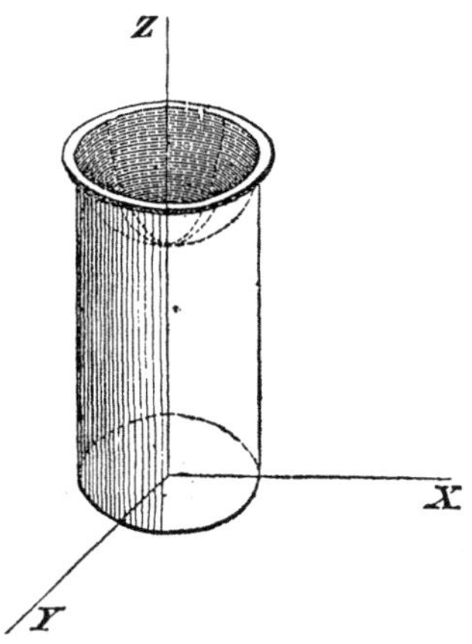

$$a^2 b = 2 \int_0^a z r . d r ;$$

replacing $r . d r$ by its value from Equation (432), and integrating between the limits $z = C$ and $z = h + C$, which are the values given by Equation (432) for $r = 0$ and $r = a$, we find

$$C = b - \tfrac{1}{2} h,$$

and the equation of the upper surface becomes

$$z = \frac{h r^2}{a^2} + b - \tfrac{1}{2} h.$$

The least and greatest values for z, are $b - \frac{1}{2} h$ and $b + \frac{1}{2} h$, obtained by making $r = 0$ and $r = a$, so that the depression of the liquid at the axis is equal to its elevation at the surface of the cylindrical vessel, and is equal to half the height due to the velocity of the latter.

§ 265.—*Example* 2.—Let the fluid elements be attracted to the centre of the mass by a force varying inversely as the square of the distance. Take the origin at the centre; denote the distance to the particle m from that point by r, and the intensity of the attractive force at the unit's distance by k. Then will

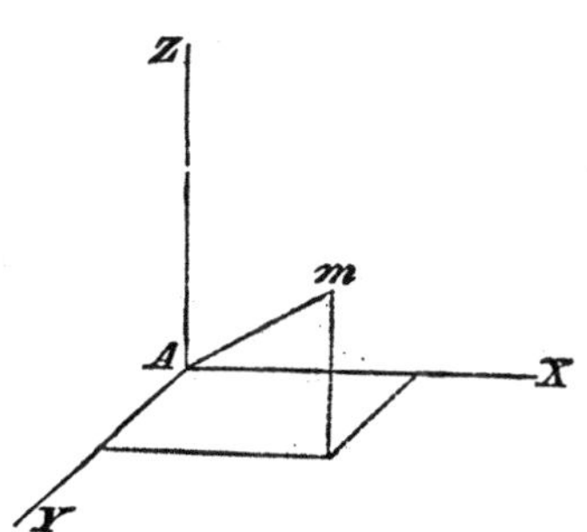

$$P = m\frac{k}{r^2}; \quad \cos\alpha = -\frac{x}{r}; \quad \cos\beta = -\frac{y}{r}; \quad \cos\gamma = -\frac{z}{r},$$

and

$$X = -\frac{kx}{r^3}; \quad Y = -\frac{ky}{r^3}; \quad Z = -\frac{kz}{r^3};$$

which in Equation (430), give

$$\frac{k}{r^3}(x\,dx + y\,dy + z\,dz) - \varphi^2(x\,dx + y\,dy) = 0,$$

or

$$\frac{k\,dr}{r^2} - \frac{\varphi^2}{2}d(x^2 + y^2) = 0,$$

and by integration,

$$\frac{k}{r} + \frac{\varphi^2}{2}(x^2 + y^2) = C;$$

making

$$x^2 + y^2 = r^2\cos^2\theta,$$

in which θ denotes the angle made by r, with the plane xy,

$$\frac{k}{r} + \frac{\varphi^2}{2}\cdot r^2\cos^2\theta = C,$$

and denoting the distance from the origin to the point in which the free surface cuts the axis z by unity, we have, by making $\theta = 90°$,

$$\frac{k}{1} = C;$$

which substituted above, and solving with respect to $\cos^2\theta$, gives

$$\tfrac{1}{2}\varphi^2 \cdot \cos^2\theta = \frac{k(r-1)}{r^3} \quad \cdots\cdots \quad (434)$$

and making $r = 1 + u$, we have

$$\tfrac{1}{2}\varphi^2 \cdot \cos^2\theta = \frac{ku}{(1+u)^3}.$$

If the angular velocity be small, then will u be very small. Developing the second member with this supposition, and limiting the terms to the first power of u, we find

$$\tfrac{1}{2}\varphi^2 \cdot \cos^2\theta = k(u - 3u^2). \quad \cdots\cdots \quad (434)'$$

Neglecting $3u^2$, and replacing u by its value, viz.: $r - 1$, we have for a first approximation,

$$r = 1 + \frac{\varphi^2}{2k} \cdot \cos^2\theta.$$

From Equation (434)′, we find

$$u = \frac{\varphi^2 \cdot \cos^2\theta}{2k} + 3u^2,$$

and this in the equation

$$r = 1 + u,$$

gives

$$r = 1 + \frac{\varphi^2}{2k} \cdot \cos^2\theta + 3u^2;$$

and replacing u^2 by its approximate value $\frac{\varphi^4 \cdot \cos^4\theta}{4k^2}$, above, by neglecting $3u^2$, we have

$$r = 1 + \frac{\varphi^2}{2k} \cdot \cos^2\theta + \frac{3\varphi^4 \cdot \cos^4\theta}{4k^2},$$

for the polar equation of the meridian section.

Comparing this with the equation

$$r = \frac{1}{\sqrt{1 - e^2 \cos^2 \theta}} = 1 + \tfrac{1}{2} e^2 \cos^2 \theta + \tfrac{3}{8} \cdot e^4 \cdot \cos^4 \theta + \&c.,$$

they become identical by neglecting the higher powers and making

$$e = \sqrt{\frac{\varphi^2}{k}}.$$

The free surface of the fluid approximates therefore very closely to an ellipsoid of revolution of which the eccentricity of its meridian section is equal to the square root of the quotient arising from dividing the centrifugal force at the unit's distance from the axis of rotation, by the force of attraction at an equal distance from the centre.

PRESSURE OF HEAVY FLUIDS.

§ 266.—When a fluid contained in any vessel is acted upon by its own weight, if the axis z be taken vertical and positive downwards, then will

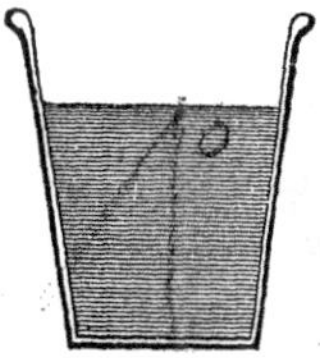

$$X = 0; \quad Y = 0; \quad Z = g;$$

and Equation (418) becomes, after integrating,

$$p = Dgz + C;$$

and assuming the plane xy to coincide with the upper surface of the fluid, which must, when in equilibrio, be horizontal, we have, by making $z = 0$,

$$p' = C;$$

in which p' denotes the pressure exerted upon the unit of the free surface. Whence,

$$p - p' = D.g.z. \quad \ldots\ldots\ldots \quad (435)$$

The first member is the pressure exerted upon a unit of surface, every point of which unit having a pressure equal to that sustained by the element whose co-ordinate is z.

If $p' = 0$, then will

$$p = Dgz; \quad \cdot \cdot \cdot \cdot \cdot \cdot \cdot \cdot \cdot (436)$$

and denoting by b the area of the surface pressed, and by db, the element of this surface, whose co-ordinate is z, we have, Equation (397), for the pressure upon this element denoted by $p_{,}$,

$$p_{,} = Dg \,.\, z \,.\, db,$$

and the same for any other element of the surface; whence, denoting the entire pressure by P, we shall have

$$P = \Sigma p_{,} = Dg \,.\, \Sigma z \,.\, db. \quad \cdot \cdot \cdot \cdot \cdot \cdot \cdot (437)$$

But if $z_{,}$ denote the co-ordinate of the centre of gravity of the entire surface b, then will, Equations (91),

$$\Sigma z \,.\, db = b z_{,},$$

and

$$P = Dg \,.\, b \,.\, z_{,}. \quad \cdot \cdot \cdot \cdot \cdot \cdot \cdot \cdot \cdot (438)$$

Now $b z_{,}$ is the volume of a right cylinder or prism, whose base is b, and altitude $z_{,}$; $Dg \,.\, b \,.\, z_{,}$ is the weight of this volume of the pressing fluid. Whence we conclude, that *the pressure exerted upon any surface by a heavy fluid is equal to the weight of a cylindrical or prismatic column of the fluid whose base is equal to the surface pressed, and whose altitude is equal to the distance of the centre of gravity of the surface below the upper surface of the fluid.*

When the surface pressed is horizontal, its centre of gravity will be at a distance from the upper surface equal to the depth of the fluid.

This result is wholly independent of the quantity of the pressing fluid, and depends solely upon the density of the fluid, its height, and the extent of the surface pressed.

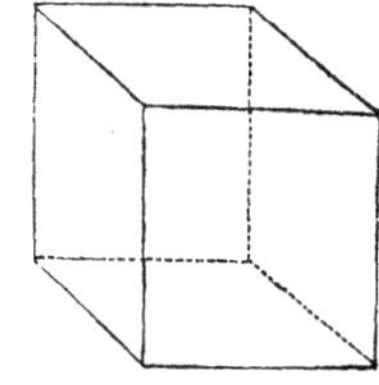

Example 1.—Required the pressure against the inner surface of a cubical vessel filled with water, one of its faces being horizontal. Call the edge of the cube a, the area of each face will be a^2, the distance of the centre of gravity of each [illegible] face below the upper surface will be $\frac{1}{2} a$, and that of the lower face a; whence, the principle of the centre of gravity gives,

$$z_{\prime} = \frac{4a^2 \times \frac{1}{2}a + a^2 \times a}{5a^2} = \frac{3}{5}a.$$

Again,

$$b = 5a^2;$$

and these, substituted in Equation (438), give

$$P = D \cdot g \cdot b \cdot z_{\prime} = D \cdot g \cdot 3a^3.$$

Now $Dg \times 1^3 = Dg$, is the weight of a cubic foot of water $= 62{,}5$ lbs., whence,

$$P = \overset{lbs.}{62{,}5} \times 3a^3.$$

Make $a = 7$ feet, then will

$$P = 62{,}5 \times 3 \times (7)^3 = \overset{lbs.}{64312{,}5}.$$

The weight of the water in the vessel is $62{,}5\ a^3$, yet the pressure is $62{,}5 \times 3a^3$, whence we see that the outward pressure to break the vessel, is three times the weight of the fluid.

Example 2.—Let the vessel be a sphere filled with mercury, and let its radius be R. Its centre of gravity is at the centre, and therefore below the upper surface at the distance R. The surface of the sphere being equal to that of four of its great circles, we have

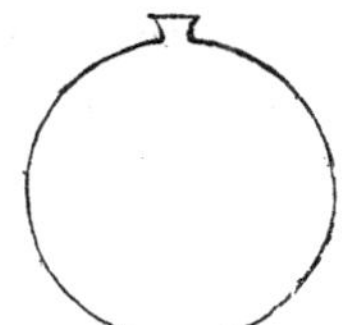

$$b = 4\pi R^2;$$

whence,

$$b \cdot z_{\prime} = 4\pi R^3;$$

and, Equation (438),

$$P = 4\pi \cdot D \cdot g \cdot R^3.$$

The quantity $Dg \times 1^3 = Dg$, is the weight of a cubic foot of mercury $= 843{,}75$ lbs., and therefore, substituting the value of $\pi = 3{,}1416$,

$$P = 4 \times 3{,}1416 \times \overset{lbs.}{843{,}75} \cdot R^3.$$

Now suppose the radius of the sphere to be two feet, then will $R^3 = 8$, and

$$P = 4 \times 3{,}1416 \times \overset{lbs.}{843{,}75} \times 8 = \overset{lbs.}{84822{,}4}.$$

The volume of the sphere is $\frac{4}{3} \pi R^3$; and the weight of the contained mercury will therefore be $\frac{4}{3} \pi R^3 g D = W$. Dividing the whole pressure by this, we find

$$\frac{P}{W} = 3;$$

whence the outward pressure is three times the weight of the fluid.

Example 3.—Let the vessel be a cylinder, of which the radius r of the base is 2, and altitude l, 6 feet. Then will

$$b \,.\, z_{\prime} = \pi r l (r + l) = 3{,}1416 \times 2 \times 6 \times 8;$$

which, substituted in Equation (438),

$$P = 301{,}5936 \times D g,$$

and

$$W = 3{,}1416 \times 2^2 \times 6 \times D g = 75{,}398 \times D g;$$

whence,

$$\frac{P}{W} = \frac{301{,}5936 \times D g}{75{,}3984 \,.\, D g} = 4;$$

that is, the pressure against this particular vessel is four times the weight of the fluid.

§ 267.—The point through which the resultant of the pressure upon all the elements of the surface passes, is called the *centre of pressure.* Let $E I F$ be any plane, and $M N$ the intersection of this plane produced with the upper surface of the fluid which presses against it. Denote the area of any elementary portion n of the plane $E I F$ by $d b$; and let m be the projection of its place upon the upper surface of the fluid; draw $m M$ perpendicular to $M N$, and join n with M by the right line $n M$, the

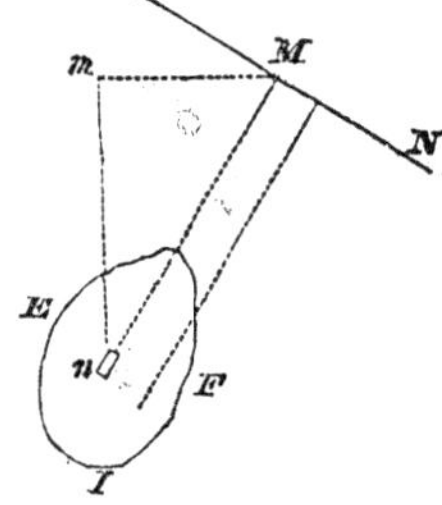

latter will also be perpendicular to MN, and the angle nMm will measure the inclination of the plane EIF to the surface of the fluid. Denote this angle by φ, the distance mn by h', and Mn by r' then will

$$h' = r' \sin \varphi;$$

the pressure upon the element db,

$$Dg \,.\, r' \sin \varphi \, db;$$

its moment with reference to the line MN,

$$D g r'^2 \sin \varphi \,.\, db;$$

and for the entire surface, the moment becomes

$$D g \,.\, \sin \varphi \,.\, \Sigma r'^2 \, db.$$

Denote by r the distance of the centre of gravity of the surface pressed from the line MN, its distance below the upper surface of the fluid will be $r \,.\, \sin \varphi$; and the pressure upon this surface will be

$$D g \,.\, r \sin \varphi \,.\, b;$$

and if l denote the distance of the centre of pressure from the line MN, then will

$$D g \,.\, r \sin \varphi \,.\, b \,.\, l = D g \,.\, \sin \varphi \,.\, \Sigma r'^2 \,.\, db,$$

from which we have,

$$l = \frac{\Sigma r'^2 \,.\, db}{r \,.\, b} = \frac{k_{\prime}^2 + r^2}{r}; \quad . \quad . \quad . \quad . \quad . \quad (439)$$

whence, Equation (238), the centre of pressure is found at the centre of percussion of the surface pressed.

§ 268.—The principles which have just been explained, are of great practical importance. It is often necessary to know the precise amount of pressure exerted by fluids against the sides of vessels and obstacles exposed to their action, to enable us so to adjust the dimensions of the latter as to give them sufficient strength to resist. Reservoirs in which considerable quantities of water are collected and retained till needed for purposes of irrigation, the supply of cities and towns, or to drive machinery; dykes to keep the sea and lakes from inundating low districts; artificial embankments constructed along the shores of rivers to protect the adjacent country in times of freshets; boilers in which elastic vapors are pent up in a high state of tension to propel boats and cars, and to give motion to machinery, are examples.

§ 269.—As a single instance, let it be required to find the thickness of a pipe of any material necessary to resist a given pressure.

Let $A\,B\,C$ be a section of pipe perpendicular to the axis, the inner surface of which is subjected to a pressure of p pounds on each superficial unit. Denote by R the radius of the interior circle, and by l the length of the pipe parallel to the axis; then will the surface pressed be measured by $2\pi R\,.\,l$; and the whole pressure by $2\pi R\,.\,l\,.\,p$.

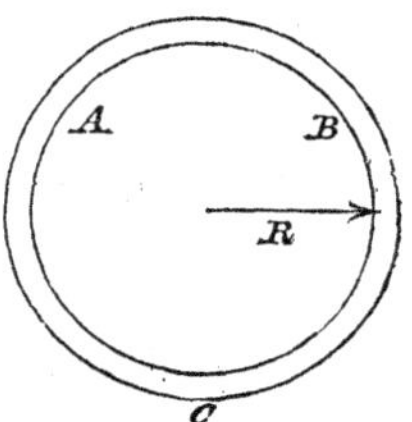

By virtue of the pressure, the pipe will stretch; its radius will become $R + d\,R$, the path described by the pressure will be $d\,R$, and its quantity of work

$$2\pi R\,.\,l\,.\,p\,d\,R.$$

The interior circumference before the pressure was $2\pi R$, afterwards $2\pi(R + d\,R)$, and the path described by resistance, $2\pi\,d\,R$. And if B denote the resistance which the material of the pipe is capable of opposing, to a stretching force, without losing its elasticity over each unit of section, t the thickness of the pipe, then, by the principle of the transmission of work, must

$$2\pi\,.\,B\,.\,l\,.\,d\,R\,.\,t = 2\pi R\,.\,l\,.\,p\,.\,d\,R;$$

whence,

$$t = \frac{R p}{B}.$$

The value of p is estimated in the case of water pressure by the rules just given. That in the case of steam or condensed gases, by rules to be given presently. The value of B is readily obtained from Table I, giving the results of experiments on the strength of materials.

EQUILIBRIUM AND STABILITY OF FLOATING BODIES.

§ 270.—When a body is immersed in a fluid it is not only acted upon by its own weight, but also by the pressure arising from the weight of the fluid, and the circumstances of its rest or motion will be made known by Equations (A) and (B).

Let ED be the body; take the plane $x\,y$ in the plane of the upper surface of the fluid, supposed at rest, and the axis of z therefore vertical. Denote by b the entire surface of the body, and by $d\,b$, one of its elements, whose co-ordinates of position are $x\,y\,z$. The pressure upon this element will be

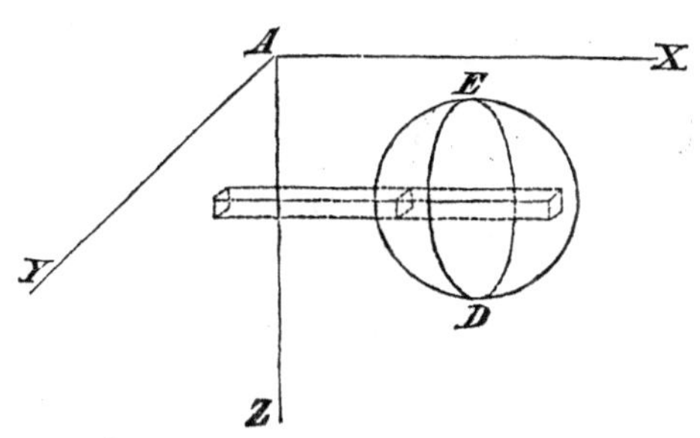

$$D\,.\,g\,.\,z\,.\,d\,b,$$

in which D is the density of the fluid, and g the force of gravity.

This pressure is, § 248, normal to the surface, and denoting by α, β and γ, the angles which this normal makes with the axes $x\,y\,z$, respectively, the components of the pressure in the direction of these axes will be

$$D\,.\,g\,.\,z\,.\,d\,b\,.\cos\alpha\,; \quad D\,.\,g\,.\,z\,.\,d\,b\,.\cos\beta\,; \quad D\,.\,g\,.\,z\,.\,d\,b\,.\cos\gamma.$$

Similar expressions being found for the components of the pressure on other elements, we have, by taking their sum,

$$D\,g\,.\,\Sigma\,z\,.\,d\,b\,.\cos\alpha\,; \quad D\,g\,.\,\Sigma\,z\,.\,d\,b\,.\cos\beta\,; \quad D\,g\,.\,\Sigma\,z\,.\,d\,b\,.\cos\gamma.$$

But $d\,b\,.\cos\alpha$, $d\,b\,.\cos\beta$, and $d\,b\,.\cos\gamma$, are the projections of the area $d\,b$ on the co-ordinate planes $z\,y$, $z\,x$ and $x\,y$, respectively; and

$\Sigma z . d b . \cos \alpha, \Sigma z . d b . \cos \beta, \Sigma z . d b . \cos \gamma$, are volumes of righ prisms whose bases are projections of the entire surface pressed upon the same co-ordinate planes, and of which the altitude of each is the depth of the common centre of gravity of the elements of its base submerged to the depths of their corresponding surface elements.

Whence we conclude, *that the component of the pressure on any surface, estimated in any direction, is equal to the pressure on so much of that surface as is equal to its projection on a plane at right angles to the given direction.*

The cylinder or prism which projects an element on one side of the body will also project an element situated on the opposite side; these projections will, therefore, be equal in extent, but will have contrary signs, for the normal to the one will make an acute, and to the other an obtuse angle with the axis of the plane of projection. When these projections are made upon any vertical plane, the value of z will be the same in both, and hence, for each positive product, $z . d b . \cos \alpha$ and $z . d b . \cos \beta$, there will be an equal negative product; therefore,

$$D g . \Sigma z . d b . \cos \alpha = \Sigma P \cos \alpha = 0; \quad D g . \Sigma z . d b . \cos \beta = \Sigma P \cos \beta = 0.$$

That is, the sum of the horizontal pressures in the directions of x and y, and therefore in *all horizontal directions*, will be zero; and the first and second of Equations (120), give

$$\Sigma m \cdot \frac{d^2 x}{d t^2} = 0; \quad \Sigma m \cdot \frac{d^2 y}{d t^2} = 0;$$

or, which is the same thing, there can be no horizontal motion of translation from the fluid pressure.

When the projections of opposite elements are made upon a horizontal plane, they will still be equal with contrary signs, the normal to the elements on the lower side making obtuse, while the normals to the elements above make acute angles with the axis z; but the corresponding values of z will differ, and by a length equal to that of the vertical filament of the body of which these elements form the opposite bases, and hence

$$D g . \Sigma z . d b . \cos \gamma = D g . \Sigma (z' - z_{\prime}) d b \cos \gamma = - D g \Sigma c \, d b \cos \gamma \cdot \cdot (440)$$

in which z' denotes the ordinate for the upper, and $z_{,}$ that for the lower element in the same vertical line, and c the distance between the elements; and the third of Equations (120) becomes

$$\Sigma \left(P \cos \gamma - m \cdot \frac{d^2 z}{d t^2}\right) = Mg - D g \cdot \Sigma c \cdot d b \cdot \cos \gamma - \Sigma m \cdot \frac{d^2 z}{d t^2} = 0.$$

But $\Sigma c . d b . \cos \gamma$ is the volume of the immersed body which is obviously equal to that of the displaced fluid; also $D g . \Sigma c . d b . \cos \gamma$ is the weight of the displaced fluid; and Mg that of the body. Denoting the volume of the body by V', its density by D', the above may be written

$$V' D' g - V' D g - \Sigma m \cdot \frac{d^2 z}{d t^2} = 0. \quad . \quad . \quad . \quad (441)$$

Now, when

$$V' D' g - V' D g = 0,$$

or

$$D = D',$$

then will

$$\Sigma m \frac{d^2 z}{d t^2} = 0;$$

and there can be no vertical motion of translation from the fluid pressure and the body's weight.

When $D' > D$, then will

$$\Sigma m \cdot \frac{d^2 z}{d t^2} = (D' - D) V' . g;$$

and the body will sink with an accelerated motion.

When $D' < D$, then will

$$\Sigma m \cdot \frac{d^2 z}{d t^2} = -(D' - D) V' . g,$$

and the body will rise with an accelerated motion till

$$\Sigma m \cdot \frac{d^2 z}{d t^2} = V' D' g - V D g = 0; \quad . \quad . \quad (442)$$

in which V denotes the volume $A\,B\,C$, of the fluid displaced. At this instant we have

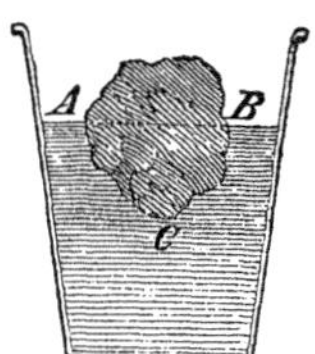

$$V'\,D'g = V\,D\,g; \quad . \quad . \quad . \quad (443)$$

and if the body be brought to rest, it will remain so. That is, the body will float at the surface when the weight of the fluid it displaces is equal to its own weight.

The action of a heavy fluid to support a body wholly or partly immersed in it, is called the *buoyant effort.* The intensity of the buoyant effort is equal to *the weight of the fluid displaced.*

Substituting the values of the horizontal and vertical components of the pressures in Equations (118), and reducing by the relations,

$$\left.\begin{array}{l} D\,g\,.\,\Sigma\,c\,.\,d\,b\,.\cos\gamma\,.\,x' = D\,g\,.\,V\,.\,\bar{x}\,; \\ D\,g\,.\,\Sigma\,c\,.\,d\,b\,.\cos\gamma\,.\,y' = D\,g\,.\,V\,.\,\bar{y}\,; \end{array}\right\} \quad . \quad . \quad . \quad (444)$$

in which $\bar{x}$ and $\bar{y}$ are the co-ordinates of the centre of gravity of the displaced fluid referred to the centre of gravity of the body, we find

$$\left.\begin{array}{l} \Sigma\,m\,.\,\dfrac{x'\,.\,d^2\,y' - y'\,.\,d^2\,x'}{d\,t^2} = 0\,; \\ \Sigma\,m\,.\,\dfrac{z'\,.\,d^2\,x' - x'\,.\,d^2\,z'}{d\,t^2} = D\,g\,.\,V\,.\,\bar{x}\,; \\ \Sigma\,m\,.\,\dfrac{y'\,.\,d^2\,z' - z'\,.\,d^2\,y'}{d\,t^2} = -\,D\,g\,.\,V\,.\,\bar{y}. \end{array}\right\} \quad . \quad . \quad (445)$$

Equations (444) show that the line of direction of the buoyant effort passes through the centre of gravity of the displaced fluid. This point is called the *centre of buoyancy.* And from Equations (445), we see that as long as $\bar{x}$ and $\bar{y}$ are not zero, there will be an angular acceleration about the centre of gravity. At the instant $\bar{x} = 0$ and $\bar{y} = 0$, that is to say, when the centres of gravity of the body and displaced fluid are on the same vertical line, this acceleration will cease, and if the body were brought to rest, it would have no tendency to rotate.

To recapitulate, we find,

1st. *That the pressures upon the surface of a body immersed in a heavy fluid have a single resultant, called the buoyant effort of the fluid, and that this resultant is directed vertically upwards.*

2d. *That the buoyant effort is equal in intensity to the weight of the fluid displaced.*

3d. *That the line of direction of the buoyant effort passes through the centre of gravity of the displaced fluid.*

4th. *That the horizontal pressures destroy one another.*

§271.—Having discussed the equilibrium, consider next the stability of a floating body. The density of the body may be homogeneous or heterogeneous. Let $A\,B\,C\,D$ be a section of the body by the upper surface of the fluid when the body is at rest, G its centre of gravity, and H that of the fluid displaced. Denote by V the volume of the displaced fluid, and by M the mass of the entire body. The body being in equilibrio, the line $G\,H$ will be vertical, and denoting the density of the fluid by D, we shall have

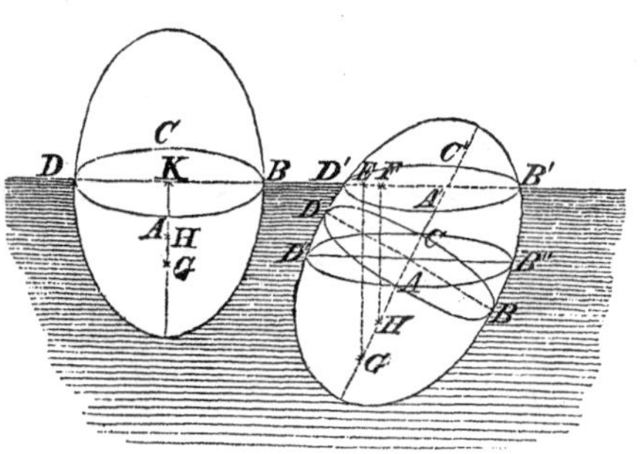

$$M = D \, . \, V. \quad . \quad . \quad . \quad . \quad . \quad . \quad . \quad . \quad . \qquad (446)$$

Suppose the section $A\,B\,C\,D$ either raised above or depressed below the surface of the fluid, and at the same time slightly careened; also suppose, when the body is abandoned, that the elements have a slight velocity denoted by u, u', &c. Now the question of stability will consist in ascertaining whether the body will return to its former position, or will depart more and more from it.

The free surface of the fluid is called the *plane of floatation,* and during the motion of the body this plane will cut from it a variable section.

Let $A'\,B'\,C'\,D'$ be one of these sections at any given instant of time; $A B'' C D''$, another variable section of the body by a horizontal plane through the centre of gravity of the primitive section $A B C D$, and $A C$ the intersection of the two. Denote by θ the inclination of these two sections, and by ζ the vertical distance of $A B'' C D''$, from the plane of floatation, which now coincides with $A' B' C' D'$, this distance being regarded as negative or positive, according as $A B'' C D''$ is below or above the plane of floatation. The variable quantities θ and ζ will be supposed very small at the instant the body is abandoned. Will they continue so during the whole time of motion?

From the principles of living force and quantity of work, we have, Equation (121),

$$\int u^2 . d M = 2 \int (X d x + Y d y + Z d z) + C.$$

The forces acting are the weights of the elements $d M$ and the vertical pressures, the horizontal pressures destroying one another; whence, $X = 0$, $Y = 0$, and

$$\int u^2 d M = 2 \int Z d z + C = 2 \Sigma Z z + C. \quad . \quad . \quad (447)$$

The force which acts upon an element above the plane of floatation is its own weight, and the force which acts upon any element below that plane is the difference between its own weight and that of the fluid it displaces; the first of these latter will be $g . d M$, and the second, $g . D . d V$, in which $d V$ is the volume of $d M$; whence,

$$\Sigma Z z = \int g . z . d M - \int g D . z . d V. \quad . \quad . \quad (448)$$

But, drawing from the centre of gravity G, of the body, the perpendicular $G E$, to the plane of floatation $A' B' C' D'$, and denoting $G E$ by $z_{,}$, we have

$$\int g . z . d M = g M z_{,}.$$

The integral $\int g D . z . d V$, will be divided into two parts, viz: one relating to the volume of the body below $A B C D$, or the volume immersed in a state of rest, and the other that comprised between

$A\,B\,C\,D$ and the plane of floatation $A'\,B'\,C'\,D'$, when the body is in motion. Denote by $g\,D\,V z'$, the value of the first, in which z' denotes the variable distance $H\,F$, of the centre of gravity H, of the volume V, from the plane of floatation $A'\,B'\,C'\,D'$. And representing for the instant by h the value of the integral $\int z\,d\,V$, comprehended between the planes $A\,B\,C\,D$ and $A'\,B'\,C'\,D'$, $g\,D\,h$ will be the second part; and Equation (447) becomes

$$\int u^2\,d\,M = 2\,g\,.\,M z_{\prime} - 2\,g\,D\,V z' - 2\,g\,D\,h + C. \quad . \quad . \quad (449)$$

The line $G\,H$, being perpendicular to the plane $A\,B\,C\,D$, the angle which it makes with the line $G\,E$ is equal to θ, and denoting the distance $G\,H$ by a, we have

$$z_{\prime} = z' \pm a \cos\theta\,;$$

the upper sign being taken when the point G is below the point H, and the lower when it is above. This value reduces Equation (449) to

$$\int u^2\,d\,M = \pm\, 2\,g\,D\,V a \cos\theta - 2\,g\,D\,h + C. \quad . \quad . \quad . \quad (450)$$

Let us now find the integral h. For this purpose, conceive the area $A\,B\,C\,D$ to be divided into indefinitely small elements denoted by $d\lambda$, and let these be projected upon the plane of floatation, $A'\,B'\,C'\,D'$. The projecting surfaces will divide the volume comprised between these two sections into an indefinite number of vertical elementary prisms, and these being cut by a series of horizontal planes indefinitely near each other, will give a series of elementary volumes, each of which will be denoted by $d\,V$, and we shall have

$$d\,V = d\,z\,.\,d\lambda\,.\cos\theta\,;$$

whence, for a single elementary vertical prism,

$$\int z\,d\,V = \int z\,d\,z\,.\,d\,\lambda\,.\cos\theta = \tfrac{1}{2}\,(z)^2\,.\cos\theta\,.\,d\,\lambda\,;$$

in which (z) denotes the mean altitude of the prism, and consequently

$$h = \tfrac{1}{2}\cos\theta\,.\int (z)^2\,.\,d\,\lambda,$$

which must be extended to embrace the entire surface $A\,B\,C\,D$.

The value of (z) is composed of two parts, viz.: one comprised between the parallel sections $A' B' C' D'$ and $A B'' C D''$, and which has been denoted by ζ; the other comprised between the base $d\lambda$ and the second of these planes, and which is equal to $l . \sin\theta$, denoting by l the distance of $d\lambda$ from the intersection $A C$; whence,

$$(z) = \zeta + l . \sin\theta,$$

in which l will be positive or negative according as $d\lambda$ happens to be below or above the plane $A B'' C D''$. Substituting this in the value of h, and recollecting that ζ and θ are constant in the integration, we find

$$h = \tfrac{1}{2}\zeta^2 . \cos\theta . \int d\lambda + \zeta \sin\theta \cos\theta \int l\, d\lambda + \tfrac{1}{2}\sin^2\theta . \cos\theta \int l^2\, d\lambda.$$

Denote by b the area of $A B C D$, or the value of $\int d\lambda$. The line $A C$ passing through the centre of gravity of $A B C D$, we have $\int l\, d\lambda = 0$. And denoting by $k_{\prime}$ the principal radius of gyration of the surface b, in reference to the axis $A C$,

$$\int l^2\, d\lambda = b\, k_{\prime}^2,$$

in which the value of $k_{\prime}$ is dependent upon the figure and extent of the surface $A B C D$, and upon the position of the line $A C$. Whence,

$$h = \tfrac{1}{2} b . \cos\theta\, (\zeta^2 + k_{\prime}^2 \sin^2\theta). \quad . \quad . \quad . \quad . \quad (451)$$

Taking

$$\sin\theta = \theta - \frac{\theta^3}{2 . 3} + \&c; \quad \cos\theta = 1 - \frac{\theta^2}{1 . 2} + \&c.$$

Neglecting all the terms of the third and higher orders, substituting in the value of h, and then in Equation (450) we find, after transposing and including the term $\pm 2 g D V a$, in the constant C,

$$\int u^2 . d M + g D \Big[b \zeta^2 + (b k_{\prime}^2 \pm V a)\, \theta^2 \Big] = C. \quad . \quad . \quad . (452)$$

Now the value of the constant C depends upon the initial values of u, θ and ζ; but these by hypothesis are very small; hence C, must also be very small. As long as the second term of the first member is positive, $\int u^2 \, dM$ must remain very small, since it is essentially positive itself, and being increased by a positive quantity, the sum is very small. Hence ζ and θ must remain very small. But when the second term is negative, which can only be when $b k_{,}^2 \pm Va$, is negative and greater than $b \frac{\zeta^2}{\theta^2}$, the value of $\int u^2 \, dM$ may increase indefinitely; for, being diminished by a quantity that increases as fast as itself, the difference may be constant and very small. Hence, ζ and θ may increase more and more after the body is abandoned to itself, and finally it may overturn.

The stability of the equilibrium depends, therefore, upon the sign of $b k_{,}^2 \pm Va$; the equilibrium is always stable when this quantity is positive; it is unstable when it is negative and greater than $b \frac{\zeta^2}{\theta^2}$. The value of $b k_{,}^2 = \int l^2 \, d\lambda$, must always be positive, since all its elements are positive; the value of $\pm Va$ becomes negative when the centre of gravity of the body is above that of the displaced fluid, in which case the stability requires that

$$b k_{,}^2 > Va, \quad \text{or,} \quad k_{,}^2 > \frac{Va}{b}.$$

When the centre of gravity of the body is below that of the displaced fluid, the sign of Va is positive.

Whence we conclude that the equilibrium of a body floating at the surface of a heavy fluid, will be stable as long as the centre of gravity of the body is below that of the displaced fluid; that it will also be stable about all lines $A\,C$, with reference to which the principal radius of gyration of the section of the body by the plane of floatation squared, is greater than the volume of the displaced fluid multiplied by the distance between the centres of gravity of the displaced fluid and that of the body, when the latter is in equilibrio, divided by the area of the section of the body by the plane of floatation. When this condition is not fulfilled, the equilibrium may be unstable. A ship whose centre of gravity is above that of the water she displaces, may overturn about her longer, but not about her shorter axis.

§ 272.—A line $B\,K$ through the centre of gravity G of the body and which is vertical when the body is in equilibrio, is called a ***line of rest.*** A vertical line $H'M$ through the centre of gravity H' of the displaced fluid, is called a *line of support.* The point M, in which the line of support cuts the line of rest, is called the *metacentre.* The body will be in equilibrio when the line of rest and of support coincide. The equilibrium will be stable if the metacentre fall above the centre of gravity ; it may be unstable if below.

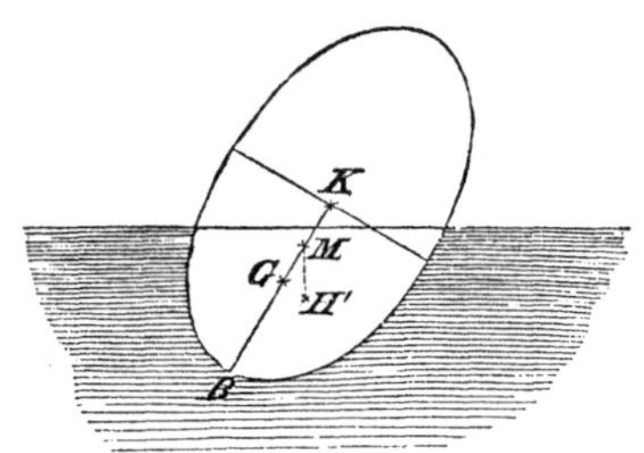

§ 273.—When the equilibrium is stable, and the body is disturbed and then abandoned to the action of its own weight and that of the fluid pressure, it will, in its efforts to regain its place of rest, oscillate about this position, and finally come to rest.

The circumstances of those oscillations about the *centre of gravity* of the body will readily result from Equations (445).

SPECIFIC GRAVITY.

§ 274.—The *specific gravity* of a body, is the weight of so much of the body, as would be contained under a unit of volume.

It is measured by the quotient arising from dividing the weight of the body by the weight of an equal volume of some other substance, assumed as a standard ; for the ratio of the weights of equal volumes of two bodies being always the same, if the unit of volume of each be taken, and one of the bodies become the standard, its weight will become the unit of weight.

The term *density* denotes the degree of proximity among the particles of a body. Thus, of two bodies, that will have the greater density which contains, under an equal volume, the greater number of particles. The force of gravity acts, within moderate limits, equally upon all elements of matter. The weight of a substance is, therefore, directly proportional to its density, and the ratio of the weights of equal volumes of two bodies is equal to the ratio of their densities. Denote the weight of the first by W, its density by D, its volume by V, and the force of gravity by g, then will

$$W = g \,.\, D \,.\, V;$$

and denoting the like elements of the other body by $W_{,}$, $D_{,}$ and $V_{,}$, we have

$$W_{,} = g \,.\, D_{,} \,.\, V_{,}.$$

Dividing the first by the second,

$$\frac{W}{W_{,}} = \frac{g\,D\,V}{g\,D_{,}\,V_{,}} = \frac{D\,V}{D_{,}V_{,}};$$

and making the volumes equal,

$$\frac{W}{W_{,}} = \frac{D}{D_{,}} \qquad \ldots \ldots \ldots \qquad (453)$$

Now suppose the body whose weight is $W_{,}$ to be assumed as the standard both for specific gravity and density, then will $D_{,}$ be unity, and

$$S = \frac{W}{W_{,}} = D \qquad \ldots \ldots \ldots \qquad (454)$$

in which S denotes the specific gravity of the body whose density is D; and from which we see, that when specific gravities and densities are referred to the same substance as a standard,. the numbers which express the one will also express the other.

§ 275.—Bodies present themselves under every variety of condition—gaseous, liquid, and solid; and in every kind of shape and of all sizes. The determination of their specific gravity, in every instance, depends upon our ability to find the weight of an equal volume of the standard. When a solid is immersed in a fluid, it loses a portion of its weight equal to that of the displaced fluid. The volume of the body and that of the displaced fluid are equal. Hence the weight of the body in vacuo, divided by its loss of weight when immersed, will give the ratio of the weights of equal volumes of the body and fluid; and if the latter be taken as the

20

standard, and the loss of weight be made to occupy the denominator, this ratio becomes the measure of the specific gravity of the body immersed. For this reason, and in view of the consideration that it may be obtained pure at all times and places, *water* is assumed as the general standard of specific gravities and densities for all bodies. Sometimes the gases and vapors are referred to atmospheric air, but the specific gravity of the latter being known as referred to water, it is very easy, as we shall presently see, to pass from the numbers which relate to one standard to those that refer to the other.

§ 276.—But water, like all other substances, changes its density with its temperature, and, in consequence, is not an invariable standard. It is hence necessary either to employ it at a constant temperature, or to have the means of reducing the apparent specific gravities, as determined by means of it at different temperatures, to what they would have been if the water had been at the standard temperature. The former is generally impracticable; the latter is easy.

Let D denote the density of any solid, and S its specific gravity, as determined at a standard temperature corresponding to which the density of the water is $D_{,}$. Then, Equation (453),

$$S = \frac{D}{D_{,}}.$$

Again, if S' denote the specific gravity of the same body, as indicated by the water when at a temperature different from the standard, and corresponding to which it has a density $D_{,,}$, then will

$$S' = \frac{D}{D_{,,}}.$$

Dividing the first of these equations by the second, we have

$$\frac{S}{S'} = \frac{D_{,,}}{D_{,}};$$

whence,

$$S = S' \cdot \frac{D_{,,}}{D_{,}}; \quad \ldots \ldots \quad (455)$$

and if the density $D_{,}$, be taken as unity,

$$S = S' \cdot D_{,,}. \quad \ldots \ldots \quad (456)$$

That is to say, *the specific gravity of a body as determined at the standard temperature of the water, is equal to its specific gravity determined at any other temperature, multiplied by the density of the water corresponding to this temperature, the density at the standard temperature being regarded as unity.*

To make this rule practicable, it becomes necessary to find the relative densities of water at different temperatures. For this purpose, take any metal, say silver, that easily resists the chemical action of water, and whose rate of expansion for each degree of Fahr. thermometer is accurately known from experiment; give it the form of a slender cylinder, that it may readily conform to the temperature of the water when immersed. Let the length of the cylinder at the temperature of 32° Fahr. be denoted by l, and the radius of its base by $m\,l$; its volume at this temperature will be,

$$\pi m^2 l^2 \times l = \pi m^2 l^3.$$

Let $n\,l$ be the amount of expansion in length for each degree of the thermometer above 32°. Then, for a temperature denoted by t, will the whole expansion in length be

$$n\,l \times (t - 32°),$$

and the entire length of the cylinder will become

$$l + n\,l\,(t - 32°) = l[1 + n\,(t - 32°)];$$

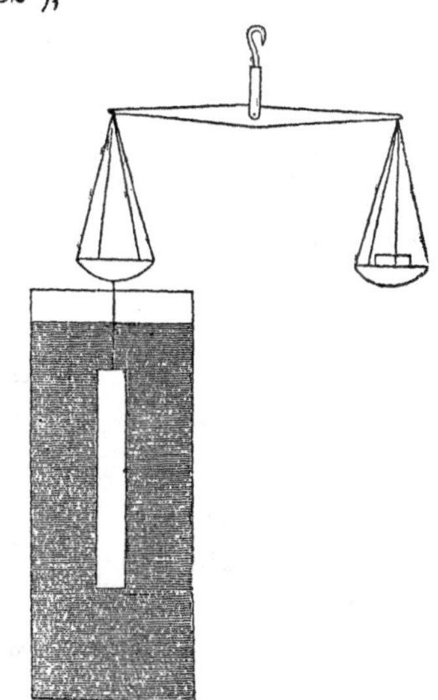

which, substituted for l in the first expression, will give the volume for the temperature t, equal to

$$\pi m^2 l^3 [1 + n\,(t - 32°)]^3.$$

The cylinder is now weighed in vacuo and in the water, at different temperatures, varying from 32° upward, through any desirable range, say to one hundred degrees. The temperature at each process being substituted above, gives the volume of the displaced fluid; the weight of the displaced fluid is known from the loss of weight of the cylinder. Dividing this weight by the volume, gives the weight of the unit of volume of the water at the temperature t. It was found by *Stampfer*, that the weight of the unit of volume is greatest when the temperature is 38°.75 Fahrenheit's scale. Taking the density of the water at this temperature as unity, and dividing the weight of the unit of volume at each of the other temperatures by the weight of the unit of volume at this, 38°.75, Table II will result.

The column under the head V, will enable us to determine how much the volume of any mass of water, at a temperature t, exceeds that of the same mass at its maximum density. For this purpose, we have but to multiply the volume at the maximum density by the tabular number corresponding to the given temperature.

§ 277.—Before proceeding to the practical methods of finding the specific gravity of bodies, and to the variations in the processes rendered necessary by the peculiarities of the different substances, it will be necessary to give some idea of the best instruments employed for this purpose. These are the *Hydrostatic Balance* and *Nicholson's Hydrometer*.

The first is similar in principle and form to the common balance. It is provided with numerous weights, extending through a wide range, from a small fraction of a grain to several ounces. Attached to the under surface of one of the basins is a small hook, from which may be suspended any body by means of a thin platinum wire, horsehair, or any other delicate thread that will neither absorb nor yield to the chemical action of the fluid in which it may be desirable to immerse it.

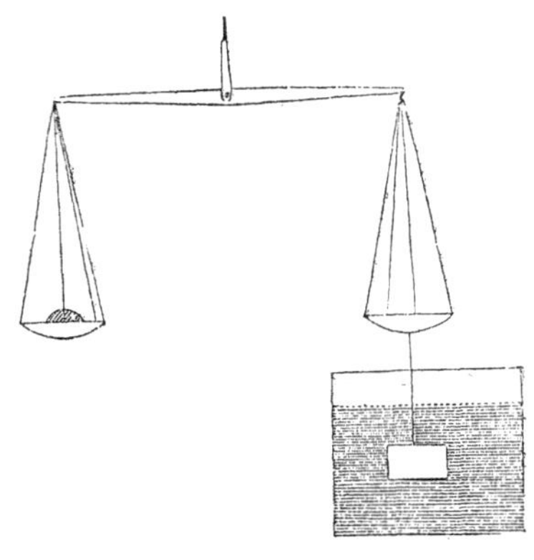

Nicholson's Hydrometer consists of a hollow metalic ball A, through the centre of which passes a metallic wire, prolonged in both directions beyond the surface, and supporting at either end a basin B and B'. The concavities of these basins are turned in the same direction, and the basin B' is made so heavy that when the instrument is placed in water the stem $C C'$ shall be vertical, and a weight of 500 grains being placed in the basin B, the whole instrument will sink till the upper surface of distilled water, at the standard temperature, comes to a point C marked on the upper stem near its middle. This instrument is provided with weights similar to those of the Hydrostatic Balance.

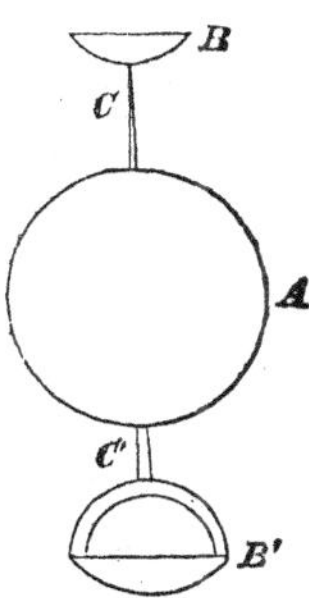

§ 278.—(1). *If the body be solid, insoluble in water, and will sink in that fluid*, attach it, by means of a hair, to the hook of the basin of the hydrostatic balance; counterpoise it by placing weights in the opposite scale; now immerse the body in water, and restore the equilibrium by placing weights in the basin above the body, and note the temperature of the water. Divide the weights in the basin to which the body is not attached by those in the basin to which it is, and multiply the quotient by the density corresponding to the temperature of the water, as given by the table; the result will be the specific gravity.

Thus denote the specific gravity by S, the density of the water by $D_{,,}$, the weight in the first case by W, and that in the scale above the solid by w, then will

$$S = D_{,,} \times \frac{W}{w}.$$

(2). *If the body be insoluble, but will not sink in water*, as would be the case with most varieties of wood, wax, and the like, attach to it some body, as a metal, whose weight in the air and loss of weight in the water are previously found. Then proceed, as in the case before, to find the weights which will counterpoise the compound in air and restore the equilibrium of the balance when it is immersed in the water. From the weight of the compound in air, subtract that of the denser body in air; from the loss of weight of the compound in water, subtract that of the denser body; divide the first difference by the second, and multiply by the density of the water answering to its temperature, and the result will be the specific gravity sought.

Example.

A piece of wax and copper in air	=	438 grs.	$= W + W'$,
Lost on immersion in water - -	=	95,8	$= w + w'$,
Copper in air - - - - - - - -	=	388	$= W'$,
Loss of copper in water - · -	=	44,2	$= w'$.

Then

$$W + W' - W' = 438 - 388 = 50, \quad = W,$$

$$w + w' - w' = 95{,}8 - 44{,}2 = 51{,}6 = w.$$

Temperature of water 43°,25,

$$D_{\prime\prime} = 0{,}999952,$$

$$S = D_{\prime\prime} \times \frac{W}{w} = 0{,}999952 \times \frac{50}{51{,}6} = 0{,}968.$$

(3). *If the body readily dissolve in water*, as many of the salts, sugar, &c., find its apparent specific gravity in some liquid in which it is insoluble, and multiply this apparent specific gravity by the density or specific gravity of the liquid referred to water at its maximum density as a standard; the product will be the true specific gravity.

If it be inconvenient to provide a liquid in which the solid is insoluble, saturate the water with the substance, and find the apparent specific gravity with the water thus saturated. Multiply this apparent specific gravity by the density of the saturated fluid, and the product will be the specific gravity referred to the standard. This is a common method of finding the specific gravity of gunpowder, the water being saturated with nitre.

(4). *If the body be a liquid*, select some solid that will resist its chemical action, as a massive piece of glass suspended from fine platinum wire; weigh it in air, then in water, and finally in the liquid; the differences between the first weight and each of the latter, will give the weights of equal volumes of water and the liquid. Divide the weight of the liquid by that of the water, and the quotient will be the specific gravity of the liquid, provided the temperature of water be at the standard. If the water have not the standard temperature, multiply this apparent specific gravity by the tabular density of the water corresponding to the actual temperature.

Example.

Loss of glass in water at 41°, $150^{grs.} = w'$,

" " sulphuric acid, $277{,}5 = w$,

$$S = \frac{277{,}5}{150} \times 0{,}999988 = 1{,}85.$$

(5). *If the body be a gas or vapor*, provide a large glass flask-shaped vessel, weigh it when filled with the gas; withdraw the gas, which may be done by means to be explained presently, fill with water, and weigh again; finally, withdraw the water and exclude the air, and weigh again. This last weight subtracted from the first, will give the weight of the gas that filled the vessel, and subtracted from the second will give the weight of an equal volume of water; divide the weight of the gas by that of the water, and multiply by the tabular density of the water answering to the actual temperature of the latter; the result will be the specific gravity of the gas.

The atmosphere in which all these operations must be performed, varies at different times, even during the same day, in respect to temperature, the weight of its column which presses upon the earth, and the quantity of moisture or aqueous vapor it contains. That is to say, its density depends upon the state of the thermometer, barometer, and hygrometer. On all these accounts corrections must be made, before the specific gravity of atmospheric air, or that of any gas exposed to its pressure, can be accurately determined. The principles according to which these corrections are made, will be discussed when we come to treat of the properties of elastic fluids.

To find the specific gravity of a solid by means of Nicholson's Hydrometer, place the instrument in water, and add weights to the upper basin until it sinks to the mark on the upper stem; remove the weights and place the solid in the upper basin, and add weights till the hydrometer sinks to the same point; the difference between the first weights and those added with the body, will give the weight of the latter in air. Take the body from the upper basin, leaving the weights behind, and place it in the lower basin; add weights to the upper basin till the instrument sinks to the same point as before, the last added weights will be the weight of the water displaced by the body; divide the weight in air by the weight of the displaced water, and multiply the quotient by the tabular density of the water answering to its actual temperature; the result will be the specific gravity of the solid.

To find the specific gravity of a fluid by this instrument, immerse it in water as before, and by weights in the upper basin sink it to the mark on the upper stem; add the weights in the basin to the weight of the instrument, the sum will be the weight of the displaced water. Place the instrument in the fluid whose specific gravity is to be found, and add weights in the upper basin till it sinks to the mark as before; add these weights to the weight of the instrument, the sum will be the weight of an equal volume of the fluid; divide this weight by the weight of the water, and multiply by the tabular density corresponding to the temperature of the water, the result will be the specific gravity.

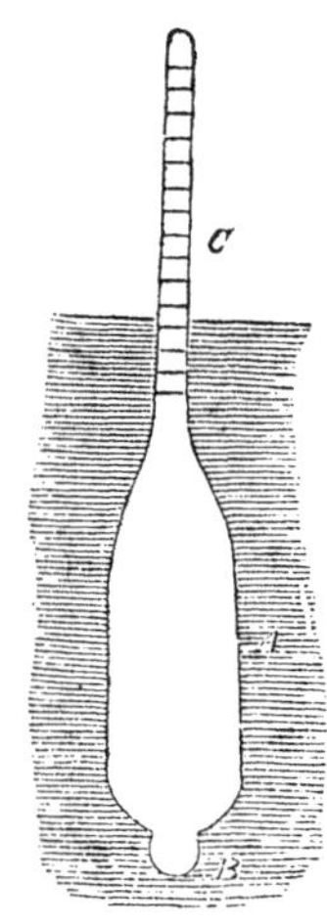

§ 279.—Besides the hydrometer of Nicholson, which requires the use of weights, there is another form of this instrument which is employed solely in the determination of the specific gravities of liquids, and its indications are given by means of a scale of equal parts. It is called the *Scale-Areometer*. It consists, generally, of a glass vial-shaped vessel A, terminating at one end in a long slender neck C, to receive the scale, and at the other in a small globe B, filled with some heavy substance, as lead or mercury to keep it upright when immersed in a fluid. The application and use of the scale depend upon this, that a body floating on the surface of different liquids, will sink deeper and deeper, in proportion as the density of the fluid approaches that of the body; for when the body is at rest its weight and that of the displaced fluid must be equal. Denoting the volume of the instrument by V, that of the displaced fluid by V', the density of the instrument by D, and that of the fluid by D', we must always have

$$g\,V D = g\,V' D';$$

in which g denotes the force of gravity, the first member the weight of the instrument, and the second that of the displaced fluid. Dividing both members by $D' V$, and omitting the common factor g we have

$$\frac{D}{D'} = \frac{V'}{V}.$$

In which, if the densities be equal, the volumes must be equal; if the density D' of the fluid be greater than D, or that of the solid, the volume V of the solid must be greater than V', or that of the displaced fluid; and in proportion as D' increases in respect to D, will V' diminish in respect to V; that is, the solid will rise higher and higher out of the fluid in proportion as the density of the latter is increased, and the reverse. The neck C of the vessel should be of the same diameter throughout. To establish the scale, the instrument is placed in distilled water at the standard temperature, and when at rest the place of the surface of the water on the neck is marked and numbered 1; the instrument is then placed in some heavy solution of salt, whose specific gravity is accurately known by means of the Hydrostatic Balance, and when at rest the place on the neck of the fluid surface is again marked and characterized by its appropriate number. The same process being repeated for rectified alcohol, will give another point towards the opposite extreme of the scale, which may be completed by graduation.

To use this instrument, it will be sufficient to immerse it in a fluid and take the number on the scale which coincides with the surface.

To ascertain the circumstances which determine the sensibility both of the Scale-Areometer and Nicholson's Hydrometer, let s denote the specific gravity of the fluid, c the volume of the vial, l the length of the immersed portion of the narrow neck, r its semi-diameter, and w the total weight of the instrument. Then will πr^2, denote the area of a section of the neck, and $\pi r^2 l$, the volume of fluid displaced by the immersed part of the neck. The weight, therefore, of the whole fluid displaced by the vial and neck will be

$$s c + s \pi r^2 l;$$

but this must be equal to the weight of the instrument, whence,

$$w = s (c + \pi r^2 l),$$

from which we deduce,

$$s = \frac{w}{c + \pi r^2 l},$$

$$l = \frac{w - s c}{\pi r^2 s} \quad . \quad . \quad . \quad . \quad . \quad . \quad . \quad . \quad . \quad (457)$$

Now, immersing the instrument in a second fluid whose specific gravity is s', the neck will sink through a distance l', and from the last equation we have

$$l' = \frac{w - s' c}{\pi r^2 s'};$$

subtracting this equation from that above and reducing, we find

$$l - l' = \frac{w}{\pi r^2} \left(\frac{s' - s}{s s'} \right).$$

The difference $l - l'$ is the distance between two points on the scale which indicates the difference $s' - s$ of specific gravities, and this we see becomes longer, and the instrument more sensible, therefore, in proportion as w is made greater and r less. Whence we conclude that the Areometer is the more valuable in proportion as the vial portion is made larger and the neck smaller.

If the specific gravity of the fluid remain the same, which is the case with Nicholson's Hydrometer, and it becomes a question to know the effect of a small weight added to the instrument, denote this weight by w', then will Equation (457) become

$$l' = \frac{w + w' - s c}{\pi r^2 s};$$

subtracting from this Equation (457), we find

$$l' - l = \frac{w'}{\pi r^2 s}.$$

From which we see that the narrower the upper stem of Nicholson's instrument, the greater its sensibility.

The knowledge of the specific gravities or densities of different substances, Table III, is of great importance, not only for scientific purposes, but also for its application to many of the useful arts. This knowledge enables us to solve such problems as the following, viz. :—

1st. The weight of any substance may be calculated, if its volume and specific gravity be known.

2d. The volume of any body may be deduced from its specific gravity and weight. Thus we have always

$$W = g D V;$$

in which g is the force of gravity, D the density, V the volume, and W the weight, of which the unit of measure is the weight of a unit of volume of water at its maximum density.

Making D and V equal to unity, this equation becomes

$$W_{,} = g;$$

but if the density be one, the substance must be water at 38°,75 Fahr. The weight of a cubic foot of water at 60° is 62,5 lbs., and, therefore, at 38°,75, it is

$$\frac{\overset{lbs.}{62{,}5}}{0{,}99914} = \overset{lbs.}{62{,}556};$$

whence, if the volume be expressed in cubit feet,

$$W = \overset{lbs.}{62{,}556} \times D V \quad . \quad . \quad . \quad . \quad . \quad . \quad . \quad (458)$$

in which W is expressed in pounds; and if the unit of volume be a cubic inch,

$$W = \frac{62{,}556}{1728} D\,V = 0{,}036201\,D\,V_{\prime} \quad . \quad . \quad . \quad (459)$$

Also,

$$V = \frac{\overset{lbs.}{W}}{62{,}556\,.\,D} \quad . \quad . \quad . \quad . \quad . \quad . \quad . \quad . \quad (460)$$

$$V_{\prime} = \frac{\overset{lbs.}{W}}{0{,}036201\,.\,D} \quad . \quad . \quad . \quad . \quad . \quad . \quad . \quad . \quad (461)$$

Example 1.—Required the weight of a block of dry fir, containing 50 cubic inches. The specific gravity or density of dry fir is 0,555, and $V = 50$; substituting these values in Equation (459),

$$W = 0{,}036201 \times 0{,}555 \times 50 = \overset{lbs.}{1{,}00457}.$$

Example 2.—How many cubic inches are there in a 12-pound cannon-ball? Here W is 12 pounds, the mean specific gravity of cast iron is 7,251, which, in Equation (461), give

$$V_{\prime} = \frac{12}{0{,}036201 \times 7{,}251} = \overset{in.}{45{,}6}.$$

ATMOSPHERIC PRESSURE.

§ 280.—The atmosphere encases, as it were, the whole earth. It has weight, else the repulsive action among its own particles would cause it to expand and extend itself through space. The weight of the upper stratum of the atmosphere is in equilibrio with the repulsive action of the strata below it, and this condition determines the exterior limit.

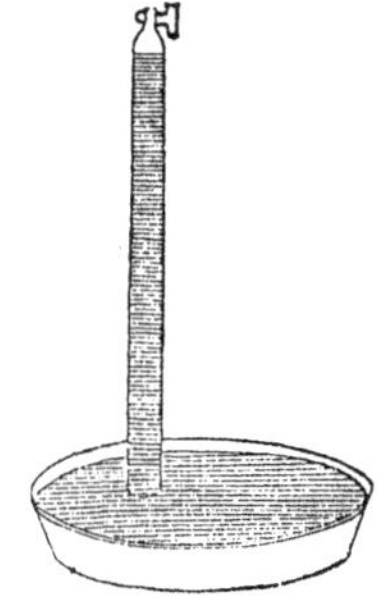

Since the atmosphere has weight, it must exert a pressure upon all bodies within it. To illustrate, fill with mercury a glass tube, about 32 or 33 inches long, and closed at one end by an iron stop-cock. Close the open end by pressing the finger against it, and invert the tube in a basin of mercury; remove the finger, the mercury will not escape, but remain apparently suspended, at the level of the ocean, nearly 30 inches above the surface of the mercury in the basin.

The atmospheric air presses on the mercury with a force sufficient to maintain the quicksilver in the tube at a height of nearly 30 inches; whence, *the intensity of its pressure must be equal to the weight of a column of mercury whose base is equal to that of the surface pressed and whose altitude is about* 30 *inches. The force thus exerted, is called the atmospheric pressure.*

The absolute amount of atmospheric pressure was first discovered by Torricelli, and the tubes employed in such experiments are called, on this account, *Torricellian tubes*, and the vacant space above the mercury in the tube, is called the *Torricellian vacuum.*

The pressure of the atmosphere at the level of the sea, supporting as it does a column of mercury 30 inches high, if we suppose the bore of the tube to have a cross-section of one square inch the atmospheric pressure up the tube will be exerted upon this extent of surface, and will support 30 cubic inches of mercury. Each cubic inch of mercury weighs 0,49 of a pound—say half a pound—from which it is apparent that *the surfaces of all bodies, at the level of the sea, are subjected to an atmospheric pressure of fifteen pounds to each square inch.*

BAROMETER.

§ 281.—The atmosphere being a heavy and elastic fluid, is compressed by its own weight. Its density cannot be the same throughout, but diminishes as we approach its upper limit where it is least, being greatest at the surface of the earth. If a vessel filled with air be closed at the base of a high mountain and afterwards opened on its summit, the air will rush out; and the vessel being closed again on the summit and opened at the base of the mountain, the air will rush in.

The evaporation which takes place from large bodies of water, the activity of vegetable and animal life, as well as vegetable decompositions, throw considerable quantities of aqueous vapor, carbonic acid, and other foreign ingredients temporarily into the permanent portions of the atmosphere. These, together with its ever-varying temperature, keep the density and elastic force of the air in a state of almost incessant change. These changes are indicated by the *Barometer*, an instrument employed to measure the intensity of atmospheric pressure, and frequently called a *weather-glass*, because of certain agreements found to exist between its indications and the state of the weather.

The barometer consists of a glass tube about thirty-four or thirty-five inches long, open at one end, partly filled with distilled mercury, and inverted in a small cistern also containing mercury. A scale of equal parts is cut upon a slip of metal, and placed against the tube to measure the height of the mercurial column, the zero being on a level with the surface of the mercury in the cistern. The elastic force of the air acting freely upon the mercury in the cistern, its pressure is transmitted to the interior of the tube, and sustains a column of mercury whose weight it is just sufficient to counterbalance. If the density and consequent elastic force of the air be increased, the column of mercury will rise till it attain a corresponding increase of weight; if, on the contrary, the density of the air diminish, the column will fall till its diminished weight is sufficient to restore the equilibrium.

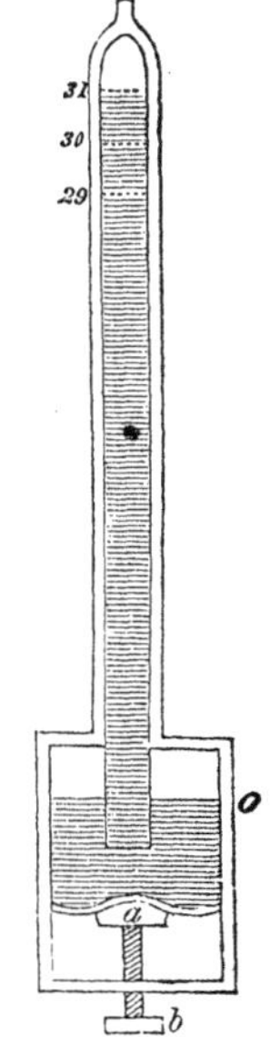

In the *Common Barometer*, the tube and its cistern are partly inclosed in a metallic case, upon which the scale is cut, the cistern, in this case, having a flexible bottom of leather, against which a plate *a* at the end of a screw *b* is made to press, in order to elevate or depress the mercury in the cistern to the zero of the scale.

De Luc's Siphon Barometer consists of a glass tube bent upward so as to form two unequal parallel legs: the longer is hermetically sealed, and constitutes the Torricellian tube; the shorter is open, and on the surface of the quicksilver the pressure of the atmosphere is exerted. The difference between the levels in the longer and shorter legs is the barometric height. The most convenient and practicable way of measuring this difference, is to adjust a movable scale between the two legs, so that its zero may be made to coincide with the level of the mercury in the shorter leg.

Different contrivances have been adopted to render the minute variations in the atmospheric pressure, and consequently in the height of the barometer, more readily perceptible by enlarging the divisions on the scale, all of which devices tend to hinder the exact measurement of the length of the column. Of these we may name Morland's Diagonal, and Hook's Wheel-Barometer, but especially Huygen's Double-Barometer.

The essential properties of a good barometer are width of tube; purity of the mercury; accurate graduation of the scale; and a good *vernier*.

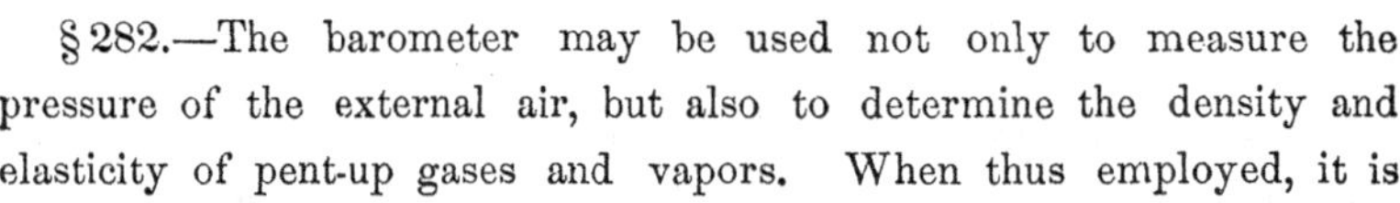

§ 282.—The barometer may be used not only to measure the pressure of the external air, but also to determine the density and elasticity of pent-up gases and vapors. When thus employed, it is called the *barometer-gauge*. In every case it will only be necessary to establish a free connection between the cistern of the barometer and the vessel containing the fluid whose elasticity is to be indicated; the height of the mercury in the tube, expressed in inches, reduced to a standard temperature, and multiplied by the known weight of a cubic inch of mercury at that temperature, will give the pressure in pounds on each square inch. In the case of the steam in the boiler of an engine, the upper end of the tube is sometimes left open. The cistern A is a steam-tight vessel, partly filled with mercury, a is a tube communicating with the boiler, and through which the steam flows and presses upon the mercury; the barometer tube $b\,c$, open at top, reaches nearly to the bottom of the vessel A.

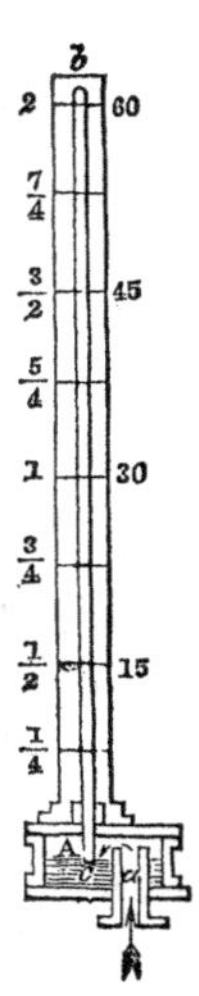

having attached to it a scale whose zero coincides with the level of the quicksilver. On the right is marked a scale of inches, and on the left a scale of atmospheres.

If a very high pressure were exerted, one of several atmospheres for example, an apparatus thus constructed would require a tube of great length, in which case *Mariotte's manometer* is considered preferable. The tube being filled with air and the upper end closed, the surface of the mercury in both branches will stand at the same level as long as no steam is admitted. The steam being admitted through d, presses on the surface of the mercury a and forces it up the branch $b\,c$, and the scale from b to c marks the force of compression in atmospheres. The greater width of tube is given at a, in order that the level of the mercury at this point may not be materially affected by its ascent up the branch $b\,c$, the point a being the zero of the scale.

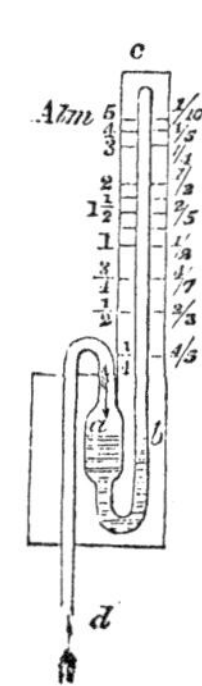

§ 283.—Another very important use of the barometer, is to find the difference of level between two places on the earth's surface, as the foot and top of a hill or mountain.

Since the altitude of the barometer depends on the pressure of the atmosphere, and as this force depends upon the height of the pressing column, a shorter column will exert a less pressure than a longer one. The quicksilver in the barometer falls when the instrument is carried from the foot to the top of a mountain, and rises again when restored to its first position: if taken down the shaft of a mine, the barometric column rises to a still greater height. At the foot of the mountain the whole column of the atmosphere, from its utmost limits, presses with its entire weight on the mercury; at the top of the mountain this weight is diminished by that of the intervening stratum between the two stations, and a shorter column of mercury will be sustained by it.

It is well known that the surface of the earth is not uniform, and does not, in consequence, sustain an equal atmospheric pressure at its different points; whence the mean altitude of the barometric column will vary at different places. This furnishes one of the best and most expeditious means of getting a profile of an extended section of the earth's surface, and makes the barometer an instrument of great value in the hands of the traveller in search of geographical information.

§ 284.—To find the relation which subsists between the altitudes of two barometric columns, and the difference of level of the points where they exist, resume Equation (427). The only extraneous force acting being that of gravity, we have, taking the axis z vertical, and counting z positive upwards,

$$X = 0; \quad Y = 0; \quad Z = -g.$$

and hence,

$$p = Ce^{-\frac{gz}{P}}. \quad \ldots \ldots \ldots \quad (462)$$

Making $z = 0$, and denoting the corresponding pressure by $p_{\prime}$, we find

$$p_{\prime} = C;$$

and dividing the last equation by this one,

$$\frac{p}{p_{\prime}} = e^{-\frac{gz}{P}},$$

whence, denoting the reciprocal of the common modulus by M,

$$z = \frac{MP}{g} \cdot \log \frac{p_{\prime}}{p}. \quad \ldots \ldots \quad (463)$$

Denote by $h_{\prime}$ and h, the barometric heights at the lower and upper stations, respectively, then will

$$\frac{p_{\prime}}{p} = \frac{h_{\prime}}{h};$$

and reducing the barometric column h to what it would have been had the temperature of the mercury at the upper not differed from that at the lower station, by Equation (394), we have

$$\frac{p_{\prime}}{p} = \frac{h_{\prime}}{h[1 + (T - T') \,.\, 0{,}0001001]};$$

in which T denotes the temperature of the mercury at the lower and T' that at the upper station.

Moreover, Equation (381),

$$g = g'(1 - 0{,}002551 \cos 2\psi);$$

in which,

$$g' = \overset{f}{32{,}1808} = \text{force of gravity at the latitude of } 45°.$$

Substituting the value of $\frac{p_{\prime}}{p}$., of g, and that of P, as given by Equation (393), in Equation (463), we find

$$z = \frac{MD_m h_{\prime\prime}}{D_{\prime}} \cdot \frac{1+(t-32°)0{,}00204}{1-0{,}002551\cos 2\psi} \times \log\left[\frac{h_{\prime}}{h} \times \frac{1}{1+(T-T')0{,}0001001}\right].$$

In this it will be remembered that t denotes the temperature of the air; but this may not be, indeed scarcely ever is, the same at both stations, and thence arises a difficulty in applying the formula. But if we represent, for a moment, the entire factor of the second member, into which the factor involving t is multiplied, by X, then we may write

$$z = [1 + (t - 32°)0{,}00204]\, X.$$

If the temperature of the *lower station* be denoted by $t_{\prime}$, and this temperature be the same throughout to the upper station, then will

$$z_{\prime} = [1 + (t_{\prime} - 32°)\, 0{,}00204]\, X.$$

And if the actual temperature of the *upper station* be denoted by t', and this be supposed to extend to the lower station, then would

$$z' = [1 + (t' - 32°)\, 0{,}00204]\, X.$$

Now if $t_{\prime}$ be greater than t', which is usually the case, then will the barometric column, or h, at the upper station, be greater than would result from the temperature t', since the air being more expanded, a portion which is actually below would pass above the upper station and press upon the mercury in the cistern; and because h enters the denominator of the value X, $z_{\prime}$ would be too small. Again, by supposing the temperature the same as that at the upper station throughout, then would the air be more condensed at the lower station, a portion of the air would sink below the upper station that before was above it, and would cease to act upon the mercurial column h, which would, in consequence, become too small.

and this would make z' too great. Taking a mean between $z_{\prime}$ and z' as the true value, we find

$$z = \frac{z_{\prime} + z'}{2} = [1 + \tfrac{1}{2}(t_{\prime} + t' - 64^{\circ})\ 0{,}00204]\,X.$$

Replacing X by its value,

$$z = \frac{MD_m h_{\prime\prime}}{D_{\prime}} \cdot \frac{1+\frac{1}{2}(t_{\prime}+t'-64^{\circ})0{,}00204}{1-0{,}002551\cos 2\psi} \times \log\left[\frac{h_{\prime}}{h} \times \frac{1}{1+(T-T')0{,}0001001}\right]$$

The factor $\frac{MD_m h_{\prime\prime}}{D_{\prime}}$, we have seen, is constant, and it only remains to determine its value. For this purpose, measure with accuracy the difference of level between two stations, one at the base and the other on the summit of some lofty mountain, by means of a Theodolite, or levelling instrument—this will give the value of z; observe the barometric column at both stations—this will give h and $h_{\prime}$; take also the temperature of the mercury at the two stations—this will give T and T'; and by a detached thermometer in the shade, at both stations, find the values of $t_{\prime}$ and t'. These, and the latitude of the place, being substituted in the formula, every thing will be known except the co-efficient in question, which may, therefore, be found by the solution of a simple equation. In this way, it is found that

$$\frac{MD_m h_{\prime\prime}}{D_{\prime}} = 60345{,}51 \text{ English feet};$$

which will finally give for z,

$$z = 60345{,}51^{ft.} \cdot \frac{1+\frac{1}{2}(t_{\prime}+t'-64^{\circ})0{,}00204}{1-0{,}002551\cos 2\psi} \times \log\left[\frac{h_{\prime}}{h} \times \frac{1}{1+(T-T')0{,}0001001}\right]$$

To find the difference of level between any two stations, the latitude of the locality must be known; it will then only be necessary to note the barometric columns, the temperature of the mercury, and that of the air at the two stations, and to substitute these observed elements in this formula.

Much labor is, however, saved by the use of a table for the computation of these results, and we now proceed to explain how it may be formed and used.

Make

$$60345,51\,[1 + (t_{,} + t' - 64^\circ)\,0,00102] = A,$$

$$\frac{1}{1 - 0,002551\,\cos 2\psi} = B,$$

$$\frac{1}{1 + (T - T')\,0,0001} = C.$$

Then will

$$z = AB \cdot \log \frac{C \cdot h_{,}}{h},$$

$$z = AB \cdot [\log C + \log h_{,} - \log h];$$

and taking the logarithms of both members,

$$\log z = \log A + \log B + \log[\log C + \log h_{,} - \log h] \quad \cdot \cdot \quad (464)$$

Making $t_{,} + t'$ to vary from 40° to 162°, which will be sufficient for all practical purposes, the logarithms of the corresponding values of A are entered in a column, under the head A, opposite the values $t_{,} + t'$, as an argument.

Causing the latitude ψ to vary from 0° to 90°, the logarithms of the corresponding values of B are entered in a column headed B, opposite the values of ψ.

The value of $T - T'$ being made, in like manner, to vary from $-30°$ to $+30°$, the logarithms of the corresponding values of C are entered under the head of C, and opposite the values of $T - T'$. In this way a table is easily constructed. Table IV was computed by Samuel Howlet, Esq., from the formula of Mr. Francis Baily, which is very nearly the same as that just described, there being but a trifling difference in the co-efficients.

Taking Equation (464) in connection with Table IV, we have this rule for finding the altitude of one station above another, viz. :—

Take the logarithm of the barometric reading at the lower station, to which add the number in the column headed C, opposite the observed value of $T - T'$, and subtract from this sum the logarithm of the barometric reading at the upper station; take the logarithm of this difference, to which add the numbers in the columns headed A and B, corresponding to the observed values of $t_{,} + t'$ and ψ; the sum will be the logarithm of the height in English feet.

Example.—At the mountain of Guanaxuato, in Mexico, M. Humboldt observed at the

	Upper Station.	Lower Station.
Detached thermometer,	$t' = 70°,4$;	$t_{,} = 77°,6$.
Attached "	$T' = 70,4$;	$T = 77,6$.
Barometric column,	$h = 23,66$;	$h_{,} = 30,05$.

What was the difference of level?

Here

$$t_{,} + t' = 148°; \quad T - T' = 7°,2; \quad \text{Latitude } 21°.$$

To log $\overset{in.}{30,05}$	= 1,4778445		
Add C for 7°,2	= 9,9996814		
	1,4775259		
Sub. log $\overset{in.}{23,66}$	= 1,3740147		
Log of - - -	0,1035112	=	− 1,0149873
Add A for 148° - - - -		=	4,8193975
Add B for 21° - - - -		=	0,0008689
$\overset{ft.}{6843,1}$ - - - - -			3,8352537;

whence the mountain is 6843,1 feet high.

It will be remembered that the final Equation (464) was deduced on the supposition that the air is in equilibrio—that is to say, when there is no wind. The barometer can, therefore, only be used for levelling purposes in calm weather. Moreover, to insure accuracy, the observations at the two stations whose difference of level is to be found, should be made simultaneously, else the temperature of the air may change during the interval between them; but with a single instrument this is impracticable, and we proceed thus, viz.: Take the barometric column, the reading of the attached and detached thermometers, and time of day at one of the stations, say the lower; then proceed to the upper station, and take the same elements there; and at an equal interval of time afterward, observe these elements at the lower station again; reduce the mercurial columns at the lower station to the same temperature by Equation (394), take a mean of these columns, and a mean of the temperatures of the air at this station, and use these means as a single set of observations made simultaneously with those at the higher station.

Example.—The following observations were made to determine the height of a hill near West Point, N. Y.

	Upper Station.	Lower Station. (1)		(2)
Detached thermometer,	$t' = 57°$;	$t_{,} = 56°$	and	$61°$.
Attached "	$T' = 57{,}5$;	$T = 56{,}5$	and	63.
Barometric column,	$h = \overset{in.}{28{,}94}$;	$h_{,} = \overset{in.}{29{,}62}$	and	$\overset{in.}{29{,}63}$.

First, to reduce 29,63 inches at 63°, to what it would have been at 56°,5. For this purpose, Equation (394) gives

$$h\,(1 + \overline{T - T'} \times 0{,}0001) = 29{,}63\,(1 - 6{,}5 \times 0{,}0001) = \overset{in.}{29{,}611}$$

Then

$$h_{,} = \frac{29{,}62 + 29{,}611}{2} = \overset{in.}{29{,}6155}.$$

$$t_{,} = \frac{56° + 61°}{2} \;\text{- -}\; = 58°{,}5,$$

$$t_{,} + t' = 58°{,}5 + 57° \;\text{- -}\; = 115°{,}5,$$

$$T - T' = 56°{,}5 - 57°{,}5 \;\text{-}\; = -1°.$$

To log $\overset{in.}{29{,}6155}$ =	1,4715191		
Add C for $-1°$ =	0,0000434		
	1,4715625		
Sub. log of $\overset{in.}{28{,}94}$ =	1,4614985		
Log of - - - -	0,0100640	=	$-$ 2,0027706
Add A for 115°,5 - - -		=	4,8048112
Add B for 41°,4 - - - -		=	0,0001465
$\overset{ft.}{642{,}28}$ - - -		- -	2,8077283;

whence the height of the hill is 642,28 English feet.

MOTION OF HEAVY INCOMPRESSIBLE FLUIDS IN VESSELS.

§ 285.—Let it now be the question to investigate the flow of a heavy, homogeneous and incompressible fluid through an opening in any part of a vessel which contains it. And for this purpose, resume Eq. (407), which is directly applicable to the case. The only incessant force being the weight of the fluid, take the axis z vertical and positive upwards; then will

$$X = 0; \; Y = 0; \text{ and } Z = -g.$$

The lateral or horizontal velocities will be insignificant in comparison with the vertical, and may be disregarded or neglected, and we have, Eq. (405),

$$\left(\frac{d\varphi}{dx}\right)^2 = 0; \; \left(\frac{d\varphi}{dy}\right)^2 = 0; \text{ and } \left(\frac{d\varphi}{dz}\right)^2 = w^2;$$

which will reduce the general Eq. (407) to

$$\frac{1}{D} dp = -g\,dz - d\frac{d\varphi}{dt} - \tfrac{1}{2} d\,(w)^2;$$

and integrating,

$$p = -Dgz - D \cdot \frac{d\varphi}{dt} - \tfrac{1}{2} D \cdot w^2 + C \quad . \quad . \quad . \quad (465)$$

Next, find the function φ, and its differential co-efficient of the first order with respect to t.

Equations (405) give

$$d\varphi = w \cdot dz,$$

$$\varphi = \int w \cdot dz. \quad . \quad . \quad . \quad . \quad . \quad . \quad . \quad . \quad (466)$$

Let $A\,B\,C\,D$ be a vertical section of the vessel, and take the following notation, viz.:

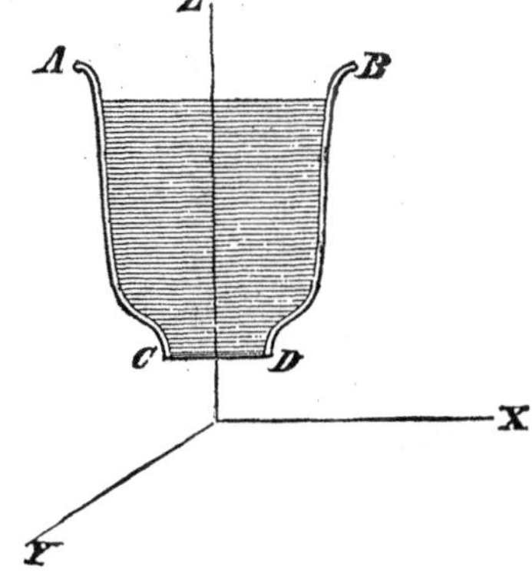

s = the variable area of the stratum whose velocity is w.

$s_{\prime}$ = the constant area of any determinate horizontal section of the vessel, as $C\,D$.

S = the area of the section of the vessel by the upper surface of the liquid; this may be constant or variable, according as the upper surface is stationary or movable.

$w_{,}$ = velocity of the stratum passing the section $s_{,}$ at $C\,D$, at the time t.

The continuity of the fluid requires that

$$w \cdot s = w_{,} \cdot s_{,} \, ,$$

because the same quantity must flow through every horizontal section in the same time: whence

$$w = w_{,} \frac{s_{,}}{s} ;$$

which, in Eq. (466), gives

$$\varphi = w_{,} \cdot s_{,} \cdot \int_{z_{,}+h}^{z_{,}} \frac{d\,z}{s} ;$$

the integration being taken with respect to the variable z, of which s is a function. This function will be given by the figure of the vessel, h being the height of the upper surface of the fluid above the opening.

It may be well to remark here, that

$$\int_{z_{,}+h}^{z_{,}} \frac{d\,z}{s} ,$$

will be constant for the same vessel and same value for h; and if the figure of the vessel be that of revolution about a vertical axis, it will only be necessary to have the equation of this vertical section to find the value of the integral. The quantity h is called the *head* of fluid.

Differentiating φ with respect to t, and recollecting that the velocity downward is negative, we find

$$\frac{d\,\varphi}{d\,t} = - s_{,} \cdot \frac{d\,w_{,}}{d\,t} \int_{z_{,}+h}^{z_{,}} \frac{d\,z}{s} ;$$

and this, with the value of w, in Eq. (465), gives

$$p = - D g z + D \cdot s_{,} \cdot \frac{d\,w_{,}}{d\,t} \int \frac{d\,z}{s} - D \frac{w_{,}^{2}}{2} \cdot \frac{s_{,}^{2}}{s^{2}} + C \cdot \cdot (467)$$

To find the value of C, let $p = P_{,}$, when $z = z_{,}$, which corresponds to the section CD of the liquid; then will

$$P_{,} = -Dgz_{,} + D \cdot s_{,} \cdot \frac{dw_{,}}{dt} \cdot \int_{z=z_{,}} \frac{dz}{s} - D\frac{w_{,}^2}{2} \cdot \frac{s_{,}^2}{s_{,}^2} + C,$$

which, subtracted from the equation above, gives

$$p - P_{,} = -Dg(z - z_{,}) + D \cdot s_{,} \cdot \frac{dw_{,}}{dt} \int_{z_{,}}^{z} \frac{dz}{s} - D\frac{w_{,}^2}{2}\left[\frac{s_{,}^2}{s^2} - 1\right] \cdot (468)$$

Also, if P' denote the pressure at the upper surface corresponding to which $z = z'$, we have

$$P' - P_{,} = -Dg(z' - z_{,}) + D \cdot s_{,} \cdot \frac{dw_{,}}{dt} \cdot \int_{z_{,}}^{z'} \frac{dz}{s} + D\frac{w_{,}^2}{2}\left[1 - \frac{s_{,}^2}{S^2}\right] \cdot (469)$$

Now $z' - z_{,} = h =$ height of the fluid surface above the section CD; whence, by substitution and transposition,

$$P' - P_{,} + Dgh - Ds_{,} \cdot \frac{dw_{,}}{dt} \cdot \int_{z_{,}}^{z'} \frac{dz}{s} - D \cdot \frac{w_{,}^2}{2}\left(1 - \frac{s_{,}^2}{S^2}\right) = 0 \cdot (470)$$

The quantity of fluid flowing through every section in the same time being equal, we also have

$$-Sdh = s_{,} . w_{,} . dt. \quad \cdots\cdots \quad (471)$$

By means of this equation, t may be eliminated from Equation (470); then knowing the quantity of the liquid, the size and figure of the vessel, we will know h, S and the integral $\int_{z_{,}}^{z'} \frac{dz}{s} = \int_{0}^{h} \frac{dz}{s}$, in which s is a function of z.

§ 287.—The value of $\frac{dw_{,}}{dt}$ being found from Equation (470), and substituted in Equation (468), this latter equation will give the value of the pressure p at any point of the fluid mass as soon as $w_{,}$ becomes known.

Two cases may arise. Either the vessel may be kept constantly full while the liquid is flowing out at the bottom, or it may be suffered to empty itself.

§ 288.—To discuss the case in which the vessel is always full, or the fluid retains the same level by being supplied at the top as fast as it flows out at the bottom, the quantity h must be constant, and Equation (471) will not be used.

And making, in Equation (470),

$$A = 2 s_{,} \int_{0}^{h} \frac{dz}{s};$$

$$B = 2g\left(h + \frac{P' - P_{,}}{Dg}\right);$$

$$C = \frac{s_{,}^{2}}{S^{2}} - 1;$$

and solving with respect to $d.t$, we have

$$dt = \frac{A \cdot dw_{,}}{B + Cw_{,}^{2}}. \quad \ldots \ldots \quad (472)$$

Now, three cases may occur.

1st. S may be less than $s_{,}$, and C will be positive.

2d. S may be equal to $s_{,}$, in which case C will be zero.

3d. S may be greater than $s_{,}$, when C will be negative, and this is usually the case in practice.

In the first case, when C is positive, we have, by integrating Equation (472), and supposing $t = 0$, when $w_{,} = 0$,

$$t = \frac{A}{\sqrt{BC}} \cdot \tan^{-1} w_{,} \sqrt{\frac{C}{B}}; \quad \ldots \ldots \quad (473)$$

whence,

$$w_{,} = \sqrt{\frac{B}{C}} \cdot \tan \frac{\sqrt{BC}}{A} \cdot t. \quad \ldots \ldots \quad (474)$$

from which we see that the velocity of egress increases rapidly with the time; it becomes infinite when

$$\frac{\sqrt{BC}}{A} \cdot t = \frac{\pi}{2},$$

or

$$t = \frac{\pi \cdot A}{2\sqrt{BC}}. \quad \ldots \ldots \quad (475)$$

When $C = 0$, then will the integration of Equation (472) give

$$t = \frac{A}{B} \cdot w_{,}, \quad \ldots \ldots \quad (476)$$

or replacing A and B by their values, and finding the value of $w_{\prime}$,

$$w_{\prime} = \frac{g\left(h + \frac{P' - P_{\prime}}{D g}\right)}{s_{\prime} \int_0^h \frac{dz}{s}} \cdot t; \quad \ldots \ldots \quad (477)$$

whence, the velocity varies directly as the time, as it should, since the whole fluid mass would fall like a solid body under the action of its own weight.

When C is negative, the integration gives

$$t = \frac{A}{2\sqrt{B C}} \cdot \log \frac{\sqrt{B} + w_{\prime}\sqrt{C}}{\sqrt{B} - w_{\prime}\sqrt{C}};$$

whence,

$$w_{\prime} = \frac{e^{\frac{2\sqrt{B C}}{A} \cdot t} - 1}{e^{\frac{2\sqrt{B C}}{A} \cdot t} + 1} \cdot \sqrt{\frac{B}{C}}; \quad \ldots \ldots \quad (478)$$

in which e is the base of the Naperian system of logarithms $= 2{,}718282$.

If the section S exceeds $s_{\prime}$ considerably, the exponent of e will soon become very great, and unity may be neglected in comparison with the corresponding power of e; whence,

$$w_{\prime} = \sqrt{\frac{B}{C}} = \sqrt{\frac{2g\left(h + \frac{P' - P_{\prime}}{D g}\right)}{1 - \frac{s_{\prime}^2}{S^2}}}; \quad \ldots \quad (479)$$

that is to say, the velocity will soon become constant.

If the pressure at the upper surface be equal to that at the place of egress, which would be sensibly the case in the atmosphere, $P' - P_{\prime} = 0$, and

$$w_{\prime} = \sqrt{\frac{2 g h}{1 - \frac{s_{\prime}^2}{S^2}}}; \quad \ldots \ldots \ldots \quad (480)$$

and if the opening below become a mere orifice, the fraction

$$\frac{s_{\prime}^2}{S^2} = 0;$$

and

$$w_{\prime} = \sqrt{2 g h}; \quad \ldots \ldots \quad (481)$$

that is to say, the velocity with which a heavy liquid will issue from a small orifice in the bottom of a vessel, when subjected to the pressure of the superincumbent mass, is equal to that acquired by a heavy body in falling through a height equal to the depth of the orifice below the upper surface of the liquid. The velocities given by Equations (479), (480), (481), are independent of the figure of the vessel.

If the velocity $w_{\prime}$ be multiplied by the area $s_{\prime}$ of the orifice, the product will be the volume of fluid discharged in a unit of time. This is called the *expense*. The expense multiplied by the time of flow will give the whole volume discharged.

§ 289.—The velocity $w_{\prime}$ being constant in the case referred to in Equation (479), we shall have

$$\frac{d w_{\prime}}{d t} = 0,$$

and Equation (468) becomes

$$p = P_{\prime} - D g (z - z_{\prime}) - D \cdot \frac{w_{\prime}^2}{2} \cdot \left(\frac{s_{\prime}^2}{s^2} - 1 \right),$$

or, substituting the value of $w_{\prime}$, given by Equation (470),

$$p = P_{\prime} - D g (z - z_{\prime}) + (D g h + P' - P_{\prime}) \cdot \frac{\frac{s_{\prime}^2}{s^2} - 1}{\frac{s_{\prime}^2}{S^2} - 1}; \quad . \quad . \quad (482)$$

whence, it appears, that when the flow has become uniform, the pressure upon any stratum is wholly independent of the figure of the vessel, and depends only upon the area s of the stratum, its distance from the upper surface of the fluid, and upon the ratio $\frac{s_{\prime}^2}{S^2}$.

§ 290.—If the vessel be not replenished, but be allowed to empty itself, h will be variable, as will also S except in the particular cases of the prism and cylinder.

Making

$$w_{\prime} = \sqrt{2 g H}, \quad . \quad . \quad . \quad . \quad . \quad . \quad . \quad . \quad (483)$$

in which H denotes the height due to the velocity of discharge: we have

$$d\,w_{,} = \frac{g \cdot d\,H}{\sqrt{2gH}}; \quad . \quad . \quad . \quad . \quad . \quad . \quad . \quad . \qquad (484)$$

and, Equation (471),

$$d\,t = -\frac{S \cdot d\,h}{s_{,}\sqrt{2\,g\,H}}; \quad . \quad . \quad . \quad . \quad . \quad . \quad . \qquad (485)$$

and by integration,

$$t = C - \frac{1}{s_{,}\sqrt{2\,g}} \cdot \int \frac{S \cdot d\,h}{\sqrt{H}} \cdot \quad . \quad . \quad . \quad . \quad . \quad . \qquad (486)$$

To effect the integration, S and H must be found in terms of h. The relation between S and h will be given by the figure of the vessel. Then to find the relation between H and h, eliminate $w_{,}$, $d\,w_{,}$, and $d\,t$ from Equation (470), by the values above, and we have

$$\left(\frac{P' - P_{,}}{D\,g} + h\right) \cdot d\,h + \frac{s_{,}^{2}\,d\,H}{S}\int_{0}^{h}\frac{d\,z}{s} - H\left(1 - \frac{s_{,}^{2}}{S^{2}}\right)d\,h = 0;$$

or, dividing by

$$\frac{s_{,}^{2}}{S} \cdot \int_{0}^{h}\frac{d\,z}{s},$$

$$\frac{S \cdot \left(\frac{P' - P_{,}}{D\,g} + h\right)}{s_{,}^{2}\int_{0}^{h}\frac{d\,z}{s}} \cdot d\,h + d\,H - \frac{S \cdot \left(1 - \frac{s_{,}^{2}}{S^{2}}\right)}{s_{,}^{2}\int_{0}^{h}\frac{d\,z}{s}} \cdot H\;d\,h = 0 \cdot (487)$$

and making

$$R = -\frac{S \cdot \left(1 - \frac{s_{,}^{2}}{S^{2}}\right)}{s_{,}^{2}\int_{0}^{h}\frac{d\,z}{s}}; \quad Q = \frac{S \cdot \left(\frac{P' - P_{,}}{D\,g} + h\right)}{s_{,}^{2}\int_{0}^{h}\frac{d\,z}{s}};$$

$$Q\,d\,h + d\,H + R\,H\,d\,h = 0. \quad . \quad . \quad . \qquad (488)$$

Multiplying by $e^{\int R\,d\,h}$,

$$d\,h \cdot Q \cdot e^{\int R\,d\,h} + d\,H \cdot e^{\int R\,d\,h} + H \cdot e^{\int R\,d\,h} \times R\,d\,h = 0;$$

or

$$dh \cdot Q \cdot e^{\int R\,dh} + d\left(He^{\int R\,dh}\right) = 0;$$

and integrating

$$\int dh \cdot Q \cdot e^{\int R\,dh} + He^{\int R\,dh} = C; \quad . \quad . \quad . \quad . \quad (489)$$

whence,

$$H = e^{-\int R\,dh} \cdot \left(C - \int dh \cdot Q \cdot e^{\int R\,dh}\right) \quad . \quad . \quad . \quad . \quad (490)$$

The constant must result from the condition, that when $H = 0$, h must be $h_{,}$, the initial height of the fluid in the vessel.

Thus H becomes known in terms of h, and its value substituted in Equation (486) will make known the time required for the fluid to reach any altitude h. The constant in Equation (486) must be determined, so that when $t = 0$, $h = h_{,}$.

§ 291.—The Equation (490) gives a direct relation between S, h, and H; the figure and dimensions of the vessel give another between S and h. From these, two of the three variables may be eliminated from the Equation (486) and the integration performed. Take, for example, the case of a right cylinder or prism. Here S will be constant, and equal to s.

$$\int_o^h \frac{dz}{s} = \frac{h}{S}.$$

Moreover, let us suppose $P' - P_{,} = 0$, which would be sensibly true were the fluid to flow into the atmosphere that surrounds the vessel. Also, for the sake of abbreviation, make $\frac{S}{s_{,}} = k$, then will

$$R = -\frac{k^2 - 1}{h} = \frac{1 - k^2}{h},$$

$$Q = k^2;$$

and

$$\int Rdh = (1 - k^2)\int \frac{dh}{h} = (1 - k^2)\log h.$$

and Eq. (490) becomes

$$H = e^{-(1-k^2)\log h} \cdot [C - \int k^2 dh \cdot e^{(1-k^2)\log h}]$$

Multiplying the last term by

$$\frac{2-k^2}{2-k^2} \cdot \frac{h}{h},$$

we may write

$$H = e^{-(1-k^2)\log h} \cdot \left[C - \frac{k^2}{2-k^2} \cdot \int d(h \cdot e^{(1-k^2)\log h})\right]$$

$$= e^{-(1-k^2)\log h} \cdot \left[C - \frac{k^2}{2-k^2} \cdot h \cdot e^{(1-k^2)\log h}\right];$$

when $H = 0$, then will $h = h_{\prime}$ and

$$C = \frac{k^2}{2-k^2} \cdot h_{\prime} \cdot e^{(1-k^2)\log h_{\prime}};$$

which substituted above, gives, after reduction,

$$H = \frac{k^2 \cdot h}{2-k^2} \cdot \left[\frac{h_{\prime}}{h} e^{(1-k^2)\log \frac{h_{\prime}}{h}} - 1\right]:$$

but,

$$e^{(1-k^2)\log \frac{h_{\prime}}{h}} = \left(\frac{h_{\prime}}{h}\right)^{1-k^2};$$

and therefore,

$$H = \frac{k^2 \cdot h}{2-k^2}\left[\left(\frac{h_{\prime}}{h}\right)^{2-k^2} - 1\right] = \frac{k^2 \cdot h}{k^2-2}\left[1 - \left(\frac{h}{h_{\prime}}\right)^{k^2-2}\right] \quad . \; . \; . \quad (492)$$

which substituted in Equation (486), gives

$$t = C - \sqrt{\frac{k^2-2}{2g}} \cdot \int \frac{dh}{\sqrt{h\left[1 - \left(\frac{h}{h_{\prime}}\right)^{k^2-2}\right]}}, \quad . \; . \quad (493)$$

in which the only variable is h.

§ 292.—The particular case in which $k^2 = 2$, gives to this value for t the form of indetermination. When this occurs, we must have recourse to the form assumed by Equation (488), which, under this supposition, becomes

$$2h\,dh + h\,dH - H\,dh = 0:$$

multiplying by h^{-2},

$$2h^{-1}dh + h^{-1}.dH - H.h^{-2}dh = 0,$$

$$2 \cdot \frac{dh}{h} + d\frac{H}{h} = 0,$$

$$2 \log h + \frac{H}{h} = C;$$

and because $H = 0$ when $h = h_{\prime}$,

$$2 \log h_{\prime} = C;$$

whence,

$$H = 2h \cdot \log \frac{h_{\prime}}{h},$$

and this, in Equation (486), gives

$$t = C - \frac{1}{\sqrt{g}} \int \frac{dh}{\sqrt{2h \cdot \log \frac{h_{\prime}}{h}}}.$$

Making $\frac{h_{\prime}}{h} = \frac{1}{x^2}$, this becomes, between the limits $x = 0$, $x = 1$,

$$t = C - \sqrt{\frac{h_{\prime}}{g}} \cdot \int \frac{dx}{\sqrt{\log \frac{1}{x}}} = \sqrt{\frac{\pi h_{\prime}}{g}}.$$

§ 293.—If the orifice be very small in comparison with a cross section of the prismatic or cylindrical vessel, then will $H = h$, and Equation (486) gives

$$t = C - \frac{2S}{s_{\prime}\sqrt{2g}} \cdot \sqrt{h}.$$

Making $t = 0$ when $h = h_{\prime}$, we have

$$t = \frac{2S}{s_{\prime}\sqrt{2g}} \cdot (\sqrt{h_{\prime}} - \sqrt{h}), \quad \dots \dots \quad (494)$$

and for the time required for the vessel to empty itself, $h = 0$, and

$$t = \frac{2S}{s_{\prime}} \cdot \sqrt{\frac{h_{\prime}}{2g}}. \quad \dots \dots \quad (495)$$

Now, with the same relation of the orifice to the cross section of the cylindrical vessel, we have, Equation (481),

$$w_{,} = \sqrt{2 g h},$$

and for the volume of fluid discharged in the time t, when the vessel is kept full,

$$w_{,} \,.\, s_{,} \,.\, t = s_{,} \,.\, t \,.\, \sqrt{2 g h},$$

and if this be equal to the contents of the vessel,

$$s_{,} \,.\, t \,.\, \sqrt{2 g h_{,}} = S \,.\, h_{,};$$

whence,

$$t = \frac{S}{s_{,}} \cdot \sqrt{\frac{h_{,}}{2 g}}.$$

That is, Equation (495), the time required for a prismatic or cylindrical vessel to discharge itself through a small orifice at the bottom is double that required to discharge an equal volume, if the vessel were kept full.

§ 294.—The orifice being still small, we obtain, from Equation (485),

$$\frac{d h}{d t} = - \frac{s_{,}}{S} \cdot \sqrt{2 g h};$$

whence it appears that, for a cylindrical or prismatic vessel, the motion of the upper surface of the fluid is uniformly retarded. It will be easy to cause S so to vary, in other words, to give the vessel such figure as to cause the motion of the upper surface to follow any law. If, for example, it were required to give such figure as to cause the motion of the upper surface to be uniform, then would the first member of the above equation be constant; and, denoting the rate of motion by α, we should have

$$\alpha = \frac{s_{,}}{S} \cdot \sqrt{2 g h};$$

whence,

$$S^2 = \frac{s_{,}^2 \,.\, 2 g h}{\alpha^2};$$

but supposing the horizontal sections circular,

$$S^2 = \pi^2 r^4 = \frac{s_{,}^2 \cdot 2 g h}{\alpha^2},$$

and, therefore,

$$r = \sqrt[4]{\frac{2g \cdot s_{,}^{2}}{\pi^{2} \cdot a^{2}}}\sqrt[4]{h};$$

whence the radii of the sections must vary as the fourth root of their distances from the bottom. These considerations apply to the construction of *Clepsydras* or *Water Clocks.*

MOTION OF ELASTIC FLUIDS IN VESSELS.

§ 295.—As in the case of incompressible, so also in that of elastic fluids, it is assumed that in their movement through vessels, they arrange themselves into parallel strata at right angles to the direction of the motion. The quantity of matter in each stratum is supposed to remain the same, while its density, which is always uniform throughout, may vary from one position of the stratum to another; hence, the volume of each stratum may vary.

All lateral velocity among the particles will be supposed zero; and as the weight of the elements of elastic fluids is insignificant in comparison to their elasticity, the former will be disregarded. The motion will, therefore, be due only to the elastic force arising from some force of compression; and as the fluid will be supposed to communicate freely with the air, or with a vessel partly filled with some other elastic fluid, this force within may be greater or less than it is on the exterior of the vessel.

§ 296.—Assuming the axis of the vessel horizontal, take that line as the axis of x. Then, by the supposition above, will

$$X = 0;$$
$$Y = 0;$$
$$Z = 0;$$
$$v = 0;$$
$$w = 0;$$

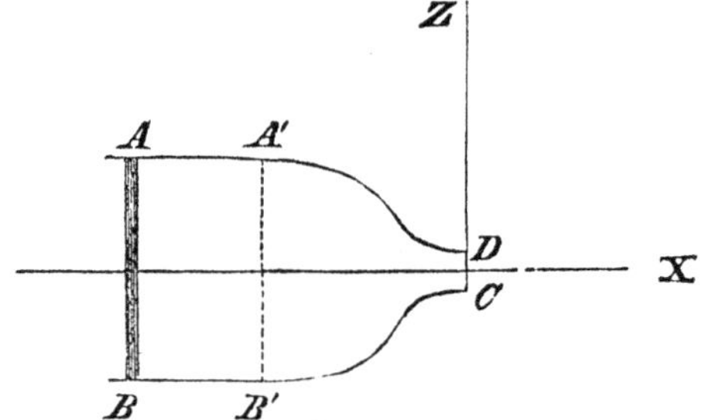

and Equations (400) give

$$\frac{1}{D} \cdot \frac{dp}{dx} = -\left(\frac{du}{dt}\right) - \frac{du}{dx} \cdot u. \quad . \quad . \quad . \quad . \quad (496)$$

Moreover, if we suppose the motion to have been established and become permanent, the velocity of a stratum as it passes any particular cross section of the vessel will always be constant, and the quantity of fluid which flows through every cross section will be the same. Hence the partial differential of u in regard to the time, that is, supposing x, y, z, to be constant, must be zero, and the above equation reduces to

$$dp = -D \, . \, u \, . \, du.$$

From Mariotte's law, Equation (389),

$$p = P \, . \, D,$$

and by division,

$$\frac{dp}{p} = -\frac{1}{P} \cdot u \, du,$$

and by integration,

$$\log p = C - \frac{1}{2P} \cdot u^2. \quad . \quad . \quad . \quad . \quad . \quad . \quad (497)$$

To determine the constant, let $p_{\prime}$ be the pressure at the opening CD, that is, the pressure of the atmosphere, and denote by $u_{\prime}$ the velocity of the fluid at this point, then will

$$\log p_{\prime} = C - \frac{1}{2P} \cdot u_{\prime}^2,$$

and by subtraction,

$$\log \frac{p}{p_{\prime}} = \frac{1}{2P} \cdot (u_{\prime}^2 - u^2). \quad . \quad . \quad . \quad . \quad . \quad (498)$$

Denote by s the area of any section of the vessel $A'B'$, at which the pressure is p and velocity u, by D the density of the fluid at this section, and by $D_{\prime}$ that at the section CD equal to $s_{\prime}$. Then, since the quantities of fluid flowing through these sections in a unit of time must be equal, we have

$$D \, . \, s \, . \, u = D_{\prime} \, . \, s_{\prime} \, . \, u_{\prime} \, ;$$

but, § 244,

$$\frac{D}{D_{,}} = \frac{p}{p_{,}};$$

whence,

$$p \, . \, s \, . \, u = p_{,} s_{,} u_{,} ,$$

or

$$u = \frac{p_{,} s_{,} u_{,}}{p \, . \, s},$$

which, in Equation (498), gives

$$\log \frac{p}{p_{,}} = \frac{u_{,}^2}{2P}\left[1 - \left(\frac{p_{,} s_{,}}{p \, . \, s}\right)^2\right]. \quad . \quad . \quad . \quad . \quad (499)$$

If p' denote the pressure exerted by the piston AB, and S denote its area, we have

$$\log \frac{p'}{p_{,}} = \frac{u_{,}^2}{2P}\left[1 - \left(\frac{p_{,} s_{,}}{p' \, S}\right)^2\right]; \quad . \quad . \quad . \quad . \quad (500)$$

whence,

$$u_{,} = \sqrt{\frac{2P \, . \log \frac{p'}{p_{,}}}{1 - \left(\frac{p_{,} s_{,}}{p' \, S}\right)^2}}. \quad . \quad . \quad . \quad . \quad . \quad (501)$$

This is the velocity with which the fluid will issue into the atmosphere or other fluid whose pressure on the unit of surface is $p_{,}$.

§ 297.—The volume discharged in a unit of time is

$$u_{,} s_{,} = s_{,} \cdot \sqrt{\frac{2P \cdot \log \frac{p'}{p_{,}}}{1 - \left(\frac{p_{,} s_{,}}{p' \cdot S}\right)^2}},$$

while under the pressure $p_{,}$; and under a pressure equal to that on the unit of surface of the piston, or top of a gasometer, and which would be indicated by a gauge, since the volumes are inversely as the pressures,

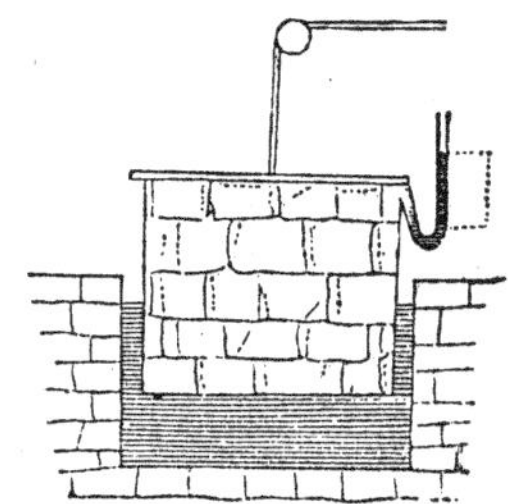

$$u_{,} s_{,} = \frac{p_{,}}{p'} \cdot s_{,} \cdot \sqrt{\frac{2P \cdot \log \frac{p'}{p_{,}}}{1 - \left(\frac{p_{,} s_{,}}{p' \, S}\right)^2}}. \quad . \quad (502)$$

§ 298.—Dividing Equation (499) by Equation (500), we have

$$\frac{\log \frac{p}{p_{\prime}}}{\log \frac{p'}{p_{\prime}}} = \frac{1 - \left(\frac{p_{\prime}\, s_{\prime}}{p \cdot s}\right)^2}{1 - \left(\frac{p_{\prime}\, s_{\prime}}{p'\, S}\right)^2}; \quad \ldots \ldots \quad (503)$$

which will give the pressure p at any section of the vessel.

§ 299.—If the opening CD is very small in reference to AB, the velocity $u_{\prime}$ will become, Equation (501),

$$u_{\prime} = \sqrt{2\, P \cdot \log \frac{p'}{p_{\prime}}}; \quad \ldots \ldots \quad (504)$$

and the volume of fluid discharged in a unit of time and of a density equal to that pressing upon the gauge,

$$\frac{p_{\prime}}{p'} \cdot s_{\prime} \cdot \sqrt{2\, P \cdot \log \frac{p'}{p_{\prime}}}; \quad \ldots \ldots \quad (505)$$

and Equation (503) becomes

$$\frac{\log \frac{p}{p_{\prime}}}{\log \frac{p'}{p_{\prime}}} = 1 - \left(\frac{p_{\prime}\, s_{\prime}}{p\, s}\right)^2.$$

§ 300.—A stream flowing through an orifice is called a *vein*. In estimating the quantity of fluid discharged, it is supposed that there are neither within nor without the vessel any causes to obstruct the free and continuous flow; that the fluid has no viscosity, and does not adhere to the sides of the vessel and orifice; that the particles of the fluid reach the upper surface with a common velocity, and also leave the orifice with equal and parallel velocities. None of these conditions are fulfilled in practice, and the theoretical discharge must, therefore, differ from the actual. Experience teaches that the former always exceeds the latter. If we take water, for example, which is far the most important of the liquids in a practical point of view, we shall find it to a certain degree viscous, and always exhibiting a tendency to adhere to ununctuous surfaces with which it may be brought in contact. When water flows through an opening, the adhesion of its particles to the surface will check their motion, and the viscosity of the fluid will transmit this effect towards the interior of the vein; the velocity will, therefore, be greatest at the axis of the latter, and least on and near its surface; the inner particles thus flowing away from those without, the vein will increase in length and diminish in thickness, till, at a certain distance from the orifice, the velocity becomes the same throughout the same cross-section, which usually takes place at a short distance from the aperture. This effect will be increased by the crowding of the particles, arising from the convergence of the paths along which they approach the aperture, every particle, which enters near the edge, tending to pass obliquely across to the opposite side. This diminution of the fluid vein is called *the veinal contraction.* The quantity of fluid discharged must depend upon the degree of veinal contraction, and the velocity of the particles at the section of greatest diminution; and any cause that will diminish the viscosity and cohesion, and draw the particles in the direction of the axis of the vein as they enter the aperture, will increase the discharge.

Experience shows that the greatest contraction takes place at a distance from the vessel varying from a half to once the greatest dimension of the aperture, and that the amount of contraction depends somewhat upon the shape of the vessel about the orifice and the head of fluid. It is further found by experiment, that if a tube of the same shape and size as the vein, from the side of the vessel to the place of greatest contraction, be inserted into the aperture, the actual discharge of fluid may be accurately computed by Equation (478), provided the smaller base of the tube be substituted for the area of the aperture; and that, generally, without the use of the tube, the actual may be deduced from the theoretical discharge, as given by that equation, by simply multiplying the theoretical discharge into a co-efficient whose numerical value depends upon the size of the aperture and head of the fluid. Moreover, all other circumstances being the same, it is ascertained that this co-efficient remains constant, whether the aperture be circular, square, or oblong, which embrace all cases of practice, provided that in comparing rectangular with circular orifices, we compare the smallest dimension of the former with the diameter of the latter. The value of this co-efficient depends, therefore, when other circumstances are the same, upon the smallest dimension of the rectangular orifice, and upon the diameter of the circle, in the case of circular orifices. But should other circumstances, such as the head of fluid, and the place of the orifice, in respect to the sides and bottom of the vessel, vary, then will the co-efficient also vary. When the flow takes place through thin plates, or through orifices whose lips are bevelled externally, the co-efficient corresponding to given heads and orifices, may be found in Table V, provided the orifices be remote from the lateral faces of the vessel. This table is deduced from the experiments of Captain Lesbros, of the French engineers, and agrees with the previous experiments of Bossut, Michelotti, and others.

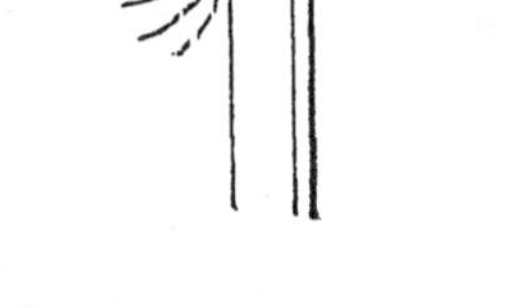

As the orifice approaches one of the lateral faces of the reservoir, the contraction on that side becomes less and less, and will ultimately become nothing, and the co-efficient will be greater than those of the table. If the orifice be near two of these faces, the contraction becomes nothing on two sides, and the co-efficient will be still greater.

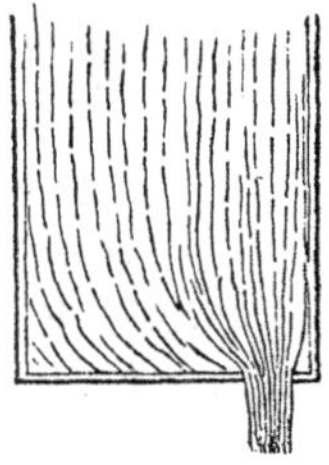

Under these circumstances, we have the following rules:—Denote by C the tabular, and by C' the true co-efficient corresponding to a given aperture and head; then, if the contraction be nothing on one side, will

$$C' = 1{,}03\ C;$$

if nothing on two sides,

$$C' = 1{,}06\ C;$$

if nothing on three sides,

$$C' = 1{,}12\ C;$$

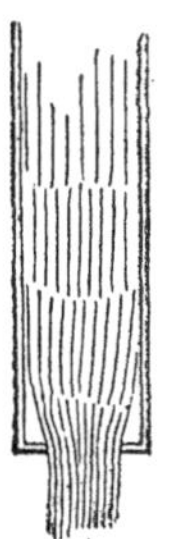

and it must be borne in mind, that these results and those of the table are applicable only when the fluid issues through holes in thin plates, or through apertures so bevelled externally that the particles may not be drawn aside by molecular action along their tubular contour.

§ 301.—When the discharge is through *thick plates without bevel*, or through cylindrical tubes whose lengths are from two to three times the smaller dimension of the orifice, the expense is increased, the mean coefficient, in such cases, augmenting, according to experiment, to about 0,815 for orifices of which the smaller dimension varies from 0,33 to 0,66 of a foot, under heads which give a coefficient 0,619 in the case of thin plates. The cause of this increase is obvious. It is within the observation of every one, that water will wet most surfaces not highly polished or covered with an unctuous coating—in other words, that there exists between the particles of the fluid and those of solids an affinity which will cause the former to spread themselves over the latter and adhere with considerable pertinacity. This affinity becoming effective between the inner surface of the tube and those particles of the fluid which enter the orifice near its edge, the latter will not only be drawn aside from their converging directions, but will take with them, by the force of viscosity, the other particles, with which they are in sensible contact. The fluid filaments leading through the tube will, therefore, be more nearly parallel than in the case of orifices through thin plates, the contraction of the vein will be less, and the discharge consequently greater.

PART III.

MECHANICS OF MOLECULES.

§ 302.—The more general circumstances attending the action of forces upon bodies of sensible magnitudes have been discussed. They constitute the subjects of Mechanics of Solids and of Fluids. Those which result from the action of forces upon the *elements* of both solids and fluids remain to be considered. They form the subject of *Mechanics of Molecules;* which comprehends the whole theory of *Electrics*, *Thermotics*, *Acoustics*, and *Optics*.

It has been seen, that all bodies are built up of elementary molecules in sensible, though not in actual, contact; that the relative places of equilibrium of these molecules are determined by the molecular forces, and that the intensities of these forces are some function of the distance between the acting molecules. A displacement of a single molecule from its position of relative rest, will break up the equilibrium of the surrounding forces, and give rise to a general and progressive disturbance throughout the body. It is proposed to investigate the nature of this disturbance, the circumstances of its progress, and the conduct of the molecules as they become involved in it.

PERIODICITY OF MOLECULAR CONDITION.

§ 303.—Molecular motions cannot, like the initial disturbances which produce them, be arbitrary; but must fulfil certain conditions imposed by the physical connections which unite the molecules into a system.

These motions are, so to speak, constrained by this connection. Let the conditions of constraint be expressed, as in § 213, Mech. of Solids, by

$$L = 0\,; \quad L' = 0\,; \quad L'' = 0\,; \quad \&c. \quad . \quad . \quad . \quad . \qquad (506)$$

L, L', L'', &c., being functions of the co-ordinates of the molecules. Denote by

$$X, Y, Z\,; \;\; X', Y', Z'\,; \;\; \&c.,$$

the accelerations impressed upon the molecules whose masses are m, m', &c., in the directions of the axes. Equation (313) will obtain for each molecule. There will be as many equations as molecules, and by addition, we find, by inverting the terms,

$$\Sigma\, m \left[\left(\frac{d^2 x}{d t^2} - X\right)\delta x + \left(\frac{d^2 y}{d t^2} - Y\right)\delta y + \left(\frac{d^2 z}{d t^2} - Z\right)\delta z\right] = 0, \quad (507)$$

There will be three co-ordinates for each molecule. Denote the number of molecules by i; the number of Equations (506) of condition by m; then will $3\,i - m = n$, be the number of co-ordinates which, being given, will reduce the number of unknown co-ordinates to the number of equations. These unknown co-ordinates may, hence, be found in functions of the known, and the places of the molecules at any instant determined.

Denote the m co-ordinates by $x\,y\,z$, $x'\,y'\,z'$, &c., and the n co-ordinates by $\alpha\,\beta\,\gamma$, $\alpha'\,\beta'\,\gamma'$, &c.: then we may write,

$$x = \varphi_x\,(\alpha\,\beta\,\gamma\,\alpha', \,\&c.) = p_x\,;$$
$$y = \varphi_y\,(\alpha\,\beta\,\gamma\,\alpha', \,\&c.) = p_y\,;$$
$$z = \varphi_z\,(\alpha\,\beta\,\gamma\,\alpha', \,\&c.) = p_z\,;$$
$$x' = \varphi_{x'}\,(\alpha\,\beta\,\gamma\,\alpha', \,\&c.) = p_{x'}\,;$$
$$\&c. = \qquad \&c. \qquad = \&c.\,;$$

also,

$$X = \psi_x\,(\alpha\,\beta\,\gamma\,\alpha', \,\&c.) = P_x\,;$$
$$Y = \psi_y\,(\alpha\,\beta\,\gamma\,\alpha', \,\&c.) = P_y\,;$$
$$Z = \psi_z\,(\alpha\,\beta\,\gamma\,\alpha', \,\&c.) = P_z\,;$$
$$X' = \psi_{x'}\,(\alpha\,\beta\,\gamma\,\alpha', \,\&c.) = P_{x'}\,;$$
$$\&c. = \qquad \&c. \qquad = \&c.\,;$$

in which φ_x, φ_y, φ_z, &c., ψ_x, ψ_y, ψ_z, &c., denote any functions of the co-ordinates $\alpha\ \beta\ \gamma$, α' &c., which result from the conditions of Equations (506) and the process of elimination.

At any time t, suppose

$$\alpha\ \beta \text{ and } \gamma \text{ to become } \alpha+\xi,\ \beta+\eta,\ \gamma+\zeta;$$
$$\alpha'\ \beta' \text{ and } \gamma' \text{ to become } \alpha'+\xi',\ \beta'+\eta',\ \gamma'+\zeta';$$
$$\cdot\quad\cdot\quad\cdot\quad\cdot\quad\cdot\quad\cdot\quad\cdot\quad\cdot\quad\cdot$$

and suppose the increments $\xi\ \eta\ \zeta$, $\xi'\ \eta'\ \zeta'$, &c., to continue so small during the entire motion as to justify the omission of all terms into which their second powers and products enter; then will

$$\left.\begin{array}{l}
x = p_x + \frac{d\,p_x}{d\,\alpha}\cdot\xi + \frac{d\,p_x}{d\,\beta}\cdot\eta + \frac{d\,p_x}{d\,\gamma}\cdot\zeta + \text{\&c., \&c.,}\\[2ex]
y = p_y + \frac{d\,p_y}{d\,\alpha}\cdot\xi + \frac{d\,p_y}{d\,\beta}\cdot\eta + \frac{d\,p_y}{d\,\gamma}\cdot\zeta + \text{\&c., \&c.,}\\[2ex]
z = p_z + \frac{d\,p_z}{d\,\alpha}\cdot\xi + \frac{d\,p_z}{d\,\beta}\cdot\eta + \frac{d\,p_z}{d\,\gamma}\cdot\zeta + \text{\&c., \&c.,}\\[2ex]
x' = p_{x'} + \frac{d\,p_{x'}}{d\,\alpha'}\cdot\xi' + \frac{d\,p_{x'}}{d\,\beta'}\cdot\eta' + \frac{d\,p_{x'}}{d\,\gamma'}\cdot\zeta' + \text{\&c., \&c.,}\\[2ex]
\cdot\quad\cdot\quad\cdot\quad\cdot\quad\cdot\quad\cdot\quad\cdot\quad\cdot\quad\cdot
\end{array}\right\}\ .\ .\ (508)$$

$$\left.\begin{array}{l}
X = P_x + \frac{d\,P_x}{d\,\alpha}\cdot\xi + \frac{d\,P_x}{d\,\beta}\cdot\eta + \frac{d\,P_x}{d\,\gamma}\cdot\zeta + \text{\&c., \&c.,}\\[2ex]
Y = P_y + \frac{d\,P_y}{d\,\alpha}\cdot\xi + \frac{d\,P_y}{d\,\beta}\cdot\eta + \frac{d\,P_y}{d\,\gamma}\cdot\zeta + \text{\&c., \&c.,}\\[2ex]
Z = P_z + \frac{d\,P_z}{d\,\alpha}\cdot\xi + \frac{d\,P_z}{d\,\beta}\cdot\eta + \frac{d\,P_z}{d\,\gamma}\cdot\zeta + \text{\&c., \&c.,}\\[2ex]
X' = P_{x'} + \quad\cdot\quad\cdot\quad\cdot\quad\cdot\quad\cdot\quad\cdot\quad\cdot
\end{array}\right\}\ .\ .\ (509)$$

From Equations (508) we have

$$\left.\begin{array}{l}
d^2 x = \frac{d\,p_x}{d\,\alpha}\cdot d^2\xi + \frac{d\,p_x}{d\,\beta}\cdot d^2\eta + \frac{d\,p_x}{d\,\gamma}\cdot d^2\zeta + \text{\&c., \&c.,}\\[2ex]
d^2 y = \frac{d\,p_y}{d\,\alpha}\cdot d^2\xi + \frac{d\,p_y}{d\,\beta}\cdot d^2\eta + \frac{d\,p_y}{d\,\gamma}\cdot d^2\zeta + \text{\&c., \&c.,}\\[2ex]
d^2 z = \frac{d\,p_z}{d\,\alpha}\cdot d^2\xi + \frac{d\,p_z}{d\,\beta}\cdot d^2\eta + \frac{d\,p_z}{d\,\gamma}\cdot d^2\zeta + \text{\&c., \&c.,}\\[2ex]
d^2 x' = \quad \text{\&c,} \qquad \text{\&c.,} \qquad \text{\&c.}
\end{array}\right\}\ .\ .\ (510)$$

also from the same

$$\left.\begin{aligned} \delta x &= \frac{d\,p_x}{d\,\alpha}\cdot\delta\xi + \frac{d\,p_x}{d\,\beta}\cdot\delta\eta + \frac{d\,p_x}{d\,\gamma}\cdot\delta\zeta + \text{\&c.,} \\ \delta y &= \frac{d\,p_y}{d\,\alpha}\cdot\delta\xi + \frac{d\,p_y}{d\,\beta}\cdot\delta\eta + \frac{d\,p_y}{d\,\gamma}\cdot\delta\zeta + \text{\&c.,} \\ \delta z &= \frac{d\,p_z}{d\,\alpha}\cdot\delta\xi + \frac{d\,p_z}{d\,\beta}\cdot\delta\eta + \frac{d\,p_z}{d\,\gamma}\cdot\delta\zeta + \text{\&c.} \end{aligned}\right\} \quad . \quad . \quad (511)$$

The Equation (507) contains three times as many terms as there are molecules, each term consisting of a variation with its coefficient. Eliminate from this equation X, Y, Z, X', &c., $d^2\,x$, $d^2\,y$, $d^2\,z$, $d^2\,x'$, &c., and $\delta\,x$, $\delta\,y$, $\delta\,z$, $\delta\,x'$, &c., by means of Equations (509), (510), and (511); collect the coefficients of $\delta\,\xi$, $\delta\,\eta$, $\delta\,\zeta$, $\delta\,\xi'$, &c.; the number of terms will reduce to n, this being the number of the co-ordinates α, β, γ, α', &c. These variations are independent of one another, since the co-ordinates α, β, γ, α', &c., are so. The coefficients of these variations must, therefore, be separately equal to zero. Performing the operation, omitting all the terms containing products and powers of ξ, η, ζ, ξ', &c., higher than the first, there will result n equations of the form,

$$\Sigma\,D\,.\frac{d^2\,\xi}{d\,t^2} + \Sigma\,E\,.\frac{d^2\,\eta}{d\,t^2} + \Sigma\,F\,.\frac{d^2\,\zeta}{d\,t^2} + \Sigma\,G\,.\,\xi + \Sigma\,H\,.\,\eta + \Sigma\,K\,.\,\zeta + A = 0\,; \quad (512)$$

in which D, E, F, G, H, &c., are functions of the differential co-efficients in Equations (509), (510), and (511); and A consists of a series of terms each composed of two factors, one of which is either P_x, P_y, P_z, or some other P with subscript co-ordinate accented.

If $\alpha\;\beta\;\gamma$, α', &c., give the places of rest of the molecules, then will $P_x = P_y = P_z =$ &c. $= 0$, and Equations (512) become

$$\Sigma\,D\,.\frac{d^2\,\xi}{d\,t^2} + \Sigma\,E\,.\frac{d^2\,\eta}{d\,t^2} + \Sigma\,F\,.\frac{d^2\,\zeta}{d\,t^2} + \Sigma\,G\,.\,\xi + \Sigma\,H\,.\,\eta + \Sigma\,K\,.\,\zeta = 0. \quad (513)$$

These equations are satisfied by making

$$\left.\begin{aligned} \xi &= R\,.\,N_\xi\,.\sin\,(t\,.\sqrt{\rho} - r), \\ \eta &= R\,.\,N_\eta\,.\sin\,(t\,.\sqrt{\rho} - r), \\ \zeta &= R\,.\,N_\zeta\,.\sin\,(t\,.\sqrt{\rho} - r), \\ \xi' &= \quad . \quad . \quad . \end{aligned}\right\} \quad . \quad . \quad . \quad . \quad . \quad (514)$$

in which R and r are arbitrary constants, and ρ, N_ξ, N_η, N_ζ, &c., are constants to be determined. For, after two differentiations, regarding ξ, η, ζ, &c., and t variable, we have

$$\frac{d^2\xi}{dt^2} = -R \,.\, N_\xi \,.\, \sin(t\sqrt{\rho} - r)\,\rho,$$

$$\frac{d^2\eta}{dt^2} = -R \,.\, N_\eta \,.\, \sin(t\sqrt{\rho} - r)\,\rho,$$

$$\frac{d^2\zeta}{dt^2} = -R \,.\, N_\zeta \,.\, \sin(t\sqrt{\rho} - r)\,\rho,$$

&c., &c.

which, substituted in Equations (513), give, after dividing out the common factor $R \,.\, \sin(t\sqrt{\rho} - r)$,

$$(\Sigma D \,.\, N_\xi + \Sigma E \,.\, N_\eta + \Sigma F N_\zeta)\rho - \Sigma G N_\xi - \Sigma H N_\eta - \Sigma K N_\zeta = 0. \quad (515)$$

Now, there being n of these equations; $n-1$ of them will give the values of N_η, N_ζ, N_ξ', &c., in terms of N_ξ; and these being substituted in the nth equation, must, from the form of the equations, give a resulting equation having N_ξ as a common factor and of the nth degree in ρ. The common factor N_ξ will divide out. The quantity ρ will have n values. The values of N_η, N_ζ, $N_{\xi'}$, &c., will be rational fractions of the $(n-1)^{\text{th}}$ degree in ρ, having a common denominator, and each multiplied by N_ξ, which is, as yet, arbitrary. Make N_ξ equal to the common denominator, and N_η, N_ζ, $N_{\xi'}$, &c., will be expressed in symmetrical functions of ρ, of the $(n-1)^{\text{th}}$ degree. Each of the quantities N_ξ, N_η, N_ζ, &c., will have as many values as ρ; and each of the increments ξ, η, ζ, ξ', &c., will also have n values, each set of which will satisfy Equations (513).

But Equations (513) are linear; not only, therefore, will each of the values of ξ, η, ζ, ξ', &c., satisfy them, but their respective sums substituted for ξ, η, ζ, ξ', &c., will also satisfy them.

Denoting the roots of the nth equation in ρ by ρ, ρ_1, ρ_2, &c., and using the subscript figures to designate the corresponding values for the other letters, the general solution of Equations (513) will be

$$
\left.\begin{array}{l}
\xi = R.N_{\xi}.\sin(t.\sqrt{\rho}-r)+R_1N_{\xi_1}.\sin(t.\sqrt{\rho_1}-r_1)+R_2N_{\xi_2}.\sin(t.\sqrt{\rho_2}-r_2)+\&c. \\
\eta = R.N_{\eta}.\sin(t.\sqrt{\rho}-r)+R_1.N_{\eta_1}.\sin(t.\sqrt{\rho_1}-r_1)+R_2.N_{\eta_2}.\sin(t.\sqrt{\rho_2}-r_2)+\&c. \\
\zeta = R.N_{\zeta}.\sin(t.\sqrt{\rho}-r)+R_1.N_{\zeta_1}.\sin(t.\sqrt{\rho_1}-r_1)+R_2.N_{\zeta_2}.\sin(t.\sqrt{\rho_2}-r_2)+\&c. \\
\ldots\ldots\ldots\ldots\ldots\ldots\ldots\ldots
\end{array}\right\} \quad (516)
$$

R, R_1, &c., and r, r_1, &c., are arbitrary constants, in these complete integrals. They must be found in terms of the initial values of ξ, η, ζ, &c., and their differential coefficients. They are small, because the original disturbance is supposed small.

§ 304.—If all the quantities R, R_1, R_2, &c., except the first, vanish, Equations (516) lose all their terms in the second members except the first, and Equations (508) become

$$
\left.\begin{array}{l}
x = p_x + \left(\frac{d\,p_x}{d\,\alpha}N_{\xi} + \frac{d\,p_x}{d\,\beta}N_{\eta} + \frac{d\,p_x}{d\,\gamma}\cdot N_{\zeta} + \ldots\right)R.\sin(t.\sqrt{\rho}-r), \\
y = p_y + \left(\frac{d\,p_y}{d\,\alpha}N_{\xi} + \frac{d\,p_y}{d\,\beta}N_{\eta} + \frac{d\,p_y}{d\,\gamma}\cdot N_{\zeta} + \ldots\right)R.\sin(t.\sqrt{\rho}-r), \\
z = p_z + \left(\frac{d\,p_z}{d\,\alpha}N_{\xi} + \frac{d\,p_z}{d\,\beta}N_{\eta} + \frac{d\,p_z}{d\,\gamma}\cdot N_{\zeta} + \ldots\right)R.\sin(t.\sqrt{\rho}-r), \\
x' = p'_{x'} + \left(\frac{d\,p_{x'}}{d\,\alpha'}N_{\xi} + \frac{d\,p_{x'}}{d\,\beta'}N_{\eta} + \frac{d\,p_{x'}}{d\,\gamma'}\cdot N_{\zeta} + \ldots\right)R.\sin(t.\sqrt{\rho}-r),
\end{array}\right\} \quad (517)
$$

§ 305.—If the roots ρ, ρ_1, ρ_2, &c., be real, the different terms in the values of ξ, η, ζ, &c., as given in Equations (516), will disappear periodically, and the precise times of disappearance of each will be found by making

$$
t\sqrt{\rho} - r = a\pi; \quad t\sqrt{\rho_1} - r_1 = a\pi; \quad t\sqrt{\rho_2} - r_2 = a\pi; \quad \&c.,
$$

or,

$$
t = \frac{a\pi + r}{\sqrt{\rho}}; \quad t = \frac{a\pi + r_1}{\sqrt{\rho_1}}; \quad t = \frac{a\pi + r_2}{\sqrt{\rho_2}}; \quad \&c.,
$$

in which a is any whole number whatever. The intervals of disappearance will be

$$
\frac{\pi + r}{\sqrt{\rho}}; \quad \frac{\pi + r_1}{\sqrt{\rho_1}}; \quad \frac{\pi + r_2}{\sqrt{\rho_2}}; \quad \&c.
$$

When these intervals are commensurable, then will ξ, η, ζ, &c., resume the values they had at some previous time, the molecules will return to their former simultaneous places, the movement will become periodical, and the period will be equal to the least common multiple of the above intervals. This phenomenon of periodical returns of molecules to their initial places, is called the *periodicity of molecular condition.*

§ 306.—From Equations (516) it is apparent that each and every individual of a system of molecules in which the connection is such as to leave n of their co-ordinates independent, may, when slightly disturbed from rest in positions of stable equilibrium, assume a number n of oscillatory movements, and that all or any number of these may take place simultaneously. And conversely, whatever be the initial derangement of such a system, the resulting motions of each molecule may be resolved into n or less than n simple components parallel to each of any three rectangular axes. Here we have, under a different form, the principle of the *coexistence of small motions.*

§ 307.—Again, let ξ_1, η_1, ζ_1, &c., be the values of ξ, η, ζ, &c., when the system is in motion by the action of one set of forces; ξ_2, η_2, ζ_2, &c., when under the action of another set, and so on—the initial condition being determined for each set of movements—then, Equation (516) being linear, will the resultant values of ξ, η, ζ, &c., be given by

$$\xi = \xi_1 + \xi_2 + \xi_3 + \text{ \&c.},$$
$$\eta = \eta_1 + \eta_2 + \eta_3 + \text{ \&c.},$$
$$\zeta = \zeta_1 + \zeta_2 + \zeta_3 + \text{ \&c.};$$

and here we also have again the *superposition of small motions.* That is, each molecule may take up simultaneously the motions due to each disturbing cause acting separately and alone.

§ 308.—Equations (513) may also be satisfied by making

$$\xi = R \, . \, N_\xi \, . \, e^{t\sqrt{\rho} - r},$$

$$\eta = R \, . \, N_\eta \, . \, e^{t\sqrt{\rho} - r},$$

$$\zeta = R \, . \, N_\zeta \, . \, e^{t\sqrt{\rho} - r},$$

which give

$$\frac{d^2 \xi}{d t^2} = \rho . R . N_{\xi} . e^{t \sqrt{\rho} - r}$$

$$\frac{d^2 \eta}{d t^2} = \rho . R . N_{\eta} . e^{t . \sqrt{\rho} - r},$$

$$\frac{d^2 \zeta}{d t^2} = \rho . R . N_{\zeta} . e^{t . \sqrt{\rho} - r};$$

and these substituted in Equations (513), give Equations (515), with the exceptions of the signs of the terms which are independent of ρ. But with this solution there would be no limit to the increase of ξ, η, ζ, ξ', &c., which is contrary to the conditions that the disturbances are to continue small. In fact, this last solution supposes the molecules to be moved from positions of *unstable equilibrium;* the other, which is the case of nature, *from stable equilibrium.*

WAVES.

§ 309.—It thus appears that every molecule, subjected to certain conditions of aggregation, may, when disturbed from its place of relative rest, describe, under the action of surrounding molecules, a closed orbit. The disturbed molecule being acted upon by its neighbors, will react upon the latter, and cause them, in turn, to take up their appropriate paths; and the same being true of the next molecules in order of distance, the disturbance will be progressive and in all directions. That is, an initial disturbance of a molecule at one time and place, becomes a cause of disturbance of another molecule at another time and place. While, therefore, any molecule A_1 is travelling over its orbit, the disturbance is being propagated on all sides, and at the instant the former completes its circuit, the latter will have reached a molecule A_2, in the distance, which will then, for the first time, begin to move; and the molecules A_1 and A_2 will, thereafter, always be at the same relative distance from their respective starting points. In the same way, a molecule A_3, still further in the distance, will begin its first circuit when A_2 begins its second and A_1 its third, and so on.

Between the molecules A_1 and A_2, as also between A_2 and A_3, &c., molecules will be found at all distances from their starting points and moving in all directions, consistently with the dimensions and shapes of their respective orbits. The term *phase* is used to express the condition of a molecule with respect to its *displacement* and the *direction of its motion.*

Molecules are said to be in *similar phases*, when moving in parallel orbital elements and in the same direction; and in *opposite phases*, when moving in parallel orbital elements and in opposite directions.

The particular form of aggregation assumed by the molecules between the nearest two concentric surfaces in which the same phases simultaneously exist throughout, is called a *wave.*

A surface which contains molecules only in similar phases, is called a *wave front.* This latter term is generally, though not always, applied to the surface upon which the molecules are just beginning to move. The velocity of a wave front will always be that of disturbance propagation. A *wave length* is the interval, measured in the direction of wave propagation, between two consecutive surfaces upon which the molecules have similar phases.

WAVE FUNCTION.

§ 310.—Denote the masses of the molecules by m, m', &c.; the coordinates of

m	by	x,	y,	z,
m'	"	$x + \Delta x$,	$y + \Delta y$,	$z + \Delta z$,
m''	"	$x + \Delta x'$,	$y + \Delta y'$,	$z + \Delta z'$,
&c.,		&c.,	&c.,	&c.,

and the distance between any two molecules, as m and m', by r; then will

$$r = \sqrt{\Delta x^2 + \Delta y^2 + \Delta z^2} \quad . \quad . \quad . \quad . \quad . \quad . \quad (518)$$

Let $f(r)$ be the intensity of the reciprocal action between m and m'; in which f denotes any function whatever. This reciprocal action will determine the elastic force of the body.

Before the system is disturbed, there will be no inertia developed, the inertia terms in Equations (A) will disappear, and we shall have for the action on any molecule as m,

$$\left.\begin{aligned}\Sigma f(r)\,.\frac{\Delta x}{r} &= 0,\\ \Sigma f(r)\,.\frac{\Delta y}{r} &= 0,\\ \Sigma f(r)\,.\frac{\Delta z}{r} &= 0.\end{aligned}\right\}\quad . \quad . \quad . \quad . \quad . \quad . \quad . \quad . \quad (519)$$

Now suppose the system slightly disturbed, and denote the displacement at the time t in the direction of the axes x, y, z, respectively, of

m	by	ξ,	η,	ζ,
m'	"	$\xi + \Delta\,\xi$,	$\eta + \Delta\,\eta$,	$\zeta + \Delta\,\zeta$,
m''	"	$\xi + \Delta\,\xi'$,	$\eta + \Delta\,\eta'$,	$\zeta + \Delta\,\zeta'$,
&c.,		&c.,	&c.,	&c.

Then, denoting the change in r by $\Delta\,r$, Equation (518) becomes

$$r + \Delta\,r = \sqrt{(\Delta\,x + \Delta\,\xi)^2 + (\Delta\,y + \Delta\,\eta)^2 + (\Delta\,z + \Delta\,\zeta)^2}\quad . \quad (520)$$

and by the principle of the superposition of small motions, Equations (A) give for the action on m,

$$\left.\begin{aligned}m\,.\frac{d^2\,\xi}{d\,t^2} &= \Sigma f(r + \Delta\,r)\,.\frac{\Delta\,x + \Delta\,\xi}{r + \Delta\,r},\\ m\,.\frac{d^2\,\eta}{d\,t^2} &= \Sigma f(r + \Delta\,r)\,.\frac{\Delta\,y + \Delta\,\eta}{r + \Delta\,r},\\ m\,.\frac{d^2\,\zeta}{d\,t^2} &= \Sigma f(r + \Delta\,r)\,.\frac{\Delta\,z + \Delta\,\zeta}{r + \Delta\,r}.\end{aligned}\right\}\quad . \quad . \quad . \quad . \quad (521)$$

But

$$\frac{1}{r + \Delta\,r} = \frac{1}{r} - \frac{\Delta\,r}{r^2} + \frac{\Delta\,r^2}{r^3} - \text{\&c.},$$

$$f(r + \Delta\,r) = f(r) + \frac{d\,f(r)}{d\,r}\Delta\,r + \text{\&c.},$$

whence, neglecting the powers of $\Delta\,r$ higher than the first,

$$\frac{f(r+\Delta r)}{r+\Delta r}\cdot(\Delta x+\Delta\xi)=\left\{\frac{f(r)}{r}+\left(\frac{d f(r)}{r\,.\,d r}-\frac{f(r)}{r^2}\right)\Delta r\right\}.(\Delta x+\Delta\xi).$$

Squaring Equation (520), neglecting the squares of Δr, $\Delta\xi$, $\Delta\eta$, and $\Delta\zeta$, and subtracting the square of Equation (518) from the result, we find

$$\Delta r=\frac{\Delta x\,.\,\Delta\xi+\Delta y\,.\,\Delta\eta+\Delta z\,.\,\Delta\zeta}{r}. \quad . \quad . \quad . \quad (522)$$

Substituting this above, and making

$$\left.\begin{aligned}\varphi(r)&=\frac{f(r)}{r},\\ \psi(r)&=\frac{d f(r)}{r^2\,.\,d r}-\frac{f(r)}{r^3};\end{aligned}\right\} \quad . \quad . \quad . \quad . \quad . \quad . \quad (523)$$

Equations (521) become

$$\left.\begin{aligned}m\,.\,\frac{d^2\xi}{d t^2}&=\Sigma\{\varphi(r)\,.\,\Delta\xi+\psi(r)(\Delta x\,.\,\Delta\xi+\Delta y\,.\,\Delta\eta+\Delta z\,.\,\Delta\zeta)\,.\,\Delta x\}\\ m\,.\,\frac{d^2\eta}{d t^2}&=\Sigma\{\varphi(r)\,.\,\Delta\eta+\psi(r)(\Delta x\,.\,\Delta\xi+\Delta y\,.\,\Delta\eta+\Delta z\,.\,\Delta\zeta)\,.\,\Delta y\}\\ m\,.\,\frac{d^2\zeta}{d t^2}&=\Sigma\{\varphi(r)\,.\,\Delta\zeta+\psi(r)(\Delta x\,.\,\Delta\xi+\Delta y\,.\,\Delta\eta+\Delta z\,.\,\Delta\zeta)\,.\,\Delta z\}\end{aligned}\right\}(524)$$

Performing the multiplication as indicated in the last term of the second members, there will result terms of the form,

$$\Sigma\,\psi(r)\,.\,\Delta\eta\,.\,\Delta x\,.\,\Delta y;\quad \Sigma\,\psi(r)\,.\,\Delta\zeta\,.\,\Delta x\,.\,\Delta z;\quad \Sigma\,\psi(r)\,.\,\Delta\xi\,.\,\Delta x\,.\,\Delta y;$$
$$\Sigma\,\psi(r)\,.\,\Delta\eta\,.\,\Delta y\,.\,\Delta z;\quad \Sigma\,\psi(r)\,.\,\Delta\zeta\,.\,\Delta z\,.\,\Delta y;\quad \Sigma\,\psi(r)\,.\,\Delta\xi\,.\,\Delta x\,.\,\Delta z;$$

and it may be shown by the process of § 164, to prove the existence of principal axes, that the co-ordinate axes may be so taken as to cause these terms to vanish. Assuming the axes to satisfy these conditions, Equations (524) become

$$\left.\begin{aligned}m\,.\,\frac{d^2\xi}{d t^2}&=\Sigma\,\{\varphi(r)+\psi(r)\,\Delta x^2\}\,\Delta\xi,\\ m\,.\,\frac{d^2\eta}{d t^2}&=\Sigma\,\{\varphi(r)+\psi(r)\,\Delta y^2\}\,\Delta\eta,\\ m\,.\,\frac{d^2\zeta}{d t^2}&=\Sigma\,\{\varphi(r)+\psi(r)\,\Delta z^2\}\,\Delta\zeta.\end{aligned}\right\} \quad . \quad . \quad . \quad . \quad (525)$$

Making

$$\left.\begin{aligned} mp' &= \varphi(r) + \psi(r) \,.\, \Delta x^2, \\ mp'' &= \varphi(r) + \psi(r) \,.\, \Delta y^2, \\ mp''' &= \varphi(r) + \psi(r) \,.\, \Delta z^2. \end{aligned}\right\} \quad . \; . \; . \; . \; . \quad (526)$$

Equations (525) take the form,

$$\left.\begin{aligned} \frac{d^2 \xi}{d t^2} &= \Sigma p' \,.\, \Delta \xi, \\ \frac{d^2 \eta}{d t^2} &= \Sigma p'' \,.\, \Delta \eta, \\ \frac{d^2 \zeta}{d t^2} &= \Sigma p''' \,.\, \Delta \zeta. \end{aligned}\right\} \quad . \; . \; . \; . \; . \; . \; . \quad (527)$$

An initial and arbitrary displacement of a molecule at one time and place, becomes, through a series of actions and reactions of the molecular forces alone, the cause of displacement of another molecule, at another time and place. In this latter displacement, which results alone from the molecular forces, the molecular motions must take place in the direction of least molecular resistance. This direction is at right angles to that of wave propagation; for, the force which resists the approach of any two strata of molecules will be much greater than that which opposes their sliding the one by the other. Indeed, this view is abundantly confirmed by many of the phenomena that result from wave transmission; and it will be taken for granted, without further remark, that the molecular orbits are in planes at right angles to the direction of wave propagation.

§ 311.—The first of Equations (527) appertains, therefore, to wave propagation in the plane $y\,z$, the second in the plane $x\,z$, and the third in the plane $x\,y$.

The integrations of Equations (527) are given by

$$\left.\begin{aligned} \xi &= \alpha_x \,.\, \sin \frac{2\pi}{\lambda_x} (V_x \,.\, t - r_x), \\ \eta &= \alpha_y \,.\, \sin \frac{2\pi}{\lambda_y} (V_y \,.\, t - r_y), \\ \zeta &= \alpha_z \,.\, \sin \frac{2\pi}{\lambda_z} (V_z \,.\, t - r_z), \end{aligned}\right\} \quad . \; . \; . \; . \; . \quad (528)$$

in which V_x, V_y, and V_z are the velocities with which the disturbance is propagated in directions perpendicular to the axes x, y, and z, respectively; λ_x, λ_y, and λ_z the shortest distances, in the same directions, between the places of rest of any two molecules that may have at the same instant the same phase; r_x, r_y, and r_z the distances of any molecule's place of rest from that of primitive disturbance, estimated in the same directions. This being understood, we have the relations,

$$\left.\begin{array}{l} r_x = \sqrt{y^2 + z^2}; \quad r_y = \sqrt{x^2 + z^2}; \quad r_z = \sqrt{x^2 + y^2}. \\ \text{Make} \\ \dfrac{2\,\pi}{\lambda_x} V_x = n_x; \quad \dfrac{2\,\pi}{\lambda_y} V_y = n_y; \quad \dfrac{2\,\pi}{\lambda_z} V_z = n_z; \\ \dfrac{2\,\pi}{\lambda_x} = k_x; \quad \dfrac{2\,\pi}{\lambda_y} = k_y; \quad \dfrac{2\,\pi}{\lambda_z} = k_z; \end{array}\right\} \quad . \quad . \quad (529)$$

and the above become

$$\left.\begin{array}{l} \xi = \alpha_x \,.\, \sin\,(n_x \,.\, t - k_x \,.\, r_x), \\ \eta = \alpha_y \,.\, \sin\,(n_y \,.\, t - k_y \,.\, r_y), \\ \zeta = \alpha_z \,.\, \sin\,(n_z \,.\, t - k_z \,.\, r_z). \end{array}\right\} \quad . \quad . \quad . \quad . \quad . \quad . \quad (530)$$

§ 312.—To show that these are the solutions of Equations (527), it will be sufficient to prove that they will satisfy those equations with real values for n_x, n_y, and n_z. Differentiate twice with respect to t, and we have

$$\left.\begin{array}{l} \dfrac{d^2\,\xi}{d\,t^2} = -\,n_x^2 \,.\, \xi, \\ \dfrac{d^2\,\eta}{d\,t^2} = -\,n_y^2 \,.\, \eta, \\ \dfrac{d^2\,\zeta}{d\,t^2} = -\,n_z^2 \,.\, \zeta. \end{array}\right\} \quad . \quad . \quad . \quad . \quad . \quad . \quad . \quad . \quad (531)$$

Give to r_x, r_y, and r_z the increments $\Delta\,r_x$, $\Delta\,r_y$, and $\Delta\,r_z$, respectively; the corresponding increments of ξ, η, and ζ are $\Delta\,\xi$, $\Delta\,\eta$, and $\Delta\,\zeta$, and Equations (530) become

$$\xi + \Delta \xi = \alpha_x . \sin (n_x . t - k_x . r_x + k_x . \Delta r_x),$$
$$\eta + \Delta \eta = \alpha_y . \sin (n_y . t - k_y . r_y + k_y . \Delta r_y),$$
$$\zeta + \Delta \zeta = \alpha_z . \sin (n_z . t - k_z . r_z + k_z . \Delta r_z).$$

Developing the second members, regarding $n_x . t - k_x . r_x$, $n_y . t - k_y . r_y$ and $n_z . t - k_z . r_z$ as single arcs; subtracting Equations (530) in order replacing $1 - \cos k_x \Delta r_x$, $1 - \cos k_y \Delta r_y$, and $1 - \cos k_z \Delta r_z$ by their respective values, we find

$$\left. \begin{aligned} \Delta \xi &= - 2 \xi . \sin^2 \frac{(k_x \Delta r_x)}{2} + \sin (k_x \Delta r_x) . \alpha_x \cos (n_x . t - k_x . r_x), \\ \Delta \eta &= - 2 \eta . \sin^2 \frac{(k_y \Delta r_y)}{2} + \sin (k_y \Delta r_y) . \alpha_y \cos (n_y . t - k_y . r_y), \\ \Delta \zeta &= - 2 \zeta . \sin^2 \frac{(k_z \Delta r_z)}{2} + \sin (k_z \Delta r_z) . \alpha_z \cos (n_z . t - k_z . r_z). \end{aligned} \right\} \quad (532)$$

Substituting these in the second members of Equations (527), we have

$$\left. \begin{aligned} \frac{d^2 \xi}{d t^2} &= - 2 \xi . \Sigma p' . \sin^2 \frac{(k_x \Delta r_x)}{2} + \Sigma p' . \sin (k_x \Delta r_x) . a_x . \cos (n_x . t - k_x . r_x), \\ \frac{d^2 \eta}{d t^2} &= - 2 \eta . \Sigma p'' . \sin^2 \frac{(k_y \Delta r_y)}{2} + \Sigma p'' . \sin (k_y \Delta r_y) . a_y . \cos (n_y . t - k_y . r_y), \\ \frac{d^2 \zeta}{d t^2} &= - 2 \zeta . \Sigma p''' . \sin^2 \frac{(k_z \Delta r_z)}{2} + \Sigma p''' . \sin (k_z \Delta r_z) . a_z . \cos (n_z . t - k_z . r_z). \end{aligned} \right\} (533)$$

In the state of equilibrium of the molecules, we may suppose their masses equal, two and two, and symmetrically disposed on either side of that whose mass is m. Indeed, this is the most general way in which we may conceive the equilibrium to exist. Then, since for every positive arc $k_x . \Delta r_x$ there will be an equal negative one, we must have

$$\left. \begin{aligned} \Sigma p' \ . \sin (k_x . \Delta r_x) . \alpha_x . \cos (n_x . t - k_x . r_x) &= 0, \\ \Sigma p'' \ . \sin (k_y . \Delta r_y) . \alpha_y . \cos (n_y . t - k_y . r_y) &= 0, \\ \Sigma p''' . \sin (k_z . \Delta r_z) . \alpha_z . \cos (n_z . t - k_z . r_z) &= 0, \end{aligned} \right\} . \ . \quad (534)$$

and therefore,

$$\left.\begin{aligned}\frac{d^2\xi}{dt^2} &= -2\xi.\Sigma p' \,.\sin^2\frac{k_x.\Delta r_x}{2},\\ \frac{d^2\eta}{dt^2} &= -2\eta.\Sigma p''.\sin^2\frac{k_y.\Delta r_y}{2},\\ \frac{d^2\zeta}{dt^2} &= -2\zeta.\Sigma p'''.\sin^2\frac{k_z.\Delta r_z}{2},\end{aligned}\right\} \quad \ldots\ldots \quad (535)$$

whence, Equations (531) and (535),

$$\left.\begin{aligned}n_x^2 &= 2\,\Sigma p' \,.\sin^2\frac{k_x.\Delta r_x}{2},\\ n_y^2 &= 2\,\Sigma p''.\sin^2\frac{k_y.\Delta r_y}{2},\\ n_z^2 &= 2\,\Sigma p'''.\sin^2\frac{k_z.\Delta r_z}{2},\end{aligned}\right\} \quad \ldots\ldots \quad (536)$$

which are, Equations (526) and (522), real values for n_x, n_y, and n_z.

§ 313.—Substituting the values of n_x, n_y, n_z, and k_x, k_y, k_z, Equations (529), there will result, after multiplying the first, second, and third by $1 = \Delta r_x^2 \div \Delta r_x^2$; $1 = \Delta r_y^2 \div \Delta r_y^2$; $1 = \Delta r_z^2 \div \Delta r_z^2$, respectively,

$$\left.\begin{aligned}V_x^2 &= \tfrac{1}{2}\Sigma p' \,.\Delta r_x^2.\frac{\sin^2\dfrac{\pi\,\Delta r_x}{\lambda_x}}{\left(\dfrac{\pi\,\Delta r_x}{\lambda_x}\right)^2},\\ V_y^2 &= \tfrac{1}{2}\Sigma p''.\Delta r_y^2.\frac{\sin^2\dfrac{\pi\,\Delta r_y}{\lambda_y}}{\left(\dfrac{\pi\,\Delta r_y}{\lambda_y}\right)^2},\\ V_z^2 &= \tfrac{1}{2}\Sigma p'''.\Delta r_z^2.\frac{\sin^2\dfrac{\pi\,\Delta r_z}{\lambda_z}}{\left(\dfrac{\pi.\Delta r_z}{\lambda_z}\right)^2}.\end{aligned}\right\} \quad \ldots\ldots \quad (537)$$

WAVE SECTION.

§ 314.—Resuming either of Equations (528), say the first, viz.:

$$\xi = \alpha_x \sin \frac{2\pi}{\lambda_x} (V_x \, . \, t - r_x),$$

it is apparent that if t be made constant and r_x variable, so as to reach in succession all the molecules in its direction between the limits

$$V_x \, . \, t - \lambda_x, \text{ and } V_x \, . \, t,$$

the displacement ξ will also vary, and from zero to zero, passing between these limits through the maximum values α_x and minimum value $-\alpha_x$; thus determining the curved line $C\,D$, of the annexed figure, to be the locus of the corresponding displaced molecules, of which the places of rest are on the straight line $A\,B$, coincident in direction with the line r_x in the plane $y\,z$. And it is also apparent that if the above value of t receive an increment, making the time equal to t', and, with this new value for the time, r_x be made to vary between the limits

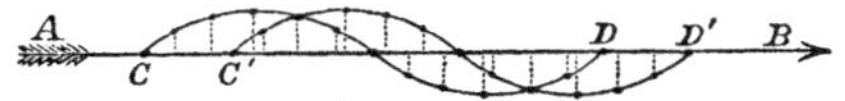

$$V_x \, . \, t' - \lambda_x, \text{ and } V_x \, . \, t',$$

the locus of the corresponding displaced molecules will be found to have shifted its place to $C'\,D'$, in the direction towards which the disturbance is propagated.

This peculiar arrangement of a series of consecutive molecules, by which the latter are made to occupy the various positions, arranged in the order of continuity about their places of rest, is, as we have seen, § 305, called a *wave*, and the functions, Equations (528), from which a section of the waves may be constructed, are called *wave functions*.

WAVE VELOCITY.

§ 315.—From either of Equations (537), say the first, it appears that the velocity of wave propagation depends upon the ratio between the arc $\frac{\pi \, . \, \Delta \, r_x}{\lambda_x}$ and its sine. If the distance $\Delta \, r_x$, between the molecules, in the direction of r_x, have any appreciable value as compared with the wave length λ_x, this ratio will be less than unity; and in proportion as the wave length increases, in the same medium, will the velocity increase. When the distance $\Delta \, r_x$ is insignificant in comparison with the wave length λ_x, the ratio of the sine to the arc will be unity, and that factor will cease to appear.

§ 316.—If the medium be homogeneous, then will

$$p' = p'' = p''' ; \quad \Delta \, r_x = \Delta \, r_y = \Delta \, r_z ;$$

and, therefore,

$$V_x = V_y = V_z .$$

That is, the velocity will be the same in all directions. Denote this velocity by V; we may write

$$V^2 = H \, . \, \frac{\sin^2 \frac{\pi \, . \, \Delta \, r}{\lambda}}{\left(\frac{\pi \, . \, \Delta \, r}{\lambda} \right)^2} \quad . \; . \; . \; . \; . \; . \; . \; . \quad (538)$$

in which the two factors that compose the second member have such average values as to give a product equal to the sum of the products which make up the second members of either of Equations (537).

Supposing, in addition to the existence of homogeneity, that the interval between the molecules is insignificant in regard to the wave length, the last factor of Equations (537) reduces to unity, and taking the axis x in the direction of the velocity to be estimated, $\Delta \, r$ becomes $\Delta \, x$, and, first of Equations (537),

$$V^2 = \tfrac{1}{2} \, \Sigma \, p' \, . \, (\Delta \, x)^2 ;$$

replacing p' by its value, Equations (526) and (523),

$$V^2 = \frac{1}{2 \, m} \, . \, \Sigma \left[\frac{f(r)}{r} \Delta \, x^2 + \left(\frac{d \, f(r)}{d \, r} \, . \, \frac{1}{r^2} - \frac{f(r)}{r^3} \right) . \, \Delta \, x^4 \right] .$$

The distances between the molecules being very small, the term of which $\Delta \, x^4$ is a factor may be neglected in comparison with that containing $\Delta \, x^2$, and the above may be written

$$V^2 = \frac{1}{2\,m} \cdot \Sigma f(r) \cdot \frac{\Delta\, x}{r} \cdot \Delta\, x.$$

Now, $f(r) \cdot \frac{\Delta\, x}{r}$ is the component of the elastic force exerted betweer two molecules whose distance is r, in the direction of the axis x; and $f(r) \cdot \frac{\Delta\, x}{r} \cdot \Delta\, x$ is the quantity of work of this component acting through a distance $\Delta\, x$. Making

$$\Sigma f(r) \cdot \frac{\Delta\, x}{r} = 2\, e_{,},$$

we may, by the principle of parallel forces, write

$$\Sigma f(r) \cdot \frac{\Delta\, x}{r} \cdot \Delta\, x = 2\, e_{,}\, x_{,};$$

in which $e_{,}$ is the sum of the component molecular forces which act on one side of the molecule m, in the direction of the axis x, or, which is the same thing, the elastic force limited to a single molecule; and $x_{,}$ the path over which this force would perform an amount of work equal to that measured by the first member. Substituting this above,

$$V^2 = \frac{e_{,}\, x_{,}}{m}.$$

Denote by i the number of molecules in a unit of length, and multiply both numerator and denominator by i^3; we have

$$V^2 = \frac{i^2 \cdot e_{,} \cdot i\, x_{,}}{i^3 \cdot m};$$

but $i^2 \cdot e_{,}$ is the elastic force extended to a unit of surface, and is the measure of the elastic force of the medium; call this e. The factor $\iota\, x_{,}$ is the number of molecules in the distance $x_{,}$; call this k. The denominator $i^3\, m$ is the quantity of matter in a unit of volume, which is the density; call this Δ, and the above becomes

$$V = \sqrt{\frac{e}{\Delta} \cdot k} \quad . \quad . \quad . \quad . \quad . \quad . \quad . \quad . \quad (539)$$

Denote by c the ratio which the contraction produced in a given vol·

ume of the medium by the pressure of a standard atmosphere A, bears to the volume without any external pressure; then will

$$e = \frac{A}{c} = \frac{g \cdot D_{,,} \cdot 30^{\text{inches}}}{c} \quad \ldots \ldots \quad (540)$$

in which g is the force of gravity and $D_{,,}$ the density of mercury at a standard temperature.

In the case of gases, c is sensibly equal to unity; for if such bodies were relieved from their atmospheric pressures they would expand indefinitely, thus making their increments of volumes sensibly equal to the volumes they would ultimately attain.

RELATION OF WAVE VELOCITY TO WAVE LENGTH.

§ 317.—Denote the resultant displacement, of which ξ, η, and ζ are the components, by σ; and the angles which σ makes with the axes x, y, and z, by α, β, and γ, respectively; then will

$$\xi = \sigma \cdot \cos \alpha; \quad \eta = \sigma \cdot \cos \beta; \quad \zeta = \sigma \cdot \cos \gamma;$$

which, substituted in the second members of Equations (531), give

$$\left.\begin{aligned} \frac{d^2 \xi}{d t^2} &= - \sigma \cdot n_x^2 \cdot \cos \alpha, \\ \frac{d^2 \eta}{d t^2} &= - \sigma \cdot n_y^2 \cdot \cos \beta, \\ \frac{d^2 \zeta}{d t^2} &= - \sigma \cdot n_z^2 \cdot \cos \gamma. \end{aligned}\right\} \quad \ldots \ldots \quad (541)$$

Squaring, adding, taking square root, and denoting the resultant by ε_m, we have

$$\varepsilon^2_m = \left(\frac{d^2 \xi}{d t^2}\right)^2 + \left(\frac{d^2 \eta}{d t^2}\right)^2 + \left(\frac{d^2 \zeta}{d t^2}\right)^2 = \sigma^2 (n^4_x \cdot \cos^2 \alpha + n^4_y \cdot \cos^2 \beta + n^4_z \cdot \cos^2 \gamma). \quad (542)$$

The first member is the square of the resultant acceleration due to the molecular action developed by the displacement σ.

Denote by $\alpha_{,}$, $\beta_{,}$, and $\gamma_{,}$ the angles which the direction of this resultant makes with the axes x, y, and z, respectively; and by ψ the inclination of this direction to that of displacement. Then will

$$\cos\psi = \cos\alpha \cdot \cos\alpha_{,} + \cos\beta \cdot \cos\beta_{,} + \cos\gamma \cdot \cos\gamma_{,} \quad . \quad . \quad (543)$$

The components of the acceleration, in the directions of the axes x, y, and z, are, respectively,

$$\varepsilon_m \cos\alpha_{,}\,; \quad \varepsilon_m \cos\beta_{,}\,; \quad \varepsilon_m \cos\gamma_{,}\,;$$

and, therefore, Equations (541),

$$\varepsilon_m \cos\alpha_{,} = -\sigma \cdot n_x^2 \cdot \cos\alpha,$$
$$\varepsilon_m \cos\beta_{,} = -\sigma \cdot n_y^2 \cdot \cos\beta,$$
$$\varepsilon_m \cos\gamma_{,} = -\sigma \cdot n_z^2 \cdot \cos\gamma.$$

Whence,

$$\left.\begin{aligned} \cos\alpha_{,} &= -\frac{\sigma \cdot n_x^2 \cdot \cos\alpha}{\varepsilon_m}, \\ \cos\beta_{,} &= -\frac{\sigma \cdot n_y^2 \cdot \cos\beta}{\varepsilon_m}, \\ \cos\gamma_{,} &= -\frac{\sigma \cdot n_z^2 \cdot \cos\gamma}{\varepsilon_m}, \end{aligned}\right\} \quad . \quad . \quad . \quad . \quad . \quad . \quad (544)$$

These, in Equation (543), give, Equations (531),

$$\varepsilon_m \cdot \cos\psi = -\sigma \cdot n_\sigma^2 = -\sigma \cdot (n_x^2 \cdot \cos^2\alpha + n_y^2 \cdot \cos^2\beta + n_z^2 \cdot \cos^2\gamma)\,;$$

and replacing n_σ, n_x, n_y, and n_z, by their values, Equations (529),

$$\frac{V_\sigma^2}{\lambda_\sigma^2} = \frac{V_x^2}{\lambda_x^2} \cdot \cos^2\alpha + \frac{V_y^2}{\lambda_y^2} \cdot \cos^2\beta + \frac{V_z^2}{\lambda_z^2} \cdot \cos^2\gamma.$$

But, because the number of waves, in a unit of time, arising from the components of a common initial disturbance must be the same, the coefficients of the circular functions above must be equal, and hence,

$$\frac{V_\sigma^2}{\lambda_\sigma^2} = \frac{V_x^2}{\lambda_x^2}\,(\cos^2\alpha + \cos^2\beta + \cos^2\gamma) = \frac{V_x^2}{\lambda_x^2} \qquad (545)$$

Whence the wave velocity is proportional to the wave length.

SURFACE OF ELASTICITY.

§ 318.—Replacing, in Equations (541), n_x, n_y, n_z, by their values in Equations (529), multiplying the first by $c \,.\, \pi \lambda_x^2 \,.\, m$, the second by $c \,.\, \pi \lambda_y^2 \,.\, m$, and the third by $c \,.\, \pi \lambda_z^2 \,.\, m$, we have

$$\left.\begin{aligned} c \,.\, \pi \,.\, \lambda_x^2 \,.\, m \,.\, \frac{d^2 \xi}{d t^2} &= - \sigma \,.\, c \,.\, 4 \pi^3 \,.\, m \, V_x^2 \,.\, \cos \alpha, \\ c \,.\, \pi \,.\, \lambda_y^2 \,.\, m \,.\, \frac{d^2 \eta}{d t^2} &= - \sigma \,.\, c \,.\, 4 \pi^3 \,.\, m \, V_y^2 \,.\, \cos \beta, \\ c \,.\, \pi \,.\, \lambda_z^2 \,.\, m \,.\, \frac{d^2 \zeta}{d t^2} &= - \sigma \,.\, c \,.\, 4 \pi^3 \,.\, m \, V_z^2 \,.\, \cos \gamma. \end{aligned}\right\} \quad . \quad (546)$$

Now, $\pi \,.\, \lambda_x^2$, $\pi \,.\, \lambda_y^2$, and $\pi \,.\, \lambda_z^2$ are the projections of the waves arising from the component displacements ξ, η, and ζ, on the planes $y z$, $x z$, and $x y$, respectively; and if every molecule in each of these waves had the same acceleration, the first members would measure the elastic forces exerted over these projections by making c equal to unity. These are, however, not equal; but if c denote a proper fractional coefficient, and ε_x, ε_y, and ε_z the actual elastic forces in the three waves, we may write,

$$\left.\begin{aligned} \varepsilon_x &= - \sigma \,.\, \varsigma \,.\, V_x^2 \,.\, \cos \alpha, \\ \varepsilon_y &= - \sigma \,.\, \varsigma \,.\, V_y^2 \,.\, \cos \beta, \\ \varepsilon_z &= - \sigma \,.\, \varsigma \,.\, V_z^2 \,.\, \cos \gamma. \end{aligned}\right\} \quad . \quad . \quad . \quad . \quad . \quad . \quad (547)$$

in which $\varsigma = 4 c \,.\, \pi^3 \,.\, m$. Squaring, adding, taking square root of sum, and denoting the resultant by ε_σ,

$$\varepsilon_\sigma = \sqrt{\varepsilon_x^2 + \varepsilon_y^2 + \varepsilon_z^2} = \sigma \,.\, \varsigma \,.\, \sqrt{V_x^4 \,.\, \cos^2 \alpha + V_y^4 \,.\, \cos^2 \beta + V_z^4 \,.\, \cos^2 \gamma}\,;$$

from which it is apparent that if the displacement be made in the direction of either axis, the elastic force will be wholly in the direction of that axis—a property possessed by these particular axes in consequence of the fact that they were assumed in directions to satisfy the conditions of symmetry in molecular arrangement, which caused Equations (524) to reduce to Equations (525). The *directions* of these special axes are called *axes of elasticity*. The resultant elastic force will not, in general, act in the direction of the displacement.

Denote the angles which ε_σ makes with the axes of elasticity by $\alpha_{,}$, $\beta_{,}$, and $\gamma_{,}$, and the angle which it makes with the displacement by ψ, then will

$$\cos\psi = \cos\alpha \,.\, \cos\alpha_{,} + \cos\beta \,.\, \cos\beta_{,} + \cos\gamma \,.\, \cos\gamma_{,},$$

$$\varepsilon_\sigma \,.\, \cos\alpha_{,} = \varepsilon_x = -\sigma \,.\, \varsigma \,.\, V_x^2 \,.\, \cos\alpha,$$

$$\varepsilon_\sigma \,.\, \cos\beta_{,} = \varepsilon_y = -\sigma \,.\, \varsigma \,.\, V_y^2 \,.\, \cos\beta,$$

$$\varepsilon_\sigma \,.\, \cos\gamma_{,} = \varepsilon_z = -\sigma \,.\, \varsigma \,.\, V_z^2 \,.\, \cos\lambda.$$

Whence

$$\left.\begin{aligned} \cos\alpha_{,} &= -\frac{\sigma \,.\, \varsigma \,.\, V_x^2 \,.\, \cos\alpha}{\varepsilon_\sigma}, \\ \cos\beta_{,} &= -\frac{\sigma \,.\, \varsigma \,.\, V_y^2 \,.\, \cos\beta}{\varepsilon_\sigma}, \\ \cos\gamma_{,} &= -\frac{\sigma \,.\, \varsigma \,.\, V_z^2 \,.\, \cos\gamma}{\varepsilon_\sigma}; \end{aligned}\right\} \quad . \quad . \quad . \quad . \quad (548)$$

which substituted above, give,

$$\varepsilon_\sigma \,.\, \cos\psi = -\sigma \,.\, \varsigma \,.\, V_\sigma^2 = -\sigma \,.\, \varsigma\,(V_x^2 \,.\, \cos^2\alpha + V_y^2 \,.\, \cos^2\beta + V_z^2 \,.\, \cos^2\gamma);$$

in which V_σ is the velocity perpendicular to the displacement. Making

$$V_\sigma = V; \quad V_x = a; \quad V_y = b; \quad V_z = c;$$

we have

$$V = \sqrt{a^2 \,.\, \cos^2\alpha + b^2 \,.\, \cos^2\beta + c^2 \,.\, \cos^2\gamma} \quad . \quad . \quad . \quad (549)$$

The quantities a, b, and c are called *definite axes of elasticity*, in contradistinction to axes of elasticity which merely give direction. The surface of which the above is the equation, is called the surface of elasticity. The value of V will measure the velocity of any point on the wave surface in a direction normal to the displacement, and being squared and multiplied by $\sigma \,.\, \varsigma$ will give the elasticity developed in the direction of the displacement itself.

The definite axes of elasticity are the geometrical axes of figure of the surface of elasticity; the general axes of elasticity are directions parallel to these, and drawn from any point in the medium taken at pleasure.

WAVE SURFACE.

§ 319.—This is the locus of those molecules which have, simultaneously, the same phase, § 309; and whatever this phase may be, the particular surface characterized by it will be concentric with that which marks, at any epoch, the exterior limits of the disturbance, or upon which the molecules are beginning to participate in the disturbance propagation.

It is now the question to determine the equation of this latter surface; for this purpose, assume the origin of co-ordinates at the point of primitive disturbance, and let

$$lx + my + nz = V \quad . \quad . \quad . \quad . \quad . \quad . \quad . \quad (550)$$

be the equation of a plane tangent to the wave front at any point, and at the end of a unit of time. The coefficients l, m, and n, will be the cosines of the angles which the normal to this plane makes with the axes $x\,y\,z$, respectively, and its length will measure the velocity V, of wave propagation in its own direction. This plane must be parallel to the displacement and its normal perpendicular thereto; hence

$$l \cos \alpha + m \cos \beta + n \cos \gamma = 0 \quad . \quad . \quad . \quad . \quad (551);$$

also

$$\cos^2 \alpha + \cos^2 \beta + \cos^2 \gamma = 1 \quad . \quad . \quad . \quad . \quad (552).$$

Equations (549), (550), (551), and (552) must exist simultaneously for real values of the cosines of α, β, and γ. To find an equation which shall express this condition, square Eq. (549), and divide it by $V^2 \cdot \cos^2 \alpha$, it becomes

$$\frac{a^2 + b^2 \cdot \dfrac{\cos^2 \beta}{\cos^2 \alpha} + c^2 \cdot \dfrac{\cos^2 \gamma}{\cos^2 \alpha}}{V^2} = \frac{1}{\cos^2 \alpha} \quad \ldots \ldots \quad (553)$$

divide Eq. (551) by $\cos \alpha$, we have

$$l + m \cdot \frac{\cos \beta}{\cos \alpha} + n \cdot \frac{\cos \gamma}{\cos \alpha} = 0 \quad \ldots \ldots \quad (554);$$

and divide Eq. (552) by $\cos^2 \alpha$, the result is

$$1 + \frac{\cos^2 \beta}{\cos^2 \alpha} + \frac{\cos^2 \gamma}{\cos^2 \alpha} = \frac{1}{\cos^2 \alpha} \quad \ldots \ldots \quad (555).$$

Equations (553) and (555) give

$$1 + \frac{\cos^2 \beta}{\cos^2 \alpha} + \frac{\cos^2 \gamma}{\cos^2 \alpha} = \frac{a^2 + b^2 \cdot \dfrac{\cos^2 \beta}{\cos^2 \alpha} + c^2 \cdot \dfrac{\cos^2 \gamma}{\cos^2 \alpha}}{V^2};$$

whence

$$V^2 - a^2 + (V^2 - b^2) \cdot \frac{\cos^2 \beta}{\cos^2 \alpha} + (V^2 - c^2) \cdot \frac{\cos^2 \gamma}{\cos^2 \alpha} = 0 \quad \ldots \ldots \quad (556).$$

From Equation (554) we have

$$\frac{\cos \gamma}{\cos \alpha} = - \frac{l + m \cdot \dfrac{\cos \beta}{\cos \alpha}}{n};$$

which in Equation (556) gives

$$[(V^2 - b^2) n^2 + (V^2 - c^2) m^2] \cdot \frac{\cos^2 \beta}{\cos^2 \alpha} + 2 (V^2 - c^2) \cdot l \cdot m \cdot \frac{\cos \beta}{\cos \alpha} = -(V^2 - a^2) n^2 - (V^2 - c^2) l^2,$$

or

$$\frac{\cos^2 \beta}{\cos^2 \alpha} + 2 \frac{(V^2 - c^2) . l . m}{(V^2 - b^2) n^2 + (V^2 - c^2) m^2} \cdot \frac{\cos \beta}{\cos \alpha} = - \frac{(V^2 - a^2) n^2 + (V^2 - c^2) l^2}{(V^2 - b^2) n^2 + (V^2 - c^2) m^2};$$

and solving with respect to $\dfrac{\cos \beta}{\cos \alpha}$, there will result,

$$\frac{\cos \beta}{\cos \alpha} = - \frac{(V^2 - c^2) . m . l \mp n \sqrt{-[(V^2 - a^2)(V^2 - b^2) n^2 + (V^2 - a^2)(V^2 - c^2) m^2 + (V^2 - c^2)(V^2 - b^2) l^2]}}{(V^2 - b^2) n^2 + (V^2 - c^2) m^2} \quad (557);$$

and this in Equation (554) gives

$$\frac{\cos\gamma}{\cos\alpha} = -\frac{(V^2-b^2).n.l \pm m\sqrt{-[(V^2-a^2)(V^2-b^2)n^2+(V^2-a^2)(V^2-c^2)m^2+(V^2-c^2)(V^2-b^2)l^2]}}{(V^2-b^2)n^2+(V^2-c^2)m^2} \quad (558).$$

For any assumed displacement, the value of V, Eq. (549), becomes known, and the values of the first members of Eqs. (557) and (558) must be real; whence l, m, and n, must, in addition to Eq. (549), also satisfy the condition

$$(V^2-a^2)(V^2-b^2)n^2 + (V^2-a^2)(V^2-c^2)m^2 + (V^2-b^2)(V^2-c^2)l^2 = 0.$$

Dividing by

$$(V^2-a^2)\ (V^2-b^2)\ (V^2-c^2),$$

and inverting the order of the terms,

$$\frac{l^2}{V^2-a^2} + \frac{m^2}{V^2-b^2} + \frac{n^2}{V^2-c^2} = 0. \quad . \quad . \quad . \quad . \quad (559)$$

From this equation, together with Equation (550), and the relation

$$l^2 + m^2 + n^2 = 1, \quad . \quad . \quad . \quad . \quad . \quad . \quad . \quad . \quad (560)$$

we have all the conditions necessary to find the equation of the wave surface; this is done by eliminating V, m, l, and n.

For this purpose, differentiate each of these equations with respect to the quantities to be eliminated. We have, from Equation (550),

(1) . . . . $x\,dl + y\,dm + z\,dn = dV;$

from Equation (560),

(2) . . . . $l\,dl + m\,dm + n\,dn = 0;$

and from Equation (559),

(3) $$\frac{l\,dl}{V^2-a^2} + \frac{m\,dm}{V^2-b^2} + \frac{n\,dn}{V^2-c^2} = V\,dV\left(\frac{l^2}{(V^2-a^2)^2} + \frac{m^2}{(V^2-b^2)^2} + \frac{n^2}{(V^2-c^2)^2}\right).$$

Multiply the first by λ, the second by $-\lambda'$, the third by -1, and add members to members, and collect the coefficients of like differentials; there will result,

$$\left.\begin{array}{r} \left(\lambda x - \lambda' l - \dfrac{l}{V^2 - a^2}\right) d l \\ + \left(\lambda y - \lambda' m - \dfrac{m}{V^2 - b^2}\right) d m \\ + \left(\lambda z - \lambda' n - \dfrac{n}{V^2 - c^2}\right) d n \\ + \left(\lambda - V\left\{\dfrac{l^2}{(V^2 - a^2)^2} + \dfrac{m^2}{(V^2 - b^2)^2} + \dfrac{n^2}{(V^2 - c^2)^2}\right\}\right) d V \end{array}\right\} = 0.$$

Taking λ and λ' of such values as to make the coefficients of $d V$ and $d n$ each zero, the equation will reduce to the first two terms; and as $d m$ and $d l$ are wholly arbitrary, Equation (560), as long as $d n$ is undetermined, we may, from the principle of indeterminate coefficients, write,

(4) . . . . . . $\lambda x - \lambda' l - \dfrac{l}{V^2 - a^2} = 0,$

(5) . . . . . . $\lambda y - \lambda' m - \dfrac{m}{V^2 - b^2} = 0,$

(6) . . . . . . $\lambda z - \lambda' n - \dfrac{n}{V^2 - c^2} = 0,$

(7) . $\lambda - V\left\{\dfrac{l^2}{(V^2 - a^2)^2} + \dfrac{m^2}{(V^2 - b^2)^2} + \dfrac{n^2}{(V^2 - c^2)^2}\right\} = 0;$

Multiply (4) by l, (5) by m, (6) by n, add and reduce by Equations (550), (560), and (559); we have

(8) . . . . $\lambda V - \lambda' = 0;$

Multiply (4) by x, (5) by y, and (6) by z; add and reduce by Equation (550) and the relation $x^2 + y^2 + z^2 = r^2$; we have

$$\lambda r^2 - \left(\lambda' V + \frac{l x}{V^2 - a^2} + \frac{m y}{V^2 - b^2} + \frac{n z}{V^2 - c^2}\right) = 0;$$

substituting for λ' its value, (8), and transposing,

(9) . . $\lambda (r^2 - V^2) = \dfrac{l x}{V^2 - a^2} + \dfrac{m y}{V^2 - b^2} + \dfrac{n z}{V^2 - c^2};$

transposing in (4), (5), and (6), squaring and adding, we have

$$\lambda^2 r^2 = \lambda'^2 + \frac{l^2}{(V^2 - a^2)^2} + \frac{m^2}{(V^2 - b^2)^2} + \frac{n^2}{(V^2 - c^2)^2};$$

substituting for λ'^2 its value, (8), and reducing by (7), we have

$$\lambda^2 (r^2 - V^2) = \frac{\lambda}{V};$$

and, therefore,

$$(10) \quad . \quad . \quad . \quad . \quad \lambda = \frac{1}{V(r^2 - V^2)}; \quad \lambda' = \frac{1}{r^2 - V^2}.$$

Substituting these in (4), we find

$$\frac{x}{V(r^2 - V^2)} = l\left(\frac{1}{r^2 - V^2} + \frac{1}{V^2 - a^2}\right);$$

whence

$$\frac{x}{r^2 - a^2} = \frac{Vl}{V^2 - a^2};$$

similarly,

$$\frac{y}{r^2 - b^2} = \frac{Vm}{V^2 - b^2};$$

$$\frac{z}{r^2 - c^2} = \frac{Vn}{V^2 - c^2};$$

multiply the first by x, the second by y, the third by z, add and reduce by (9) and (10); we have

$$\frac{x^2}{r^2 - a^2} + \frac{y^2}{r^2 - b^2} + \frac{z^2}{r^2 - c^2} = 1 \quad . \quad . \quad . \quad . \quad . \quad (561)$$

From this, which is one form of the equation of the wave surface, subtract

$$\frac{x^2 + y^2 + z^2}{r^2} = 1,$$

and we have

$$\frac{a^2 x^2}{r^2 - a^2} + \frac{b^2 y^2}{r^2 - b^2} + \frac{c^2 z^2}{r^2 - c^2} = 0 \quad . \quad . \quad . \quad . \quad (562)$$

which is a second form of the equation of the wave surface.

Clearing the fractions, it becomes, after substituting for r^2 its value $x^2 + y^2 + z^2$,

$$\left.\begin{array}{l}(x^2+y^2+z^2)(a^2x^2+b^2y^2+c^2z^2)\\ \quad -a^2(b^2+c^2)x^2\\ \quad -b^2(a^2+c^2)y^2\\ \quad -c^2(a^2+b^2)z^2\\ \quad +a^2b^2c^2\end{array}\right\} = 0 \quad . \quad . \quad . \quad (563)$$

DOUBLE WAVE VELOCITY.

§ 320.—The radius vector r measures the velocity of the point of the wave to which it belongs; and denoting by $l_{,}$, $m_{,}$, and $n_{,}$ the cosines of the angles which r makes with x, y, and z, respectively, we have

$$x = r \, . \, l_{,} \, ; \quad y = r m_{,} \, ; \quad z = r n_{,} \, ;$$

and writing V_r for r, we have, by substituting in Equation (563), and dividing by $V_r^4 \, . \, a^2 \, . \, b^2 \, . \, c^2$,

$$\frac{1}{V_r^4} - \left[\left(\frac{1}{b^2}+\frac{1}{c^2}\right)l_{,}^2 + \left(\frac{1}{c^2}+\frac{1}{a^2}\right) . \, m_{,}^2 + \left(\frac{1}{a^2}+\frac{1}{b^2}\right) . \, n_{,}^2\right] . \, \frac{1}{V_r^2} + \frac{l_{,}^2}{b^2c^2} + \frac{m_{,}^2}{a^2c^2} + \frac{n_{,}^2}{a^2b^2} = 0, \quad (564)$$

a trinomial equation, of which the second powers of the equal roots are

$$\frac{1}{V_r^2} = \tfrac{1}{2}\left(\frac{1}{c^2}+\frac{1}{a^2}\right) + \tfrac{1}{2}\left(\frac{1}{c^2}-\frac{1}{a^2}\right)\left[A' . A'' \pm \sqrt{1-A'^2} \times \sqrt{1-A''^2}\right], \quad (565)$$

and in which,

$$A' = l_{,} \, . \sqrt{\frac{\frac{1}{b^2}-\frac{1}{a^2}}{\frac{1}{c^2}-\frac{1}{a^2}}} + n_{,} \, . \sqrt{\frac{\frac{1}{c^2}-\frac{1}{b^2}}{\frac{1}{c^2}-\frac{1}{a^2}}} \, ; \quad . \quad . \quad . \quad (566)$$

$$A'' = l_{,} \, . \sqrt{\frac{\frac{1}{b^2}-\frac{1}{a^2}}{\frac{1}{c^2}-\frac{1}{a^2}}} - n_{,} \, . \sqrt{\frac{\frac{1}{c^2}-\frac{1}{b^2}}{\frac{1}{c^2}-\frac{1}{a^2}}} \, ; \quad . \quad . \quad . \quad (567)$$

If $a > b > c$, the values of A' and A'' will be real, and there will, in general, be two real values for $\frac{1}{V_r^2}$; and with this condition, Equation (565) will give two pairs of real and equal roots with contrary signs.

The positive roots give two velocities in any one direction, and the negative in a direction contrary to this.

Through the origin, conceive two lines to be drawn, making with the axis a, angles whose cosines are $\alpha_{\prime}$ and $\alpha_{\prime\prime}$; with the axis b, angles whose cosines are $\beta_{\prime}$ and $\beta_{\prime\prime}$; with the axis c, angles whose cosines are $\gamma_{\prime}$ and $\gamma_{\prime\prime}$; and such that

$$\alpha_{\prime} = \alpha_{\prime\prime} = \sqrt{\frac{\frac{1}{b^2} - \frac{1}{a^2}}{\frac{1}{c^2} - \frac{1}{a^2}}}; \quad \beta_{\prime} = \beta_{\prime\prime} = 0; \quad \gamma_{\prime} = \gamma_{\prime\prime} = \sqrt{\frac{\frac{1}{c^2} - \frac{1}{b^2}}{\frac{1}{c^2} - \frac{1}{a^2}}}; \quad (568)$$

and denote the angle which r makes with the first of these lines by $u_{\prime}$, and that which it makes with the second by $u_{\prime\prime}$; then will

$$A' = l_{\prime}\alpha_{\prime} + n_{\prime} . \gamma_{\prime} = \cos u_{\prime},$$

$$A'' = l_{\prime}\alpha_{\prime} - n_{\prime}\gamma_{\prime} = \cos u_{\prime\prime}.$$

$$\sqrt{1 - A'^2} = \sin u_{\prime}; \quad \sqrt{1 - A''^2} = \sin u_{\prime\prime}.$$

These, in Equation (565), give for the two values of $\frac{1}{V_r^2}$,

$$\frac{1}{V_{r_1}^2} = \tfrac{1}{2}\left(\frac{1}{c^2} + \frac{1}{a^2}\right) + \tfrac{1}{2}\left(\frac{1}{c^2} - \frac{1}{a^2}\right) . (\cos u_{\prime} . \cos u_{\prime\prime} + \sin u_{\prime} . \sin u_{\prime\prime}) \;.\;.\; (569)$$

$$\frac{1}{V_{r_2}^2} = \tfrac{1}{2}\left(\frac{1}{c^2} + \frac{1}{a^2}\right) + \tfrac{1}{2}\left(\frac{1}{c^2} - \frac{1}{a^2}\right) . (\cos u_{\prime} . \cos u_{\prime\prime} - \sin u_{\prime} . \sin u_{\prime\prime}) \;.\;.\; (570)$$

and by subtraction,

$$\frac{1}{V_{r_2}^2} - \frac{1}{V_{r_1}^2} = \left(\frac{1}{c^2} - \frac{1}{a^2}\right) . \sin u_{\prime} . \sin u_{\prime\prime} \;.\;.\;.\;.\; (571)$$

Now,

$$\frac{1}{V_{r_1}} \text{ and } \frac{1}{V_{r_2}}$$

are the retardations of wave velocity. As long as a and c differ, the second member can only reduce to zero, when $u_{\prime}$ or $u_{\prime\prime}$ is zero; whence it appears that, as a general rule, every direction except two is distinguished by transmitting two waves, one in advance of the other. The two directions which form the exceptions are in the plane of the axes of greatest and least elasticity, and make with these axes the angles of which the cosines are $\alpha_{,}$ and $\gamma_{,}$, $\alpha_{,,}$ and $\gamma_{,,}$, Equations (568). In these directions the waves will travel with equal velocities.

Any direction along which the component waves travel with equal velocities is called an *axis of equal wave velocity.* All bodies in which the elasticities in three rectangular directions differ, possess, Equation (571), two of these axes, and are called biaxial bodies. The retardation of one component wave over that of the other, will vary with the inclination of the direction of its motion to the axis of equal wave velocity; and Equation (571) shows that the loci of equal retardations will be arranged in the form of *spherical lemniscates* about the poles of the axes.

§ 321.—The form of the wave surface and its properties become better known from its principal sections and singular points.

Its sections by the planes yz, xz, and xy give, respectively,

$$\left.\begin{array}{l} x = 0\,; \quad (y^2 + z^2 - a^2)\,(b^2 y^2 + c^2 z^2 - b^2 c^2) = 0, \\ y = 0\,; \quad (z^2 + x^2 - b^2)\,(c^2 z^2 + a^2 x^2 - c^2 a^2) = 0, \\ z = 0\,; \quad (x^2 + y^2 - c^2)\,(a^2 x^2 + b^2 y^2 - a^2 b^2) = 0, \end{array}\right\} \quad . \quad (572)$$

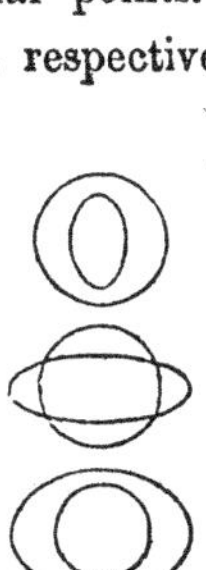

If a be greater, and c less than b, then will the first give a circle and an ellipse, the latter lying wholly within the former; the third will give the same kind of curves, but the ellipse will wholly envelop the circle; the second will give the same kind of curves, intersecting one another in four points. This last is the most important. It is the section *parallel to the axes of greatest and least elasticities.*

§ 322.—If $b = c$, then, Equations (568),

$$\alpha_{,} = 1\,; \quad \gamma_{,} = 0\,;$$

the axes will coincide with one another and with the axis a, that is, with x; $u_{,}$ will equal $u_{,,}$, and, Equation (571),

$$\frac{1}{V_{r_2}^2} - \frac{1}{V_{r_1}^2} = \left(\frac{1}{c^2} - \frac{1}{a^2}\right) \cdot \sin^2 u_{,} \quad . \quad . \quad . \quad . \quad . \qquad (573)$$

Also, Equation (563),

$$(x^2 + y^2 + z^2 - c^2)\left[a^2 x^2 + c^2 (y^2 + z^2) - a^2 c^2\right] = 0 \quad . \quad . \qquad (574)$$

and the wave surface will be resolved into the surface of a sphere, and that of an ellipsoid of revolution. Making $u_{,} = 0$, it will be seen from Equation (571) that these waves travel with equal velocities in the direction of the axis a. For any other value for u, since $u_{,} = u_{,,}$, $\cos u_{,} \cos u_{,,} + \sin u_{,} \sin u_{,,} = 1$, Equations (569) and (570) become

$$\frac{1}{V_{r_1}^2} = \frac{1}{c^2}; \quad \frac{1}{V_{r_2}^2} = \frac{1}{c^2} - \left(\frac{1}{c^2} - \frac{1}{a^2}\right) \cdot \sin^2 u_{,}; \quad . \quad . \qquad (575)$$

and it hence appears, that the velocity of one of the component waves will be constant throughout its entire extent, while that of the other will be variable from one point to another. The first is called the *ordinary*, the second the *extra-ordinary wave.*

If c be greater than a, then will the ellipsoid be prolate; if less than a, it will be oblate. There is but one direction which will make $V_{r_1}^2 = V_{r_2}^2$, and that is coincident with the axis a. Bodies in which this is true have but one axis of equal wave velocity, and are called *Uniaxial bodies.*

From Equation (571) it appears that the loci of equal retardations are concentric surfaces, of which the common axis is on the axis of equal wave velocity, and common vertex at the origin.

UMBILIC POINTS.

§ 323.—Let $L = 0$ represent Equation (563), and take

$$\cos A = \frac{1}{w} \cdot \frac{dL}{dz}; \quad \cos B = \frac{1}{w} \cdot \frac{dL}{dy}; \quad \cos C = \frac{1}{w} \cdot \frac{dL}{dx}; \qquad (576)$$

in which A, B, and C are the angles which a tangent plane to the surface makes with the co-ordinate planes xy, xz, and yz, respectively, and,

$$\frac{1}{w} = \frac{1}{\sqrt{\left(\frac{dL}{dx}\right)^2 + \left(\frac{dL}{dy}\right)^2 + \left(\frac{dL}{dz}\right)^2}} \quad . \quad . \quad . \quad . \quad (577)$$

Performing the operation here indicated on Equation (563), we have

$$\frac{dL}{dz} = 2z(a^2x^2 + b^2y^2 + c^2z^2) + 2c^2z(x^2 + y^2 + z^2 - a^2 - b^2),$$

$$\frac{dL}{dy} = 2y(a^2x^2 + b^2y^2 + c^2z^2) + 2b^2y(x^2 + y^2 + z^2 - a^2 - c^2);$$

$$\frac{dL}{dx} = 2x(a^2x^2 + b^2y^2 + c^2z^2) + 2a^2x(x^2 + y^2 + z^2 - b^2 - c^2).$$

Making $y = 0$, brings the tangential point in the plane ac, and the above become

$$\left.\begin{aligned} \frac{dL}{dz} &= 2z(a^2x^2 + c^2z^2) + 2c^2z(x^2 + z^2 - a^2 - b^2), \\ \frac{dL}{dy} &= 0, \\ \frac{dL}{dx} &= 2x(a^2x^2 + c^2z^2) + 2a^2x(x^2 + z^2 - b^2 - c^2). \end{aligned}\right\} \quad . \quad . \quad (578)$$

the second of which shows the tangent plane to be normal to the plane ac.

But $y = 0$ gives, Equations (572),

$$x^2 + z^2 - b^2 = 0; \quad a^2x^2 + c^2z^2 - a^2c^2 = 0,$$

whence we have

$$\left.\begin{aligned} z &= \pm a\sqrt{\frac{b^2 - c^2}{a^2 - c^2}} \\ x &= \pm c\sqrt{\frac{a^2 - b^2}{a^2 - c^2}} \end{aligned}\right\} \quad . \quad . \quad . \quad . \quad . \quad . \quad . \quad (579)$$

for the co-ordinates of the points in which the circle and ellipse intersect, and which are real as long as $a > b > c$. Substituting these in Equations (576), (577), and (578), we have

$$\cos A = \frac{0}{0}; \quad \cos B = \frac{0}{0}; \quad \cos C = \frac{0}{0};$$

hence the points of intersection of the ellipse and circle in the plane of the axis $a\,c$, are the vertices of conoidal cusps, each having a tangent cone. If a line be drawn tangent both to the ellipse and the circle in the plane $a\,c$, the tangential points will belong to the circumference of a circle along which a plane through this line may be drawn tangent to the wave surface. This circumference is in fact the margin of the conoidal or umbilic cusp, determined by the surface of the tangent cone reaching its limit by becoming a plane in the gradual increase of the inclination of its elements, as the tangential circumference recedes from the cusp point. A narrow annular plane wave, starting from this circle, will contract to a point in one direction; and, conversely, an element of a plane wave starting in the opposite direction will expand into a ring.

It thus appears that the general wave surface, and of which (563) is the equation, consists of two *nappes*, the one wholly within the other, except at four points, where they unite, and at each of which they form a double umbilic, somewhat after the manner of the opposite *nappes* of a very obtuse cone. The figure represents a model of the wave surface, so cut, by three rectangular planes, as to show two of the umbilic points, as well as the general course of the nappes, by the removal of a pair of the resulting diedral quadrantal fragments.

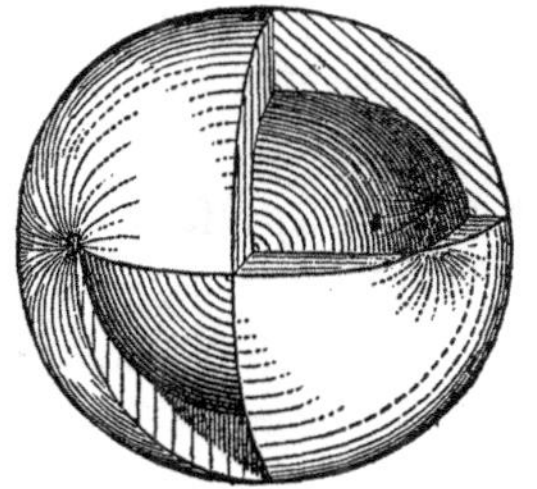

MOLECULAR VELOCITY.

§ 324.—Multiply the first of Equations (531) by $2\,d\,\xi$, the second by $2\,d\,\eta$, the third by $2\,d\,\zeta$, and integrate; there will result, recollecting that the molecule is moved from its place of rest,

$$\left.\begin{aligned}\frac{d\,\xi^2}{d\,t^2} &= -\,n_x^2\,.\,\xi^2,\\ \frac{d\,\eta^2}{d\,t^2} &= -\,n_y^2\,.\,\eta^2,\\ \frac{d\,\zeta^2}{d\,t^2} &= -\,n_z^2\,.\,\zeta^2.\end{aligned}\right\} \quad \ldots\ldots\ldots \quad (580)$$

whence it appears that the velocity of a molecule in the direction of either axis is proportional to its displacement in that direction, from its place of rest. The place of rest is only relative. When a molecule is in a position such that its neighbors are symmetrically disposed around it, it is in its place of rest, and its displacement therefrom will be directly proportional to the excess of condensation on one side over that on the other. This excess and the molecule's motion will reduce to zero simultaneously, and a single displacement, not repeated, can only give rise to what is called a *pulse.*

These equations also show that the *living force of the molecule is proportional to the square of the displacement.*

MOLECULAR ORBITS.

§ 325.—The molecular orbits are on the wave front. Suppose the wave due to the displacement ξ to be superposed upon that due to η, and take a molecule of which the place of rest is on the axis z. The first and second of Equations (528), will be sufficient to find the orbit of this molecule under the simultaneous action of both waves. From these two equations we find, after writing z for r_x and r_y,

$$(1) \ldots \quad \frac{2\,\pi}{\lambda_x}\,.\,(V_x\,.\,t - z) = \sin^{-1}\frac{\xi}{\alpha_x},$$

$$(2) \ldots \quad \frac{2\,\pi}{\lambda_y}\,.\,(V_y\,.\,t - z) = \sin^{-1}\frac{\eta}{\alpha_y},$$

$$(3) \ldots \quad \frac{2\,\pi}{\lambda_x}\,.\,(V_x\,.\,t - z) = \cos^{-1}\sqrt{1 - \frac{\xi^2}{\alpha_x^2}},$$

$$(4) \ldots \quad \frac{2\,\pi}{\lambda_y}\,.\,(V_y\,.\,t - z) = \cos^{-1}\sqrt{1 - \frac{\eta^2}{\alpha_y^2}}.$$

Subtracting (2) from (1),

$$2\pi\left(\frac{V_x \cdot t - z}{\lambda_x} - \frac{V_y \cdot t - z}{\lambda_y}\right) = \sin^{-1}\frac{\xi}{\alpha_x} - \sin^{-1}\frac{\eta}{\alpha_y};$$

in which $V_x \cdot t - z$, is the distance of the wave front due to ξ from the molecule's place of rest, and $V_y \cdot t - z$, that of the wave front due to η from the same point. Make

t_x = time required for the wave front due to			ξ	to travel over	$V_x \cdot t - z$;
τ_x =	"	"	"	"	"	λ_x;
t_y =	"	"	"	η	"	$V_y \cdot t - z$;
τ_y =	"	"	"	"	"	λ_y;

then will

$$\frac{V_x \cdot t - z}{\lambda_x} - \frac{V_y \cdot t - z}{\lambda_y} = \frac{t_x}{\tau_x} - \frac{t_y}{\tau_y} = \frac{t_x \cdot \tau_y - t_y \cdot \tau_x}{\tau_x \cdot \tau_y} = \frac{t_{,}}{\tau_{,}}; \quad (581)$$

which substituted above gives, after taking cosine of both members,

$$\cos 2\pi\frac{t_{,}}{\tau_{,}} = \sqrt{1 - \frac{\xi^2}{\alpha_x^2}} \cdot \sqrt{1 - \frac{\eta^2}{\alpha_y^2}} + \frac{\xi}{\alpha_x} \cdot \frac{\eta}{\alpha_y}.$$

Clearing the radical and reducing,

$$\frac{\xi^2}{\alpha_x^2} + \frac{\eta^2}{\alpha_y^2} - 2\cos 2\pi\frac{t_{,}}{\tau_{,}} \cdot \frac{\xi}{\alpha_x} \cdot \frac{\eta}{\alpha_y} - \sin^2 2\pi\frac{t_{,}}{\tau_{,}} = 0 \quad . \quad . \quad (582)$$

which is the equation of an ellipse referred to its centre.

§ 326.—To find the position of the transverse axis, take the usual formulas for the transformation of co-ordinates from one set, which are rectangular, to another, also rectangular. They are,

$$\xi = \xi' \cos\varphi - \eta' \sin\varphi,$$

$$\eta = \xi' \sin\varphi + \eta' \cos\varphi;$$

in which φ is the angle which the axis ξ' makes with that of ξ.

Substituting these values of ξ and η in Equation (582), collecting the coefficients, and placing that of the rectangle $\xi' \eta'$, equal to zero, we have

$$2 \sin \varphi . \cos \varphi (\alpha_x^2 - \alpha_y^2) - 2 (\sin^2 \varphi - \cos^2 \varphi) . \alpha_x . \alpha_y . \cos 2 \pi \frac{t_{,}}{\tau_{,}} = 0 ;$$

and because

$$\sin^2 \varphi - \cos^2 \varphi = \cos 2 \varphi,$$

$$2 \sin \varphi . \cos \varphi = \sin 2 \varphi,$$

the above becomes,

$$\tan 2 \varphi = 2 . \frac{\alpha_x . \alpha_y}{\alpha_x^2 - \alpha_y^2} . \cos 2 \pi . \frac{t_{,}}{\tau_{,}} \quad . \quad . \quad . \quad . \quad (583)$$

§ 327.—Now, if the successive pairs of component waves which disturb the molecule, reach it with a variable difference of phase, then will $\cos 2 \pi \frac{t_{,}}{\tau_{,}}$ be variable, and the transverse axis of the elliptical orbit be continually shifting its place. A wave in which the molecular motions fulfil this condition is called a *common wave;* being far the most frequent in nature. When the successive pairs of component waves are such as to make the second member of Equation (583) constant, the transverse axes of the molecular orbits will retain the same direction, and the wave is said to be *elliptically polarized.*

§ 328.—If $\frac{t_{,}}{\tau_{,}}$ equal $\frac{1}{4}$, or any odd multiple of $\frac{1}{4}$, and $\alpha_x = \alpha_y$, then will, Equation (582),

$$\xi^2 + \eta^2 - \alpha_x^2 = 0, \quad . \quad . \quad . \quad . \quad . \quad . \quad . \quad . \quad (584)$$

and the orbit becomes a circle. When this happens, the wave is said to be *circularly polarized.*

§ 329.—If $\frac{t_{,}}{\tau_{,}}$ be equal to any even multiple of $\frac{1}{4}$, then will

$$\cos 2 \pi . \frac{t_{,}}{\tau_{,}} = 1 ; \quad \sin^2 2 \pi \cdot \frac{t_{,}}{\tau_{,}} = 0 ;$$

and, Equation (582),

$$\frac{\xi}{\alpha_x} - \frac{\eta}{\alpha_y} = 0, \quad . \quad . \quad . \quad . \quad . \quad . \quad . \quad . \quad . \quad . \quad (585)$$

and the orbit is a straight line through the molecule's place of rest. The motion of the molecule will take place in a plane normal to the wave front, and the wave is said to be *plane polarized;* and a plane normal to the wave front and in the molecular paths, is called the *plane of polarization.*

§ 330.—Referring the curve to the new axes, and omitting the accents from ξ' and η', Equation (582) may be written,

$$\frac{\xi^2}{\alpha_x^2} + \frac{\eta^2}{\alpha_y^2} - \sin^2 2\pi \cdot \frac{t_{\prime}}{\tau_{\prime}} = 0, \quad . \quad . \quad . \quad . \quad . \quad (586)$$

in which α_x and α_y will take new values.

REFLEXION AND REFRACTION OF WAVES.

§ 331.—The elastic force which the molecules in the surface of one body exert upon those in the surface of another, in sensible contact, must, when the molecules are at relative rest, be equal to that exerted by the molecules in the interior of either body; else these surface molecules would be urged in opposite directions by unequal forces, and relative repose would be impossible. But, for equal displacements, the elastic forces developed in different bodies are in general unequal, and this is one of the most common of the causes that produce a resolution of primitive into secondary or component waves.

The velocity of a wave molecule varies, Equations (580), directly as the molecule's distance from its place of rest. If, therefore, a wave, in its progress through any medium, meet with a constitutional change of elasticity or density, the elastic force developed at the place of change will either be greater or less than that which determined the places of rest in the interior of either body. In the first case, the condensation in front cannot, by the forward movement, reduce to an equality with that behind; the surface molecules will first be checked, and then partly driven back upon those behind, and a return and an onward pulse will proceed in opposite directions from the surface which marks the change of structure, as from a primitive disturbance. In the second case, the molecules, meeting with less opposition, will go beyond their neutral limits with reference to those behind, the latter will close up in succession, and thus a return and transmitted pulse will arise as before, but with this difference, viz.: in the latter case, the molecular motions in the return pulse will continue in the same direction as before, whereas, in the former case, those motions will be reversed. The return pulse is said to be *reflected;* that transmitted, *refracted.* The primitive pulse, and of which these are the components, is called the *incident pulse.* A change of density or of elasticity will, Equation (537), produce a change in the velocity of wave propagation. A surface which is the locus of a change of density or of elasticity, is called a *deviating surface.* Two planes which are tangent, the one to the deviating surface, the other to the wave front, at a point common to both, will intersect in a line parallel to that of the nodes of the molecular orbits, which are in the deviating surface and near the common tangential point. This line of intersection is called the *line of nodes.* A plane through the tangential point and perpendicular to the line of nodes, is called the *plane of incidence.* The medium through which the wave moves before it meets the deviating surface, is called the *medium of incidence;* that into which it enters on passing this surface, the *medium of intromittance.*

§ 332.—Let A be a point common both to the wave and deviating surface. $A\,C$ a linear element of the former, and $A\,B$ a like element of the latter, both lying in the plane of incidence. Denote by V and λ the velocity and length of the wave in the medium of incidence; by $V_{\prime}$ and $\lambda_{\prime}$ the same in that of intromittance; and by t the time. Now, supposing the wave to proceed in the direction CB, and taking $A\,B = d\,s$, we have $CB = V \, . \, d\,t$.

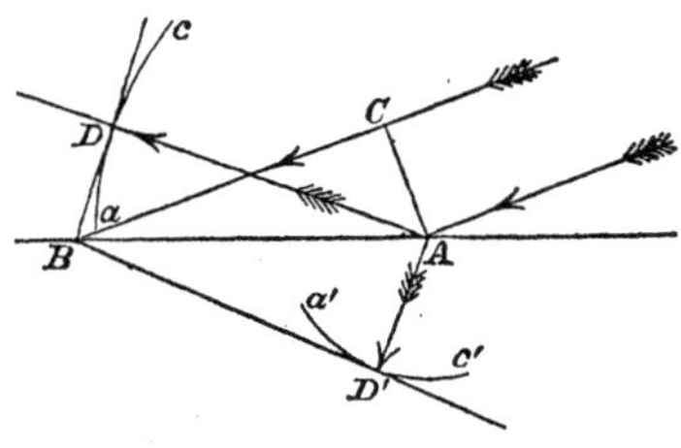

But while the point C, in the incident wave front, is moving from C to B, the reflected pulse, proceeding from A as a centre of disturbance, will move over a distance equal to $V\ d\,t$ in the medium of incidence; the refracted pulse over a distance equal to $V_{\prime} \, . \, d\,t$ in that of intromittance. With A as a centre, and radius $V.dt$, describe the arc ac, and with the radius $V_{,}dt$, the arc $a'c'$; and from B draw the tangents BD and BD'; the first will be the front of the new wave element in the medium of incidence, the second in that of intromittance.

§ 333.—Denote the angle $CAB = ABD$ by φ; the angle ABD' by φ'; then will

$$ds.\sin\varphi = Vdt; \quad ds.\sin\varphi' = V_{,}dt \quad . \quad . \quad . \quad . \qquad (587)$$

and by division, denoting the ratio of the velocities by m,

$$\frac{\sin\varphi}{\sin\varphi'} = \frac{V}{V_{,}} = m \quad . \quad . \quad . \quad . \quad . \quad . \quad . \quad . \qquad (588)$$

whence

$$\sin\varphi = m\sin\varphi' \quad . \quad . \quad . \quad . \quad . \quad . \quad . \quad . \qquad (589)$$

The angle φ measures the inclination of the incident, and φ' that of the refracted wave to the deviating surface. These are equal, respectively, to the angles which the normals to the incident and refracted waves make with the normal to the deviating surface, at the point of incidence. The first is called the *angle of incidence*, the second the *angle of refraction*. The inclination of the reflected wave to the deviating surface, is called the *angle of reflexion*. The normals to the incident and reflected waves fall on opposite sides of the normal to the deviating surface; and because the velocity of the reflected wave is equal to that of the incident, with contrary sign, Equation (589) becomes applicable to the reflected wave, by making $m = -1$.

LIVING FORCE AND QUANTITY OF MOTION IN A PLANE POLARIZED WAVE.

§ 334.—Take either of Equations (528), say the first, and which relates to a wave plane polarized, the plane of polarization being perpendicular to the co-ordinate plane yz, differentiate with respect to ξ and t, dropping the subscripts—we get

$$\frac{d\xi}{dt} = \alpha.V.\cos\frac{2\pi}{\lambda}(Vt - r)\frac{2\pi}{\lambda}.$$

Denote the density of the medium by Δ, and the area of any portion of the wave-front by a, then will the mass between two consecutive positions of this area be $a.\Delta.dr$, and the living force within a quarter of a wave-length be

$$\int_{r+\frac{1}{4}\lambda}^{r}\Delta.a.dr.\frac{d\xi^2}{dt^2} = \int_{Vt-r=\frac{1}{4}\lambda}^{Vt-r=0}\Delta.a.\frac{2\pi}{\lambda}.a^2.V^2.\cos^2\frac{2\pi}{\lambda}(Vt-r)\frac{2\pi}{\lambda}dr = \tfrac{1}{2}\pi^2.\frac{V}{\lambda}.\Delta.a.V.a^2, \qquad (590)$$

Dividing by the volume $a . V$, and recalling that π and $\frac{V}{\lambda}$ are constant, we shall find that the quantity of living force in a unit of volume of the medium will vary directly as the product of the density and square of the greatest displacement; and the relation of these products, in the case of any two waves, will determine the relation of the effects of these waves upon the organs of sense upon which they act.

Again, the quantity of motion in this quarter of wave-length will be

$$\int_{r+\frac{1}{4}\gamma}^{r} \Delta . a . dr . \frac{d\xi}{dt} = \int_{Vt-r=\frac{1}{4}\lambda}^{Vt-r=0} \Delta . a . a . V . \cos \frac{2\pi}{\lambda}(Vt-r)\frac{2\pi}{\lambda} dr = \Delta . a . a . V \quad . \quad (591)$$

RESOLUTION OF LIVING FORCE AND OF MOTION, BY DEVIATING SURFACES.

§ 335.—Take the co-ordinate plane xz in the plane of incidence, and the axis z in the direction of the normal to the incident wave, the axis y will be parallel to the line of the nodes of the molecular orbit in the deviating surface, at the place of incidence. Then, preserving the notation of § 332, will the element of the deviating surface at the place of incidence be $ds . dy$, and its projections upon the incident, reflected and refracted wave-fronts, respectively, be $ds . dy . \cos\varphi$, $ds . dy \cos\varphi$, and $ds . dy . \cos\varphi'$. These will take the place of a in Equations (590) and (591), in computing the living force and quantity of motion in the incident, reflected and refracted waves. The living force in the incident must be equal to the sum of the living forces of its reflected and refracted components. First take the wave in which the molecular motions are parallel to the axis x, and employ the subscripts i, r and t to denote the incident, reflected and refracted or transmitted waves, respectively. The living force in a quarter of each of these waves will, omitting the common factors, Equations (529), (545) and (590), give

$$\Delta . \cos\varphi . V . \alpha^2_{xr} + \Delta_{,} . \cos\varphi' . V_{,} . \alpha^2_{xt} - \Delta . \cos\varphi . V . \alpha^2_{xi} = 0;$$

or, Equations (588) and (589),

$$\alpha^2_{xr} + \frac{\Delta_{,}}{\Delta} \cdot \frac{\cos\varphi'}{\cos\varphi} \cdot \frac{\sin\varphi'}{\sin\varphi} \cdot \alpha^2_{xt} - \alpha^2_{xi} = 0 \quad . \quad . \quad . \quad . \quad (592)$$

in which Δ and $\Delta_{,}$ are the densities of the medium of incidence and of intromittance.

The molecular motions are all parallel to the plane of incidence, and at the same time normal to the directions of their respective wave motions;

they, therefore, make with one another angles equal to those made by the directions of these latter motions, and we obtain two more equations from the relations of Equations (59) for the resolution and composition of oblique forces. The angles made by the direction of the motion in the incident with the directions of the motions in the reflected and refracted waves, are $180° - 2\varphi$ and $360° - (\varphi - \varphi')$, respectively; and the angles under which the directions of the motions in the latter waves are inclined to one another, is $180° - (\varphi + \varphi')$. Whence

$$\Delta . \cos\varphi . V . \alpha_{xr} = -\Delta . \cos\varphi . V . \alpha_{xi} . \frac{\sin(\varphi - \varphi')}{\sin(\varphi + \varphi')};$$

$$\Delta_{,} . \cos\varphi' . V_{,} . \alpha_{xt} = \Delta . \cos\varphi . V . \alpha_{xi} . \frac{\sin 2\varphi}{\sin(\varphi + \varphi')};$$

or,

$$\alpha_{xr} = -\alpha_{xi} . \frac{\sin(\varphi - \varphi')}{\sin(\varphi + \varphi')} \quad \ldots \ldots \ldots \quad (593)$$

$$\alpha_{xt} = \alpha_{xi} . \frac{\Delta}{\Delta_{,}} . \frac{\cos\varphi}{\cos\varphi'} . \frac{\sin\varphi}{\sin\varphi'} . \frac{\sin 2\varphi}{\sin(\varphi + \varphi')} \quad \ldots \ldots \quad (594)$$

Substituting these in Equation (592), we readily find,

$$\frac{\Delta}{\Delta_{,}} = \frac{4\cos^2\varphi' . \sin^2\varphi'}{\sin^2 2\varphi} = \frac{\cos^2\varphi' . \sin^2\varphi'}{\cos^2\varphi . \sin^2\varphi};$$

whence,

$$\sqrt{\Delta_{,}} = \sqrt{\Delta} . \frac{\sin 2\varphi}{2\cos\varphi' . \sin\varphi'} = \sqrt{\Delta} . \frac{\cos\varphi . \sin\varphi}{\cos\varphi' . \sin\varphi'} \quad \ldots \quad (595)$$

Substituting the above ratio of the densities in the equation just preceding, we get

$$\alpha_{xt} = \alpha_{xi} . \frac{2\cos\varphi' . \sin\varphi'}{\sin(\varphi + \varphi')}; \quad \ldots \ldots \quad (596)$$

multiplying this by Equation (595), member by member, and the equation giving the value of α_{xr} by $\sqrt{\Delta}$, and taking

$$\sqrt{\Delta} . \alpha_{xi} = 1; \quad \sqrt{\Delta}\, \alpha_{xr} = v; \quad \sqrt{\Delta_{,}} . \alpha_{xt} = u,$$

we find

$$v = -\frac{\sin(\varphi - \varphi')}{\sin(\varphi + \varphi')} \quad \ldots \ldots \quad (597)$$

$$u = \frac{\sin 2\varphi}{\sin(\varphi + \varphi')} \quad \ldots \ldots \ldots \quad (598)$$

To which may be added the relations, Equation (589),

$$\sin\varphi' = \frac{\sin\varphi}{m}; \quad \cos\varphi' = \sqrt{1 - \frac{\sin^2\varphi}{m^2}}$$

Transposing the term of which α_{zt} is a factor to the second member in Equation (592), subtracting Equation (593) from $\alpha_{zt} = \alpha_{zi}$, dividing the first result by the second, and multiplying the quotient by Equation (593)′, we readily find

$$\frac{\alpha_{zi} + \alpha_{zr}}{\cos\varphi} = \frac{\alpha_{zt}}{\cos\varphi'} \quad . \quad . \quad . \quad . \quad . \quad . \quad . \quad . \qquad (599)$$

That is, the projection in the direction of wave propagation and on the deviating surface, of the greatest displacement in the incident, increased by that in the reflected wave, is equal to like projection of the greatest displacement in the refracted wave.

Next, take the wave in which the molecular motions are parallel to the axis y; these are parallel to the deviating surface. The motions in the incident, reflected and refracted waves are parallel to one another, and, by the principles of parallel forces, the sum of the motions in the reflected and refracted waves must be equal to that in the incident. The equation for the living force will be the same as before. Whence Equations (529), (545) and (590), omitting the common factors,

$$\Delta \,.\, \cos\varphi \,.\, V \,.\, \alpha^2_{yr} + \Delta_{,} \,.\, \cos\varphi' \,.\, V_{,} \,.\, \alpha^2_{yt} - \Delta \,.\, \cos\varphi \,.\, V \,.\, \alpha^2_{yi} = 0\,;$$

$$\Delta \,.\, \cos\varphi \,.\, V \,.\, \alpha_{yr} + \Delta_{,} \,.\, \cos\varphi' \,.\, V_{,} \,.\, \alpha_{yt} - \Delta \,.\, \cos\varphi \,.\, V \,.\, \alpha_{yi} = 0 \quad . \qquad (600)$$

In which Δ and $\Delta_{,}$ are, as before, the densities of the medium of incidence and of intromittance, respectively; or, Equations (588) and (589),

$$\alpha^2_{yr} + \frac{\Delta_{,}}{\Delta} \cdot \frac{\sin\varphi'}{\sin\varphi} \cdot \frac{\cos\varphi'}{\cos\varphi} \cdot \alpha^2_{yt} - \alpha^2_{yi} = 0$$

$$\alpha_{yr} + \frac{\Delta'}{\Delta} \cdot \frac{\sin\varphi'}{\sin\varphi} \cdot \frac{\cos\varphi'}{\cos\varphi} \cdot \alpha_{yt} - \alpha_{yi} = 0 \quad . \quad . \quad . \qquad (601)$$

Transposing the terms containing α_{yr} and α_{yi} to the second members, and dividing the first by the second, we find

$$\alpha_{yr} + \alpha_{yi} = \alpha_{yt} \,. \quad . \quad . \quad . \quad . \quad . \quad . \quad . \quad . \qquad (602)$$

That is, the greatest displacement in the refracted is equal to the sum of the greatest displacements in the incident and reflected waves.

Substituting the value of $\frac{\Delta'}{\Delta}$, as given by Equation (597), in Equation (601), we have

$$\alpha_{yr} + \frac{\sin\varphi \,.\, \cos\varphi}{\sin\varphi' \,.\, \cos\varphi'} \cdot \alpha_{yt} - \alpha_{yi} = 0 \,. \quad . \quad . \quad . \qquad (603)$$

Substituting in this, first the value of α_{yt}, and then of α_{yr}, deduced from Equation (602), we readily get

$$\alpha_{yr} = -\alpha_{yi} \cdot \frac{\tan(\varphi - \varphi')}{\tan(\varphi + \varphi')}; \quad . \quad . \quad . \quad . \quad . \quad . \quad (604)$$

$$\alpha_{yt} = \alpha_{yi} \cdot \frac{4 \cos \varphi' . \sin \varphi'}{\sin 2\varphi + \sin 2\varphi'} \quad . \quad . \quad . \quad . \quad . \quad . \quad (605)$$

Multiplying the first of these by $\sqrt{\Delta}$, and the second by Equation (595), and making

$$\sqrt{\Delta} . \alpha_{yi} = 1; \quad \sqrt{\Delta} . \alpha_{yr} = v'; \quad \sqrt{\Delta_{\prime}} . \alpha_{yt} = u';$$

there will result,

$$v' = -\frac{\tan(\varphi - \varphi')}{\tan(\varphi + \varphi')} \quad . \quad . \quad . \quad . \quad . \quad . \quad . \quad (606)$$

$$u' = \frac{\sin 2\varphi}{\sin(\varphi + \varphi') . \cos(\varphi - \varphi')} \quad . \quad . \quad . \quad . \quad (607)$$

§ 336.—Divide Equation (598) by Equation (597), and Equation (607) by Equation (606), replace v, u, v' and u' by their values, and substitute for the ratio of the square roots of the densities its value as given in Equation (595), we find

$$\frac{\alpha_{xt} . \cos \varphi'}{\alpha_{xr} . \cos \varphi} = -\frac{\cos \varphi'}{\cos \varphi} \cdot \frac{\sin 2\varphi'}{\sin(\varphi - \varphi')},$$

$$\frac{\alpha_{yt}}{\alpha_{yr}} = -\frac{\sin 2\varphi'}{\sin(\varphi - \varphi') . \cos(\varphi + \varphi')} \quad . \quad . \quad . \quad (608)$$

But $\alpha_{xt} . \cos \varphi'$ and $\alpha_{xr} . \cos \varphi$, are the components parallel to the deviating surface of the displacements which are in the plane of incidence; α_{yt} and α_{yr} are already parallel to the deviating surface; whence, as long as $\varphi > \varphi'$, that is, as long as the velocity of wave-motion in the medium of incidence exceeds that in the medium of intromittance, the molecular phases in the refracted and reflected waves will be opposite, and conversely.

§ 337.—Denote the living force in the original incident wave, supposed common, by unity; that in each of its two original components will be denoted by one half of unity, and the total living force of the reflected wave will, Equations (597), (606), be

$$v^2 + v'^2 = \tfrac{1}{2} \cdot \frac{\sin^2(\varphi - \varphi')}{\sin^2(\varphi + \varphi')} + \tfrac{1}{2} \cdot \frac{\tan^2(\varphi - \varphi')}{\tan^2(\varphi + \varphi')} \quad . \quad . \quad (609)$$

and that of the refracted,

$$u^2 + u'^2 = \tfrac{1}{2}\left(1 - \frac{\sin^2(\varphi - \varphi')}{\sin^2(\varphi + \varphi')}\right) + \tfrac{1}{2}\left(1 - \frac{\tan^2(\varphi - \varphi')}{\tan^2(\varphi + \varphi')}\right) \quad (610)$$

POLARIZATION BY REFLEXION AND REFRACTION.

§ 338.—The first term in the second member of Equation (609), measures the living force in that portion of the reflected wave which is due to vibrations parallel to the plane of incidence; the second, that due to vibrations perpendicular to this plane. The former exceeds the latter. These living forces being proportional to the squares of the greatest displacements, the former may be represented by α_x^2, and the latter by α_y^2, in Equation (582). The factor $\frac{t_{\prime}}{\tau_{\prime}}$, in this equation, determines the difference of phase simultaneously impressed by both waves upon the same molecule, and when the waves have passed from one medium to another, its value will depend not only upon the nature of both media, but also upon the action to which the waves may have been subjected while crossing the space wherein the physical changes occur that constitute the transition from one medium to another. The amount of this action, in any particular case, can only be known from experience. The resultant waves, both in the medium of incidence and of intromittance, will be elliptically polarized.

When $\varphi + \varphi' = 90°$, then, Equation (589), will $\sin \varphi' = \cos \varphi$, and

$$m = \frac{\sin \varphi}{\cos \varphi} = \tan \varphi \, ; \quad . \quad . \quad . \quad . \quad . \quad . \quad . \quad . \qquad (611)$$

the second, term of Equation (609) will disappear, and the reflected wave will be wholly polarized in the plane of incidence. This angle, of which the tangent is equal to the index of refraction, is called the *polarizing angle.*

The index of refraction varies with the wave length, Eqs. (588), (545), and it will, therefore, be impossible wholly to polarize, by a single reflexion, a wave compounded of several components, having different wave lengths.

Of the terms of the second member of Equation (610), the last is the greater, because

$$\frac{\sin^2(\varphi - \varphi')}{\sin^2(\varphi + \varphi')} = \frac{\tan^2(\varphi - \varphi')}{\tan^2(\varphi + \varphi')} \cdot \frac{\cos^2(\varphi - \varphi')}{\cos^2(\varphi + \varphi')};$$

and the excess will measure the preponderance of that part of the refracted wave due to vibrations perpendicular over that due to vibrations parallel to the plane of incidence. This excess is exactly equal to the excess in the reflected wave which arises from vibrations parallel over those perpendicular to the plane of incidence.

§ 338′.—If the wave velocity in the medium of incidence be less than in that of intromittance, then will m be less than unity, and the values of v and v' become imaginary for all angles of incidence greater than that whose sine is equal to m, and at this limit the problem changes its nature. In fact, this is the limit of refraction, according to the law of the sines, Equation (589), and for any increase of the angle of incidence beyond this, the wave will be wholly reflected.

§ 339.—If the wave be plane polarized, and its plane of polarization inclined to that of incidence, under any angle denoted by α, then will the reflected component displacements parallel and perpendicular to the plane of incidence be, respectively, Equations (597) and (606),

$$-\frac{\sin(\varphi - \varphi')}{\sin(\varphi + \varphi')} \cdot \cos\alpha, \text{ and } -\frac{\tan(\varphi - \varphi')}{\tan(\varphi + \varphi')} \cdot \sin\alpha.$$

The component waves due to these displacements will proceed onwards, and may satisfy the condition of $\frac{t_{\prime}}{\tau_{\prime}}$ being an even multiple of $\frac{1}{4}$; in which case the resultant will, Equation (585), be a plane polarized wave. Denote the inclination of its plane of polarization to that of reflexion by α', then will

$$\tan\alpha' = \frac{v'}{v} = \frac{\dfrac{\tan(\varphi - \varphi')}{\tan(\varphi + \varphi')} \cdot \sin\alpha}{\dfrac{\sin(\varphi - \varphi')}{\sin(\varphi + \varphi')} \cdot \cos\alpha} = \frac{\cos(\varphi + \varphi')}{\cos(\varphi - \varphi')} \cdot \tan\alpha \quad . \quad (612)$$

If $\varphi + \varphi' = 90°$, then will $\alpha' = 0°$, whatever be α; also if $\alpha = 0°$, then will $\alpha' = 0°$; finally, if $\varphi = 0°$, then will $\varphi' = 0$, and $\alpha' = \alpha$. That is, when a plane polarized wave is incident under the polarizing angle, it is reflected polarized in the plane of reflexion. Where an incident wave is polarized in the plane of incidence, the reflected wave preserves its plane of polarization unchanged under all angles of incidence. Finally, under a perpendicular incidence, the plane of polarization of the incident and that of the reflected wave coincide.

Equation (612) shows that α' is always less than α, and that the plane of polarization approaches that of incidence at each reflexion, and may be made, by a sufficient number of reflexions, ultimately to coincide with it.

§ 340.—Still, supposing the velocity of the wave less in the medium of incidence than in that of intromittance, or $\varphi' > \varphi$, let the wave be plane polarized, and its plane of polarization inclined to that of incidence. The vibrations will be resolved into their components, respectively parallel and perpendicular to this latter plane; and as long as $\sin \varphi < m$, two components will be reflected and two refracted. If $\frac{t_{\prime}}{T_{\prime}}$ be any even multiple of $\frac{1}{4}$, in both sets of components, the reflected and intromitted resultant waves will be plane polarized.

The inclinations, denoted by α' and $\alpha_{\prime}$, of the planes of polarization of the reflected and refracted waves, respectively, to the plane of incidence, will be given by

$$\tan \alpha' = \frac{v'}{v} \cdot \tan \alpha ; \quad \tan \alpha_{\prime} = \frac{u'}{u} \cdot \tan \alpha ;$$

in which v, v', u and u', are to be found by Equations (597), (606), (598), and (607).

If $\alpha = 45°$ and $\sin \varphi = m$, then will

$$\tan \alpha' = -1, \text{ and } \tan \alpha_{\prime} = \frac{1}{m}.$$

At this limit, the refracted wave takes the direction of the deviating surface. An infinitesimal increment to φ will cause this wave to be reflected and make $m = -1$, $\tan \alpha_{\prime} = -1$, and give to $\tan \alpha'$ the form of indetermination. But, retaining the limiting value of this function above, we have,

$$1 + \tan \alpha' . \tan \alpha_{\prime} = 1 - 1 = 0 ;$$

and since the planes of polarization pass through the same line, viz., a normal to the wave front, they will make with one another an angle of 90°, and the whole reflected wave will be compounded of two equal components polarized in planes at right angles to each other. If these waves reach the molecules in their common path, so as to satisfy the condition that $\frac{t_{\prime}}{\tau_{\prime}}$ shall be an even multiple of $\frac{1}{4}$, the resultant wave will be plane polarized; if an odd multiple, then circularly polarized; and if between these limits, then elliptically polarized.

§ 341.—If the polarization be circular, then will $\alpha_x = \alpha_y = \alpha_{\prime}$, be equal to the radius vector of the circular orbit. Denote the angle which this radius makes with the axis x, at any instant, by θ; then will

$$\alpha_{\prime} . \cos\theta = \xi = \alpha_x . \sin 2\pi \frac{V_x . t - z}{\lambda_x},$$

$$\alpha_{\prime} . \sin\theta = \eta = \alpha_y . \sin 2\pi \frac{V_y . t - z}{\lambda_y}.$$

Denote the time required for the first wave to describe $V_x . t - z$, by t_x, that for the second to describe $V_y . t - z$ by t_y, and the periodic time of a molecule in both waves by τ; then, because the wave velocity is constant, and the wave length and orbit are described in the same time,

$$\frac{V_x . t - z}{\lambda_x} = \frac{t_x}{\tau}; \quad \frac{V_y . t - z}{\lambda_y} = \frac{t_y}{\tau},$$

which, in the above, give,

$$\cos\theta = \sin 2\pi . \frac{t_x}{\tau},$$

$$\sin\theta = \sin 2\pi . \frac{t_y}{\tau};$$

and making

$$t_y = t_x \pm t', \quad . \quad . \quad . \quad . \quad . \quad . \quad . \quad . \quad (613)$$

in which t' denotes the time the wave due to vibrations parallel to one axis is in advance of that due to those parallel to the other; we have,

$$\cos\theta = \sin 2\pi\,\frac{t_x}{\tau}; \quad . \; . \; . \; . \; . \; . \; . \quad (614)$$

$$\sin\theta = \sin 2\pi\left(\frac{t_x}{\tau} \pm \frac{t'}{\tau}\right). \quad . \; . \; . \quad (615)$$

Differentiating, regarding $\frac{t'}{\tau}$ as constant, we find,

$$\frac{d\theta}{d t_x} = \frac{2\pi}{\tau\,.\,\cos\theta}\,.\,\cos 2\pi\left(\frac{t_x}{\tau} \pm \frac{t'}{\tau}\right);$$

and, developing the last factor,

$$\frac{d\theta}{d t_x} = \frac{2\pi}{\tau\,.\,\cos\theta}\cdot\left[\cos 2\pi\cdot\frac{t_x}{\tau}\cdot\cos 2\pi\cdot\frac{t'}{\tau} \mp \sin 2\pi\cdot\frac{t_x}{\tau}\cdot\sin 2\pi\,\frac{t'}{\tau}\right];$$

and making $\frac{t'}{\tau} = \frac{1}{4}$,

$$\cos\theta\,.\,\frac{d\theta}{d t_x} = \mp\,\frac{2\pi}{\tau}\,.\,\sin 2\pi\,.\,\frac{t_x}{\tau} \quad . \; . \; . \; . \; . \; . \quad (616)$$

Differentiating (614), we find,

$$\sin\theta\,.\,\frac{d\theta}{d t_x} = -\,\frac{2\pi}{\tau}\,.\,\cos 2\pi\,\frac{t_x}{\tau} \quad . \; . \; . \; . \; . \quad (617)$$

Squaring, adding to the square of Equation (616), and taking square root,

$$\frac{d\theta}{d t_x} = \mp\,\frac{2\pi}{\tau} \quad . \; . \; . \; . \; . \; . \; . \; . \; . \quad (618)$$

whence the velocity is constant.

The first member of Equation (616) is the velocity in the direction of the axis y, and Equation (617) in the direction of the axis x, and these equations show that the upper sign must be taken in Equation (618) when t' is positive in Equation (613), and the lower when t' is negative. Whence it appears, that two waves plane polarized will, by their simultaneous action upon a molecule, cause it to move uniformly in a circle, provided they be of the same length, and one wave lag, as it were, behind the other, by a distance equal to $\frac{1}{4}$ of a wave length; and the motion will be from right to left, or the converse, according to wave precedence.

Two waves distinguished by these peculiarities are said to be *oppositely polarized.* The plane perpendicular to the wave front, and through that diameter of the orbit into which the molecule would be brought at the same instant by the separate action of the two waves, is called the *plane of crossing.*

§ 342.—Let

$$(1) \quad . \quad . \quad . \quad . \quad . \quad . \quad \alpha_{\prime} \cos \theta = \xi = \alpha_{\prime} \sin 2\pi \frac{t_{\prime}}{\tau},$$

$$(2) \quad . \quad . \quad . \quad . \quad . \quad . \quad \alpha_{\prime} \sin \theta = \eta = \alpha_{\prime} \sin \left(2\pi . \frac{t_{\prime}}{\tau} + \frac{t'}{\tau}\right),$$

$$(3) \quad . \quad . \quad . \quad . \quad . \quad . \quad \alpha_{\prime} \cos \theta = \xi = \alpha_{\prime} \sin \left(2\pi . \frac{t_{\prime}}{\tau} + \frac{t'}{\tau}\right),$$

$$(4) \quad . \quad . \quad . \quad . \quad . \quad . \quad \alpha_{\prime} \sin \theta = \eta = \alpha_{\prime} \sin 2\pi \frac{t_{\prime}}{\tau},$$

be the displacements in two oppositely circularly polarized waves. The union of (1) and (4) gives a resultant wave plane polarized; that of (2) and (3) also a wave plane polarized, the equation of the path being

$$\xi = \eta$$

in the plane of crossing. It thus appears that the union of two circularly polarized waves, polarized in opposite directions, gives a plane polarized wave, of which the intensity is double of either. Conversely, a wave plane polarized may be resolved into two components of equal intensity, circularly polarized in opposite directions.

§ 343.—Because the time of describing the wave length is equal to the molecular periodic time, we have, denoting the velocity of wave propagation by V,

$$\lambda = V\tau,$$

whence

$$\tau = \frac{\lambda}{V};$$

which, in Equation (618), gives, after multiplying by t_x and dividing by 2π,

$$\frac{\frac{d\theta}{dt_x} \cdot t_x}{2\pi} = \frac{Vt_x}{\lambda} \quad \ldots \ldots \quad (619)$$

The first member is the arc, expressed in circumferences, described by the molecule while the wave is moving through a thickness $V \cdot t_x$ of the medium. So that a wave, compounded of many components having different wave lengths, but all polarized, on entering a medium, may emerge with the planes of polarization of its several components so twisted through different angles as to diverge from a common line perpendicular to the wave front. The department of optics furnishes some fine examples of this. A piece of quartz, of a peculiar kind, is known to twist the extreme red wave through an angle of 17° 29′ 47″, and the extreme violet, 44° 04′ 58″, for each millimetre of thickness.

DIFFUSION AND DECAY OF LIVING FORCE.

§ 344.—The living force of any molecule whose mass is m and velocity $v_{,}$, is

$$m v_{,}^2;$$

and denoting by n the number of molecules on a superficial unit of the wave front, the living force on this unit will be

$$n \cdot m \cdot v_{,}^2;$$

and on the surface of a sphere of which the radius is $r_{,}$,

$$4\pi \cdot r_{,}^2 \cdot n \cdot m v_{,}^2;$$

and for another sphere, of which the radius is $r_{,,}$, and molecular velocity $v_{,,}$,

$$4\pi \cdot r_{,,}^2 \cdot n m v_{,,}^2.$$

If these spherical surfaces occupy the same relative positions in a diverging wave, in any two of its positions, their molecular living forces must be equal; whence, suppressing the common factors,

$$r_{,}^2 \cdot m v_{,}^2 = r_{,,}^2 m v_{,,}^2. \quad \ldots \ldots \quad (620)$$

The molecules describe elliptical orbits, and under the action of molecular forces directed to the centres of these curves. The periodic time will, therefore, § 207, Equation (286), be constant, however the dimensions of these orbits may vary; and the average velocities of the molecules will be proportional to the lengths of their respective orbits, or, in similar orbits, to any homologous dimensions of the same—as their transverse axes or greatest molecular displacements. Denoting the latter by c' and c'' in the two waves, then will

$$\frac{v_{\prime}}{v_{\prime\prime}} = \frac{c'}{c''};$$

which, with Equation (620), gives

$$c'' r_{\prime\prime} = c' r_{\prime} \quad \text{. . . . .} \quad (621)$$

Whence it appears, *that the living force of the molecules of any wave varies inversely as the second, and the greatest displacement inversely as the first power of the distance to which the wave has been propagated from its place of primitive disturbance.*

INTERFERENCE.

§ 345.—Resuming Equation (586), viz.,

$$\frac{\xi^2}{\alpha_{\prime}^2} + \frac{\eta^2}{\alpha_{\prime\prime}^2} - \sin^2 2\pi\frac{t}{\tau} = 0;$$

denote the radius vector of the molecular orbit by ρ', and the angle it makes with the axis of ξ by θ', then will

$$\xi = \rho' . \cos\theta'; \quad \eta = \rho' . \sin\theta';$$

which, in the above, give

$$\rho' = \frac{\alpha_{\prime} . \alpha_{\prime\prime}}{\sqrt{\alpha_{\prime\prime}^2 \cos^2\theta' + \alpha_{\prime}^2 \sin^2\theta'}} . \sin 2\pi . \frac{t}{\tau};$$

and making

$$\frac{\alpha_{\prime} . \alpha_{\prime\prime}}{\sqrt{\alpha_{\prime\prime}^2 . \cos^2\theta' + \alpha_{\prime}^2 . \sin^2\theta'}} = c',$$

we have

$$\rho' = c' . \sin 2\pi . \frac{t}{\tau} \quad . \quad . \quad . \quad . \quad . \quad . \quad . \qquad (622)$$

In this equation, ρ' is the actual displacement of the molecule from its place of rest, and becomes a maximum when $\frac{t}{\tau}$ is any odd multiple of $\frac{1}{4}$. If, however, there be added to the arc $2\pi\frac{t}{\tau}$, an arbitrary arc a', this latter may be so taken as to make the maximum or any other displacement occur at such time and place as we please, and, therefore, to give to the molecule any particular phase at pleasure, at the time t. We may write, then, generally,

$$\rho' = c' . \sin\left(2\pi . \frac{t}{\tau} + a'\right); \quad . \quad . \quad . \quad . \qquad (623)$$

and for a second resultant wave,

$$\rho'' = c'' . \sin\left(2\pi . \frac{t}{\tau} + a''\right); \quad . \quad . \quad . \quad . \quad . \qquad (624)$$

and if these waves act simultaneously upon the same molecules, the resultant displacement, denoted by ρ, will, § 306, be given by

$$\rho = \rho' + \rho'' = c' . \sin\left(2\pi . \frac{t}{\tau} + a'\right) + c'' . \sin\left(2\pi . \frac{t}{\tau} + a''\right).$$

Developing the circular functions and collecting the coefficients of like factors,

$$\rho = (c' \cos a' + c'' \cos a'') . \sin 2\pi\frac{t}{\tau} + (c' \sin a' + c'' \sin a'') . \cos 2\pi\frac{t}{\tau};$$

and making

$$\left.\begin{aligned} c \cos a &= c' . \cos a' + c'' \cos a'', \\ c \sin a &= c' \sin a' + c'' \sin a'', \end{aligned}\right\} \quad . \quad . \quad . \quad . \quad . \qquad (625)$$

we have

$$\rho = c . \cos a . \sin 2\pi . \frac{t}{\tau} + c \sin a . \cos 2\pi . \frac{t}{\tau};$$

or,

$$\rho = c \sin \left(2 \pi . \frac{t}{\tau} + a\right). \quad . \quad . \quad . \quad . \quad . \quad . \quad . \quad . \quad . \quad . \qquad (626)$$

Squaring Equations (625), and adding,

$$c^2 = c'^2 + c''^2 + 2\, c'\, c'' \cos (a' - a''), \quad . \quad . \quad . \qquad (627)$$

and dividing the second by the first,

$$\tan a = \frac{c' . \sin a' + c'' . \sin a''}{c' \cos a' + c'' \cos a''}. \quad . \quad . \quad . \quad . \quad . \quad . \quad . \qquad (628)$$

From Equation (626) we see that the resultant wave is of the same length as that of the component waves to which Equations (623) and (624) appertain; the length being determined by the molecular periodic time τ; but the value of a in that equation differing from a' and a'' in Equations (623) and (624), shows that the maximum displacement of a given molecule does not take place in the resultant wave at the same time as in either of its components.

§ 346.—The maximum displacement in the resultant wave is given by

$$c = \sqrt{c'^2 + c''^2 + 2\, c'\, c'' . \cos (a' - a'')}\,; \quad . \quad . \quad . \qquad (629)$$

which will be the greatest possible when $a' - a'' = 0$, and least possible when $a' - a'' = 180°$; the maximum in the former case being given by

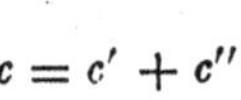

$$c = c' + c''$$

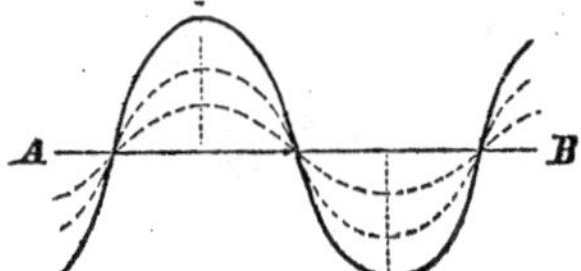

and the minimum, by

$$c = c' - c''.$$

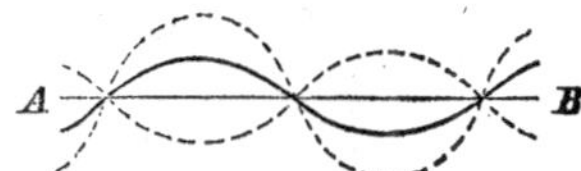

In the first case, Equation (628),

$$\tan a = \frac{(c' + c'') . \sin a'}{(c' + c'') . \cos a'} = \tan a'.$$

Whence $a = a' = a''$, and the maximum displacement will occur at the same place and time in the resultant and component waves.

In the second case, Equation (628), if we make $a' = 180° + a''$,

$$\tan a = \frac{(c' - c'') \,.\, \sin a''}{(c' - c'') \,.\, \cos a''} = \tan a'' = \tan (a' - 180°) = \tan a' ;$$

that is, a will be equal to one at least of the arcs a' and a'', and the greatest displacement will occur at the same time and place in the resultant wave as in one of its components.

If $c' = c''$, then, Equation (629),

$$c = c' \sqrt{2 \left[1 + \cos (a' - a'')\right]} ;$$

and because

$$1 + \cos (a' - a'') = 2 \cos^2 \frac{a' - a''}{2},$$

$$c = 2\, c' \,.\, \cos \frac{a' - a''}{2}, \quad . \quad . \quad . \quad . \quad . \quad . \quad . \quad (630)$$

and, Equation (628),

$$\tan a = \frac{\sin a' + \sin a''}{\cos a' + \cos a''} = \tan \frac{a' + a''}{2} \quad . \quad . \quad . \quad . \quad (631)$$

If, while c' and c'' continue equal, we also have $a' - a'' = 180°$, then, Equation (630),

$$c = 0.$$

Thus it appears that two equal waves may reach the same molecules in such relative condition as to keep them in their places of rest; in other words, two equal waves may destroy one another.

§ 347.—To ascertain the precise relation of two waves which will cause this mutual destruction, make, in Equation (623),

$$a' = a'' \pm \pi = a'' \pm \frac{2\pi \,.\, \tau}{2\tau},$$

and that equation becomes,

$$\rho' = c' \,.\, \sin \left(2\pi \frac{t}{\tau} + a'' \pm \frac{2\pi \,.\, \tau}{2\tau}\right),$$

$$\rho' = c' \,.\, \sin \left(2\pi \frac{t \pm \frac{1}{2}\tau}{\tau} + a''\right); \quad . \quad . \quad . \quad . \quad (632)$$

which becomes identical with Equation (624) by making

$$c' = c'',$$

and

$$t = t \pm \tfrac{1}{2}\tau. \quad . \quad . \qquad . \quad . \quad . \quad . \qquad (633)$$

Now, the same value for t, in Equations (623) and (624), will, for equal values of the arbitrary arcs a' and a'', determine the component waves to give to a molecule subjected to their simultaneous action, *similar phases;* and a value for t, in the one, which differs from that in the other, by one-half, or any odd multiple of one-half, of the molecular periodic time, *opposite phases.* And, because the waves progress by a wave length during each molecular revolution, the above result shows that, *when two waves meet, after having travelled over routes, estimated from points at which their molecular phases are similar, and which routes differ by half, or any odd multiple of half a wave length, they will destroy one another, provided the waves have the same length and equal maximum molecular displacements.* This act, by which one wave destroys another, is called wave *interference.*

The same process of combination will equally apply to three or more wave functions in which τ is the same in all; that is, wherein the wave lengths are the same; for, in that case, $\sin 2\pi . \frac{t}{\tau}$ and $\cos 2\pi . \frac{t}{\tau}$ being common factors, after developing each function in the sum, the resultant displacement ρ becomes,

$$\rho = \sin 2\pi . \frac{t}{\tau} . \Sigma c' \cos a' + \cos 2\pi . \frac{t}{\tau} . \Sigma c' \sin a',$$

and assuming

$$c . \cos a = \Sigma c' \cos a',$$

$$c . \sin a = \Sigma c' \sin a';$$

$$\rho = c . \sin (2\pi \frac{t}{\tau} + a), \quad . \quad . \quad (634)$$

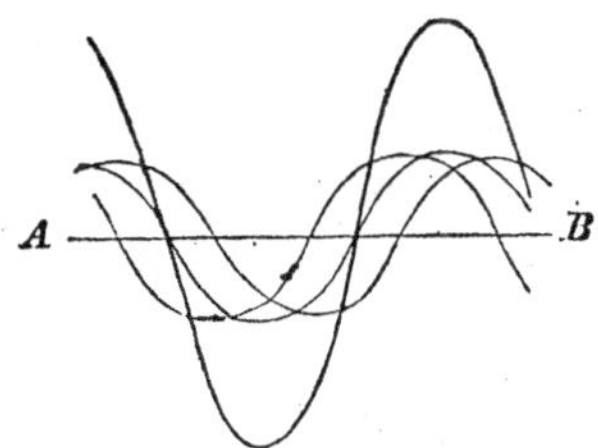

thus making the resultant wave of the same length as that of either of its components.

But, if the component waves be not of equal lengths, the sum of the corresponding functions cannot reduce to the form of Equation (634), because of the absence of common factors, arising from a change in the value of τ from one component to another. Such components can never destroy one another.

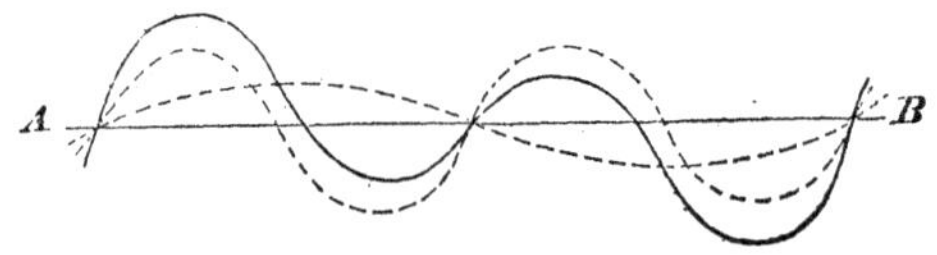

INFLEXION.

§ 348.—Make, in Equation (621), $r'' = 1$, and that equation becomes

$$c' = \frac{c''}{r_{,}};$$

and this value being substituted for c', in Equation (622), gives,

$$\rho' = \frac{c''}{r} \cdot \sin 2\pi \cdot \frac{t}{\tau};$$

and making

$$\frac{t}{\tau} = \frac{Vt - r_{,}}{\lambda},$$

we have, omitting all the accents,

$$\rho = \frac{c}{r} \cdot \sin 2\pi \frac{Vt - r}{\lambda}, \quad \ldots \ldots \quad (635)$$

which is of the same form as Equations (528), and in which V is the velocity of wave propagation; t, the time of its motion from primitive disturbance; λ, the wave length; $\frac{c}{r}$, the maximum displacement of a molecule of which the distance of the place of rest from the point of primitive disturbance is r; and ρ the actual displacement, at the time t, of this same molecule. And from which it is apparent that the displacements will always be the same for equal distances, $Vt - r$, behind the wave front.

Every disturbance of a molecule, at one time, becomes a cause of disturbance to another molecule at some subsequent time. All the molecules in a wave front, when they first begin to move, become, therefore, centres of disturbance for every molecule in advance; and if the primitive disturbance be kept up, secondary waves proceeding from these centres will reach a molecule in advance simultaneously, and determine, § 307, at any instant t, its displacement $\Sigma \rho$.

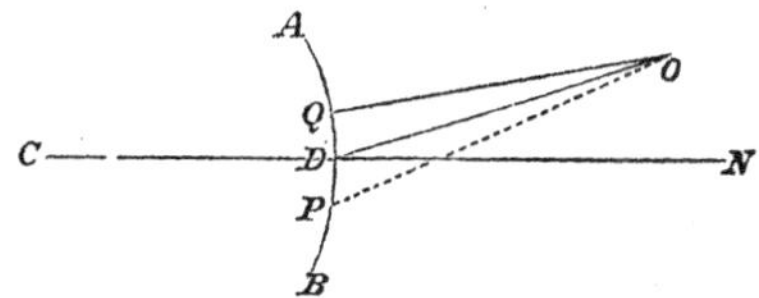

Suppose a wave, whose centre of disturbance is C, to have reached the position AB, so remote from C that a small portion, AB, may be regarded as sensibly plane: What is the displacement of a molecule at O, produced by the simultaneous action of the secondary waves proceeding from the molecules in any portion, as AB, of a section of this wave front? Draw the normal CDN, through the middle of PQ; denote the variable distance DQ by z, and QO by r. The displacement of the molecule O, by the secondary waves from the arc $AB = 2b$, will, Eq. (635), be given by

$$\Sigma \rho = \int_{-b}^{+b} \rho \, dz = \int_{-b}^{+b} \frac{c\,dz}{r} \cdot \sin 2\pi \cdot \frac{Vt - r}{\lambda} \quad . \quad . \quad (636)$$

Here r and z are variable. To eliminate the former, join O with the middle of AB by the line DO, and denote its length by l, and the angle QDO, which it makes with the wave front, by θ. Then will

$$r = \sqrt{l^2 + z^2 - 2lz\cos\theta};$$

and by Maclaurin's formula,

$$r = l - \cos\theta \cdot z + \frac{\sin^2\theta}{2l} \cdot z^2 - \&c. \quad . \quad . \quad . \quad . \quad (637)$$

If the greatest value of z be small as compared to l, we may take

$$r = l - \cos\theta \cdot z, \quad . \quad . \quad . \quad . \quad . \quad . \quad (638)$$

and regard the displacements of the molecule O, by the partial waves from z to be equal. Whence, substituting the value of r, with this restriction, in Equation (636), we have,

$$\Sigma \rho = \int_{-b}^{+b} \rho \, . \, dz = \frac{c}{l} \, . \int_{-b}^{+b} \sin \frac{2\pi}{\lambda} \left(Vt - l + \cos\theta \, . \, z \right) dz,$$

and, performing the integration without regard to limits,

$$\Sigma \rho = - \frac{c\lambda}{2\pi l \cos\theta} \cdot \cos \frac{2\pi}{\lambda} \left(Vt - l + \cos\theta \, . \, z \right),$$

and between the limits $-b$ and $+b$,

$$\Sigma \rho = \frac{c\lambda}{2\pi \, . \, l \, . \, \cos\theta} \cdot \left[\cos \frac{2\pi}{\lambda} \left(Vt - l - b \cos\theta \right) - \cos \frac{2\pi}{\lambda} \left(Vt - l + b \, . \, \cos\theta \right) \right],$$

or,

$$\Sigma \rho = \frac{c\lambda}{\pi \, . \, l \, . \, \cos\theta} \cdot \sin \frac{2\pi \, . \, b \, . \, \cos\theta}{\lambda} \cdot \sin 2\pi \cdot \frac{Vt - l}{\lambda}; \quad . \; . \; . \; . \quad (639)$$

so that the function whose value gives the resultant displacement, is of the same form as that of the function which determines either of the partial displacements.

The maximum value of the resultant displacement is given by

$$\Sigma \rho = \frac{c\lambda}{\pi \, . \, l \, . \, \cos\theta} \cdot \sin \frac{2\pi \, . \, b \, . \, \cos\theta}{\lambda}; \quad . \; . \; . \quad (640)$$

and this will become zero for such values of θ as make $b \, . \cos\theta$ equal to either of the following values, viz.,

$$\tfrac{1}{2}\lambda, \quad \tfrac{3}{2}\lambda, \quad \tfrac{5}{2}\lambda, \quad \tfrac{7}{2}\lambda, \quad \&c.$$

Conceiving the figure to be revolved about the normal CN, and all the wave except the circular portion whose diameter is $2b = AB$, to be intercepted, the space in advance of the wave will, when the above values obtain, find itself divided by the secondary waves into a series of concentric cone-like zones around the normal CN, as an axis, and of which the alternate ones, beginning with that immediately about the axis, will be filled with molecules in motion, while the molecules in the others will be at rest. A section in advance of the primitive wave will cut from these zones a series of concentric circular rings distinguished by the same peculiarities.

But if λ be very great as compared with b, then will the arc

$$\frac{2\pi . b . \cos\theta}{\lambda}$$

be so small as to justify the substitution of the arc for its sine and for the maximum value of resultant displacement,

$$(\Sigma\rho)_{,} = \frac{c\lambda}{\pi . l . \cos\theta} \cdot \frac{2\pi . b . \cos\theta}{\lambda} = \frac{2cb}{l}; \quad . \quad . \quad . \quad (641)$$

and this result being independent of θ, the conic zones cannot exist, and the effect of the secondary waves will be diffused in all directions to the front. This lateral action of secondary waves proceeding from a small portion of a primitive wave, is called *wave inflection.*

When θ approaches nearly to 90°, $\cos\theta$ will be exceedingly small, and the arc

$$\frac{2\pi . b . \cos\theta}{\lambda}$$

may again be substituted for its sine; again Equation (641) suits the case, and determines the maximum displacement immediately about the normal.

The maximum of the maxima displacements will occur when, in Equation (640),

$$\sin . \frac{2\pi . b . \cos\theta}{\lambda} = \pm 1;$$

and which would reduce that equation to

$$(\Sigma\rho)_{,,} = \frac{c\lambda}{\pi . l . \cos\theta};$$

and as the living forces are proportional to the squares of the greatest displacements, we have

$$m . v_{,}^{2} : m . v_{,,}^{2} :: \frac{4c^2 b^2}{l^2} : \frac{c^2\lambda^2}{\pi^2 . l^2 \cos^2\theta}.$$

Whence

$$m \cdot v_{\prime\prime}^2 = m \cdot v_{\prime}^2 \cdot \frac{\lambda^2}{4\pi^2 b^2 \cdot \cos^2\theta}, \quad . \quad . \quad . \quad . \quad (642)$$

in which $v_{\prime}$ is the velocity of the molecule on the normal, and $v_{\prime\prime}$ that at the angular distance θ from it. When the waves are very short, as compared with b, it is obvious that the living force of the molecules would be sensibly nothing, except immediately about the normal. When the waves are long, as compared with b, the living force will be appreciable for every value of θ, and, therefore, in every direction in front of the primitive wave. The importance of this discussion will be apparent in the subjects of sound and light.

PART IV.

APPLICATION OF THE PRECEDING PRINCIPLES TO SIMPLE MACHINES, PUMPS, ETC.

§ 349.—Any device by which the action of a force may be received at one place and transmitted to another is called a *Machine.*

There are usually seven elementary machines discussed in *Mechanics;* viz., the *Cord*, *Lever*, *Inclined Plane*, *Pulley*, *Screw*, *Wheel and Axle*, and *Wedge.* The Cord, Lever, and Inclined Plane are called Simple Machines; the others, being combinations of these, are called Compound Machines.

§ 350.—In Machines, as in all other bodies, every action is accompanied by an equal and contrary reaction. A force which acts upon a Machine to impress or preserve motion is called a *Power.* A force which reacts to prevent or destroy motion, is called a *Resistance.* The Agent which is the source of power, is, § 38, called a *Motor.*

§ 351.—Resuming Equation (30), and supposing the displacement, which in that equation was wholly arbitrary, to conform in every respect to that caused by the powers and resistances, we shall have $\delta s = d s$, s being the path described by the elementary mass m; and hence,

$$\Sigma P \delta p - \Sigma m \cdot \frac{d^2 s}{d t^2} . d s = 0;$$

but

$$\frac{d^2 s}{d t^2} d s = \frac{d s}{d t} \cdot \frac{d^2 s}{d t} = v\, d v = \tfrac{1}{2} d\,(v^2);$$

whence,

$$\Sigma P \delta p - \tfrac{1}{2} \Sigma m . d\,(v^2) = 0. \quad \cdot \quad \cdot \quad \cdot \quad \cdot \quad \cdot \quad \cdot \quad (643)$$

Denoting by Q, Q', &c. the resistances, by P, P', &c. the powers, δq, &c. and δp, &c. the projections of their respective virtual velocities; the first term, which embraces all the forces except inertia in action on the machine, may be replaced by $\Sigma P \delta p - \Sigma Q \delta q$, and we have

$$\Sigma P \delta p - \Sigma Q \delta q = \tfrac{1}{2} \Sigma m \,.\, d v^2. \quad . \quad . \quad . \quad . \quad (644)$$

Integrating,

$$\int \Sigma P \delta p - \int \Sigma Q \delta q = \tfrac{1}{2} \Sigma m v^2 + C;$$

and denoting by $v_{\prime}$ the initial velocity, and taking the integral so as to vanish when $t = 0$,

$$\int \Sigma P \delta p - \int \Sigma Q \delta q = \tfrac{1}{2} \Sigma m v^2 - \tfrac{1}{2} \Sigma m v_{\prime}^2. \quad . \quad . \quad . \quad (645)$$

The products $P \delta p$ and $Q \delta q$ are the elementary quantities of work performed by a power and a resistance respectively, in the element of time dt; the product $\frac{1}{2} m d v^2$ is the elementary quantity of work performed by the inertia, or one half the increment of living force of the mass m in this time. And Equation (645) shows that in any machine, in motion, the increment of the half sum of the living forces of all its parts is always equal to the excess of the work of the powers or motors over that of the resistances

§ 352.—If the machine start from rest, Equation (645) becomes

$$\int \Sigma P \delta p - \int \Sigma Q \delta q = \tfrac{1}{2} \Sigma m v^2, \quad . \quad . \quad . \quad . \quad (646)$$

and as the second member is essentially positive, the work of the motors must exceed that of the resistances embraced in the term $\int \Sigma Q \delta q$; in other words, the inertia will oppose the motor and act as a resistance. When the motion becomes uniform, the second member will be constant; from that instant inertia will cease to act, and the subsequent work of the motor will be equal to that of the resistances as long as this motion continues. If the motion be now retarded, the second member will decrease, the inertia will act with the power, and this will continue till the machine comes to rest, and the excess of work of the *Resistance* during retardation will be exactly equal to that of the *Power* during acceleration. Generally, then, when a machine is at rest or is moving uniformly, inertia does not act; when the motion is variable, it does, and opposes or aids the motor according as the motion is accelerated or retarded.

§ 353.—The essential parts of every machine are those which receive directly the action of the motor, those which act directly upon the body to be moved or transformed, and those which serve to transmit the action. The arrangement of the latter is often a source of resistance, arising from *Friction*, *Adhesion*, *Stiffness of Cordage*, &c., whose work enters largely into the general term $\int \Sigma\, Q\, \delta q$.

FRICTION.

§ 354.—When two bodies are pressed together, experience shows that a certain effort is always required to cause one to roll or slide along the other. This arises almost entirely from the inequalities in the surfaces of contact interlocking with each other, thus rendering it necessary, when motion takes place, either to break them off, compress them, or force the bodies to separate far enough to allow them to pass each other. This cause of resistance to motion is called *friction*, of which we distinguish two kinds, according as it accompanies a sliding or rolling motion. The first is denominated *sliding,* and the second *rolling friction.* They are governed by the same laws; the former is much greater in amount than the latter under given circumstances, and being of more importance in machines, will principally occupy our attention.

The intensity of friction, in any given case, is measured by the force exerted in the direction of the surface of contact, which will place the bodies in a condition to resist, during a change of state, in respect to motion or rest, only by their inertia.

§ 355.—The friction between two bodies may be measured directly by means of the spring balance. For this purpose, let the surface

CD of one of the bodies M be made perfectly level, so that the other body M', when laid upon it, may press with its entire weight. To some point, as E, of the body M', attach a cord with a spring balance in the manner indicated in the figure, and apply to the latter a force F of such intensity as to produce in the body M' a uniform motion. The motion being uniform, the accelerating and retarding forces must be equal and contrary; that is to say, the friction must be equal and contrary to the force F, of which the intensity is indicated by the balance.

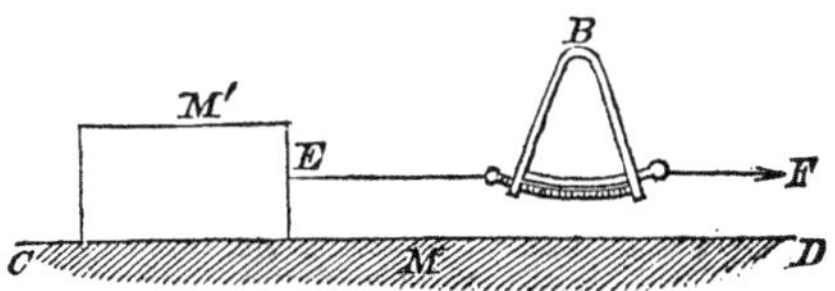

The experiments on friction which seem most entitled to conf. dence are those performed at Metz by M. Morin, under the order of the French government, in the years 1831, 1832, and 1833. They were made by the aid of a contrivance, first suggested by M. Poncelet, which is one of the most beautiful and valuable contributions that theory has ever made to practical mechanics. Its details are given in a work by M. Morin, entitled "*Nouvelles Expériences sur le Frottement.*" Paris, 1833.

The following conclusions have been drawn from these experiments, viz.:

The friction of two surfaces which have been for a considerable time in contact and at rest is not only different in amount, but also in nature, from the friction of surfaces in continuous motion; especially in this, that the friction of quiescence is subjected to causes of variation and uncertainty from which the friction during motion is exempt. This variation does not appear to depend upon the *extent* of the surface of contact; for, with different pressures, the ratio of the friction to the pressure varied greatly, although the surfaces of contact were the same.

The slightest jar or shock, producing the most imperceptible movement of the surfaces of contact, causes the friction of quiescence to pass to that which accompanies *motion.* As every machine may be regarded as being subject to slight shocks, producing imper ceptible motions in the surfaces of contact, the kind of friction to be employed in all questions of equilibrium, as well as of motions of machines, should obviously be this last mentioned, or that which accompanies continuous motion.

The LAWS of friction which accompanies continuous motion are remarkably *uniform* and *definite*. These laws are:

1st. Friction accompanying continuous motion of two surfaces, between which no unguent is interposed, bears a constant proportion to the force by which those surfaces are pressed together, whatever be the intensity of the force.

2d. Friction is wholly independent of the *extent* of the surfaces in contact.

3d. Where *unguents* are interposed, a distinction is to be made between the case in which the surfaces are simply *unctuous* and in intimate contact with each other, and that in which the surfaces are wholly *separated* from one another by an *interposed stratum of the unguent.* The friction in these two cases is not the same in amount under the same pressure, although the law of the independence of extent of surface obtains in each. When the pressure is increased sufficiently to *press out* the unguent so as to bring the unctuous surfaces in contact, the latter of these cases passes into the first; and this fact may give rise to an *apparent* exception to the law of the independence of the extent of surface, since a diminution of the surface of contact may so concentrate a given pressure as to remove the unguent from between the surfaces. The exception is, however, but apparent, and occurs at the passage from one of the cases above-named to the other. To this extent, the law of independence of the extent of surface is, therefore, to be received with restriction.

There are, then, three conditions in respect to friction, under which the surfaces of bodies in contact may be considered to exist, viz.: 1st, that in which no unguent is present; 2d, that in which the surfaces are simply *unctuous;* 3d, that in which there is an interposed stratum of the unguent. Throughout each of these states the friction which accompanies motion is always proportional to the pressure, but for the same pressure in each, very different in amount.

4th. The friction which accompanies motion is always independ ent of the *velocity* with which the bodies move; and this, whether the surfaces be without unguents or lubricated with water, oils, grease, glutinous liquids, syrups, pitch, &c., &c.

The variety of the circumstances under which these laws obtain, and the accuracy with which the phenomena of motion accord with them, may be inferred from a single example taken from the first set of Morin's experiments upon the friction of surfaces of oak, whose fibres were parallel to the direction of the motion. The surfaces of contact were made to vary in extent from 1 to 84; the forces which pressed them together from 88 to 2205 pounds; and the velocities from the slowest perceptible motion to 9,8 feet a second, causing them to be at one time accelerated, at another uniform, and at another retarded; yet, throughout all this wide range of variation, in no instance did the ratio of the friction to the pressure differ from its mean value of 0,478 by more than $\frac{1}{24}$ of this same fraction.

Denote the constant ratio of the entire friction F, to the normal pressure P, by f; then will the first law of friction be expressed by the following equation,

$$\frac{F}{P} = f; \quad . \quad . \quad . \quad . \quad . \quad . \quad . \quad . \quad . \quad (647)$$

whence,

$$F = f . P.$$

This constant ratio f is called the *co-efficient of friction*, because, when multiplied by the total normal pressure, the product gives the entire friction.

Assuming the first law of friction, the co-efficient of friction may easily be obtained by means of the inclined plane. Let W denote the weight of any body placed upon the inclined plane $A B$. Resolve this weight $G G'$ into two components, one $G M$ perpendicular to the plane, and the other $G N$ par-

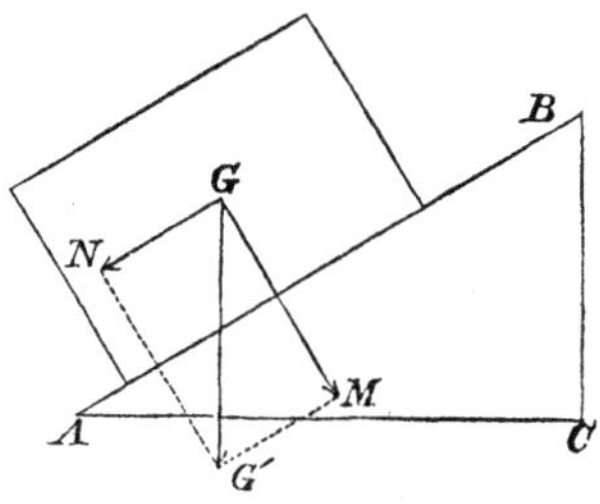

allel to it. Because the angles $G' G M$ and $B A C$ are equal, the first of these components will be

$$G M = W . \cos A,$$

and the second,

$$G N = W . \sin A,$$

in which A denotes the angle $B A C$.

The first of these components determines the total pressure upon the plane, and the friction due to this pressure will be

$$F = f . W \cos A.$$

The second component urges the body to move down the plane. If the inclination of the plane be gradually increased till the body move with uniform motion, the total friction and this component must be equal and opposed; hence,

$$f . W . \cos A = W . \sin A;$$

whence,

$$f = \frac{\sin A}{\cos A} = \tan A.$$

We, therefore, conclude, that the *unit* or *co-efficient* of friction between any two surfaces, is equal to the tangent of the angle which one of the surfaces must make with the horizon in order that the other may slide over it with a uniform motion, the body to which the moving surface belongs being acted upon by its own weight alone. This angle is called the *angle of friction* or *limiting angle of resistance.*

The values of the *unit* of friction and of the *limiting angles* for many of the various substances employed in the art of construction, are given in Tables VI, VII and VIII.

The distinction between the friction of surfaces to which no unguent is applied, those which are merely unctuous, and those between which a uniform stratum of the unguent is interposed, appears first to have been remarked by M. Morin; it has suggested to him what appears to be the true explanation of the difference between his results and those of Coulomb. He conceives, that in the experiments of this celebrated Engineer, the requisite precautions had not been taken to exclude unguents from the surfaces of contact. The slightest unctuosity, such as might present itself accidentally, unless expressly guarded against—such, for instance, as might have been left by the hands of the workman who had given the last polish to the surfaces of contact—is sufficient materially to affect the co-efficient of friction.

Thus, for instance, surfaces of oak having been rubbed with hard dry soap, and then thoroughly wiped, so as to show no traces whatever of the unguent, were found by its presence to have lost $\frac{2}{3}$ds of their friction, the co-efficient having passed from 0,478 to 0,164.

This effect of the unguent upon the friction of the surfaces may be traced to the fact, that their motion upon one another without unguents was always found to be attended by a wearing of both the surfaces; small particles of a dark color continually separated from them, which it was found from time to time necessary to remove, and which manifestly influenced the friction: now, with the presence of an unguent the formation of these particles, and the consequent wear of the surfaces, completely ceased. Instead of a new surface of contact being continually presented by the wear, the same surface remained, receiving by the motion continually a more perfect polish.

A comparison of the results enumerated in Table VIII, leads to the following remarkable conclusion, easily fixing itself in the memory, *that with the unguents, hogs' lard and olive oil interposed in a continuous stratum between them, surfaces of wood on metal, wood on wood, metal on wood, and metal on metal, when in motion, have all of them very nearly the same co-efficient of friction, the value of that co-efficient being in all cases included between* 0,07 *and* 0,08, *and the limiting angle of resistance therefore between* 4° *and* 4° 35′.

For the unguent tallow the co-efficient is the same as the above in every case, except in that of metals upon metals; this unguent seems less suited to metallic surfaces than the others, and gives for the mean value of its co-efficient 0,10, *and for its limiting angle of resistance* 5° 43′.

356.—Besides friction, there is another cause of resistance to the motion of bodies when moving over one another. The same forces which hold the elements of bodies together, also tend to keep the bodies themselves together, when brought into sensible contact. The effort by which two bodies are thus united, is called the force of *Adhesion.*

Familiar illustrations of the existence of this force are furnished by the pertinacity with which sealing-wax, wafers, ink, chalk and black-lead cleave to paper, dust to articles of dress, paint to the surface of wood, whitewash to the walls of buildings, and the like.

The intensity of this force, arising as it does from the affinity of the elements of matter for each other, must vary with the number of attracting elements, and therefore with the *extent of the surface of contact.*

This law is best verified, and the actual amount of adhesion between different substances determined, by means of a delicate spring-balance. For this purpose, the surfaces of solids are reduced to polished planes, and pressed together to exclude the air, and the efforts necessary to separate them noted by means of this instrument. The experiment being often repeated with the same substances, having different extent of surfaces in contact, it is found that the effort necessary to produce the separation divided by the area of the surface gives a constant ratio. Thus, let S denote the area of the surfaces of contact expressed in square feet, square inches, or any other superficial unit; A the effort required to separate them, and a the constant ratio in question, then will

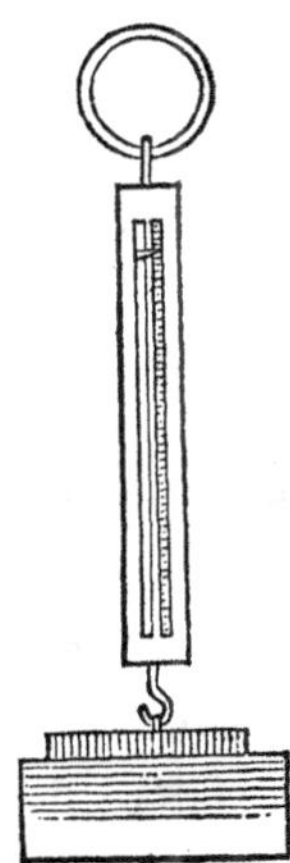

$$\frac{A}{S} = a,$$

or,

$$A = a \, . \, S.$$

The constant a is called the *unit* or *co-efficient of adhesion,* and obviously expresses the value of adhesion on each unit of surface, for making

$$S = 1,$$

we have

$$A = a.$$

To find the adhesion between solids and liquids, suspend the solid from the balance, with its polished surface downward and in a horizontal position; note the weight of the solid, then bring it in contact with the horizontal surface of the fluid and note the indication of the balance when the separation takes place, on drawing the balance up; the difference between this indication and that of the weight will give the adhesion; and this divided by the extent of surface, will give, as before, the co-efficient a. But in this experiment two opposite conditions must be carefully noted, else the cohesion of the elements of the liquid for each other may be mistaken for the adhesion of the solid for the fluid. If the solid on being removed take with it a layer of the fluid; in other words, if the solid has been wet by the fluid, then the attraction of the elements of the solid for those of the liquid is stronger than that of the elements of the liquid for each other, and a will be the unit of adhesion of two surfaces of the fluid. If, on the contrary, the solid on leaving the fluid be perfectly dry, the elements of the fluid will attract each other more powerfully than they will those of the solid, and a will denote the unit of adhesion of the solid for the liquid.

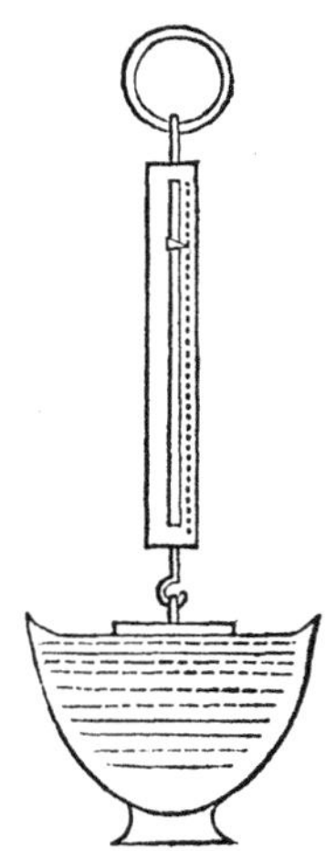

It is easy to multiply instances of this diversity in the action of solids and fluids upon each other. A drop of water or spirits of wine, placed upon a wooden table or piece of glass, loses its globular form and spreads itself over the surface of the solid; a drop of mercury will not do so. Immerse the finger in water, it becomes wet; in quicksilver, it remains dry. A tallow candle, or a feather from any species of water-fowl, remains dry though dipped in water. Gold, silver, tin, lead, &c., become moist on being immersed in quicksilver, but iron and platinum do not. Quicksilver when poured into a gauze bag will not run through; water will: place the gauze containing the quicksilver in contact with water, and the metal will also flow through.

It is difficult to ascertain the precise value of the force of adhesion between the rubbing surfaces of machinery, apart from that of friction. But this is attended with little practical inconvenience, as long as a machine is in motion. The experiments of which the results are given in Tables VI, VII and VIII, and which are applicable to machinery, were made under considerable pressures, such as those with which the parts of the larger machines are accustomed to move upon one another. Under such pressures, the adhesion of unguents to the surfaces of contact, and the opposition to motion presented by their viscosity, are causes whose influence may be safely disregarded as compared with that of friction. In the cases of lighter machinery, however, such as watches, clocks, and the like, these considerations rise into importance, and cannot be neglected.

STIFFNESS OF CORDAGE.

§ 357.—Conceive a wheel turning freely about an axle or trunnion, and having in its circumference a groove to receive a cord or rope. A weight W, being suspended from one end of the rope, while a force F, is applied to the other extremity to draw it up, the latter will experience a resistance in consequence of the rigidity of the rope, which opposes every effort to bend it around the wheel. This resistance must, of necessity, consume a portion of the work of the force F. The measure of the resistance due to the rigidity of cordage has been made the

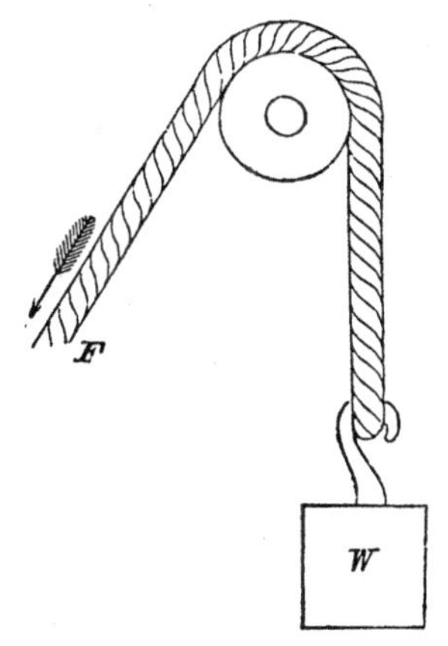

subject of experiment by Coulomb; and, according to him, it results that for the same cord and same wheel, this measure is composed of two parts, of which one remains constant, while the other varies with the weight W, and is directly proportional to it; so that, designating the constant part by K, and the ratio of the variable part to the weight W by I, the measure will be given by the expression

$$K + I \cdot W;$$

in which K represents the stiffness arising from the natural torsion or tension of the threads, and I the stiffness of the same cord due to a tension resulting from one unit of weight; for, making $W = 1$, the above becomes

$$K + I.$$

Coulomb also found that on changing the wheel, the stiffness varied in the inverse ratio of its diameter; so that if

$$K + I \cdot W$$

be the measure of the stiffness for a wheel of one foot diameter, then will

$$\frac{K + I \cdot W}{2R}$$

be the measure when the wheel has a diameter of $2R$. A table giving the values of K and I for all ropes and cords employed in practice, when wound around a wheel of one foot diameter, and subjected to a tension arising from a unit of weight, would, therefore, enable us to find the stiffness answering to any other wheel and weight whatever.

But as it would be impossible to anticipate all the different sizes of ropes used under the various circumstances of practice, Coulomb also ascertained the law which connects the stiffness with the diameter of the cross-section of the rope. To express this law in all cases, he found it necessary to distinguish, 1st, *new white rope*, either dry or moist; 2d, *white ropes partly worn*, either dry or moist; 3d, *tarred ropes;* 4th, *packthread*. The stiffness of the first class he found nearly proportional to the square of the diameter of the cross-section; that of the second, to the square root of the cube of this diameter, nearly; that of the third, to the number of yarns in the rope; and that of the fourth, to the diameter of the cross-section. So that, if S denote the resistance due to the stiffness of any given rope; d the ratio of its diameter to that of the table; and n the ratio of the number of yarns in any tarred rope to that of the table, we shall have for

New white rope, dry or moist.

$$S = d^2 \cdot \frac{K + I \cdot W}{2R} \quad \ldots \ldots \ldots \quad (648)$$

Half worn white rope, dry or moist.

$$S = d^{\frac{3}{2}} \cdot \frac{K + I \cdot W}{2R} \quad \ldots \ldots \ldots \quad (649)$$

Tarred rope.

$$S = n \cdot \frac{K + I \cdot W}{2R} \quad \ldots \ldots \ldots \quad (650)$$

Packthread.

$$S = d \cdot \frac{K + I \cdot W}{2R} \quad \ldots \ldots \ldots \quad (651)$$

For packthread, it will always be sufficient to use the tabular values given, corresponding to the least tabular diameters, and substitute them in Equation (651). An example or two will be sufficient to illustrate the use of these tables.

Example 1*st*. Required the resistance due to the stiffness of a new dry white rope, whose diameter is 1,18 inches, when loaded with a weight of 882 pounds, and wound about a wheel 1,64 feet in diameter.

Seek in No. 1, Table X, the diameter nearest that of the given rope; it is 0,79; hence,

$$d = \frac{1{,}18}{0{,}79} = 1{,}5 \text{ nearly};$$

and from the table at the side,

$$d^2 = 2{,}25.$$

From No. 1, opposite 0,79, we find

$$K = 1{,}6097,$$

$$I = 0{,}03195;$$

which, together with the weight $W = 882$ lbs., and $2R = \overset{ft.}{1,64}$, substituted in Equation (648), give

$$S = 2,25 \cdot \frac{\overset{lb.}{1,6097} + \overset{lb.}{0,03195} \times 882}{1,64} = \overset{lbs.}{40,817},$$

which is the true resistance due to the stiffness of the rope in question.

Example 2d. What is the resistance due to the stiffness of a white rope, half worn and moistened with water, having a diameter equal to 1,97 inches, wound about a wheel 0,82 of a foot in diameter, and loaded with a weight of 2205 pounds?

The tabular diameter in No. 4, Table X, next less than 1,97, is 1,57, and hence,

$$d = \frac{1,97}{1,57} = 1,3 \text{ nearly};$$

the square root of the cube of which is, by the table at the side,

$$d^{\frac{3}{2}} = 1,482.$$

In No. 4 we find, opposite 1,57,

$$K = 6,4324,$$

$$I = 0,06387;$$

which values, together with $W = 2205$ lbs., and $2R = \overset{ft.}{0,82}$, in Equation (649), give

$$S = 1,482 \times \frac{\overset{lbs.}{6,4324} + \overset{lbs.}{0,06387} \times 2205}{0,82} = \overset{lbs.}{266,109},$$

which is the required resistance.

Example 3d. What is the resistance due to the stiffness of a tarred rope of 22 yarns, when subjected to the action of a weight equal to 4212 pounds, and wound about a wheel 1,3 feet diameter, the weight of one running foot of the rope being about 0,6 of a pound?

By referring to No. 5, Table X, we find the tabular number of yarns next less than 22 to be 15, and hence,

$$n = \frac{22}{15} = 1,466 \text{ nearly}.$$

In the same table, opposite 15, we find

$$K = 0{,}7664,$$

$$I = 0{,}019879;$$

which, together with $W = 4212$, and $2\,R = \overset{ft.}{1{,}3}$, in Equation (650), give

$$S = 1{,}466\,\frac{0{,}7664 + 0{,}019879 \times 4212}{1{,}3} = \overset{lbs.}{95{,}188}.$$

Example 4th. Required the resistance due to the stiffness of a new white packthread, whose diameter is 0,196 inches, when moistened or wet with water, wound about a wheel 0,5 of a foot in diameter, and loaded with a weight of 275 pounds.

The lowest tabular diameter is 0,39 of an inch, and hence

$$d = \frac{0{,}196}{0{,}390} = 0{,}5 \text{ nearly}.$$

In No. 2, Table X, we find, opposite 0,39,

$$K = \overset{lb.}{0{,}8048},$$

$$I = 0{,}00798;$$

which, with $W = 275$, and $2\,R = 0{,}5$, we find, after substituting in Equation (651),

$$S = 0{,}5\,\frac{0{,}8048 + 0{,}00798 \times 275}{0{,}5} = \overset{lbs.}{2{,}999}.$$

§ 358.—The resistance just found is expressed in pounds, and is the amount of weight which would be necessary to bend any given rope around a vertical wheel, so that the portion $A\,E$, between the first point of contact A, and the point E, where the rope is attached to the weight, shall be perfectly straight. The entire process of bending takes place at this first or tangential point A; for, if motion be com-

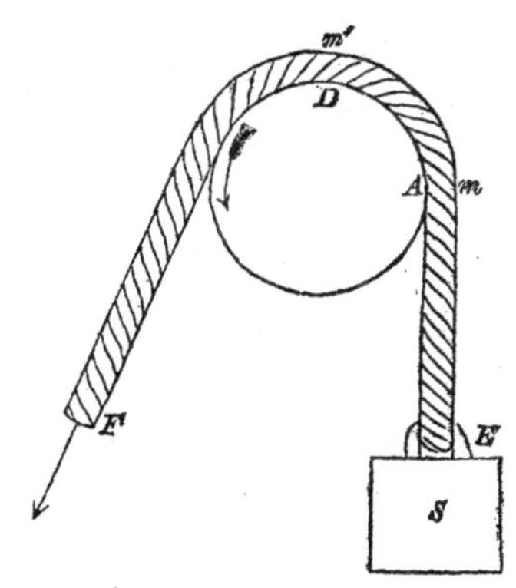

municated to the wheel in the direction indicated by the arrow-head, the rope, supposed not to slide, will, at this point, take and retain the constant curvature of the wheel, till it passes from the latter on the side of the power F. When, therefore, by the motion of the wheel, the point m of the rope, now at the tangential point, passes to m', the working point of the force S will have described in its own direction the distance AD. Denoting the arc described by a point at the unit's distance from the centre of the wheel by $s_{\prime}$, and the radius of the wheel by R, we shall have

$$AD = Rs_{\prime};$$

and representing the quantity of work of the force S by L, we get

$$L = S \cdot Rs_{\prime};$$

replacing S by its value in Equations (648) to (651),

$$L = Rs_{\prime} \cdot d_{\prime} \cdot \frac{K + I \cdot W}{2R} \quad \ldots \ldots \quad (652)$$

in which $d_{\prime}$ represents the quantity d^2, $d^{\frac{3}{2}}$, n, or d, in Equations (648) to (651), according to the nature of the rope.

Example.—Taking the 2d example of § 357, and supposing a portion of the rope, equal to 20 feet in length, to have been brought in contact with the wheel, after the motion begins, we shall have

$$L = 20 \times 266{,}109 = 5322{,}18 \text{ units of work};$$

that is, the quantity of work consumed by the resistance due to the stiffness of the rope, while the latter is moving over a distance of 20 feet, would be sufficient to raise a weight of 5322,18 pounds through a vertical height of one foot.

FRICTION ON PIVOTS, AND TRUNNIONS.

§ 359.—All rotating pieces, such as wheels supported upon other pieces, give rise by their motion to friction. This is an important element in all computations relating to the performance of machinery. It seems to be different according as the rotating pieces are kept in place by *trunnions* or by *pivots*. By *trunnions* are meant cylindrical projections $a\,a$ from the ends of the arbor $A\,B$ of a wheel. The trunnions rest on the concave surfaces of cylindrical boxes $C\,D$, with which they usually have a small surface of contact m, the linear elements of both being parallel. *Pivots* are shaped like the trunnions, but support the weight of the wheel and its arbor upon their circular end, which rests against the bottom of cylindrical sockets $F\,G\,H\,I$.

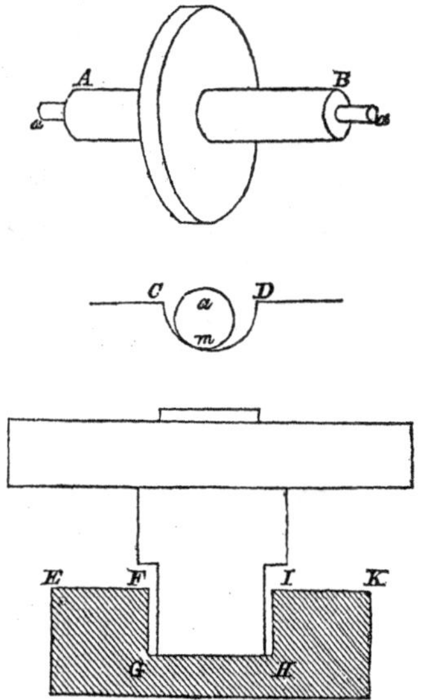

PIVOTS.

Let N denote the force, in the direction of the axis, by which the pivot is pressed against the bottom of the socket. This force may be regarded as passing through the centre of the circular end of the pivot, and as the resultant of the partial pressures exerted upon all the elementary surfaces of which this circle is composed. Denote by A the area of the entire circle, then will the pressure sustained by each unit of surface be

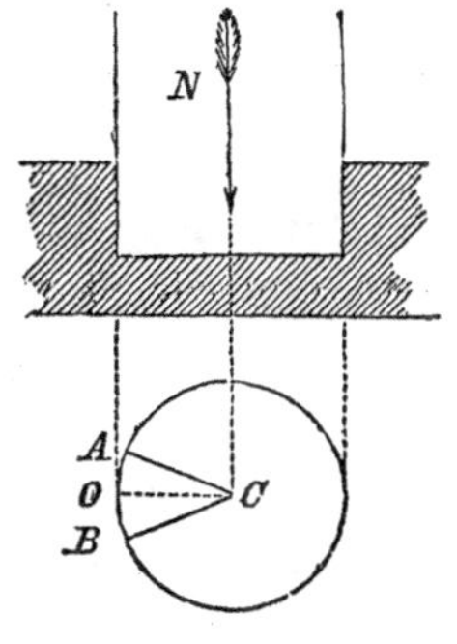

$$\frac{N}{A};$$

and the pressure on any small portion of the surface denoted by a, will obviously be

$$\frac{a\,.\,N}{A};$$

and the friction on the same will be

$$\frac{f \cdot a \cdot N}{A}.$$

This friction may be regarded as applied to the centre of the elementary surface a; it is opposed to the motion, and the direction of its action is tangent to the circle described by the centre of the element. Denote the radius of this circle by x, then will the moment of the friction be

$$f \cdot \frac{a \cdot N}{A} \cdot x.$$

Now, if s denote the length of any variable portion of the circumference at the unit's distance from the centre C, then will

$$a = x \cdot ds \cdot dx;$$

also,

$$A = \pi R^2;$$

which substituted above give

$$f \cdot N \cdot \frac{x^2 \cdot dx \cdot ds}{\pi \cdot R^2},$$

and by integration,

$$f \cdot N \cdot \frac{\int_0^R x^2\, dx \int_0^{2\pi} ds}{\pi R^2} = f \cdot N \cdot \tfrac{2}{3} R; \quad . \quad . \quad (653)$$

whence we conclude, that, in the friction of a pivot, *we may regard the whole friction due to the pressure as acting in a single point, and at a distance from the centre of motion equal to two-thirds of the radius of the base of the pivot.* This distance is called the *mean lever* of friction.

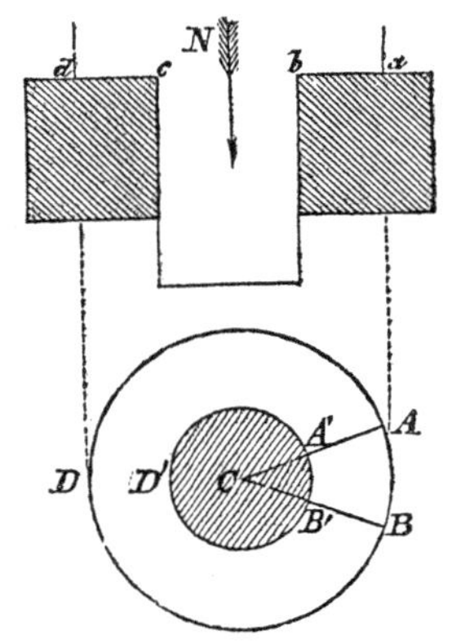

§ 360.—If the extremity of the pivot, instead of rubbing upon an entire circle, is only in contact with a ring or surface comprised between two concentric circles, as when the arbor of a wheel is urged in the direction of its length by the force N against a shoulder $d\,c\,b\,a$; then will

$$A = \pi\,(R^2 - R'^2);$$

and the integration will give

$$f \cdot N \cdot \frac{\int_{R'}^{R} x^2\,dx \int_{0}^{2\pi} ds}{\pi\,(R^2 - R'^2)} = \tfrac{2}{3} f \cdot N \cdot \frac{R^3 - R'^3}{R^2 - R'^2};$$

in which R denotes the radius of the larger, and R' that of the smaller circle.

Finally, denote by l the breadth of the ring, that is, the distance $A'A$; by r, its mean radius or distance from C to a point half way between A' and A, and we shall have

$$R = r + \tfrac{1}{2}\,l,$$

$$R' = r - \tfrac{1}{2}\,l;$$

substituting these values above and reducing, we have

$$f \cdot N \times \left[r + \tfrac{1}{12} \cdot \frac{l^2}{r}\right]; \quad \cdot \cdot \cdot \cdot \cdot \cdot \quad (654)$$

and making

$$r + \frac{l^2}{12\,r} = r_{,},$$

we obtain, for the moment of the friction on the entire ring,

$$f \,.\, N \,.\, r_{,} \,. \quad . \; . \; . \; . \; . \; . \; . \; . \; . \quad (655)$$

The quantity $r_{,}$ is called the *mean lever* of friction for a ring. Since the whole friction fN may be considered as applied at a point whose distance from the centre is $\frac{2}{3}R$, or $r_{,} = r + \frac{l^2}{12\,r}$, according as the friction is exerted over an entire circle or over a ring, and since the path described by this point lies always in the direction in which the friction acts, the quantity of work consumed by it will be equal to the product of its intensity fN into this path. Designating the length of the arc described at the unit's distance from C by $s_{,}$, the path in question will be either

$$\tfrac{2}{3}\,R\,s_{,}, \quad \text{or} \quad r_{,}\,s_{,};$$

and the quantity of work either

$$\tfrac{2}{3} R \,.\, s_{,} \,.\, f \,.\, N$$

for an entire circle, or

$$f \cdot N \left(r + \frac{l^2}{12\, r} \right) s_{,}$$

for a ring. Let Q denote the quantity of work consumed by friction in the unit of time, and n the number of revolutions performed by the pivot in the same time; then will

$$s_{,} = 2\pi \times n\,;$$

and we shall have

$$Q = \tfrac{4}{3}\pi \,.\, R \,.\, f \,.\, N \,.\, n \quad \cdots\cdots \quad (656)$$

for the circle, and

$$Q = 2\pi \cdot f \cdot N \cdot \left(r + \frac{l^2}{12\, r} \right) \cdot n \quad \cdots\cdots \quad (657)$$

for a ring; in which $\pi = 3{,}1416$.

The co-efficient of friction f, when employed in either of the foregoing cases, must be taken from Table VI, VII, or VIII.

Example.—Required the moment of the friction on a pivot of cast iron, working into a socket of brass, and which supports a weight of 1784 pounds, the diameter of the circular end of the pivot being 6 inches. Here

$$R = \tfrac{6}{2} = \overset{in.}{3} = \overset{ft.}{0{,}25},$$

$$N = \overset{lbs.}{1784},$$

$$f = 0{,}147\,;$$

which, substituted in Equation (653), gives

$$0{,}147 \times \overset{lbs.}{1784} \times \tfrac{2}{3} \times \overset{ft.}{0{,}25} = 43{,}708.$$

And to obtain the quantity of work in one unit of time, say a minute, there being 20 revolutions in this unit, we make $n = 20$, and $\pi = 3{,}1416$ in Equation (656), and find

$$Q = \tfrac{4}{3} \times 3{,}1416 \times 0{,}25 \times 0{,}147 \times 1784 \times 20 = 5492{,}80\,;$$

that is to say, during each unit of time, there is a quantity of work lost which would be sufficient to raise a weight of 5492,80 pounds through a vertical distance of one foot.

Example.—Required the moment of friction, when the pivot supports a weight of 2046 pounds, and works upon a shoulder whose exterior and interior diameters are respectively 6 and 4 inches; the pivot and socket being of cast iron, with water interposed.

$$l = \frac{6 - 4}{2} = 1 \text{ inch},$$

$$r = 2 + 0,5 = 2,5 \text{ inches},$$

$$r_{\prime} = 2,5 + \frac{(1)^2}{12 \times 2,5} = \overset{in.}{2,5333} = \overset{ft.}{0,2111},$$

$$N = 2046 \text{ pounds},$$

$$f = 0,314;$$

which, substituted in Expression (655), gives for the moment of friction,

$$0,314 \times \overset{lbs.}{2046} \times \overset{ft.}{0,2111} = 135,62.$$

The quantity of work consumed in one minute, there being supposed 10 revolutions in that unit, will be found by making in Equation (657), $\pi = 3,1416$ and $n = 10$,

$$Q = 2 \times 3,1416 \times 0,314 \times 2046 \times 0,211 \times 10 = 8517,24;$$

that is to say, friction will, in one unit of time, consume a quantity of work which would raise 8517,24 pounds through a vertical distance of one foot. The quantity of work consumed in any given time would result from multiplying the work above found, by the time reduced to minutes.

TRUNNIONS.

§ 361.—The friction on trunnions and axles, which we now proceed to consider, gives a considerably less co-efficient than that which accompanies the kinds of motion referred to in § 355. This will appear from Table IX, which is the result of careful experiment.

The contact of the trunnion with its box is along a linear element, common to the surfaces of both. A section perpendicular to its length would cut from the trunnion and its box, two circles tangent to each other internally. The trunnion being acted on only by its weight, would, when at rest, give this tangential point at o, the lowest point of the section $p\,o\,q$ of the box. If the trunnion be put in motion by the application of a force, it would turn around the point of contact and roll indefinitely along the surface of the box, if the latter were level; but this not being the case, it will ascend along the inclined surface $o\,p$ to some point as m, where the inclination of the tangent $u\,m\,v$ is such, that the friction is just sufficient to prevent the trunnion from sliding. Here let the trunnion be in equilibrio. But the equilibrium requires that the resultant of all the forces which act, friction included, shall pass through the point m and be normal to the surface of the trunnion at that point. The friction is applied at the point m; hence the resultant N of all the other forces must pass through m in some direction as $m\,d$; the friction acts in the direction of the tangent; and hence, in order that the resultant of the friction and the force N shall be normal to the surface, the tangential component of the latter must, when the other component is normal, be equal and directly opposed to the friction.

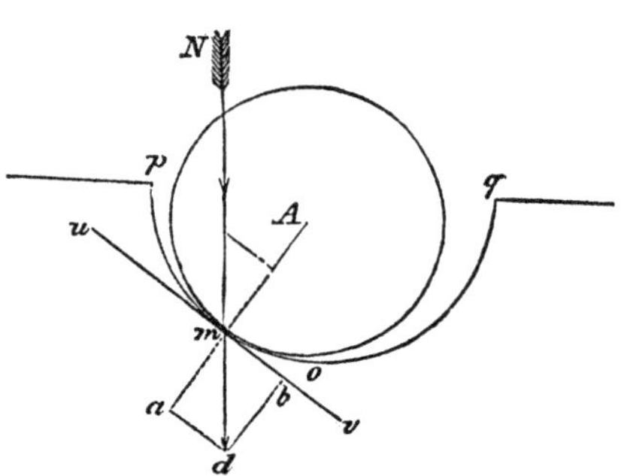

Take upon the direction of the force N the distance $m\,d$ to represent its intensity, and form the rectangle $a\,d\,b\,m$, of which the side $m\,b$ shall coincide with the tangent, then, denoting the angle $d\,m\,a$ by φ, will the component of N perpendicular to the tangent be

$$N \,.\, \cos \varphi ;$$

and the friction due to this pressure will be

$$f \,.\, N \,.\, \cos \varphi .$$

The component of N, in the direction of the tangent, will be

$$N \cdot \sin \varphi ;$$

and as this must be equal to the friction, we have

$$f \cdot N \cdot \cos \varphi = N \cdot \sin \varphi ; \quad . \quad . \quad . \quad . \quad . \quad (658)$$

whence,

$$f = \tan \varphi ;$$

that is to say, *the ratio of the friction to the pressure on the trunnion is equal to the tangent of the angle which the direction of the resultant N, of all the forces except the friction, makes with the normal to the surface of the trunnion at the point of contact.* This gives an easy method of finding the point of contact. For this purpose, we have but to draw through the centre A a line $A\,Z$, parallel to the direction of N, and through A the line $A\,m$, making with $A\,Z$ an angle of which the tangent is f; the point m, in which this line cuts the circular section of the trunnion, will be the point of contact.

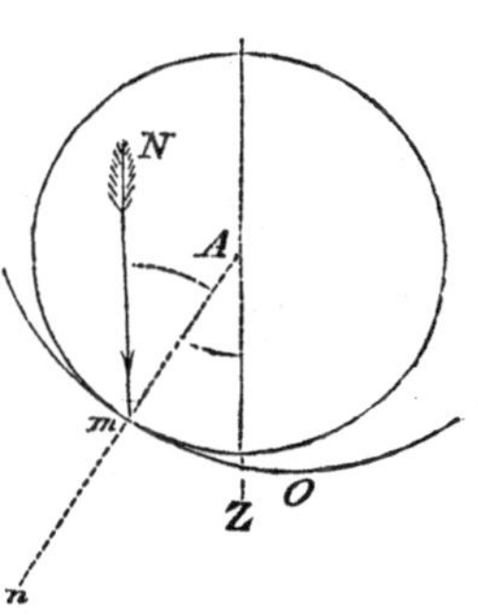

Because $m\,a\,d\,b$, last figure, is a rectangle, we have

$$N^2 = N^2 \cos^2 \varphi + N^2 \sin^2 \varphi ;$$

and, substituting for $N^2 \sin^2 \varphi$ its equal $f^2 N^2 \cos^2 \varphi$, we have

$$N^2 = N^2 \cos^2 \varphi + f^2 N^2 \cos^2 \varphi = N^2 \cos^2 \varphi \, (1 + f^2) ;$$

whence,

$$N \cos \varphi = N \times \frac{1}{\sqrt{1 + f^2}} ;$$

and multiplying both members by f,

$$f \cdot N \cdot \cos \varphi = N \cdot \frac{f}{\sqrt{1 + f^2}} ; \quad . \quad . \quad . \quad . \quad (659)$$

but the first member is the total friction; whence we conclude that *to find the friction upon a trunnion, we have but to multiply the resultant of the forces which act upon it by the unit of friction, found in Table IX, and divide this product by the square root of the square of this same unit increased by unity.*

This friction acting at the extremity of the radius R of the trunnion and in the direction of the tangent, its moment will be

$$N \cdot \frac{f}{\sqrt{1+f^2}} \times R. \quad \ldots \ldots \quad (660)$$

And the path described by the point of application of the friction being denoted by $R s_{,}$, the quantity of work of the friction will be

$$N \,.\, R \,.\, s_{,} \times \frac{f}{\sqrt{1+f^2}}; \quad \ldots \ldots \quad (661)$$

in which $s_{,}$ denotes the path described by a point at the unit's distance from the centre of the trunnion. Denoting, as in the case of the pivot, the number of revolutions performed by the trunnion in a unit of time, say a minute, by n; the quantity of work performed by friction in this time by $Q_{,}$; and making $\pi = 3{,}1416$, we have

$$s_{,} = 2\pi \,.\, n;$$

and

$$Q_{,} = 2\pi \,.\, R \,.\, n \,.\, N \,.\, \frac{f}{\sqrt{1+f^2}}. \quad \ldots \ldots \quad (662)$$

When the trunnion remains fixed and does not form part of the rotating body, the latter will turn about the trunnion, which now becomes an axle, having the centre of motion at A, the centre of the eye of the wheel; in this case, the lever of friction becomes the radius of the eye of the wheel. As the quantity of work consumed by friction is the greater, Equation (662), in proportion as this radius is greater, and as the radius of the eye of the wheel must be greater than that of the axle, the trunnion has the advantage, in this respect over the axle.

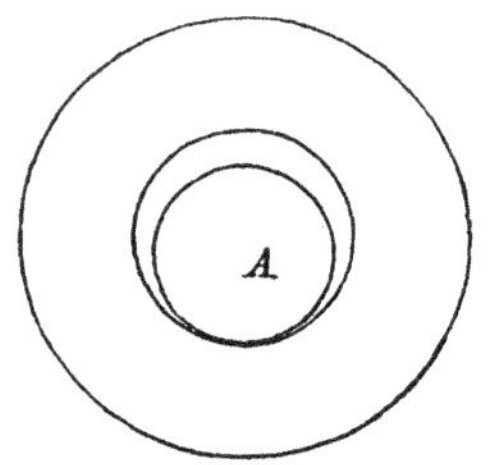

The value of the quantity of work consumed by friction is wholly independent of the length of the trunnion or axle, and no advantage is therefore gained by making it shorter or longer.

THE CORD.

§ 362.—The cord and its properties have been considered in part at § 58. It is now proposed to discuss its action under the operation of forces applied to it in any manner whatever.

Let the points A', A'', A''', be connected with each other by means of two perfectly flexible and inextensible cords $A' A''$, $A'' A'''$, the first point being acted upon by the forces P', P'', &c.; the second by the forces Q', Q'', &c.; and the third by the forces S', S'', &c.; and suppose these forces to be in equilibrio. Denote the coordinates of A' by $x' y' z'$, A'' by $x'' y'' z''$, and A''' by $x''' y''' z'''$. Also, the algebraic sum of the components of the forces acting at A' in the direction of $x y z$, by $X' Y' Z'$, at A'' by $X'' Y'' Z''$, and at A''' by $X''' Y''' Z'''$. Then will, § 101,

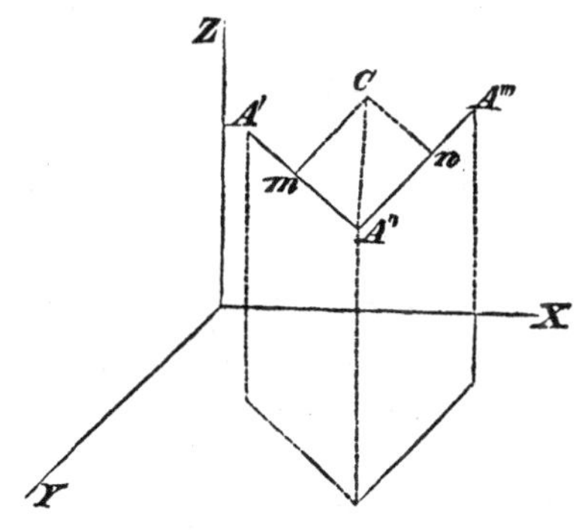

$$\left.\begin{array}{l} X' \, \delta x' + Y' \, \delta y' + Z' \, \delta z' \\ + X'' \, \delta x'' + Y'' \, \delta y'' + Z'' \, \delta z'' \\ + X''' \, \delta x''' + Y''' \, \delta y''' + Z''' \, \delta z''' \end{array}\right\} = 0. \quad . \quad . \quad (663)$$

Denote the length $A' A''$ by f, and $A'' A'''$ by g; then will

$$\left.\begin{array}{l} L = f - \sqrt{(x'' - x')^2 + (y'' - y')^2 + (z'' - z')^2} = 0; \\ H = g - \sqrt{(x''' - x'')^2 + (y''' - y'')^2 + (z''' - z'')^2} = 0. \end{array}\right\} \quad . \quad (664)$$

The displacement by which we obtain the virtual velocities whose projections are $\delta x'$, $\delta y'$, $\delta z'$, &c., is not wholly arbitrary; but must be made so as to satisfy the condition

$$\delta f = 0 \quad \text{and} \quad \delta g = 0. \quad \cdots \quad (665)$$

Differentiating Equations (664), and writing for dx', dy', dz', $\delta x'$, $\delta y'$, $\delta z'$, &c., we find

$$\frac{(x'' - x')(\delta x'' - \delta x') + (y'' - y')(\delta y'' - \delta y') + (z'' - z')(\delta z'' - \delta z')}{f} = 0;$$

$$\frac{(x''' - x'')(\delta x''' - \delta x'') + (y''' - y'')(\delta y''' - \delta y'') + (z''' - z'')(\delta z''' - \delta z'')}{g} = 0.$$

These being multiplied respectively by λ' and λ''', and added to Equation (663), we obtain by reduction, and by the principle of indeterminate co-efficients, exactly as in § 213,

$$\left.\begin{array}{l} X' - \lambda' \cdot \dfrac{x'' - x'}{f} = 0; \\ Y' - \lambda' \cdot \dfrac{y'' - y'}{f} = 0; \\ Z' - \lambda' \cdot \dfrac{z'' - z'}{f} = 0; \end{array}\right\} \quad \cdots \quad (666)$$

$$\left.\begin{array}{l} X'' + \lambda' \cdot \dfrac{x'' - x'}{f} - \lambda''' \cdot \dfrac{x''' - x''}{g} = 0; \\ Y'' + \lambda' \cdot \dfrac{y'' - y'}{f} - \lambda''' \cdot \dfrac{y''' - y''}{g} = 0; \\ Z'' + \lambda' \cdot \dfrac{z'' - z'}{f} - \lambda''' \cdot \dfrac{z''' - z''}{g} = 0; \end{array}\right\} \quad \cdots \quad (667)$$

$$\left.\begin{array}{l} X''' + \lambda''' \cdot \dfrac{x''' - x''}{g} = 0; \\ Y''' + \lambda''' \cdot \dfrac{y''' - y''}{g} = 0; \\ Z''' + \lambda''' \cdot \dfrac{z''' - z''}{g} = 0; \end{array}\right\} \quad \cdots \quad (668)$$

Taking from each group its first equation and adding, and doing the same for the second and third, we have

$$\left.\begin{array}{l} X' + X'' + X''' = 0; \\ Y' + Y'' + Y''' = 0; \\ Z' + Z'' + Z''' = 0. \end{array}\right\} \quad \cdots \quad (669)$$

That is, the conditions of equilibrium of the forces are, § 80, the same as though they had been applied to a single point.

To find the position of the points, eliminate the factors λ' and λ''', and for this purpose add the first, second and third equations of group (667) to the corresponding equations of group (668), and there will result

$$X'' + X''' + \frac{\lambda'}{f}(x'' - x') = 0;$$

$$Y'' + Y''' + \frac{\lambda'}{f}(y'' - y') = 0;$$

$$Z'' + Z''' + \frac{\lambda'}{f}(z'' - z') = 0.$$

from which we find by elimination,

$$\left.\begin{aligned} Y'' + Y''' - \frac{y'' - y'}{x'' - x'}(X'' + X''') = 0; \\ Z'' + Z''' - \frac{z'' - z'}{x'' - x'}(X'' + X''') = 0. \end{aligned}\right\} \dots (670)$$

From group (666), by eliminating λ',

$$\left.\begin{aligned} Y' - \frac{y'' - y'}{x'' - x'}X' = 0; \\ Z' - \frac{z'' - z'}{x'' - x'}X' = 0; \end{aligned}\right\} \dots\dots (671)$$

and finally from group (668) we obtain, by eliminating λ''',

$$\left.\begin{aligned} Y''' - \frac{y''' - y''}{x''' - x''}\cdot X''' = 0; \\ Z''' - \frac{z''' - z''}{x''' - x''}\cdot X''' = 0. \end{aligned}\right\} \dots\dots (672)$$

Equations (669), (670), (671) and 672), involve all the conditions necessary to the equilibrium, and the last three groups, in connection with group (664), determine the positions of the points A', A'' and A''', in space.

§ 363.—The reactions in the system which impose conditions on the displacement will be made known by Equation (331), which, because

$$\left[\frac{dL}{d(x''-x')}\right]^2+\left[\frac{dL}{d(y''-y')}\right]^2+\left[\frac{dL}{d(z''-z')}\right]^2=1\,;$$

$$\left[\frac{dH}{d(x'''-x'')}\right]^2+\left[\frac{dH}{d(y'''-y'')}\right]^2+\left[\frac{dH}{d(z'''-z'')}\right]^2=1\,;$$

becomes for the cord $A'A''$,

$$\lambda' = N'\,;$$

and for the cord $A''A'''$,

$$\lambda''' = N'''\,;$$

from which we conclude, that λ' and λ''' are respectively the tensions of the cords $A'A''$ and $A''A'''$.

This is also manifest from Equations (666) and (668); for, by transposing, squaring, adding and reducing by the relations,

$$\frac{(x''-x')^2+(y''-y')^2+(z''-z')^2}{f^2}=1,$$

$$\frac{(x'''-x'')^2+(y'''-y'')^2+(z'''-z'')^2}{g^2}=1,$$

we have

$$\left.\begin{aligned}\lambda' &= \sqrt{X'^2+Y'^2+Z'^2}=R',\\ \lambda''' &= \sqrt{X'''^2+Y'''^2+Z'''^2}=R''',\end{aligned}\right\}\quad\ldots\ldots\quad(673)$$

in which R' and R''' are the resultants of the forces acting upon the points A' and A''' respectively.

Substituting these values in Equations (666) and (668), we have

$$\frac{X'}{R'}=\frac{x''-x'}{f}\,;\quad \frac{Y'}{R'}=\frac{y''-y'}{f}\,;\quad \frac{Z'}{R'}=\frac{z''-z'}{f}\,;$$

$$\frac{X'''}{R'''}=-\frac{x'''-x''}{g}\,;\quad \frac{Y'''}{R'''}=-\frac{y'''-y''}{g}\,;\quad \frac{Z'''}{R'''}=-\frac{z'''-z''}{g}\,;$$

whence the resultants of the forces applied at the points A' and A''', act in the directions of the cords connecting these points with the point A'', and will be equal to, indeed determine the tensions of these cords.

§ 364.—From Equations (669), we have by transposition,

$$X'' = -(X''' + X'); \quad Y'' = -(Y''' + Y'); \quad Z'' = -(Z''' + Z').$$

Squaring, adding and denoting the resultant of the forces applied at A'' by R'', we have

$$R'' = \sqrt{(X''' + X')^2 + (Y''' + Y')^2 + (Z''' + Z')^2} \quad . . (674)$$

and dividing each of the above equations by this one

$$\left.\begin{aligned} \frac{X''}{R''} &= -\frac{X''' + X'}{R''}; \\ \frac{Y''}{R''} &= -\frac{Y''' + Y'}{R''}; \\ \frac{Z''}{R''} &= -\frac{Z''' + Z'}{R''}; \end{aligned}\right\} \quad . \quad . \quad . \quad . \quad . \quad (675)$$

whence, Equation (674), the resultant of the forces applied at A'' is equal and immediately opposed to the resultant of all the forces applied both at A' and A'''

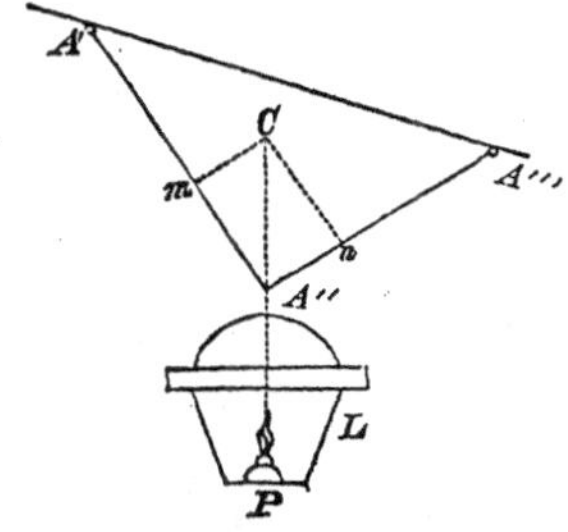

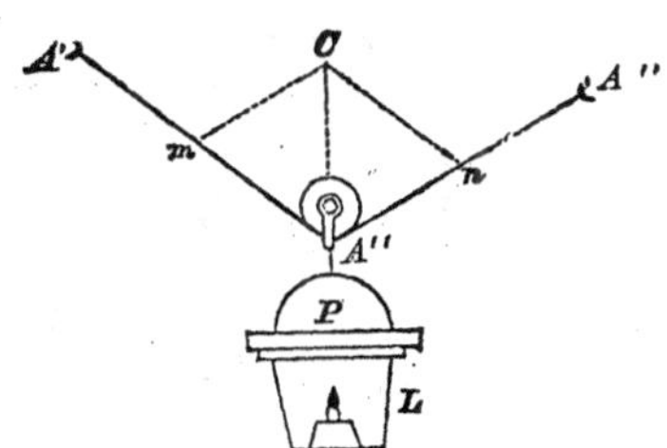

If, therefore, from the point A'', distances $A''m$ and $A''n$ be taken proportional to R' and R''' respectively, and a parallelogram $A''\,m\,C\,n$ be constructed, $A''\,C$ will represent the value of R''. If $A'\,A''\,A'''$ be a continuous cord, and the point A'' capable of sliding thereon, the tension of the cord would be the same throughout, in which case R' would be equal to R''', and the direction of R'' would bisect the angle $A'\,A''\,A'''$.

The same result is shown if, instead of making $\delta f = 0$ and $\delta g = 0$ separately, we make

$\delta(f + g) = 0$, multiply by a single indeterminate quantity λ, and proceed as before.

§ 365.—Had there been four points, A', A'', A''' and A^{iv}, connected by the same means, the general equation of equilibrium would become, by calling h the distance between the points, A''' and A^{iv},

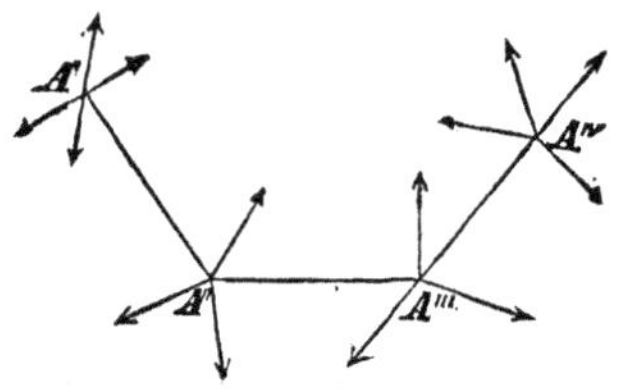

$$\left.\begin{array}{l} X'\,\delta x' + X''\,\delta x'' + X'''\,\delta x''' + X^{iv}\,\delta x^{iv} \\ + Y'\,\delta y' + Y''\,\delta y'' + Y'''\,\delta y''' + Y^{iv}\,\delta y^{iv} \\ + Z'\,\delta z' + Z''\,\delta z'' + Z'''\,\delta z''' + Z^{iv}\,\delta z^{iv} \\ + \lambda'\,\delta f + \lambda''\,\delta g + \lambda'''\,\delta h \end{array}\right\} = 0;$$

and from which, by substituting the values of δf, δg and δh, the following equations will result, viz.:

$$\left.\begin{array}{l} X' - \lambda' \cdot \dfrac{x'' - x'}{f} = 0, \\ Y' - \lambda' \cdot \dfrac{y'' - y'}{f} = 0, \\ Z' - \lambda' \cdot \dfrac{z'' - z'}{f} = 0, \end{array}\right\} \quad . \quad . \quad . \quad . \quad . \quad (676)$$

$$\left.\begin{array}{l} X'' + \lambda' \cdot \dfrac{x'' - x'}{f} - \lambda'' \cdot \dfrac{x''' - x''}{g} = 0, \\ Y'' + \lambda' \cdot \dfrac{y'' - y'}{f} - \lambda'' \cdot \dfrac{y''' - y''}{g} = 0, \\ Z'' + \lambda' \cdot \dfrac{z'' - z'}{f} - \lambda'' \cdot \dfrac{z''' - z''}{g} = 0, \end{array}\right\} \quad . \quad . \quad (677)$$

$$\left.\begin{array}{l} X''' + \lambda'' \cdot \dfrac{x''' - x''}{g} - \lambda''' \cdot \dfrac{x^{iv} - x'''}{h} = 0, \\ Y''' + \lambda'' \cdot \dfrac{y''' - y''}{g} - \lambda''' \cdot \dfrac{y^{iv} - y'''}{h} = 0, \\ Z''' + \lambda'' \cdot \dfrac{z''' - z''}{g} - \lambda''' \cdot \dfrac{z^{iv} - z'''}{h} = 0, \end{array}\right\} \quad . \quad (678)$$

$$\left.\begin{aligned} X^{\text{iv}} + \lambda''' \cdot \frac{x^{\text{iv}} - x'''}{h} &= 0, \\ Y^{\text{iv}} + \lambda''' \cdot \frac{y^{\text{iv}} - y'''}{h} &= 0, \\ Z^{\text{iv}} + \lambda''' \cdot \frac{z^{\text{iv}} - z'''}{h} &= 0, \end{aligned}\right\} \quad \ldots \quad (679)$$

Eliminating the indeterminate quantities λ', λ'' and λ''', we obtain eight equations, from which, and the three equations of conditions expressive of the lengths of f, g, and h, the position of the points A', A'', A''', and A^{iv} may be determined.

If there be n points, connected in the same way and acted upon by any forces, the law which is manifest in the formation of Equations (676), (677), (678), and (679), plainly indicates the following n equations of equilibrium:

$$\left.\begin{aligned} X' - \lambda' \cdot \frac{x'' - x'}{f} &= 0, \\ Y' - \lambda' \cdot \frac{y'' - y'}{f} &= 0, \\ Z' - \lambda' \cdot \frac{z'' - z'}{f} &= 0, \end{aligned}\right\} \quad \ldots \quad (680)$$

$$\left.\begin{aligned} X'' + \lambda' \cdot \frac{x'' - x'}{f} - \lambda'' \cdot \frac{x''' - x''}{g} &= 0, \\ Y'' + \lambda' \cdot \frac{y'' - y'}{f} - \lambda'' \cdot \frac{y''' - y''}{g} &= 0, \\ Z'' + \lambda' \cdot \frac{z'' - z'}{f} - \lambda'' \cdot \frac{z''' - z''}{g} &= 0, \end{aligned}\right\} \quad (681)$$

$$\left.\begin{aligned} X''' + \lambda'' \cdot \frac{x''' - x''}{g} - \lambda''' \cdot \frac{x^{\text{iv}} - x'''}{h} &= 0, \\ Y''' + \lambda'' \cdot \frac{y''' - y''}{g} - \lambda''' \cdot \frac{y^{\text{iv}} - y'''}{h} &= 0, \\ Z''' + \lambda'' \cdot \frac{z''' - z''}{g} - \lambda''' \cdot \frac{z^{\text{iv}} - z'''}{h} &= 0, \end{aligned}\right\} \quad (682)$$

. . . .

$$\left.\begin{aligned} X_{n-1} + \lambda_{n-2} \cdot \frac{x_{n-1} - x_{n-2}}{k} - \lambda_{n-1} \cdot \frac{x_n - x_{n-1}}{l} &= 0, \\ Y_{n-1} + \lambda_{n-2} \cdot \frac{y_{n-1} - y_{n-2}}{k} - \lambda_{n-1} \cdot \frac{y_n - y_{n-1}}{l} &= 0, \\ Z_{n-1} + \lambda_{n-2} \cdot \frac{z_{n-1} - z_{n-2}}{k} - \lambda_{n-1} \cdot \frac{z_n - z_{n-1}}{l} &= 0, \end{aligned}\right\} \cdot (683)$$

$$\left.\begin{aligned} X_n + \lambda_{n-1} \cdot \frac{x_n - x_{n-1}}{l} &= 0, \\ Y_n + \lambda_{n-1} \cdot \frac{y_n - y_{n-1}}{l} &= 0, \\ Z_n + \lambda_{n-1} \cdot \frac{z_n - z_{n-1}}{l} &= 0. \end{aligned}\right\} \quad . \; . \; . \; . \quad (684)$$

In which λ, with its particular accent, denotes the tension of the cord into the difference of whose extreme co-ordinates it is multiplied.

Adding together the equations containing the components of the forces parallel to the same axis, there will result

$$\left.\begin{aligned} X' + X'' + X''' + X^{\text{iv}} \; . \; . \; . \; X_n &= 0, \\ Y' + Y'' + Y''' + Y^{\text{iv}} \; . \; . \; . \; Y_n &= 0, \\ Z' + Z'' + Z''' + Z^{\text{iv}} \; . \; . \; . \; Z_n &= 0, \end{aligned}\right\} \; . \; . \quad (685)$$

from which we infer, that the conditions of equilibrium are the same as though the forces were all applied to a single point.

From group (680), we find by transposing, squaring, adding and extracting square root,

$$\sqrt{X'^2 + Y'^2 + Z'^2} = \lambda' = R'$$

and dividing each of the equations found after transposing in group (680) by this one,

$$\frac{X'}{R'} = \frac{x'' - x'}{f};$$

$$\frac{Y'}{R'} = \frac{y'' - y'}{f};$$

$$\frac{Z'}{R'} = \frac{z'' - z'}{f}.$$

Treating the equations of group (684) in the same way, we have

$$\frac{X_n}{R_n} = - \frac{x_n - x_{n-1}}{l};$$

$$\frac{Y_n}{R_n} = - \frac{y_n - y_{n-1}}{l};$$

$$\frac{Z_n}{R_n} = - \frac{z_n - z_{n-1}}{l};$$

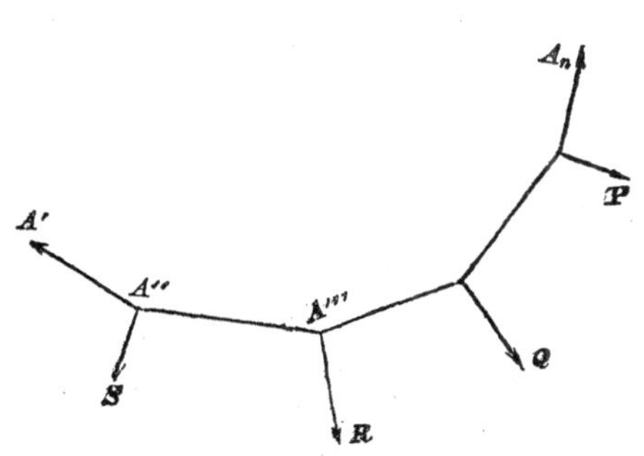

whence, the resultants of the forces applied to the extreme points A' and A_n, act in the direction of the extreme cords. And from Equations (685) it appears that the resultant of these two resultants is equal and contrary to that of all the forces applied to the other points.

§ 366.—If the extreme points be fixed, X', Y', Z' and X_n, Y_n, Z_n, will be the components of the resistances of these points in the directions of the axes; these resistances will be equal to the tensions λ' and λ_n of the cords which terminate in them. Taking the sum of the equations in groups (680) to (684), stopping at the point whose co-ordinates are x_{n-m}, y_{n-m}, z_{n-m}, we have

$$\left.\begin{aligned} X' + \Sigma X - \lambda_{n-m-1} \cdot \frac{x_{n-m} - x_{n-m-1}}{l_{n-m}} &= 0; \\ Y' + \Sigma Y - \lambda_{n-m-1} \cdot \frac{y_{n-m} - y_{n-m-1}}{l_{n-m}} &= 0; \\ Z' + \Sigma Z - \lambda_{n-m-1} \cdot \frac{z_{n-m} - z_{n-m-1}}{l_{n-m}} &= 0; \end{aligned}\right\} \quad . \quad . \quad (686)$$

in which ΣX, ΣY, ΣZ, denote the algebraic sums of the components in the directions of the axes of the active forces; λ_{n-m-1} the tension on the side of which the extreme co-ordinates are x_{n-m}, y_{n-m}, z_{n-m}, and x_{n-m-1}, y_{n-m-1}, z_{n-m-1}; and l_{n-m} the length of this side.

§ 367.—Now, suppose the length of the sides diminished and

their number increased indefinitely; the polygon will become a curve; also, making $\lambda_{n-m-1}=t$, we have

$$x_{n-m} - x_{n-m-1} = dx,$$

$$y_{n-m} - y_{n-m-1} = dy,$$

$$z_{n-m} - z_{n-m-1} = dz,$$

$$l_{n-m} = ds,$$

s being any length of the curve; and Equations (686) become

$$\left.\begin{aligned} X' + \Sigma X - t \cdot \frac{dx}{ds} &= 0; \\ Y' + \Sigma Y - t \cdot \frac{dy}{ds} &= 0; \\ Z' + \Sigma Z - t \cdot \frac{dz}{ds} &= 0; \end{aligned}\right\} \quad \cdot \quad \cdot \quad \cdot \quad \cdot \quad \cdot \quad \cdot \quad (687)$$

which will give the curved locus of a rope or chain, fastened at its ends, and acted upon by any forces whatever, as its own weight, the weight of other materials, the pressure of winds, currents of water, &c., &c.

This arrangement of several points, connected by means of flexible cords, and subjected to the action of forces, is called a *Funicular Machine.*

§ 368.—If the only forces acting be pressure from weights, we have, by taking the axis of z vertical,

$$X'' = X''' = X^{\text{v}} \text{ \&c.} = 0; \quad Y'' = Y''' \text{ \&c.} = 0;$$

and from Equations (680) to (684),

$$X' = \lambda' \cdot \frac{x'' - x'}{f} = \lambda'' \cdot \frac{x''' - x''}{g} = \cdot \cdot \cdot \cdot \lambda_{n-1} \frac{x_n - x_{n-1}}{l_n};$$

whence, the tensions on all the cords, estimated in a horizontal direction, are equal to one another. Moreover, we obtain from the same equations, by division,

$$\frac{y'' - y'}{x'' - x'} = \frac{y''' - y''}{x''' - x''} = \cdot \cdot \cdot \cdot \cdot \cdot \frac{y_n - y_{n-1}}{x_n - x_{n-1}}.$$

These are the tangents of the angles which the projections of the sides on the plane $x y$ make with the axis x. The polygon is therefore contained in a vertical plane.

THE CATENARY.

§ 369.—If a single rope or chain cable be taken, and subjected only to the action of its own weight, it will assume a curvilinear shape called the *Catenary curve.* It will lie in a vertical plane. Take the axes z and x in this plane, and z positive upwards, then will

$$\Sigma X = 0; \quad \Sigma Y = 0; \quad Y' = 0; \quad \Sigma Z = -W;$$

in which W denotes the weight of the cable, and Equations (687) become

$$\left. \begin{aligned} X' - t \cdot \frac{dx}{ds} &= 0, \\ Z' - W - t \cdot \frac{dz}{ds} &= 0. \end{aligned} \right\} \quad \cdots \cdots \quad (688)$$

These are the differential equations of the curve. The origin may be taken at any point. Let it be at the bottom point of the curve. The curve being at rest, will not be disturbed by taking any one of its points fixed at pleasure. Suppose the lowest point for a moment to become fixed. As the curve is here horizontal, $Z' = 0$, § 366, and from the second of Equations (688), we have

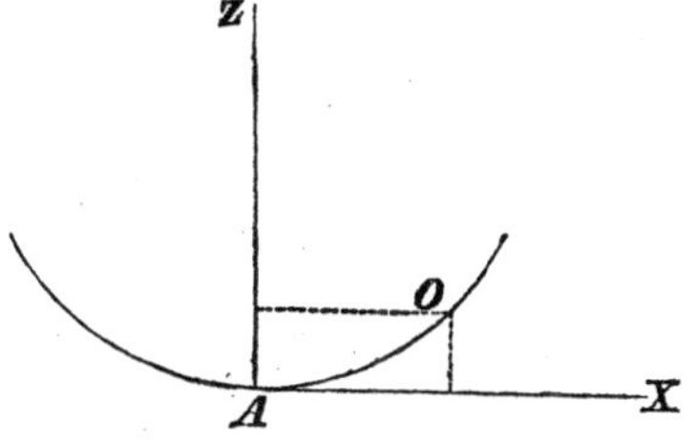

$$W = -t \cdot \frac{dz}{ds}; \quad \cdots \cdots \quad (689)$$

whence, the vertical component of the tension at any point as O of the curve, is equal to the weight of that part of the cable between this point and the lowest point. The first of Equations (688) shows that the horizontal component of the tension at O is equal to the tension at the lowest point, as it should be, since the horizontal tensions are equal throughout.

Taking the unit of length of the cable to give a unit of weight, which would give the common catenary, we have $W = s$; and, denoting the tension at the lowest point by c, we have

$$t = \pm \sqrt{s^2 + c^2},$$

and from Equation (689),

$$dz = \mp \frac{s \cdot ds}{\sqrt{c^2 + s^2}}.$$

Taking the positive sign, because z and s increase together, integrating, and finding the constant of integration such that when $z = 0$, we have $s = 0$,

$$z + c = \sqrt{c^2 + s^2};$$

whence,

$$s^2 = z^2 + 2cz.$$

Also, dividing the first of Equations (688) by Equation (689),

$$\frac{dx}{dz} = \frac{c}{s} = \frac{c}{\sqrt{z^2 + 2cz}};$$

and integrating, and taking the constant such that x and z vanish together,

$$x = c \cdot \log \frac{z + c + \sqrt{z^2 + 2cz}}{c} \quad . \quad . \quad . \quad (690)$$

which is the equation of the catenary.

This equation may be put under another form. For we may write the above,

$$c e^{\frac{x}{c}} = z + c + \sqrt{(z + c)^2 - c^2};$$

transposing $z + c$ and squaring,

$$c^2 \cdot e^{\frac{2x}{c}} - 2ce^{\frac{x}{c}}(z + c) = -c^2;$$

whence,

$$z + c = \tfrac{1}{2} c \cdot (e^{\frac{x}{c}} + e^{-\frac{x}{c}}). \quad . \quad . \quad . \quad . \quad . \quad (691)$$

Also,

$$s = \sqrt{(z + c)^2 - c^2},$$

and by substitution,

$$s = \tfrac{1}{2} c \cdot (e^{\frac{x}{c}} - e^{-\frac{x}{c}}). \quad . \quad . \quad . \quad . \quad . \quad . \quad (692)$$

§ 370.—If the length of the portion of the cable which gives a unit of weight were to vary, the variation might be made such as to cause the area of the cross section to be proportional to the tension at the point where the section is made. The general Equations (688) will give the solution for every possible case.

FRICTION BETWEEN CORDS AND CYLINDRICAL SOLIDS.

§ 371.—When a cord is wrapped around a solid cylinder, and motion is communicated by applying the power F at one end while a resistance W acts at the other, a pressure is exerted by the cord upon the cylinder; this pressure produces friction, and this acts as a resistance. To estimate its amount, denote the radius of the cylinder by R, the arc of contact by s, the tension of the cord at any point by t.

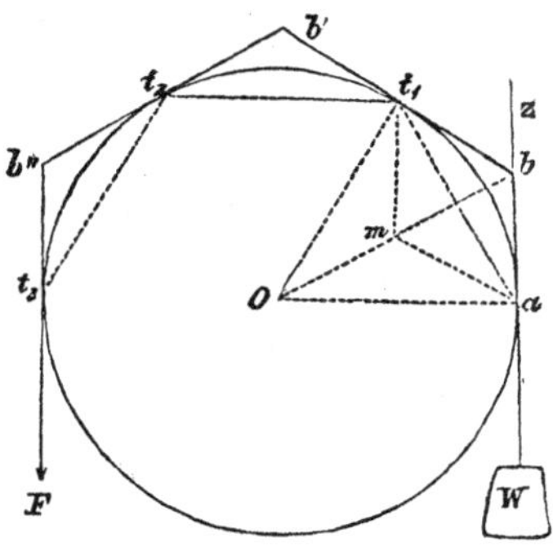

The tension t being the same throughout the length $ds = a t_{,}$ of the cord, this element will be pressed against the cylinder by two forces each equal to t, and applied at its extremities a and $t_{,}$, the first acting from a towards W, the second from $t_{,}$ towards b'. Denoting by θ the angle $a b t_{,}$, and by p the resultant $b m$ of these forces, which is obviously the pressure of ds against the cylinder, we have, Equation (56),

$$p = \sqrt{t^2 + t^2 + 2 t \,.\, t \cos \theta} = t \sqrt{2 (1 + \cos \theta)}\,;$$

but

$$1 + \cos \theta = 2 \cos^2 \tfrac{1}{2} \theta\,; \qquad (\pi - \theta) = \frac{ds}{R}\,;$$

and taking the arc for its sine, because $\pi - \theta$ is very small, we have

$$p = t \cdot \frac{ds}{R};$$

and hence, § 355, the friction on ds will be

$$f \cdot p = f \cdot t \cdot \frac{ds}{R}.$$

The element $t_{,} t_2$ of the cord which next succeeds $a t_{,}$, will have its tension increased by this friction before the latter can be overcome; this friction is therefore the differential of the tension, being the difference of the tensions of two consecutive elements; whence,

$$dt = f \cdot t \cdot \frac{ds}{R};$$

dividing by t and integrating,

$$\log t = f \cdot \frac{s}{R} + \log C,$$

or,

$$t = C e^{\frac{fs}{R}} \quad \dots\dots\dots \quad (693)$$

making $s = 0$, we have $t = W = C$; whence,

$$t = W \cdot e^{\frac{fs}{R}}; \quad \dots\dots\dots \quad (694)$$

and making $s = S = a t_1 t_2 t_3$, we have $t = F$; and

$$F = W \cdot e^{\frac{fS}{R}} \quad \dots\dots\dots \quad (695)$$

Suppose, for example, the cord to be wound around the cylinder three times, and $f = \frac{1}{3}$; then will

$$S = 3\pi \, . \, 2R = 6 \, . \, 3{,}1416 \, . \, R = 18{,}849 \, R,$$

and

$$F = W \times e^{\frac{1}{3} \times 18{,}849} = W \times (2{,}71825)^{6{,}2832};$$

or,

$$F = W \, . \, 535{,}3;$$

that is to say, one man at the end W could resist the combined effort of 535 men, of the same strength as himself, to put the cord in motion when wound three times round the cylinder.

THE INCLINED PLANE.

§ 372.—The inclined plane is used to support, in part, the weight of a body while at rest or in motion upon its surface.

Suppose a body to rest with one of its faces on an inclined plane of which the Equation is

$$L = \cos a\, x + \cos b\, y + \cos c\, z - d = 0\,; \quad \dots \quad (a)$$

in which d denotes the distance of the plane from the origin of co-ordinates, and a, b, c, the angles which a normal to the plane makes with the axes x, y, z, respectively.

Denote the weight of the body by W; the power by F; the normal pressure by N; the angles which the power makes with the axes x, y, z, by $\alpha_{\prime}$, $\beta_{\prime}$, $\gamma_{\prime}$, respectively; and the path described by the point of application of the resultant friction by s. Then, taking the axis z vertical and positive upwards, and supposing the force to produce a uniform motion of simple translation, will, Eq. (645),

$$\left.\begin{array}{l} \left(F\cos\alpha_{\prime} + f N \dfrac{dx}{ds}\right)\delta x \\ + \left(F\cos\beta_{\prime} + f N \dfrac{dy}{ds}\right)\delta y \\ + \left(F\cos\gamma_{\prime} + f N \dfrac{dz}{ds} - W\right)\delta z \end{array}\right\} = 0\,; \quad \dots \quad (b)$$

and, Equation (a),

$$\cos a\,\delta x + \cos b\,\delta y + \cos c\,\delta z = 0$$

Multiplying this last by λ, adding and proceeding as in § 213,

$$\left.\begin{array}{l} F\cos\alpha_{\prime} + f N \dfrac{dx}{ds} + \lambda\cos a = 0, \\ F\cos\beta_{\prime} + f N \dfrac{dy}{ds} + \lambda\cos b = 0, \\ F\cos\gamma_{\prime} + f N \dfrac{dz}{ds} + \lambda\cos c - W = 0\,; \end{array}\right\} \quad \dots \quad (c)$$

and, Eq. (331),

$$N = \lambda\sqrt{\left(\frac{dL}{dx}\right)^2 + \left(\frac{dL}{dy}\right)^2 + \left(\frac{dL}{dz}\right)^2} = \lambda. \quad \dots \quad (d)$$

Substituting the value of λ in Equations (c), the first two give by eliminating N,

$$-\frac{f\frac{dx}{ds}+\cos a}{f\frac{dy}{ds}+\cos b}\cdot\frac{\cos\beta_{\prime}}{\cos\alpha_{\prime}}+1=0; \quad \cdot \cdot \cdot \cdot \cdot \cdot (e)$$

and the first and third, by eliminating N,

$$F\left[f\left(\cos\gamma_{\prime}\frac{dx}{ds}-\cos\alpha_{\prime}\frac{dz}{ds}\right)+\cos\gamma_{\prime}\cos a-\cos\alpha_{\prime}\cos c\right]=W\left(f\frac{dx}{ds}+\cos a\right) \quad (g)$$

If there be no friction, then will $f=0$, and, Eq. (e),

$$-\frac{\cos a}{\cos b}\cdot\frac{\cos\beta_{\prime}}{\cos\alpha_{\prime}}+1=0;$$

whence, Eqs. (45) and (a), the power must be applied in a plane normal both to the inclined plane and to the horizon.

If without disregarding friction, the power be applied in a plane fulfilling the above condition, and also containing the centre of gravity, the resultant friction may be regarded as acting in this plane, and we may take it as the co-ordinate plane $z\,x$, in which case

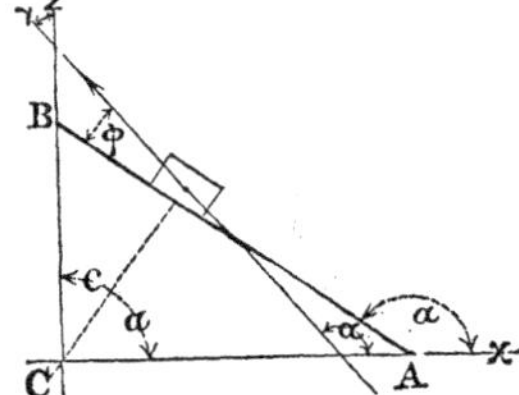

$$\cos b=0;\ \cos\beta_{\prime}=0;\ \frac{dy}{ds}=0;$$

and denoting the inclination of the plane to the horizon by α, and that of the power to the inclined plane by φ;

$$\cos a=\sin\alpha;\ \cos c=-\cos\alpha;\ \cos\gamma_{\prime}=\sin\alpha_{\prime};$$

$$\cos\gamma_{\prime}\frac{dx}{ds}-\cos\alpha_{\prime}\frac{dz}{ds}=-\sin\alpha_{\prime}\cos\alpha+\cos\alpha_{\prime}\sin\alpha=\sin(\alpha-\alpha_{\prime})=\sin\varphi;$$

$$\cos\gamma_{\prime}\cos a-\cos\alpha_{\prime}\cos c=\sin\alpha_{\prime}\sin\alpha+\cos\alpha_{\prime}\cos\alpha=\cos(\alpha-\alpha_{\prime})=\cos\varphi;$$

which, in Eq. (g), give

$$F=\frac{W(\sin\alpha+f\cos\alpha)}{\cos\varphi+f\sin\varphi}. \quad \cdot \cdot \cdot \cdot \cdot \cdot (696)$$

This supposes motion to take place *up* the plane; if the power F be just sufficient to permit the body to move uniformly *down* the plane, then will f change its sign, and we shall have

$$F=\frac{W(\sin\alpha-f\cos\alpha)}{\cos\varphi-f\sin\varphi}. \quad \cdot \cdot \cdot \cdot \cdot \cdot (697)$$

And the power may vary between the limits given by these two values without moving the body.

§ 373. If the power be zero, or $F = 0$, then will

$$\sin \alpha - f \cos \alpha = 0,$$

or

$$\tan \alpha = f,$$

which is the angle of friction, § 355.

§ 374.—If the power act parallel to the plane, then will $\varphi = 0$, and

$$F = W(\sin \alpha \pm f \cos \alpha). \quad . \quad . \quad . \quad . \quad . \quad (698)$$

the upper sign answering to the case of motion up, and the lower, down the plane; the difference of the two values being

$$2 f W \cos \alpha.$$

If $f = 0$, then will

$$\frac{F}{W} = \sin \alpha = \frac{BC}{AB};$$

that is, the power is to the weight as the height of the plane is to its length; and there will be a gain of power.

§ 375.—If the power be applied horizontally, then will φ be negative and equal to α, and we have, by including the motion in both directions,

$$F = \frac{W(\sin \alpha \pm f \cos \alpha)}{\cos \alpha \mp f \sin \alpha}; \quad . \quad . \quad . \quad . \quad . \quad (699)$$

the difference of the limiting values being

$$\frac{2 f . W}{\cos^2 \alpha - f^2 \sin^2 \alpha}.$$

If the friction be zero, or $f = 0$, then will

$$\frac{F}{W} = \tan \alpha = \frac{BC}{AC}.$$

That is, the power will be to the resistance as the height of the plane is to its base; and there may be gain or loss of power.

§ 376.—To find under what angle the power will act to greatest advantage, make the denominator in Equation (696) a maximum. For this purpose, we have, by differentiating,

$$- \sin \varphi + f \cos \varphi = 0;$$

whence,

$$\tan \varphi = f.$$

That is, the angle should be positive, and equal to that of the friction.

§ 377.—If the power act parallel to any inclined surface to move a body up, the elementary quantity of work of the power and resistances will give the relation, Equation (698),

$$F\,d\,s = W\,d\,s \sin \alpha + W f\,d\,s \cos \alpha.$$

But, denoting the whole horizontal distance passed over by $l = A\,C$, and the vertical height by $h = B\,C$, we have

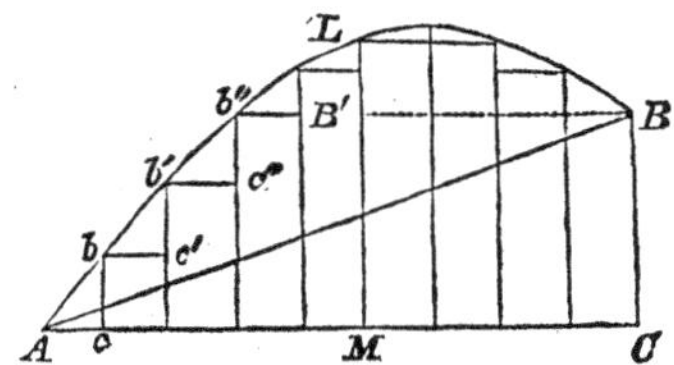

$$d\,s\,.\sin \alpha = d\,h,$$

$$d\,s\,.\cos \alpha = d\,l;$$

whence, substituting, and integrating, and supposing the body to be started from rest and brought to rest again, in which case the work of inertia will balance itself, we have

$$F s = W h + f\,.\,W\,.\,l, \quad \cdots\cdots \quad (700)$$

in which there is no trace of the path actually passed over by the body. The work is that required to raise the body through a vertical height $B\,C$, and to overcome the friction due to its weight over a horizontal distance $A\,C$.

The resultant of the weight and the power must intersect the inclined plane within the polygon, formed by joining the points of contact of the body, else the body will roll, and not slide.

THE LEVER.

§ 378.—The *Lever* is a solid bar $A\,B$, of any form, supported by a fixed point O, about which it may freely turn, called the *fulcrum.* Sometimes it is supported upon trunnions, and frequently

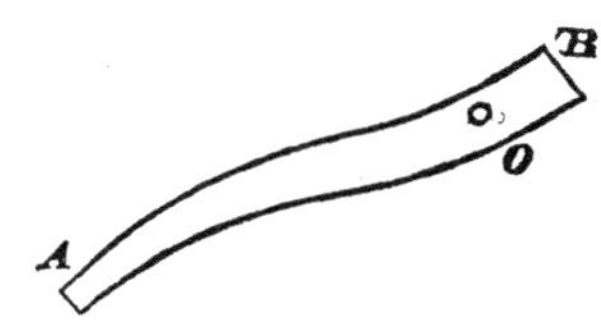

upon a knife-edge. Levers have been divided into three different classes, called orders.

In levers of the *first order*, the power F and resistance Q are applied on opposite sides of the fulcrum O; in levers of the *second order*, the resistance Q is applied to some point between the fulcrum O and the point of application of the power F; and in the *third order* of levers, the power F is applied between the fulcrum O and point of application of the resistance Q.

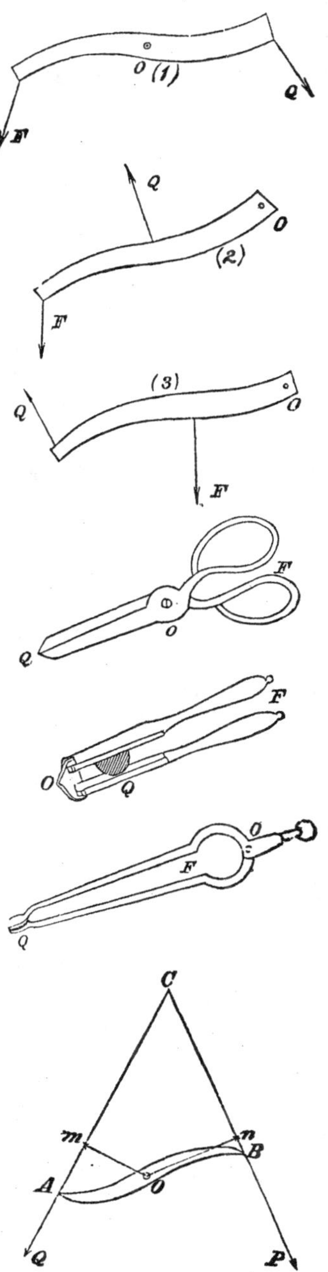

The common shears furnishes an example of a pair of levers of the first order; the nut-crackers of the second; and fire-tongs of the third. In all orders, the conditions of equilibrium are the same.

These divisions are wholly arbitrary, being founded in no difference of principle. The relation of the power to the resistances, is the same in all.

Let AB be a lever supported upon a trunnion at O, and acted upon by the power P and resistance Q, applied in a plane perpendicular to the axis of the trunnion. Draw from the axis of the trunnion, the lever arms On and Om, being the perpendicular distances of the power and resistance from the axis of motion, and denote them respectively by l_p and l_r; also denote the resultant of P and Q by N, the radius of the trunnion by r, the co-efficient of friction by f, and the arc described at the unit's distance from the axis by s_1.

Then,

$$\delta p = l_p' . d s_1 ; \quad \delta q = l_r . d s_1 ,$$

$$N = \sqrt{P^2 + Q^2 + 2 P Q \cos \theta},$$

in which θ is the angle of inclination $A C B$ of the power to the resistance. Then, supposing the lever to have attained a uniform motion, will, Equations (645) and (661),

$$P . l_p . d s_1 - Q . l_r . d s_1 - \sqrt{P^2 + Q^2 + 2 P Q \cos \theta} \cdot \frac{r . d s_1 . f}{\sqrt{1 + f^2}} = 0. \quad (701)$$

Omitting the common factor $d s_1$, and making

$$\frac{f}{\sqrt{1 + f^2}} = f' ; \quad m = \frac{l_r}{l_p} ; \quad n = \frac{r}{l_p},$$

we have,

$$P - m Q - \sqrt{P^2 + Q^2 + 2 P Q . \cos \theta} \cdot f' n = 0.$$

Transposing, squaring, and solving, with respect to P, we find,

$$P = Q \cdot \frac{m + f' n \left(f' n \cos \theta \pm \sqrt{1 + 2 m \cos \theta + m^2 - f'^2 n^2 \sin^2 \theta} \right)}{1 - f'^2 n^2} \quad (702)$$

If the fraction n be so small as to justify the omission of every term into which it enters as a factor, or if the co-efficient of friction be sensibly zero, then would

$$\frac{P}{Q} = m = \frac{l_r}{l_p} \quad . \quad . \quad . \quad . \quad . \quad . \quad . \quad . \quad (703)$$

That is, the power and the resistance are to each other inversely as the lengths of their respective lever arms.

If the power or the resistance, or both, be applied in a plane oblique to the axis of the trunnion, each oblique action must be replaced by its components, one of which is perpendicular, and the other parallel to the axis of the trunnion. The perpendicular components must be treated as above. The parallel components will, if the friction arising from the resultant of the normal components be not too great, give motion to the whole body of the lever along the trunnion; and if this be prevented by a shoulder, the friction upon this shoulder becomes an additional resistance, whose elementary quantity of work may be computed by means of Eq. (657) and made another term in Equation (701).

WHEEL AND AXLE.

§ 379.—This machine consists of a wheel mounted upon an arbor, supported at either end by a trunnion resting in a box or trunnion bed. The plane of the wheel is at right angles to the arbor; the power P is applied to a rope wound round the wheel, the resistance to another rope wound in the opposite direction about the arbor, and both act in planes at right angles to the axis of motion. Let us suppose the arbor to be horizontal and the resistance Q to be a weight.

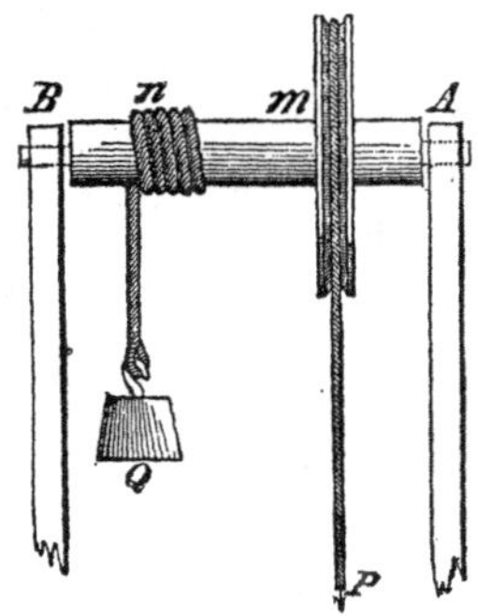

Make

N and N' = pressures upon the trunnion boxes at A and B;

R = radius of the wheel;

r = radius of the arbor;

ρ and ρ' = radii of the trunnions at A and B;

$$f' = \frac{f}{\sqrt{1+f^2}}$$

s_1 = arc described at unit's distance from axis of motion.

Then, the system being retained by a fixed axis, we have

$$P\,\delta p = P\,R\,d\,s_1;$$

$$Q\,\delta q = Q\,r\,d\,s_1.$$

The elementary work of the friction will, Eq. (661), be

$$f'\,(N\rho + N'\rho')\,d\,s_1;$$

and the elementary work of the stiffness of cordage, Equation (652),

$$d_{,} \cdot \frac{K + I \cdot Q}{2\, r} \cdot r \cdot d\, s_1 ;$$

and when the machine is moving uniformly,

$$P R\, d s_1 - Q\, r\, d s_1 - f'(N \rho + N' \rho')\, d s_1 - d_{,} \cdot \frac{K + I \cdot Q}{2\, r} \cdot r \cdot d s_1 = 0 ; \cdot (704)$$

The pressures N and N' arise from the action of the power P, the weight of the machine, and the reaction of the resistance Q, increased by the stiffness of cordage. To find their values, resolve each of these forces into two parallel components acting in planes which are perpendicular to the axis of the arbor at the trunnion beds; then resolve each of these components which are oblique to the components of Q into two others, one parallel and the other perpendicular to the direction of Q.

Make

w = weight of the wheel and axle,

g = the distance of its centre of gravity from A,

p = the distance $m\, A$,

q = the distance $n\, A$,

l = length of the arbor $A\, B$,

φ = the angle which the direction of P makes with the vertical or direction of the resistance Q.

Then the force applied in the plane perpendicular to the trunnion A, and acting parallel to the resistance Q, will, § 95, be,

$$w \cdot \frac{l - g}{l} + Q \cdot \frac{l - q}{l} + P \cdot \frac{l - p}{l} \cdot \cos \varphi ;$$

and the force applied in this plane and acting at right angles to the direction of Q, will be

$$P \cdot \frac{l - p}{l} \cdot \sin \varphi .$$

The vertical force applied in the plane at B will be

$$w \cdot \frac{g}{l} + Q \cdot \frac{q}{l} + P \cdot \frac{p}{l} \cdot \cos \varphi ,$$

and the horizontal force in this plane will be

$$P \cdot \frac{p}{l} \cdot \sin \varphi ;$$

whence,

$$N = \frac{1}{l} \cdot \sqrt{[w(l-g) + Q(l-q) + P(l-p)\cos\varphi]^2 + P^2(l-p)^2 \cdot \sin^2\varphi} ; \quad (705)$$

$$N' = \frac{1}{l} \cdot \sqrt{[w \cdot g + Q \cdot q + P \cdot p \cdot \cos\varphi]^2 + P^2 \cdot p^2 \cdot \sin^2\varphi} ; \quad (706)$$

If θ and θ' be the angles which the directions of N and N' make with that of the resistance Q, we have

$$\sin\theta = \frac{P\,(l-p)}{N \cdot l} \cdot \sin\varphi ; \quad \sin\theta' = \frac{P\,p}{N' l} \cdot \sin\varphi .$$

Equations (704), (705), and (706) are sufficient to determine the relation between P and Q to preserve the motion uniform, or an equilibrium without the aid of inertia. The values of N and N' being substituted in Equation (704), and that equation solved with reference to P, will give the relation in question.

§ 380.—If the power P act in the direction of the resistance Q, then will $\cos\varphi = 1$, $\sin\varphi = 0$, and Equation (704) would, after substituting the corresponding values of N and N', transposing, omitting the common factor $d s_1$, and supposing $\rho = \rho'$, become

$$P R = Q r + f' \rho\,(w + Q + P) + d_{\prime} \cdot \frac{K + I\,Q}{2\,r} \cdot r . \quad (707)$$

And omitting the terms involving the friction and stiffness of cordage,

$$\frac{P}{Q} = \frac{r}{R} ;$$

that is, the power is to the resistance as the radius of the arbor is to that of the wheel; which relation is exactly the same as that of the common lever.

FIXED PULLEY.

§ 381.—The pulley is a small wheel having a groove in its circumference for the reception of a rope, to one end of which the power P is applied, and to the other the resistance Q. The pulley may turn either upon trunnions or about an axle, supported in what

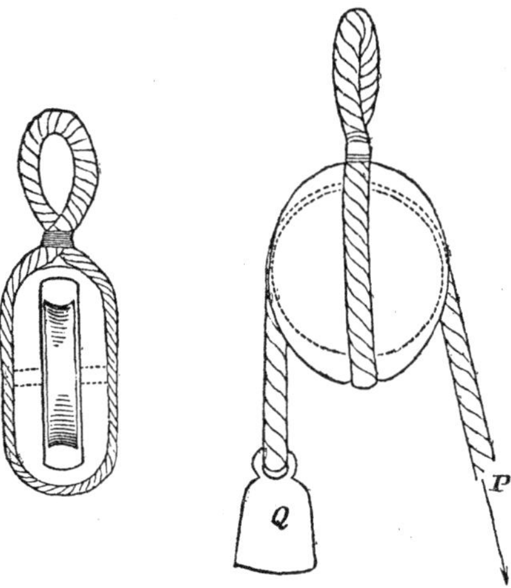

is called a *block*. This is usually a solid piece of wood, through which is cut an opening large enough to receive the pulley, and allow it to turn freely between its cheeks. Sometimes the block is a simple framework of metal. When the block is stationary, the pulley is said to be *fixed*. The principle of this machine is obviously the same as that of the wheel and axle.

The friction between the rope and pulley will be sufficient to give the latter motion.

Making, in Equations (705) and (706),

$$g = q = p = \tfrac{1}{2} l,$$

we have

$$N = \tfrac{1}{2} \sqrt{(w + Q + P \cos \varphi)^2 + P^2 \sin^2 \varphi} = N' \quad \cdot \cdot \ (708)$$

Making $R = r$, and $\rho = \rho'$, in Equation (704), and substituting the above values of N and N', we have, after omitting the common factor $d s_1$,

$$PR - QR - f'\rho \sqrt{(w + Q + P \cos \varphi)^2 + P^2 \sin^2 \varphi} - d_{,} \cdot \frac{K + IQ}{2R} \cdot R = 0. \cdot (709)$$

Solving this equation with respect to P, we find the value of the latter in terms of the different sources of resistance. But this direct process would be tedious; and it will be sufficient in all cases of practice to employ an approximate value for P under the radical, obtained by first neglecting the terms involving friction and stiffness of cordage.

Thus, dividing by R and transposing, we find

$$P = Q + f' \frac{\rho}{R} \sqrt{(w + Q + P \cos \varphi)^2 + P^2 \sin^2 \varphi} + d_{,} \cdot \frac{K + I Q}{2 R}.$$

Now $f' \cdot \frac{\rho}{R}$ is usually a small fraction; an erroneous value assumed for P under the radical, will involve but a trifling error in the result. We may therefore write Q for P in the second member; and neglecting the weight of the pulley, which is always insignificant in comparison to Q, we have

$$P = Q \left[1 + f' \cdot \frac{\rho}{R} \sqrt{2 (1 + \cos \varphi)}\right] + d_{,} \cdot \frac{K + I Q}{2 R}; \quad \cdots (710)$$

but

$$1 + \cos \varphi = 2 \cos^2 \tfrac{1}{2} \varphi ;$$

whence,

$$P = Q \left(1 + 2 f' \frac{\rho}{R} \cdot \cos \tfrac{1}{2} \varphi\right) + d_{,} \cdot \frac{K + I Q}{2 R} \quad \ldots \quad (711)$$

In which φ denotes the angle $A\ M\ B$, which is the supplement of the angle $A\ C\ B$, and denoting this latter angle by θ, we have

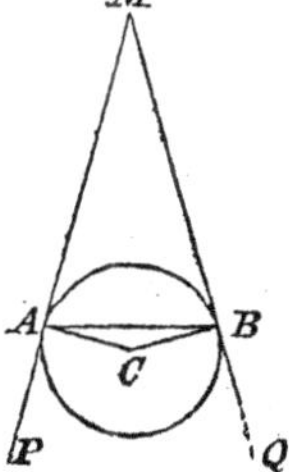

$$\cos \tfrac{1}{2} \varphi = \sin \tfrac{1}{2} \theta ,$$

whence

$$P = Q \left(1 + 2 f' \frac{\rho}{R} \sin \tfrac{1}{2} \theta\right) + d_{,} \frac{K + I Q}{2 R} \quad . \quad . \quad (712)$$

If the arc of the pulley, enveloped by the rope, be 180°, then will

$$P = Q \left(1 + 2 f' \cdot \frac{\rho}{R}\right) + d_{,} \cdot \frac{K + I Q}{2 R} \quad . \quad . \quad . \quad . \quad (713)$$

If the friction and stiffness of cordage be so small as to justify their omission, then will

$$P = Q.$$

That is, the power must be equal to the resistance, and the only office of the cord or rope is to change the direction of the power.

MOVABLE PULLEY.

§ 382.—In the fixed pulley, the resultant action of the power and resistance is thrown upon the trunnion boxes. If one end of the rope be attached to a fixed hook A, while the power P is applied to the other, and the pulley is left free to roll along the rope, the resistance W to be overcome may be connected with its trunnion, after the manner of the figure; the pulley is then said to be *movable*, and the relation between the power and resistance is still given by Eq. (704,) in which the principal resistance becomes $N + N'$, and the tension of the rope between the fixed point A, and the tangential point H, becomes Q.

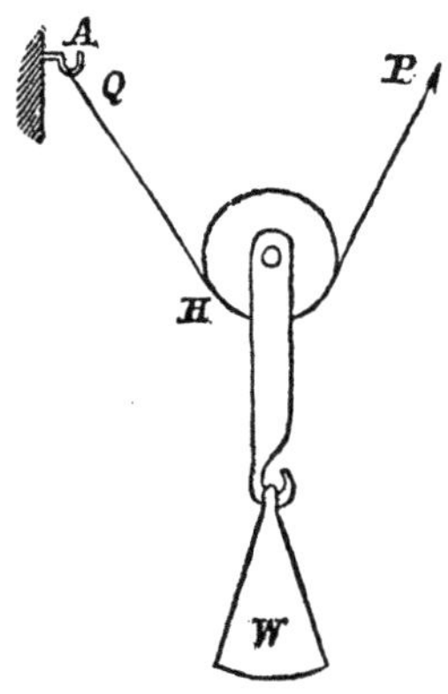

Making in Equation (704), $R = r$, $\rho = \rho'$, and $W = N + N' = 2N$, we have

$$P R - Q R - f' \rho W - d_{,} \cdot \frac{K + I Q}{2 R} \cdot R = 0 \quad . \quad . \quad . \quad . \quad (714)$$

dividing by R, and transposing

$$P = Q + f' \frac{\rho}{R} \cdot W + d_{,} \cdot \frac{K + I Q}{2 R} \quad . \quad . \quad . \quad . \quad . \quad . \quad . \quad . \quad (715)$$

Eliminating Q by means of Equation (708), and solving the resulting equation with respect to P, the value of the power will be known in terms of the resistances. The process may be much abridged by limiting the solution to an approximation, which will be found sufficient in practice.

Neglecting the weight of the pulley, which is always insignificant in comparison with P or Q, and making $Q = P$, which would be the case if we neglect friction and stiffness of cordage, Equation (708), gives

$$N = \tfrac{1}{2} W = \tfrac{1}{2} Q \sqrt{2(1 + \cos \varphi)};$$

and because

$$1 + \cos \varphi = 2 \cos^2 \tfrac{1}{2} \varphi = 2 \sin^2 \tfrac{1}{2} \theta,$$

$$W = 2\, Q \,.\, \sin \tfrac{1}{2} \theta;$$

or,

$$Q = \frac{W}{2 \sin \frac{1}{2} \theta};$$

which, in Equation (715), gives

$$P = W \left(\frac{1}{2 \sin \frac{1}{2} \theta} + f' \cdot \frac{\rho}{R} \right) + d_{,} \cdot \frac{K + I \cdot \dfrac{W}{2 \sin \frac{1}{2} \theta}}{2R}. \qquad (716)$$

The quantity of work is found by multiplying both members by $R s_1$, in which s_1 is the arc described at the unit's distance.

If the arc enveloped by the rope be 180°, then will $\frac{1}{2} \theta = 90°$, $\sin \frac{1}{2} \theta = 1$, and

$$P = W \left(\tfrac{1}{2} + f' \cdot \frac{\rho}{R} \right) + d_{,} \cdot \frac{K + \frac{1}{2} I \,.\, W}{2R}. \quad . \quad . \quad (717)$$

If the friction and stiffness of cordage be neglected, then will, Equation (716),

$$W = 2\, P \sin \tfrac{1}{2} \theta,$$

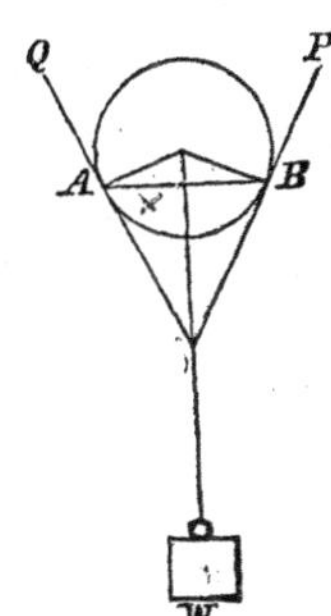

and multiplying by R,

$$R\, W = P \,.\, 2 R \,.\, \sin \tfrac{1}{2} \theta;$$

but

$$2 R \sin \tfrac{1}{2} \theta = A B;$$

whence,

$$R \,.\, W = P \,.\, A B;$$

that is, *the power is to the resistance as the radius of the pulley is to the chord of the arc enveloped by the rope.*

§ 383.—The *Muffle* is a collection of pulleys in two separate blocks or frames. One of these blocks is attached to a fixed point A, by which all of its pulleys become *fixed*, while the other block is attached to the resistance W, and its pulleys thereby made *movable*. A rope is attached at one end to a hook h at the extremity of the fixed block, and is passed around one of the movable pulleys, then about one of the fixed pulleys, and so on, in order, till the rope is made to act upon each pulley of the combination. The power P is applied to the other end of the rope, and the pulleys are so proportioned that the parts of the rope between them, when stretched, are parallel. Now, suppose the power P to maintain in uniform motion the point of application of the resistance W; denote the tension of the rope between the hook of the fixed block and the point where it comes in contact with the first movable pulley by t_1; the radius of this pulley by R_1; that of its eye by r_1; the co-efficient of friction on the axle by f; the constant and co-efficient of the stiffness of cordage by K and I, as before; then, denoting the tension of the rope between the last point of contact with the first movable, and first point of contact with the first fixed pulley, by t_2, the quantity of work of the tension t_2 will, Equation (652), be

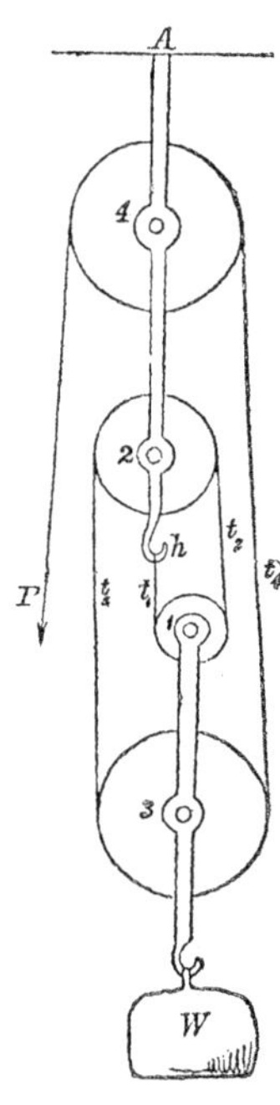

$$t_2 R_1 s_1 = t_1 R_1 s_1 + d_1 \frac{K + I t_1}{2 R_1} R_1 s_1 + f' (t_1 + t_2) r_1 s_1 ;$$

in which

$$f' = \frac{f}{\sqrt{1 + f^2}} ;$$

dividing by s_1,

$$t_2 R_1 = t_1 R_1 + d_1 \cdot \frac{K + I t_1}{2 R_1} \cdot R_1 + f' (t_1 + t_2) r_1 . \quad (718)$$

Again, denoting the tension of that part of the rope which passes from the first fixed to the second movable pulley by t_3, the radius of the first fixed pulley by R_2, and that of its eye by r_2, we shall, in like manner, have

$$t_3 R_2 = t_2 R_2 + d_, \frac{K + I t_2}{2 R_2} R_2 + f' (t_2 + t_3) r_2. \quad . \quad (719)$$

And denoting the tensions, in order, by t_4 and t_5, this last being equal to P, we shall have

$$t_4 R_3 = t_3 R_3 + d_, \frac{K + I t_3}{2 R_3} \cdot R_3 + f' (t_3 + t_4) r_3. \quad . \quad (720)$$

$$P R_4 = t_4 R_4 + d_, \frac{K + I t_4}{2 R_4} R_4 + f' (t_4 + P) r_4. \quad . \quad (721)$$

so that we finally arrive at the power P, through the tensions which are as yet unknown. The parts of the rope being parallel, and the resistance W being supported by their tensions, the latter may obviously be regarded as equal in intensity to the components of W; hence,

$$t_1 + t_2 + t_3 + t_4 = W; \quad . \quad . \quad . \quad . \quad . \quad (722)$$

which, with the preceding, gives us five equations for the determination of the four tensions and power P. This would involve a tedious process of elimination, which may be avoided by contenting ourselves with an approximation which is found, in practice, to be sufficiently accurate.

If the friction and stiffness be supposed zero, for the moment, Equations (718) to (721) become

$$t_2 R_1 = t_1 R_1,$$

$$t_3 R_2 = t_2 R_2,$$

$$t_4 R_3 = t_3 R_3,$$

$$P R_4 = t_4 R_4;$$

from which it is apparent, dividing out the radii R_1, R_2, R_3, &c.,

that $t_2 = t_1$, $t_3 = t_2$, $t_4 = t_3$, $P = t_4$; and hence, Equation (722) becomes

$$4\,t_1 = W;$$

whence,

$$t_1 = \frac{W}{4};$$

the denominator 4 being the whole number of pulleys, movable and fixed. Had there been n pulleys, then would

$$t_1 = \frac{W}{n}.$$

With this approximate value of t_1, we resort to Equations (718) to (721), and find the values of t_2, t_3, t_4, &c. Adding all these tensions together, we shall find their sum to be greater than W, and hence we infer each of them to be too large. If we now suppose the true tensions to be proportional to those just found, and whose sum is $W_1 > W$, we may find the true tension corresponding to any erroneous tension, as t_1, by the following proportion, viz.:

$$W_1 : W :: t_1 : \frac{W}{W_1}\,t_1;$$

or, which is the same thing, multiply each of the tensions found by the constant ratio $\frac{W}{W_1}$, the product will be the true tensions, very nearly. The value of t_4 thus found, substituted in Equation (721), will give that of P.

Example.—Let the radii R_1, R_2, R_3 and R_4, be respectively 0,26, 0,39, 0,52, 0,65 feet; the radii $r_1 = r_2 = r_3 = r_4$ of the eyes = 0,06 feet; the diameter of the rope, which is white and dry, 0,79 inches, of which the constant and co-efficient of rigidity are, respectively, $K = 1{,}6097$ and $I = 0{,}0319501$; and suppose the pulley of brass, and its axle of wrought iron, of which the co-efficient $f = 0{,}09$, and the resistance W a weight of 2400 pounds.

Without friction and stiffness of cordage,

$$t_1 = \frac{2400}{4} = 600.^{lbs.}$$

Dividing Equation (718) by R_1, it becomes, since $d_{,} = 1$,

$$t_2 = t_1 + \frac{K + I t_1}{2 R_1} + \frac{r_1}{R_1} f' (t_1 + t_2).$$

Substituting the value of R_1, and the above value of t_1, and regarding in the last term t_2 as equal to t_1, which we may do, because of the small co-efficient $\frac{r_1}{R_1} f'$, we find

$$t_2 = \left\{ \begin{array}{l} 600 \\ + \dfrac{1,6097 + 0,0319501 \times 600}{2 \times (0,26)} \\ + \dfrac{0,06}{0,26} \times 0,09 \times (600 + 600) \end{array} \right\} = 628,39.$$

Again, dividing Equation (719) by R_2, and substituting this value of t_2 and that of R_2, we find

$$t_3 = \overset{lbs.}{673,59}.$$

Dividing Equation (720) by R_3, and substituting this value of t_3, as well as that of R_3, there will result

$$t_4 = \overset{lbs.}{709,82};$$

whence,

$$W_1 = t_1 + t_2 + t_3 + t_4 = \left\{ \begin{array}{l} 600 \\ + 628,39 \\ + 673,59 \\ + 709,82 \end{array} \right\} = 2611\text{ lbs}$$

and

$$\frac{W}{W_1} = \frac{2400}{2611,80} = 0,919;$$

which will give for the true values of

$$\begin{aligned} t_1 &= 0,919 \times 600 &&= 551,400 \\ t_2 &= 0,919 \times 628,39 &&= 577,490 \\ t_3 &= 0,919 \times 673,59 &&= 619,029 \\ t_4 &= 0,919 \times 709,82 &&= 652,324 \\ & && \overline{2400,243} \end{aligned}$$

The above value for $t_4 = 652{,}324$, in Equation (721), will give, after dividing by R_4, and substituting its numerical value,

$$P = \begin{cases} 652{,}324 \\ + \dfrac{1{,}6097 + 0{,}03195 \times 652{,}324}{2 \times 0{,}65} \\ + \dfrac{0{,}06}{0{,}65} \times 0{,}09 \times (652{,}324 + P)\,; \end{cases}$$

and making in the last factor $P = t_4 = 652{,}324$, we find

$$P = \overset{lbs.}{652{,}324} + \overset{lbs.}{17{,}270} + \overset{lbs.}{10{,}831} = \overset{lbs.}{680{,}425}.$$

Thus, without friction or stiffness of cordage, the intensity of P would be 600 lbs.; with both of these causes of resistance, which cannot be avoided in practice, it becomes 680,425 lbs., making a difference of 80,425 lbs., or nearly one-seventh; and as the quantity of work of the power is proportional to its intensity, we see that to overcome friction and stiffness of rope, in the example before us, the motor must expend nearly a seventh more work than if these sources of resistance did not exist.

THE WEDGE.

§ 384.—The wedge is usually employed in the operation of cutting, splitting, or separating. It consists of an acute right triangular prism $A\,B\,C$. The acute dihedral angle $A\,C\,b$ is called the *edge*; the opposite plane face $A\,b$ the *back*; and the planes $A\,c$ and $C\,b$, which terminate in the edge, the *faces*. The more common application of the wedge consists in driving it, by a blow upon its back, into any substance which we wish to split or divide into parts, in such manner that after each advance it shall be supported against the faces of the opening till the work is accomplished.

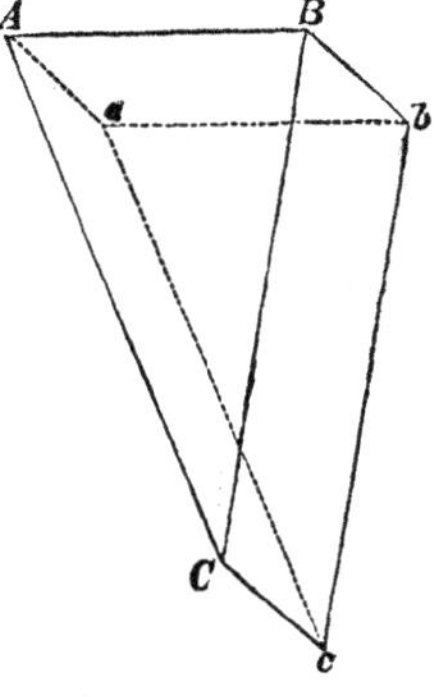

§ 385.—The blow by which the wedge is driven forward will be supposed perpendicular to its back, for if it were oblique, it would only tend to impart a rotary motion, and give rise to complications which it would be unprofitable to consider: and to make the case conform still further to practice, we will suppose the wedge to be isosceles.

The wedge $A\,C\,B$ being inserted in the opening $a\,h\,b$, and in contact with its jaws at a and b, we know that the resistance of the latter will be perpendicular to the faces of the wedge. Through the points a and b draw the lines $a\,q$ and $b\,p$ normal to the faces $A\,C$ and $B\,C$; from their point of intersection O lay off the distances $O\,q$ and $O\,p$ equal, respectively, to the resistances at a and b. Denote the first by Q, and the second by P. Completing the parallelogram $O\,q\,m\,p$, $O\,m$ will represent the resultant of the resistances Q and P. Denote this resultant by R', and the angle $A\,C\,B$ of the wedge by θ, which, in the quadrilateral $a\,O\,b\,C$, will be equal to the supplement of the angle $a\,O\,b = p\,O\,q$, the angle made by the directions of Q and P. From the parallelogram of forces, we have,

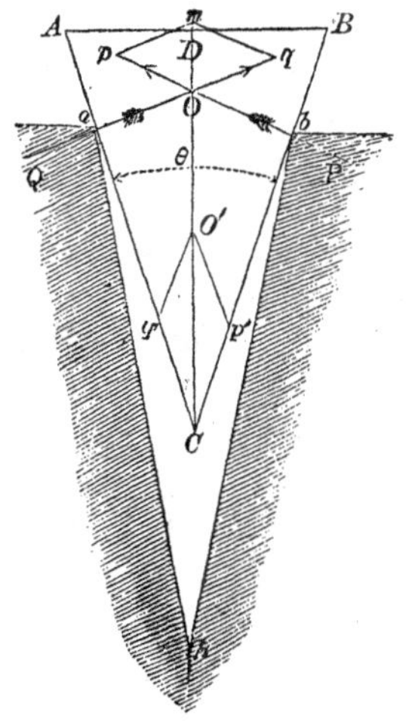

$$R'^2 = P^2 + Q^2 + 2\,P\,Q \cos p\,O\,q = P^2 + Q^2 - 2\,P\,Q \cos \theta;$$

or,

$$R' = \sqrt{P^2 + Q^2 - 2\,P\,Q \cos \theta}.$$

The resistance Q will produce a friction on the face $A\,C$ equal to $f\,Q$, and the resistance P will produce on the face $B\,C$ the friction $f\,P$: these act in the directions of the faces of the wedge. Produce them till they meet in C, and lay off the distances $C\,q'$ and $C\,p'$ to represent their intensities, and complete the parallelogram

$C q' O' p'$; $C O'$ will represent the resultant of the frictions. Denote this by R'', and we have, from the parallelogram of forces,

$$R''^2 = f^2 Q^2 + f^2 P^2 + 2 f^2 P Q \cos \theta;$$

or,

$$R'' = f \sqrt{P^2 + Q^2 + 2 P Q \cos \theta}.$$

The wedge being isosceles, the resistances P and Q will be equal, their directions being normal to the faces will intersect on the line $C D$, which bisects the angle $C = \theta$, and their resultant will coincide with this line. In like manner the frictions will be equal, and their resultant will coincide with the same line. Making Q and P equal, we have, from the above equations,

$$R' = P \sqrt{2 (1 - \cos \theta)},$$

$$R'' = f P \sqrt{2 (1 + \cos \theta)}.$$

But,

$$1 - \cos \theta = 2 \sin^2 \tfrac{1}{2} \theta,$$

$$1 + \cos \theta = 2 \cos^2 \tfrac{1}{2} \theta;$$

whence we obtain, by substituting and reducing,

$$R' = 2 P . \sin \tfrac{1}{2} \theta,$$

$$R'' = 2 f . P . \cos \tfrac{1}{2} \theta;$$

and further,

$$\sin \tfrac{1}{2} \theta = \tfrac{1}{2} \frac{A B}{A C},$$

$$\cos \tfrac{1}{2} \theta = \frac{C D}{A C};$$

therefore,

$$R' = P \cdot \frac{A B}{A C},$$

$$R'' = 2 f \cdot P \cdot \frac{C D}{A C}.$$

Denote by F the intensity of the blow on the back of the wedge. If this blow be just sufficient to produce an equilibrium bordering on motion forward, call it F'; the friction will oppose it, and we must have,

$$F' = R' + R'' = P \cdot \frac{AB}{AC} + 2f \cdot P \cdot \frac{CD}{AC} \cdot \quad . \quad . \quad . \quad (723)$$

If, on the contrary, the blow be just sufficient to prevent the wedge from flying back, call it F''; the friction will aid it, and we must have,

$$F'' = P \cdot \frac{AB}{AC} - 2f \cdot P \cdot \frac{CD}{AC} \quad . \quad . \quad . \quad . \quad (724)$$

The wedge will not move under the action of any force whose intensity is between F' and F''. Any force less than F'', will allow it to fly back; any force greater than F', will drive it forward. The range through which the force may vary without producing motion, is obviously,

$$F' - F'' = 4fP \cdot \frac{CD}{AC} \quad . \quad . \quad . \quad . \quad . \quad (725)$$

which becomes greater and greater, in proportion as CD and AC become more nearly equal; that is to say, in proportion as the wedges becomes more and more acute.

The ordinary mode of employing the wedge requires that it shall retain of itself whatever position it may be driven to. This makes it necessary that F'' should be zero or negative, Eq. (724), whence

$$P \cdot \frac{AB}{AC} = 2f \cdot P \cdot \frac{CD}{AC}, \quad \text{or} \quad P\frac{AB}{AC} < 2f \cdot P \cdot \frac{CD}{AC};$$

or, omitting the common factors and dividing both members of the equation and inequality by $2\,CD$,

$$\frac{\frac{1}{2}AB}{CD} = f, \quad \text{or} \quad \frac{\frac{1}{2}AB}{CD} < f;$$

but $\frac{\frac{1}{2}AB}{CD}$ is the tangent of the angle ACD; hence we conclude, that the wedge will retain its place when its semi-angle does not exceed that whose tangent is the co-efficient of friction between the surface of the wedge and the surface of the opening which it is intended to enlarge.

Resuming Eq. (724), and supposing the last term of the second member greater than the first term, F'' becomes negative, and will represent the intensity of the force necessary to withdraw the wedge; which will obviously be the greatest possible when $A\ B$ is the least possible. This explains why it is that nails retain with such pertinacity their places when driven into wood, &c.

THE SCREW.

§ 386.—The *Screw*, regarded as a mechanical power, is a device by which the principles of the inclined plane are so applied as to produce considerable pressures with great steadiness and regularity of motion.

To form an idea of the figure of a screw and its mode of action, conceive a right cylinder, $a\ k$, with circular base, and a rectangle, or other plane figure, $a\,b\,c\,m$, having one of its sides $a\,b$ coincident with a surface element, while its plane passes through the axis of this cylinder. Next, suppose the plane of the generatrix to rotate uniformly about the axis, and the generatrix itself to move also uniformly in the direction of that line; and let this twofold motion of rotation and of translation be so regulated, that in one entire revolution of the plane, the generatrix shall progress in the direction of the axis over a distance greater than the side $a\,b$, which is in the surface of the cylinder. The generatrix will thus generate a projecting and winding solid called a *fillet*, leaving between its turns a groove called the *channel*. Each point as m in the perimeter of the generatrix, will generate a curve called a *helix*, and it is obvious, from what has been said, that every helix will enjoy this property, viz.: any one of its points as m, being taken as an origin of reference, as well for the curve itself as for its projection on a plane through this point and at right angles to the axis, the distances $d'\,m'$, $d''\,m''$, &c., of the several points of the helix from this plane,

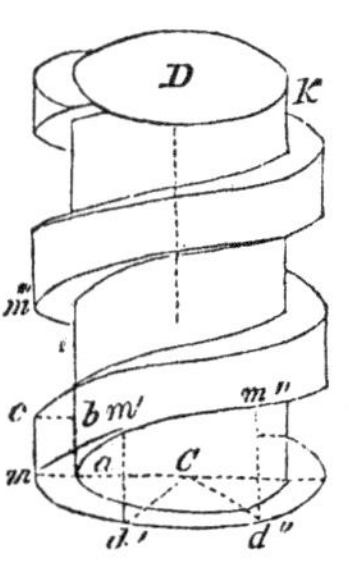

are respectively proportioned to the circular arcs $m d'$, $m d''$, &c., into which the portions $m m'$, $m m''$, &c., of the helix, between the origin and these points, are projected.

The solid cylinder about which the fillet is wound, is called the *newel* of the screw; the distance $m m'''$, between the consecutive turns of the same helix, estimated in the direction of the axis, is called the *helical interval.*

The fillet is often generated by the motion of a triangle with one of its sides coincident with $a b$; and as the discussion will be more general by considering this mode of generation, we shall adopt it. The surfaces of the fillet, which are generated by the inclined faces of the triangle, are each made up of an infinite number of helices, all of which have the same interval, though the helices themselves are at different distances from the axis, and have different inclinations to that line.

The inclination of the different helices to the axis of the screw, increases from the newel to the exterior surface of the fillet, the same helix preserving its inclination unchanged throughout. The screw is received into a hole in a solid piece B of metal or wood, called a *nut* or *burr.* The surface of the hole through the nut is furnished with a winding fillet of the same shape and size as the channel of the screw, so that the surfaces of the screw and nut are brought into accurate contact.

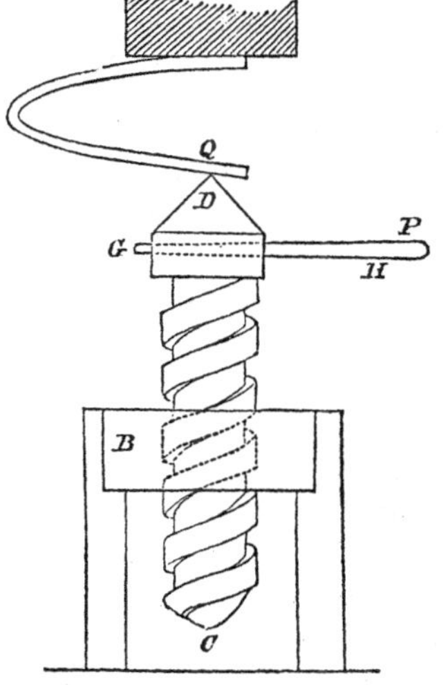

From this arrangement it is obvious that when the nut is stationary, and a rotary motion is communicated to the screw, the latter will move in the direction of its axis; also, when the screw is stationary and the nut is turned, the nut must also move in the direction of the axis. In the first case, one entire revolution of the screw will carry it longitudinally through a distance equal to the helical interval, and any fractional portion of an entire revolution will carry it through a proportional distance; the same of the nut, when the latter is movable and the screw stationary. The resistance Q is applied either to the head of the screw, or to the nut, depending upon which is the movable element; in either case it acts in the direction $D\,C$ of the axis. The power P is applied at the extremity of a bar $G\,H$ connected with the screw or nut, and acts in a plane at right angles to the axis of the screw.

From the description of the screw and its mode of generation, we may find the equation of its fillet or helicoidal surface. For this purpose, take the axis z to coincide with the axis of the newel, and the initial position of the generatrix in the plane $y\,z$. Make

$s =$ any definite portion $C\,C'$ of an assumed helix;

$\varphi =$ the angle $Y\,A\,t$, through which the rotating plane has turned during the generation of s;

$r =$ the distance $C\,D$ of this helix from the axis z;

$\alpha =$ the angle which this helix makes with the plane $x\,y$;

$\epsilon =$ the angle $C\,B\,D$ which the generatrix of the helicoidal surface makes with the axis z;

$\gamma =$ the co-ordinate $A\,B$ of the point in which the generatrix, in its initial position, intersects the axis z.

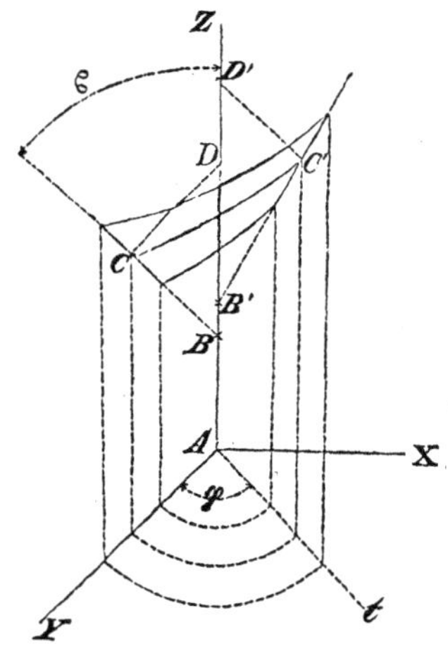

Then, for any point as C of the generatrix in its initial position, we have

$$z = A\,D = A\,B + B\,D = \gamma + r \,.\, \text{cotan}\, \epsilon,$$

and for any subsequent position, as $C'\,B'$,

$$z = \gamma + r \,.\, \text{cotan}\, \epsilon + r \,.\, \varphi \,.\, \tan \alpha, \quad \cdot \quad \cdot \quad \cdot \quad \cdot \qquad (726)$$

which is the equation sought, and in which α and r are constant for the same helix, and variable from one helix to another

The power P acts in a direction perpendicular to the axis of the newel. Denote by l its lever arm; its virtual moment will be

$$P l d \varphi.$$

The resistance Q acts in the direction of the axis of the newel; its virtual moment will be

$$Q d z.$$

The friction acts in the direction of the helicoidal surface and parallel to the helices. Conceive it to be concentrated upon a mean helix, of which the distance from the newel axis is r, and length s: denote the normal pressure by N, and co-efficient of friction by f. The virtual moment of friction will be

$$f \,.\, N \,.\, d s;$$

and Equation (645),

$$P l d \varphi - Q d z - f \,.\, N \,.\, d s = 0. \quad . \quad . \quad . \quad . \quad (727)$$

But the displacement must satisfy Equation (726), or, as in § 213, the condition,

$$L = z - r \,.\, \varphi \,.\, \tan \alpha - r \,.\, \text{cotan}\, \beta - \gamma = 0; \quad . \quad (728)$$

and also,

$$r = \text{constant}. \quad . \quad . \quad . \quad . \quad . \quad . \quad . \quad . \quad (729)$$

Differentiating, we have,

$$d z - \text{cotan}\, \beta \,.\, d r - r \tan \alpha \, d \varphi = 0,$$

$$d r = 0.$$

Multiplying the first by λ, the second by λ', adding to Equation (727), and eliminating $d s$ by the relation

$$d s = r \,.\, d \varphi \,.\, \cos \alpha + d z \,.\, \sin \alpha, \quad . \quad . \quad . \quad . \quad (730)$$

we find,

$$(P l - f.N.\cos \alpha . r - \lambda \tan \alpha . r) d \varphi + (\lambda - Q - f.N.\sin \alpha) \, d z + (\lambda' - \lambda \,\text{cotan}\, \beta) d r = 0$$

and, from the principle of indeterminate co-efficients,

$$Pl - f \,.\, N \,.\, \cos\alpha \,.\, r - \lambda \,.\, \tan\alpha \,.\, r = 0 ; \quad . \quad . \quad (731)$$

$$Q + f N \,.\, \sin\alpha - \lambda = 0 ; \quad . \quad . \quad . \quad . \quad . \quad (732)$$

$$\lambda' - \lambda \operatorname{cotan} \epsilon = 0. \quad . \quad . \quad . \quad . \quad . \quad . \quad . \quad (732)'$$

The variables dz, dr, and $rd\varphi$, are rectangular; whence, Equation (331),

$$N = \lambda \sqrt{\left(\frac{dL}{dz}\right)^2 + \left(\frac{dL}{dr}\right)^2 + \left(\frac{dL}{rd\varphi}\right)^2} = \lambda \sqrt{1 + \tan^2\alpha + \operatorname{cotan}^2 \epsilon}.$$

Substituting this in Equations (731) and (732), and eliminating λ, there will result

$$P = Q \cdot \frac{r}{l} \cdot \frac{\tan\alpha + f \,.\, \cos\alpha \,.\, \sqrt{1 + \tan^2\alpha + \operatorname{cotan}^2 \epsilon}}{1 - f \,.\, \sin\alpha \,.\, \sqrt{1 + \tan^2\alpha + \operatorname{cotan}^2 \epsilon}}. \quad (733)$$

Substituting the value of λ from Equation (732), in Equation (732)$'$, we find,

$$\lambda' = Q \cdot \frac{\operatorname{cotan} \epsilon}{1 - f \,.\, \sin\alpha \sqrt{1 + \tan^2\alpha + \operatorname{cotan}^2 \epsilon}} ; \quad . \quad (734)$$

in which λ' is, § 217, the value of the force acting in the direction of r.

§ 387.—If the fillet be rectangular, $\epsilon = 90°$, $\operatorname{cotan} \epsilon = 0$, and

$$P = Q \cdot \frac{r}{l} \cdot \frac{\tan\alpha + f \,.\, \cos\alpha \,.\, \sqrt{1 + \tan^2\alpha}}{1 - f \,.\, \sin\alpha \,.\, \sqrt{1 + \tan^2\alpha}} ; \quad . \quad (735)$$

and

$$\lambda' = 0.$$

§ 388.—If we neglect the friction, $f = 0$; and

$$Pl = Q \,.\, r \,.\, \tan\alpha,$$

multiplying both members by 2π,

$$P \,.\, 2\pi l = Q \,.\, 2\pi r \,.\, \tan\alpha. \quad . \quad . \quad . \quad . \quad . \quad (736)$$

That is, *the power is to the resistance as the helical interval is to the circumference described by the end of the lever arm of the power.*

PUMPS.

§ 389.—Any machine used for raising liquids from one level to a higher, in which the agency of atmospheric pressure is employed, is called a *Pump*. There are various kinds of pumps; the more common are the *sucking*, *forcing*, and *lifting* pumps.

§ 390.—The *Sucking-Pump* consists of a cylindrical body or barrel B, from the lower end of which a tube D, called the sucking-pipe, descends into the water contained in a reservoir or well. In the interior of the barrel is a movable piston C, surrounded with leather to make it water-tight, yet capable of moving up and down freely. The piston is perforated in the direction of the bore of the barrel, and the orifice is covered by a valve F called the *piston-valve*, which opens upward; a similar valve E, called the *sleeping-valve*, at the bottom of the barrel, covers the upper end of the sucking-pipe. Above the highest point ever occupied by the piston, a discharge-pipe P is inserted into the barrel; the piston is worked by means of a lever H, or other contrivance, attached to the piston-rod

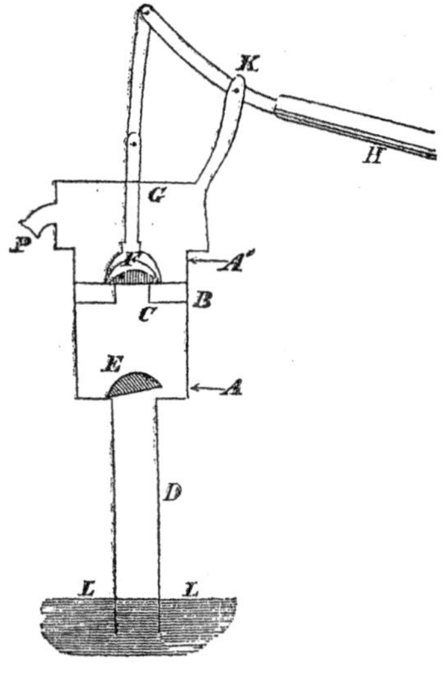

G. The distance $A\,A'$, between the highest and lowest points of the piston, is called the *play*. To explain the action of this pump, let the piston be at its lowest point A, the valves E and F closed by their own weight, and the air within the pump of the same density and elastic force as that on the exterior. The water of the reservoir will stand at the same level $L\,L$ both within and without the sucking-pipe. Now suppose the piston raised to its highest point A', the air contained in the barrel and sucking-pipe will tend by its elastic force to occupy the space which the piston leaves void, the valve E will, therefore, be forced open, and air will pass from the pipe to the barrel, its elasticity diminishing in proportion as it fills a larger space. It will, therefore, exert a less pressure on the water below it in the sucking-pipe than the exterior air does on that in the reservoir, and the excess of pressure on the part of the exterior air, will force the water up the pipe till the weight of the suspended column, increased by the elastic force of the internal air, becomes equal to the pressure of the exterior air. When this takes place, the valve E will close of its own weight; and if the piston be depressed, the air contained between it and this valve, having its density augmented as the piston is lowered, will at length have its elasticity greater than that of the exterior air; this excess of elasticity will force open the valve F, and air enough will escape to reduce what is left to the same density as that of the exterior air. The valve F will then fall of its own weight; and if the piston be again elevated, the water will rise still higher, for the same reason as before. This operation of raising and depressing the piston being repeated a few times, the water will at length enter the barrel, through the valve F, and be delivered from the discharge-pipe P. The valves E and F, closing after the water has passed them, the latter is prevented from returning, and a cylinder of water equal to that through which the piston is raised, will, at each upward motion, be forced out, provided the discharge-pipe is large enough. As the ascent of the water to the piston is produced by the difference of pressure of the internal and external air, it is plain that the lowest point to which the piston may reach, should never have a greater altitude above the water in the reservoir than that of the column of this fluid which the atmospheric pressure may support, in vacuo, at the place.

§ 391.—It will readily appear that the rise of water, during each ascent of the piston after the first, depends upon the expulsion of air through the piston-valve in its previous descent. But air can only issue through this valve when the air below it has a greater density and therefore greater elasticity than the external air; and if the piston may not descend low enough, for want of sufficient play, to produce this degree of compression, the water must cease to rise, and the working of the piston can have no other effect than alternately to compress and dilate the same air between it and the surface of the water. To ascertain, therefore, the relation which the play of the piston should bear to the other dimensions, in order to make the pump effective, suppose the water to have reached a stationary level X, at some one ascent of the piston to its highest point A', and that, in its subsequent descent, the piston-valve will not open, but the air below it will be compressed only to the same density with the external air when the piston reaches its lowest point A. The piston may be worked up and down indefinitely, within these limits for the play, without moving the water. Denote the play of the piston by a; the greatest height to which the piston may be raised above the level of the water in the reservoir, by b, which may also be regarded as the altitude of the discharge pipe; the elevation of the point X, at which the water stops, above the water in the reservoir, by x; the cross-section of the interior of the barrel by B. The volume of the air between the level X and A will be

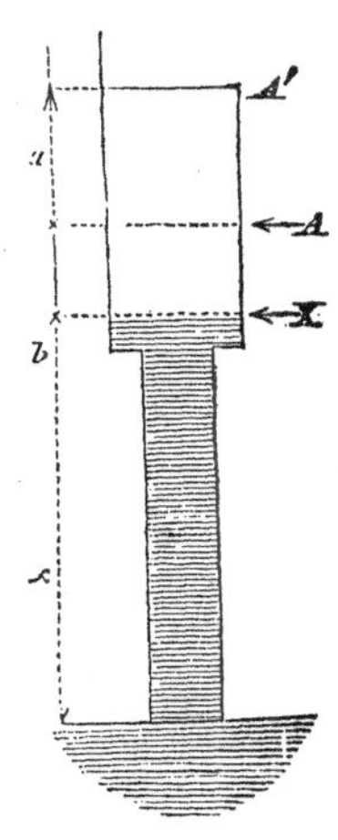

$$B \times (b - x - a);$$

the volume of this same air, when the piston is raised to A', provided the water does not move, will be

$$B(b - x).$$

Represent by h the greatest height to which water may be supported in vacuo at the place. The weight of the column of water which the elastic force of the air, when occupying the space between the limits X and A, will support in a tube, with a bore equal to that of the barrel is measured by

$$Bh \,.\, g \,.\, D;$$

in which D is the density of the water, and g the force of gravity. The weight of the column which the elastic force of this same air will support, when expanded between the limits X and A', will be

$$Bh' . g . D;$$

in which h' denotes the height of this new column. But, from Mariotte's law, we have

$$B(b - x - a) : B(b - x) :: Bh'gD : BhgD;$$

whence,

$$h' = h \cdot \frac{b - x - a}{b - x}.$$

But there is an equilibrium between the pressure of the external air and that of the rarefied air between the limits X and A', when the latter is increased by the weight of the column of water whose altitude is x. Whence, omitting the common factors B, D and g,

$$x + h' = x + h \cdot \frac{b - x - a}{b - x} = h;$$

or, clearing the fraction and solving the equation in reference to x, we find

$$x = \tfrac{1}{2} b \pm \tfrac{1}{2} \sqrt{b^2 - 4ah}. \quad \cdot \quad \cdot \quad \cdot \quad \cdot \quad \cdot \quad (737)$$

When x has a real value, the water will cease to rise, but x will be real as long as b^2 is greater than $4ah$. If, on the contrary, $4ah$ is greater than b^2, the value of x will be imaginary, and the water cannot cease to rise, and the pump will always be effective when its dimensions satisfy this condition, viz.:—

$$4ah > b^2,$$

or,

$$a > \frac{b^2}{4h};$$

that is to say, *the play of the piston must be greater than the square of the altitude of the upper limit of the play of the piston above the surface of the water in the reservoir, divided by four times the height to which the atmospheric pressure at the place, where the pump is used, will support water in vacuo.* This last height is easily found by means of the barometer. We have but to notice the altitude of the barometer at the place, and multiply its column, reduced to feet, by $13\frac{1}{2}$, this being the specific gravity of mercury referred to water as a standard, and the product will give the value of h in feet.

Example.—Required the least play of the piston in a sucking-pump intended to raise water through a height of 13 feet, at a place where the barometer stands at 28 inches.

Here $b = 13$, and $b^2 = 169$.

Barometer, $$\frac{\overset{in.}{28}}{12} = 2,333 \text{ feet.}$$

$$h = \overset{ft.}{2,333} \times 13,5 = 31,5 \text{ feet.}$$

Play $$= a > \frac{b^2}{4h} = \frac{169}{4 \times 31,5} = \overset{ft.}{1,341} + ;$$

that is, the play of the piston must be greater than one and on third of a foot.

§ 392.—The quantity of work performed by the motor during the delivery of water through the discharge-pipe, is easily computed. Suppose the piston to have any position, as M, and to be moving upward, the water being at the level LL in the reservoir, and at P in the pump. The pressure upon the upper surface of the piston will be equal to the entire atmospheric pressure denoted by A, increased by the weight of the column of water MP', whose height is c', and whose base is the area B of the piston; that is, the pressure upon the top of the piston will be

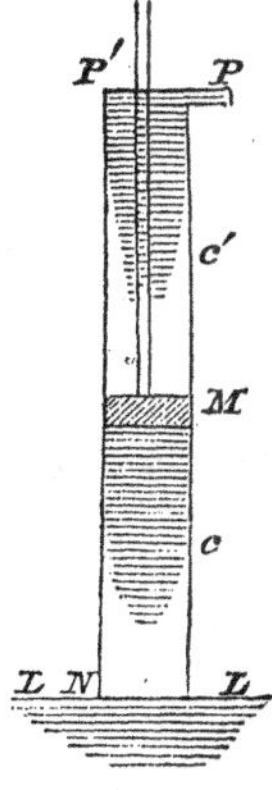

$$A + Bc'gD,$$

in which g and D are the force of gravity and density of the water, respectively. Again, the pressure upon the under surface of the piston is equal to the atmospheric pressure A, transmitted through the water in the reservoir and up the suspended column, diminished by the weight of the column of water NM below the piston, and of which the base is B and altitude c; that is, the pressure from below will be

$$A - B c g D,$$

and the difference of these pressures will be

$$A + B c' g D - (A - B c g D) = B g D (c + c');$$

but, employing the notation of the sucking-pump just described,

$$c + c' = b;$$

whence, the foregoing expression becomes

$$B b . g . D;$$

which is obviously the weight of a column of the fluid whose base is the area of the piston and altitude the height of the discharge-pipe above the level of the water in the reservoir. And adding to this the effort necessary to overcome the friction of the parts of the pump when in motion, denoted by φ, we shall have the resistance which the force F, applied to the piston-rod, must overcome to produce any useful effect; that is,

$$F = B b g D + \varphi.$$

Denote the play of the piston by p, and the number of its double strokes, from the beginning of the flow through the discharge-pipe till any quantity Q is delivered, by n; the quantity of work will, by omitting the effort necessary to depress the piston, be

$$F n p = n p \, [B b \, . \, g D + \varphi];$$

or estimating the volume in cubic feet, in which case p and b must be expressed in linear feet and B in square feet, and substituting for $g D$ its value 62,5 pounds, we finally have for the quantity of work necessary to deliver a number of cubic feet of water $Q = B n p$,

$$F n p = n p \, [62{,}5 \, . \, B b + \varphi]; \quad . \quad . \quad . \quad . \quad (738)$$

in which φ must be expressed in pounds, and may be determined either by experiment in each particular pump, or computed by the rules already given.

It is apparent that the action of the sucking-pump must be very irregular, and that it is only during the ascent of the piston that it produces any useful effect; during the descent of the piston, the force is scarcely exerted at all, not more than is necessary to overcome the friction.

§ 393.—The *Lifting-Pump* does not differ much from the sucking-pump just described, except that the barrel and sleeping-valve E are placed at the bottom of the pipe, and some distance below the surface of the water $L\,L$ in the reservoir; the piston may or may not be below this same surface when at the lowest point of its play. The piston and sleeping-valves open upward. Supposing the piston at its lowest point, it will, when raised, lift the column of water above it, and the pressure of the external air, together with the head of fluid in the reservoir above the level of the sleeping-valve, will force the latter open; the water will flow into the barrel and follow the piston. When the piston reaches the upper limit of its play, the sleeping-valve will close and prevent the return of the water above it. The piston being depressed, its valves F will open and the water will flow through them till the piston reaches its lowest point. The same operation being repeated a few times, a column of water will be lifted to the mouth of the discharge-pipe P, after which every elevation of the piston will deliver a volume of the fluid equal to that of a cylinder whose base is the area of the piston and whose altitude is equal to its play.

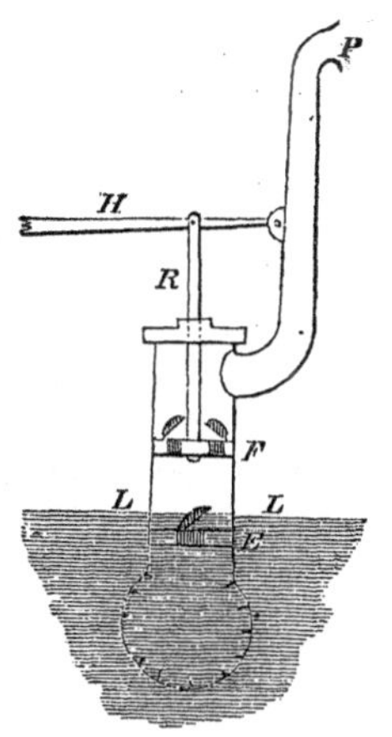

As the water on the same level within and without the pump will be in equilibrio, it is plain that the resistance to be overcome by the power will be the friction of the rubbing surfaces of the pump, augmented by the weight of a column of fluid whose base is the area of the piston, and altitude the difference of level between the surface of the water in the reservoir and the discharge-pipe. Hence the quantity of work is estimated by the same rule, Equation (738). If we omit for a moment the consideration of friction, and take but a single elevation of the piston after the water has reached the discharge-pipe, n will equal one, φ will be zero, and that equation reduces to

$$Fp = 62{,}5\ Bp \times b;$$

but $62{,}5 \times Bp$ is the quantity of fluid discharged at each double stroke of the piston, and b being the elevation of the discharge-pipe above the water in the reservoir, we see that the work will be the same as though that amount of fluid had actually been lifted through this vertical height, which, indeed, is the useful effect of the pump for every double stroke.

§ 394.—The *Forcing-Pump* is a further modification of the simple sucking-pump. The barrel B and sleeping-valve E are placed upon the top of the sucking-pipe M. The piston F is without perforation and valve, and the water, after being forced into the barrel by the atmospheric pressure without, as in the sucking-pump, is driven by the depression of the piston through a lateral pipe H into an air-vessel N, at the bottom of which is a second sleeping-valve E', opening, like the first, upward. Through the top of the air-vessel a discharge-pipe K passes, air-tight, nearly to the

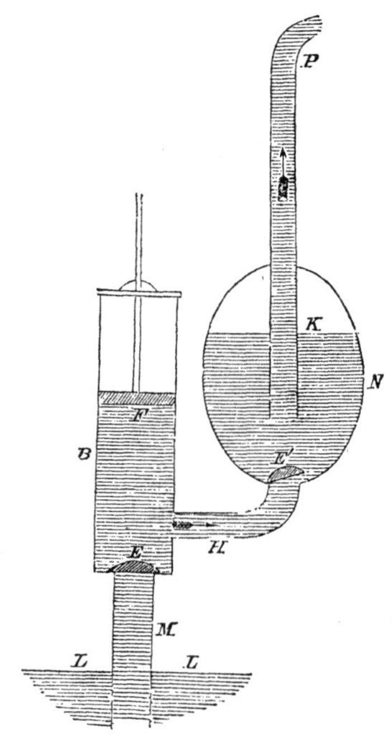

bottom. The water, when forced into the air-vessel by the descent of the piston, rises above the lower end of this pipe, confines and compresses the air, which, reacting by its elasticity, forces the water up the pipe, while the valve E' is closed by its own weight and the pressure from above, as soon as the piston reaches the lower limit of its play. A few strokes of the piston will, in general, be sufficient to raise water in the pipe K to any desired height, the only limit being that determined by the power at command and the strength of the pump.

§ 395.—During the ascent of the piston, the valve E' is closed and E is open; the pressure upon the upper surface of the piston is that exerted by the entire atmosphere; the pressure upon the lower surface is that of the entire atmosphere transmitted from the surface of the reservoir through the fluid up the pump, diminished by the weight of the column of water whose base is the area of the piston and altitude the height of the piston above the surface of the water in the reservoir; hence, the resistance to be overcome by the power will be the difference of these pressures, which is obviously the weight of this column of water. Denote the area of the piston by B, its height above the water of the reservoir at one instant by y, and the weight of a unit of volume of the fluid by w, then will the resistance to be overcome at this point of the ascent be

$$w \,.\, B \,.\, y;$$

and the elementary quantity of work will be

$$w \,.\, B \,.\, y\,dy;$$

and the whole work during the ascent will be

$$w \cdot B \int_{y_{,}}^{y'} y\,dy = w \cdot B \cdot \frac{y' + y_{,}}{2}\,(y' - y_{,});$$

in which y' and $y_{,}$ are the distances of the upper and lower limits of the play of the piston from the water in the reservoir.

But $B\,.\,(y' - y_{,})$ is the volume of the barrel within the limits of the play of the piston, and $\frac{1}{2}\,(y' + y_{,})$ is the height of its centre of gravity above the level of the fluid in the reservoir.

Denoting the play by p, and making $\frac{y' + y_{\prime}}{2} = z'$, we have for the quantity of work during the ascent,

$$w \,.\, B \,.\, p \,.\, z'.$$

During the descent of the piston, the valve E is closed, and E' open, and as the columns of the fluid in the barrel and discharge-pipe, below the horizontal plane of the lower surface of the piston, will maintain each other in equilibrio, the resistance to be overcome by the power will be the weight of a column of fluid whose base is the area of the piston and altitude the difference of level between the piston and point of delivery P; and denoting by $z_{\prime}$ the distance of the central point of the play below the point P, we shall find, by exactly the same process,

$$w\,B\,p\,z_{\prime}\,,$$

for the quantity of work of the motor during the descent of the piston; and hence the quantity of work during an entire double stroke will be the sum of these, or

$$w\,B\,p\,(z' + z_{\prime}).$$

But $z' + z_{\prime}$ is the height of the point of delivery P above the surface of the water in the reservoir; denoting this, as before, by b, we have

$$w\,B\,p\,b\,;$$

and calling the number of double strokes n, and the whole quantity of work Q, we finally have

$$Q = n\,w\,B\,p\,b. \quad \cdots\cdots\cdots \quad (739)$$

If we make $z_{\prime} = z'$, or $b = 2z_{\prime}$, which will give $z_{\prime} = \frac{b}{2}$, the quantity of work during the ascent will be equal to that during the descent, and thus, in the forcing-pump, the work may be equalized and the motion made in some degree regular. In the lifting and sucking-pumps the motor has, during the ascent of the piston, to overcome the weight of the entire column whose base is equal to the area of the piston and altitude the difference of level between the water in the reservoir and point of delivery, and being wholly relieved during the descent, when the load is thrown upon the sleeping-valve and its box, the work becomes variable, and the motion irregular.

THE SIPHON.

§ 396.—The *Siphon* is a bent tube of unequal branches, open at both ends, and is used to convey a liquid from a higher to a lower level, over an intermediate point higher than either. Its parallel branches being in a vertical plane and plunged into two liquids whose upper surfaces are at $L\ M$ and $L'\ M'$, the fluid will stand at the same level both within and without each branch of the tube when a vent or small opening is made at O. If the air be withdrawn from the siphon through this vent, the water will rise in the branches by the atmospheric pressure without, and when the two columns unite and the vent is closed, the liquid will flow from the reservoir A to A', as long as the level $L'\ M'$ is below $L\ M$, and the end of the shorter branch of the siphon is below the surface of the liquid in the reservoir A.

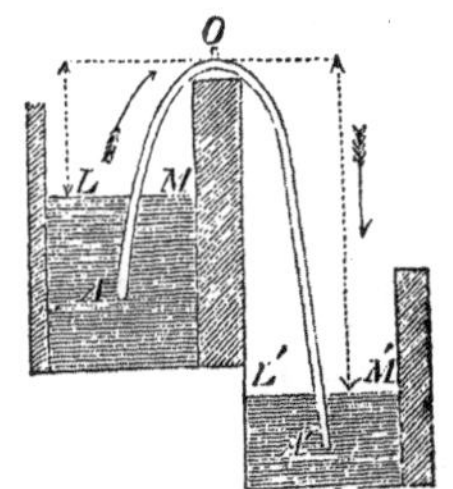

The atmospheric pressures upon the surfaces $L\ M$ and $L'\ M'$, tend to force the liquid up the two branches of the tube. When the siphon is filled with the liquid, each of these pressures is counteracted in part by the pressure of the fluid column in the branch of the siphon that dips into the fluid upon which the pressure is exerted. The atmospheric pressures are very nearly the same for a difference of level of several feet, by reason of the slight density of air. The pressures of the suspended columns of water will, for the same difference of level, differ considerably, in consequence of the greater density of the liquid. The atmospheric pressure opposed to the weight of the longer column will therefore be more counteracted than that opposed to the weight of the shorter, thus leaving an excess of pressure at the end of the shorter branch, which will produce the motion. Thus, denote by A the intensity of the atmospheric pressure upon a surface a equal to that of a cross-section of the tube; by h the difference of level between the surface $L\,M$ and the bend O; by h' the difference of level between the same point O and the level $L'\,M'$; by D the density of the liquid; and by g the force of gravity: then will the pressure, which tends to force the fluid up the branch which dips below $L\,M$, be

$$A - a\,h\,D\,g;$$

and that which tends to force the fluid up the branch immersed in the other reservoir, be

$$A - a\,h'\,D\,g;$$

and subtracting the first from the second, we find

$$a\,D\,g\,(h' - h),$$

for the intensity of the force which urges the fluid within the siphon, from the upper to the lower reservoir.

Denote by l the length of the siphon from one level to the other. This will be the distance over which the above force will be instantly transmitted, and the quantity of its work will be measured by

$$a\,D\,g\,(h' - h)\,l.$$

The mass moved will be the fluid in the siphon which is measured by $a\,l\,D$; and if we denote the velocity by V, we shall have, for the living force of the moving mass,

$$a\,l\,D\,.\,V^2;$$

whence,

$$a\,D\,g\,(h' - h)\,l = \frac{a\,D\,l\,V^2}{2};$$

and,

$$V = \sqrt{2\,g\,(h' - h)};$$

from which it appears, that *the velocity with which the liquid will flow through the siphon, is equal to the square root of twice the force of gravity, into the difference of level of the fluid in the two reser-*

voirs. When the fluid in the reservoirs comes to the same level, the flow will cease, since, in that case, $h' - h = 0$.

§ 397.—The siphon may be employed to great advantage to drain canals, ponds, marshes, and the like. For this purpose, it may be made flexible by constructing it of leather, well saturated with grease, like the common *hose*, and furnished with internal hoops to prevent its collapsing by the pressure of the external air. It is thrown into the water to be drained, and filled; when, the ends being plugged up, it is placed across the ridge or bank over which the water is to be conveyed; the plugs are then removed, the flow will take place, and thus the atmosphere will be made literally to press the water from one basin to another, over an intermediate ridge.

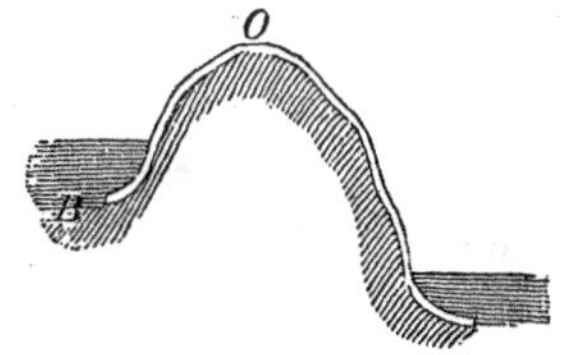

It is obvious that the difference of level between the bottom of the basin to be drained and the highest point O, over which the water is to be conveyed, should never exceed the height to which water may be supported in vacuo by the atmospheric pressure at the place.

THE AIR-PUMP.

§ 398.—Air expands and tends to diffuse itself in all directions when the surrounding pressure is lessened. By means of this property, it may be rarefied and brought to almost any degree of tenuity. This is accomplished by an instrument called the *Air-Pump* or *Exhausting Syringe.* It will be best understood by describing one of the simplest kind. It consists, essentially, of

1st. A *Receiver R*, or chamber from which the exterior air is excluded, that the air within may be rarefied. This is commonly a bell-shaped glass vessel, with ground edge, over which a small quantity of grease is smeared, that no air may pass through any remaining inequalities on its surface, and a ground glass plate $m\,n$ imbedded in a metallic table, on which it stands.

2d. A *Barrel B*, or chamber into which the air in the reservoir is to expand itself. It is a hollow cylinder of metal or glass, connected with the receiver *R* by the communication $o\,f\,g$. An air-tight piston *P* is made to move back and forth in the barrel by means of the handle *a*.

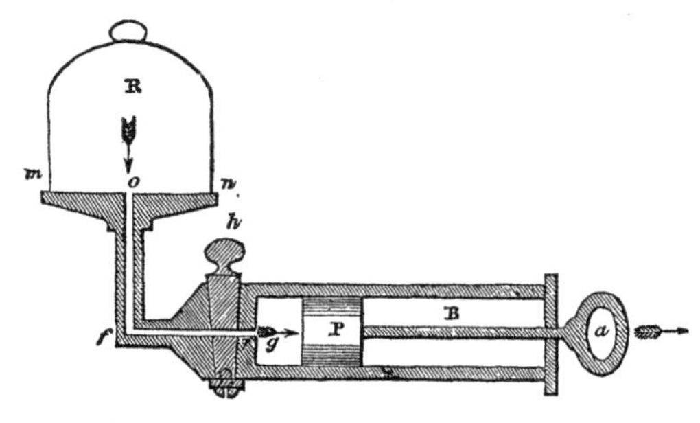

3d. A *Stop-cock h*, by means of which the communication between the barrel and receiver is established or cut off at pleasure. This cock is a conical piece of metal fitting air-tight into an aperture just at the lower end of the barrel, and is pierced in two directions; one of the perforations runs transversely through, as shown in the first figure, and when in this position the communication between the barrel and receiver is established; the second perforation passes in the direction of the axis from the smaller end, and as it approaches the first, inclines sideways, and runs out at right angles to it, as indicated in the second figure. In this position of the cock, the communication between the receiver and barrel is cut off, whilst that with the external air is opened.

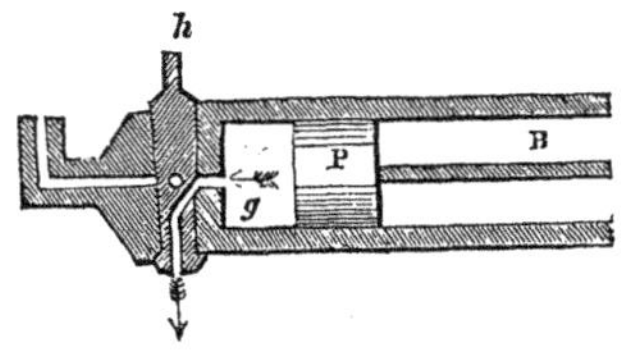

Now, suppose the piston at the bottom of the barrel, and the communication between the barrel and the receiver established; draw the piston back, the air in the receiver will rush out in the direction indicated by the arrow-head, through the communication ofg, into the vacant space within the barrel. The air which now occupies both the barrel and receiver is less dense than when it occupied the receiver alone. Turn the cock a quarter round, the communication between the receiver and barrel is cut off, and that between the latter and the open air is established; push the piston to the bottom of the barrel again, the air within the barrel will be delivered into the external air. Turn the cock a quarter back, the communication between the barrel and receiver is restored; and the same operation as before being repeated, a certain quantity of air will be transferred from the receiver to the exterior space at each double stroke of the piston.

To find the degree of exhaustion after any number of double strokes of the piston, denote by D the density of the air in the receiver before the operation begins, being the same as that of the external air; by r the capacity of the receiver, by b that of the barrel, and by p that of the pipe. At the beginning of the operation, the piston is at the bottom of the barrel, and the internal air occupies the receiver and pipe; when the piston is withdrawn to the opposite end of the barrel, this same air expands and occupies the receiver, pipe, and barrel; and as the density of the same body is inversely proportional to the space it occupies, we shall have

$$r + p + b \quad : \quad r + p \quad :: \quad D \quad : \quad x;$$

in which x denotes the density of the air after the piston is drawn back the first time. From this proportion, we find

$$x = D \cdot \frac{r + p}{r + p + b}.$$

The cock being turned a quarter round, the piston pushed back to the bottom of the barrel, and the cock again turned to open the communication with the receiver, the operation is repeated upon the air whose density is x, and we have

$$r + p + b \quad : \quad r + p \quad :: \quad D \cdot \frac{r + p}{r + p + b} \quad : \quad x';$$

in which x' is the density after the second backward motion of the piston, or after the second double stroke; and we find

$$x' = D \cdot \left(\frac{r + p}{r + p + b}\right)^2;$$

and if n denote the number of double strokes of the piston, and x_n the corresponding density of the remaining air, then will

$$x_n = D \cdot \left(\frac{r + p}{r + p + b}\right)^n.$$

From which it is obvious, that although the density of the air will become less and less at every double stroke, yet it can never be reduced to nothing, however great n may be; in other words, the air cannot be wholly removed from the receiver by the air-pump. The exhaustion will go on rapidly in proportion as the barrel is large as compared with the receiver and pipe, and after a few double strokes, the rarefaction will be sufficient for all practical purposes. Suppose, for example, the receiver to contain 19 units of volume, the pipe 1, and the barrel 10; then will

$$\frac{r + p}{r + p + b} = \frac{20}{30} = \tfrac{2}{3}:$$

and suppose 4 double strokes of the piston; then will $n = 4$, and

$$\left(\frac{r + p}{r + p + b}\right)^n = (\tfrac{2}{3})^4 = \frac{16}{81} = 0{,}197, \text{ nearly};$$

that is, after 4 double strokes, the density of the remaining air will be but about two tenths of the original density. With the best machines, the air may be rarefied from four to six hundred times.

The degree of rarefaction is indicated in a very simple manner by what are called *gauges*. These not only indicate the condition of the air in the receiver, but also warn the operator of any leakage that may take place either at the edge of the receiver or in the joints of the instrument. The mode in which the gauge acts, will be readily understood from the discussion of the barometer; it will be sufficient here simply to indicate its construction. In its more perfect form, it consists of a glass tube, about 60 inches long, bent in the middle till the straight portions are parallel to each other; one end is closed, and the branch terminating in this end is

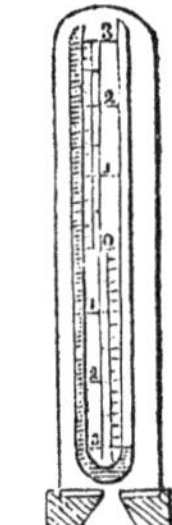

filled with mercury. A scale of equal parts is placed between the branches, having its zero at a point midway from the top to the bottom, the numbers of the scale increasing in both directions. It is placed so that the branches of the tube shall be vertical, with its ends upward, and inclosed in an inverted glass vessel, which communicates with the receiver of the air-pump.

Repeated attempts have been made to bring the air pump to still higher degrees of perfection since its first invention. Self-acting valves, opening and shutting by the elastic force of the air, have been used instead of cocks. Two barrels have been employed instead of one, so that an uninterrupted and more rapid rarefaction of the air is brought about, the piston in one barrel being made to ascend while that of the other descends. The most serious defect

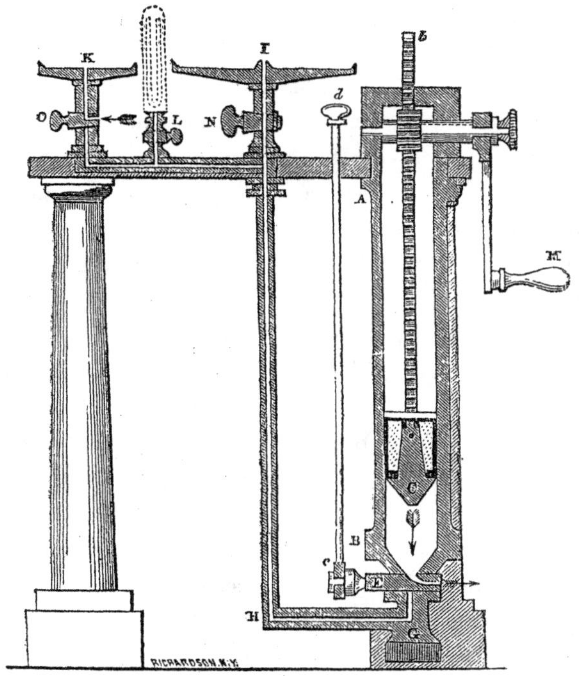

was that by which a portion of the air was retained between the piston and the bottom of the barrel. This intervening space is filled with air of the ordinary density at each descent of the piston;

when the cock is turned, and the communication re-established with the receiver, this air forces its way in and diminishes the rarefaction already attained. If the air in the receiver is so far rarefied, that one stroke of the piston will only raise such a quantity as equals the air contained in this space, it is plain that no further exhaustion can be effected by continuing to pump. This limit to rarefaction will be arrived at the sooner, in proportion as the space below the piston is larger; and one chief point in the improvements has been to diminish this space as much as possible. $A B$ is a highly polished cylinder of glass, which serves as the barrel of the pump; within it the piston works perfectly air-tight. The piston consists of washers of leather soaked in oil, or of cork covered with a leather cap, and tied together about the lower end C of the piston-rod by means of two parallel metal plates. The piston-rod Cb, which is toothed, is elevated and depressed by means of a cog-wheel turned by the handle M. If a thin film of oil be poured upon the upper surface of the piston the friction will be lessened, and the whole will be rendered more air-tight. To diminish to the utmost the space between the bottom of the barrel and the piston-rod, the form of a truncated cone is given to the latter, so that its extremity may be brought as nearly as possible into absolute contact with the cock E; this space is therefore rendered indefinitely small, the oozing of the oil down the barrel contributing still further to lessen it. The exchange-cock E has the double bore already described, and is turned by a short lever, to which motion is communicated by a rod $c d$. The communication $G H$ is carried to the two plates I and K, on one or both of which receivers may be placed; the two cocks N and O below these plates, serve to cut off the rarefied air within the receivers when it is desired to leave them for any length of time. The cock O is also an exchange-cock, so as to admit the external air into the receivers.

Pumps thus constructed have advantages over such as work with valves, in that they last longer, exhaust better, and may be employed as condensers when suitable receivers are provided, by merely reversing the operations of the exchange valve during the motion of the piston.

TABLES.

TABLE I.

THE TENACITIES OF DIFFERENT SUBSTANCES, AND THE RESISTANCES WHICH THEY OPPOSE TO DIRECT COMPRESSION.—See § 269.

SUBSTANCES EXPERIMENTED ON.	Tenacity in Tons per Square Inch.	Name of Experimenter.	Crushing Force in Tons per Sq. Inch.	Name of Experimenter.
Wrought-iron, in wire from 1-20th to 1-30th of an inch in diameter	60 to 91	Lamé		
in wire, 1-10th of an inch . . .	36 to 43	Telford		
in bars, Russian (mean) . . .	27	Lamé		
English (mean) . . .	25½	—		
hammered	30	Brunel		
rolled in sheets, and cut lengthwise	14	Mitis		
ditto, cut crosswise . . .	18	—		
in chains, oval links 6 in. clear, iron 1½ in. diameter	21½	Brown		
ditto, Brunton's, with stay across link	25	Barlow		
Cast Iron, quality No. 1	6 to 7¾	Hodgkinson	38 to 41	Hodgkinson
2	6 to 8	—	37 to 48	—
3*	6 to 9¾	—	51 to 65	—
Steel, cast	44	Mitis		
cast and tilted	60	Rennie		
blistered and hammered . . .	59½	—		
shear	57	—		
raw	50	Mitis		
Damascus	31	—		
ditto, once refined	36	—		
ditto, twice refined	44	—		
Copper, cast	8½	Rennie	52	Rennie
hammered	15	—	46	—
sheet	21	Kingston		
wire	27½			
Platinum wire	17	Guyton		
Silver, cast	18	—		
wire	17	—		
Gold, cast	9	—		
wire	14	—		
Brass, yellow (fine)	8	Rennie	73	—
Gun metal (hard)	16	—		
Tin, cast	2	—	7	—
wire	3	—		
Lead, cast	4-5ths	—	3½	—
milled sheet	1½	Tredgold		
wire	1,1	Guyton		

*The strongest quality of cast iron, is a Scotch iron known as the Devon Hot Blast, No. 3: its tenacity is 9¾ tons per square inch, and its resistance to compression 65 tons. The experiments of Major Wade on the gun iron at West Point Foundry, and at Boston, give results as high as 10 to 16 tons, and on small cast bars, as high as 17 tons.—See Ordnance Manual, 1850, p. 402

TABLE I—*continued.*

SUBSTANCES EXPERIMENTED ON.		Tenacity in Tons per Square Inch	Name of Experimenter.	Crushing Force in Tons per Sq. Inch.	Name of Experimenter.
Stone, slate (Welsh)		5,7	. .		
Marble (white)		4	. .	1,4	Rennie
Givry		1	. .		
Portland		½	. .	1,6	—
Craigleith freestone		. .	. .	2,4	—
Bramley Fall sandstone		. .	. .	2,7	—
Cornish granite		. .	. .	2,8	—
Peterhead ditto		. .	. .	3,7	—
Limestone (compact blk)		. .	. .	4	—
Purbeck		. .	. .	4	—
Aberdeen granite		. .	. .	5	—
Brick, pale red		,13	. .	,56	—
red		. .	. .	,8	—
Hammersmith (pavior's)		. .	. .	1	—
ditto (burnt)		. .	. .	1,4	—
Chalk		. .	. .	,22	—
Plaster of Paris		,03	. .		
Glass, plate		4	. .		
Bone (ox)		2,2			
Hemp fibres glued together		41			
Strips of paper glued together		13			
Wood, Box, spec. gravity	,862	9	Barlow		
Ash	,6	8	—		
Teak	,9	7	—		
Beech	,7	5	—		
Oak	,92	5	—	1,7	—
Ditto	,77	4	—		
Fir	,6	5	—		
Pear	,646	4½	—		
Mahogany	,637	3½	—		
Elm		6	. .	,57	—
Pine, American		6	. .	,73	—
Deal, white		6	. .	,86	—

TABLE II.

OF THE DENSITIES AND VOLUMES OF WATER AT DIFFERENT DEGREES OF HEAT, (ACCORDING TO STAMPFER), FOR EVERY 2¼ DEGREES OF FAHRENHEIT'S SCALE.—See § 276.

(*Jahrbuc des Polytechnischen Institutes in Wein, Bd.* 16, *S.* 70).

t Temperature.	$D_{//}$ Density.	Diff.	V Volume.	Diff.
°				
32,00	0,999887		1,000113	
34,25	0,999950	63	1,000050	63
36,50	0,999988	38	1,000012	38
38,75	1,000000	12	1,000000	12
41,00	0,999988	12	1,000012	12
43,25	0,999952	35	1,000047	35
45,50	0,999894	58	1,000106	59
47,75	0,999813	81	1,000187	81
50,00	0,999711	102	1,000289	102
52,25	0,999587	124	1,000413	124
54,50	0,999442	145	1,000558	145
56,75	0,999278	164	1,000723	165
59,00	0,999095	183	1,000906	183
61,25	0,998893	202	1,001108	202
63,50	0,998673	220	1,001329	221
65,75	0,998435	238	1,001567	238
68,00	0,998180	255	1,001822	255
70,25	0,997909	271	1,002095	273
72,50	0,997622	287	1,002384	289
74,75	0,997320	302	1,002687	303
77,00	0,997003	317	1,003005	318
79,25	0,996673	330	1,003338	333
81,50	0,996329	344	1,003685	347
83,75	0,995971	358	1,004045	360
86,00	0,995601	370	1,004418	373
88,25	0,995219	382	1,004804	386
90,50	0,994825	394	1,005202	398
92,75	0,994420	405	1,005612	410
95,00	0,994004	416	1,006032	420
97,25	0,993579	425	1,006462	430
99,50	0,993145	434	1,006902	440

With this table it is easy to find the specific gravity by means of water at any temperature. Suppose, for example, the specific gravity S' in Equation (456), had been found at the temperature of 59°, then would $D_{//}$ in that equation be 0,999095, and the specific gravity of the body referred to water at its greatest density, would be given by

$$S = S' \times 0{,}999095.$$

TABLE III.

IF THE SPECIFIC GRAVITIES OF SOME OF THE MOST IMPORTANT BODIES.

[The density of distilled water is reckoned in this Table at its maximum 38¾° F. = 1,000].

Name of the Body.	Specific Gravity.
I. SOLID BODIES.	
(1) METALS.	
Antimony (of the laboratory)	4,2 — 4,7
Brass	7,6 — 8,8
Bronze for cannon, according to Lieut. Matzka	8,414 — 8,974
Ditto, mean	8,758
Copper, melted	7,788 — 8,726
Ditto, hammered	8,878 — 8,9
Ditto, wire-drawn	8,78
Gold, melted	19,238 — 19,253
Ditto, hammered	19,361 — 19,6
Iron, wrought	7,207 — 7,788
Ditto, cast, a mean	7,251
Ditto, gray	7,2
Ditto, white	7,5
Ditto for cannon, a mean	7,21 — 7,30
Lead, pure melted	11,3303
Ditto, flattened	11,388
Platinum, native	16,0 — 18,94
Ditto, melted	20,855
Ditto, hammered and wire-drawn	21,25
Quicksilver, at 32° Fahr.	13,568 — 13,598
Silver, pure melted	10,474
Ditto, hammered	10,51 — 10,622
Steel, cast	7,919
Ditto, wrought	7,840
Ditto, much hardened	7,818
Ditto, slightly	7,833
Tin, chemically pure	7,291
Ditto, hammered	7,299 — 7,475
Ditto, Bohemian and Saxon	7,312
Ditto, English	7,291
Zinc, melted	6,861 — 7,215
Ditto, rolled	7,191
(2) BUILDING STONES.	
Alabaster	2,7 — 3,0
Basalt	2,8 — 3,1
Dolerite	2,72 — 2,93
Gneiss	2,5 — 2,9
Granite	2,5 — 2,66
Hornblende	2,9 — 3,1
Limestone, various kinds	2,64 — 2,72
Phonolite	2,51 — 2,69
Porphyry	2,4 — 2,6
Quartz	2,56 — 2,75
Sandstone, various kinds, a mean	2,2 — 2,5
Stones for building	1,66 — 2,62
Syenite	2,5 — 3,
Trachyte	2,4 — 2,6
Brick	1,41 — 1,86

TABLE III—*Continued*

Name of the Body.	Specific Gravity.	
I. SOLID BODIES.		
(3) Woods.	Fresh-felled.	Dry.
Alder	0,8571	0,5001
Ash	0,9036	0,6440
Aspen	0,7654	0,4302
Birch	0,9012	0,6274
Box	0,9822	0,5907
Elm	0,9476	0,5474
Fir	0,8941	0,5550
Hornbeam	0,9452	0,7695
Horse-chestnut	0,8614	0,5749
Larch	0,9206	0,4735
Lime	0,8170	0,4390
Maple	0,9036	0,6592
Oak	1,0494	0,6777
Ditto, another specimen	1,0754	0,7075
Pine, *Pinus Abies Picea*	0,8699	0,4716
Ditto, *Pinus Sylvestris*	0,9121	0,5502
Poplar (Italian)	0,7634	0,3931
Willow	0,7155	0,5289
Ditto, white	0,9859	0,4873
(4) Various Solid Bodies.		
Charcoal, of cork	0,1	
Ditto, soft wood	0,28	— 0,44
Ditto, oak	1,573	
Coal	1,232	— 1,510
Coke	1,865	
Earth, common	1,48	
rough sand	1,92	
rough earth, with gravel	2,02	
moist sand	2,05	
gravelly soil	2,07	
clay	2,15	
clay or loam, with gravel	2,48	
Flint, dark	2,542	
Ditto, white	2,741	
Gunpowder, loosely filled in		
coarse powder	0,886	
musket ditto	0,992	
Ditto, slightly shaken down		
musket-powder	1,069	
Ditto, solid	2,248	— 2,563
Ice	0,916	— 0,9268
Lime, unslacked	1,842	
Resin, common	1,089	
Rock-salt	2,257	
Saltpetre, melted	2,745	
Ditto, crystallized	1,900	
Slate-pencil	1,8	— 2,24
Sulphur	1,92	— 1,99
Tallow	0,942	
Turpentine	0,991	
Wax, white	0,969	
Ditto, yellow	0,965	
Ditto, shoemaker's	0,897	

TABLE III—*Continued.*

Name of the Body.	Specific Gravity.	
II. LIQUIDS.		
Acid, acetic	1,063	
Ditto, muriatic	1,211	
Ditto, nitric, concentrated	1,521 — 1,522	
Ditto, sulphuric, English	1,845	
Ditto, concentrated (Nordh.)	1,860	
Alcohol, free from water	0,792	
Ditto, common	0,824 — 0,79	
Ammoniac, liquid	0,875	
Aquafortis, double	1,300	
Ditto, single	1,200	
Beer	1,023 — 1,034	
Ether, acetic	0,866	
Ditto, muriatic	0,845 — 0,874	
Ditto, nitric	0,886	
Ditto, sulphuric	0,715	
Oil, linseed	0,928 — 0,953	
Ditto, olive	0,915	
Ditto, turpentine	0,792 — 0,891	
Ditto, whale	0,923	
Quicksilver	13,568 — 13,598	
Water, distilled	1,000	
Ditto, rain	1,0013	
Ditto, sea	1,0265 — 1,028	
Wine	0,992 — 1,038	
III. GASES.	Water = 1. Temp. 38¾° F.	Barometer 30 In. Tem.=32°
Atmospheric air = $\frac{1}{770}$ =	0,00130	1,0000
Carbonic acid gas	0,00198	1,5240
Carbonic oxide gas	0,00126	0,9569
Carbureted hydrogen, a maximum	0,00127	0 9784
Ditto, from Coals	0,00039 0,00085	0,3000 0,5596
Chlorine	0,00321	2,4700
Hydriodic gas	0,00577	4,4430
Hydrogen	0,0000895	0,0688
Hydrosulphuric acid gas	0.00155	1,1912
Muriatic acid gas	0.00162	1,2474
Nitrogen	0,00127	0,9760
Oxygen	0,00143	1,1026
Phosphureted hydrogen gas	0,00113	0,8700
Steam at 212° Fahr.	0,00082	0.6235
Sulphurous acid gas	0,00292	2,2470

TABLE IV.

TABLE FOR FINDING ALTITUDES.—See § 284.

Detached Thermometer.							
$t_{,}+t'$	A	$t_{,}+t'$	A	$t_{,}+t'$	A	$t_{,}+t'$	A
40	4,7689067	75	4,7859208	110	4,8022936	145	4,8180714
41	,7694021	76	,7863973	111	,8027525	146	,8185140
42	,7698971	77	,7868733	112	,8032109	147	,8189559
43	,7703911	78	,7873487	113	,8036687	148	,8193975
44	,7708851	79	,7878236	114	,8041261	149	,8198387
45	,7713785	80	,7882979	115	,8045830	150	,8202794
46	,7718711	81	,7887719	116	,8050395	151	,8207196
47	,7723633	82	,7892451	117	,8054953	152	,8211594
48	,7728548	83	,7897180	118	,8059509	153	,8215988
49	,7733457	84	,7901903	119	,8064058	154	,8220377
50	,7738363	85	,7906621	120	,8068604	155	,8224761
51	,7743261	86	,7911335	121	,8073144	156	,8229141
52	,7748153	87	,7916042	122	,8077680	157	,8233517
53	,7753042	88	,7920745	123	,8082211	158	,8237888
54	,7757925	89	,7925441	124	,8086737	159	,8242256
55	,7762802	90	,7930135	125	,8091258	160	,8246618
56	,7767674	91	,7934822	126	,8095776	161	,8250976
57	,7772540	92	,7939504	127	,8100287	162	,8255331
58	,7777400	93	,7944182	128	,8104795	163	,8259680
59	,7782256	94	,7948854	129	,8109298	164	,8264024
60	,7787105	95	,7953521	130	,8113796	165	,8268365
61	,7791949	96	,7958184	131	,8118290	166	,8272701
62	,7796788	97	,7962841	132	,8122778	167	,8277034
63	,7801622	98	,7967493	133	,8127263	168	,8281362
64	,7806450	99	,7972141	134	,8131742	169	,8285685
65	,7811272	100	,7976784	135	,8136216	170	.8290005
66	,7816090	101	,7981421	136	,8140688	171	,8294319
67	,7820902	102	,7986054	137	,8145153	172	,8298629
68	,7825709	103	,7990681	138	,8149614	173	,8302937
69	,7830511	104	,7995303	139	,8154070	174	,8307238
70	,7835306	105	,7999921	140	,8158523	175	,8311536
71	,7840098	106	,8004533	141	,8162970	176	,8315830
72	,7844883	107	,8009142	142	,8167413	177	,8320119
73	,7849664	108	,8013744	143	,8171852	178	,8324404
74	4,7854438	109	4,8018343	144	4,8176285	179	4,8328686

TABLE IV—*continued.*

WITH THE BAROMETER.—See § 284.

Latitude.		Attached Thermometer.		
Ψ	B	T—T′	C	C
0°	0,0011689		−	+
3	,0011624	0°	0,0000000	0,0000000
6	,0011433	1	,0000434	9,9999566
9	,0011117	2	,0000869	,9999131
12	,0010679	3	,0001303	,9998697
15	,0010124	4	,0001738	,9998263
18	,0009459	5	,0002172	,9997829
21	,0008689	6	,0002607	,9997395
24	,0007825	7	,0003041	,9996961
27	,0006874	8	,0003476	,9996527
30	,0005848	9	,0003910	,9996093
33	,0004758	10	,0004345	,9995659
36	,0003615	11	,0004780	,9995225
39	,0002433	12	,0005215	,9994792
42	,0001223	13	,0005650	,9994358
45	,0000000	14	,0006084	,9993924
48	9,9998775	15	,0006519	,9993490
49	,9998372	16	,0006954	,9993057
50	,9997967	17	,0007389	,9992623
51	,9997566	18	,0007824	,9992190
52	,9997167	19	,0008259	,9991756
53	,9996772	20	,0008695	,9991323
54	,9996381	21	,0009130	,9990889
55	,9995995	22	,0009565	,9990456
56	,9995613	23	,0010000	,9990023
57	,9995237	24	,0010436	,9989589
58	,9994866	25	,0010871	,9989156
59	,9994502	26	,0011306	,9988723
60	,9994144	27	,0011742	,9988290
63	,9993115	28	,0012177	,9987857
66	,9992161	29	,0012613	,9987424
69	,9991293	30	,0013048	,9986991
75	,9989852	31	0,0013484	9,9986558
81	,9988854			
90	9,9988300			

TABLE V.

COEFFICIENT VALUES, FOR THE DISCHARGE OF FLUIDS THROUGH THIN PLATES, THE ORIFICES BEING REMOTE FROM THE LATERAL FACES OF THE VESSEL.—See § 300.

Head of fluid above the centre of the orifice, in feet.	Values of the coefficients for orifices whose smallest dimensions or diameters are—					
	ft. 0,66	*ft.* 0,33	*ft.* 0,16	*ft.* 0,08	*ft.* 0,07	*ft.* 0,03
0,05						0,700
0,07				0,627	0,660	0,696
0,13			0,618	0,632	0,657	0,685
0,20		0,592	0,620	0,640	0,656	0,677
0,26		0,602	0,625	0,638	0,655	0,672
0,33	0,593	0,608	0,630	0,637	0,655	0,667
0,66	0,596	0,613	0,631	0,634	0,654	0,655
1,00	0,601	0,617	0,630	0,632	0,644	0,650
1,64	0,602	0,617	0,628	0,630	0,640	0,644
3,28	0,605	0,615	0,626	0,628	0,633	0,632
5,00	0,603	0,612	0,620	0,620	0,621	0,618
6,65	0,602	0,610	0,615	0,615	0,610	0,610
32,75	0,600	0,600	0,600	0,600	0,600	0,600

In the instance of gas, the generating head is always greater than 6,65 ft., and the coefficient 0,6, or 0,61, is taken in all cases.

For orifices larger than 0,66 ft., the coefficients are taken as for this dimension; for orifices smaller than 0,03 ft., the coefficients are the same as for this latter; finally, for orifices between those of the table, we take coefficients whose values are a mean between the latter, corresponding to the given head.

TABLE VI.

EXPERIMENTS ON FRICTION, WITHOUT UNGUENTS. BY M. MORIN.

The surfaces of friction were varied from 0,03336 to 2,7987 square feet, the pressures from 88 lbs. to 2205 lbs., and the velocities from a scarcely perceptible motion to 9,84 feet per second. The surfaces of wood were planed, and those of metal filed and polished with the greatest care, and carefully wiped after every experiment. The presence of unguents was especially guarded against.—See § 355.

SURFACES OF CONTACT.	Friction of Motion.*		Friction of Quiescence.†	
	Coefficient of Friction.	Limiting Angle of Resistance.	Coefficient of Friction.	Limiting Angle of Resistance.
Oak upon oak, the direction of the fibres being parallel to the motion . . .	0,478	25° 33′	0,625	32° 1′
Oak upon oak, the directions of the fibres of the moving surface being perpendicular to those of the quiescent surface and to the direction of the motion‡	0,324	17 58	0,540	28 23
Oak upon oak, the fibres of the both surfaces being perpendicular to the direction of the motion	0,336	18 35		
Oak upon oak, the fibres of the moving surface being perpendicular to the surface of contact, and those of the surface at rest parallel to the direction of the motion	0,192	10 52	0,271	15 10
Oak upon oak, the fibres of both surfaces being perpendicular to the surface of contact, or the pieces end to end . .	. .	. .	0,43	23 17
Elm upon oak, the direction of the fibres being parallel to the motion . . .	0,432	23 22	0,694	34 46
Oak upon elm, ditto§	0,246	13 50	0,376	20 37
Elm upon oak, the fibres of the moving surface (the elm) being perpendicular to those of the quiescent surface (the oak) and to the direction of the motion .	0,450	24 16	0,570	29 41
Ash upon oak, the fibres of both surfaces being parallel to the direction of the motion	0,400	21 49	0,570	29 41
Fir upon oak, the fibres of both surfaces being parallel to the direction of the motion	0,355	19 33	0,520	27 29
Beach upon oak, ditto	0,360	19 48	0,53	27 56
Wild pear-tree upon oak, ditto . . .	0,370	20 19	0,440	23 45
Service-tree upon oak, ditto	0,400	21 49	0,570	29 41
Wrought iron upon oak, ditto‖ . . .	0,619	31 47	0,619	31 47

* The friction in this case varies but very slightly from the mean.

† The friction in this case varies considerably from the mean. In all the experiments the surfaces had been 15 minutes in contact.

‡ The dimensions of the surfaces of contact were in this experiment ,947 square feet, and the results were nearly uniform. When the dimensions were diminished to ,043, a tearing of the fibre became apparent in the case of motion, and there were symptoms of the combustion of the wood; from these circumstances there resulted an irregularity in the friction indicative of excessive pressure.

§ It is worthy of remark that the friction of oak upon elm is but five-ninths of that of elm upon oak.

‖ In the experiments in which one of the surfaces was of metal, small particles of the metal began, after a time, to be apparent upon the wood, giving it a polished metallic appearance; these were at every experiment wiped off; they indicated a wearing of the metal. The friction of motion and that of quiescence, in these experiments, coincided. The results were remarkably uniform.

TABLE VI—*continued.*

SURFACES OF CONTACT.	Friction of Motion.		Friction of Quiescence.	
	Coefficient of Friction.	Limiting Angle of Resistance.	Coefficient of Friction.	Limiting Angle of Resistance.
Wrought iron upon oak, the surfaces being greased and well wetted . . .	0,256	14° 22′	0,649	33° 0′
Wrought iron upon elm	0,252	14 9	. .	. .
Wrought iron upon cast iron, the fibres of the iron being parallel to the motion	0,194	10 59	0,194	10 59
Wrought iron upon wrought iron, the fibres of both surfaces being parallel to the motion	0,138	7 52	0,137	7 49
Cast iron upon oak, ditto	0,490	26 7		
Ditto, the surfaces being greased and wetted	. .	. .	0,646	32 52
Cast iron upon elm	0,195	11 3		
Cast iron upon cast iron	0,152	8 39	0,162	9 13
Ditto, water being interposed between the surfaces	0,314	17 26		
Cast iron upon brass	0,147	8 22		
Oak upon cast iron, the fibres of the wood being perpendicular to the direction of the motion	0,372	20 25		
Hornbeam upon cast iron—fibres parallel to motion	0,394	21 31		
Wild pear-tree upon cast iron—fibres parallel to the motion	0,436	23 34		
Steel upon cast iron	0,202	11 26		
Steel upon brass	0,152	8 39		
Yellow copper upon cast iron	0,189	10 49		
Ditto oak	0,617	31 41	0,617	31 41
Brass upon cast iron	0,217	12 15		
Brass upon wrought iron, the fibres of the iron being parallel to the motion .	0,161	9 9		
Wrought iron upon brass	0,172	9 46		
Brass upon brass	0,201	11 22		
Black leather (curried) upon oak* . .	0,265	14 51	0,74	36 31
Ox hide (such as that used for soles and for the stuffing of pistons) upon oak, rough	0,52	27 29	0,605	31 11
Ditto ditto ditto smooth .	0,335	18 31	0,43	23 17
Leather as above, polished and hardened by hammering	0,296	16 30	. .	. .
Hempen girth, or pulley-band, (sangle de chanvre,) upon oak, the fibres of the wood and the direction of the cord being *parallel* to the motion . . .	0,52	27 29	0,64	32 38
Hempen matting, woven with small cords, ditto.	0,32	17 45	0,50	26 34
Old cordage, 1½ inch in diameter, ditto†	0,52	27 29	0,79	38 19

* The friction of motion was very nearly the same whether the surface of contact was the inside or the outside of the skin.—The *constancy* of the coefficient of the friction of motion was equally apparent in the rough and the smooth skins.

† All the above experiments, except that with curried black leather, presented the phenomenon of a change in the polish of the surfaces of friction—a state of their surfaces necessary to, and dependent upon, their motion upon one another.

TABLE VI—*continued.*

SURFACES OF CONTACT.	FRICTION OF MOTION.		FRICTION OF QUIESCENCE.	
	Coefficient of Friction.	Limiting Angle of Resistance.	Coefficient of Friction.	Limiting Angle of Resistance.
Calcareous oolitic stone, used in building, of a moderately hard quality, called stone of Jaumont—upon the same stone	0,64	32° 38′	0,74	36° 31′
Hard calcareous stone of Brouck, of a light gray color, susceptible of taking a fine polish, (the muschelkalk,) moving upon the same stone	0,38	20 49	0,70	35 0
The soft stone mentioned above, upon the hard	0,65	33 2	0,75	36 53
The hard stone mentioned above upon the soft	0,67	33 50	0,75	36 53
Common brick upon the stone of Jaumont	0,65	33 2	0,65	33 2
Oak upon ditto, the fibres of the wood being perpendicular to the surface of the stone	0,38	20 49	0,63	32 13
Wrought iron upon ditto, ditto . . .	0,69	34 37	0,49	26 7
Common brick upon the stone of Brouck	0,60	30 58	0,67	33 50
Oak as before (endwise) upon ditto . .	0,38	20 49	0,64	32 38
Iron, ditto ditto . .	0,24	13 30	0,42	22 47

TABLE VII.

EXPERIMENTS ON THE FRICTION OF UNCTUOUS SURFACES. BY M. MORIN.—See § 355.

In these experiments the surfaces, after having been smeared with an unguent, were wiped, so that no interposing layer of the unguent prevented their intimate contact.

SURFACES OF CONTACT.	Friction of Motion.		Friction of Quiescence.	
	Coefficient of Friction.	Limiting Angle of Resistance.	Coefficient of Friction.	Limiting Angle of Resistance.
Oak upon oak, the fibres being parallel to the motion	0,108	6° 10′	0,390	21° 19′
Ditto, the fibres of the moving body being perpendicular to the motion	0,143	8 9	0,314	17 26
Oak upon elm, fibres parallel	0,136	7 45		
Elm upon oak, ditto	0,119	6 48	0,420	22 47
Beech upon oak, ditto	0,330	18 16		
Elm upon elm, ditto	0,140	7 59		
Wrought iron upon elm, ditto	0,138	7 52		
Ditto upon wrought iron, ditto	0,177	10 3		
Ditto upon cast iron, ditto	. .	. .	0,118	6 44
Cast iron upon wrought iron, ditto	0,143	8 9		
Wrought iron upon brass, ditto	0,160	9 6		
Brass upon wrought iron	0,166	9 26		
Cast iron upon oak, ditto	0,107	6 7	0,100	5 43
Ditto upon elm, ditto, the unguent being tallow	0,125	7 8		
Ditto, ditto, the unguent being hog's lard and black lead	0,137	7 49		
Elm upon cast iron, fibres parallel	0,135	7 42	0,098	5 36
Cast iron upon cast iron	0,144	8 12		
Ditto upon brass	0,132	7 32		
Brass upon cast iron	0,107	6 7		
Ditto upon brass	0,134	7 38	0,164	9 19
Copper upon oak	0,100	5 43		
Yellow copper upon cast iron	0,115	6 34		
Leather (ox hide) well tanned upon cast iron, wetted	0,229	12 54	0,267	14 57
Ditto upon brass, wetted	0,244	13 43		

TABLE VIII.

EXPERIMENTS ON FRICTION WITH UNGUENTS INTERPOSED. BY M. MORIN.

The extent of the surfaces in these experiments bore such a relation to the pressure, as to cause them to be separated from one another throughout by an interposed stratum of the unguent.—See § 355.

SURFACES OF CONTACT.	FRICTION OF MOTION. Coefficient of Friction.	FRICTION OF QUIESCENCE. Coefficient of Friction.	UNGUENTS.
Oak upon oak, fibres parallel	0,164	0,440	Dry soap.
Ditto ditto	0,075	0,164	Tallow.
Ditto ditto	0,067	. .	Hog's lard.
Ditto, fibres perpendicular	0,083	0,254	Tallow.
Ditto ditto	0,072	. .	Hog's lard.
Ditto ditto	0,250	. .	Water.
Ditto upon elm, fibres parallel	0,136	. .	Dry soap.
Ditto ditto	0,073	0,178	Tallow.
Ditto ditto	0,066	. .	Hog's lard.
Ditto upon cast iron, ditto	0,080	. .	Tallow.
Ditto upon wrought iron, ditto	0,098	. .	Tallow.
Beech upon oak, ditto	0,055	. .	Tallow.
Elm upon oak, ditto	0,137	0,411	Dry soap.
Ditto ditto	0,070	0,142	Tallow.
Ditto ditto	0,060	. .	Hog's lard.
Ditto upon elm, ditto	0,139	0,217	Dry soap.
Ditto upon cast iron, ditto	0,066	. .	Tallow.
Wrought iron upon oak, ditto	0,256	0,649	Greased, and saturated with water.
Ditto ditto ditto	0,214	. .	Dry soap.
Ditto ditto ditto	0,085	0,108	Tallow.
Ditto upon elm, ditto	0,078	. .	Tallow.
Ditto ditto ditto	0,076	. .	Hog's lard.
Ditto ditto ditto	0,055	. .	Olive oil.
Ditto upon cast iron, ditto	0,103	. .	Tallow.
Ditto ditto ditto	0,076	. .	Hog's lard.
Ditto ditto ditto	0,066	0,100	Olive oil.
Ditto upon wrought iron, ditto	0,082	. .	Tallow.
Ditto ditto ditto	0,081	. .	Hog's lard.
Ditto ditto ditto	0,070	0,115	Olive oil.
Wrought iron upon brass, fibres parallel	0,103	. .	Tallow.
Ditto ditto ditto	0,075	. .	Hog's lard.
Ditto ditto ditto	0,078	. .	Olive oil.
Cast iron upon oak, ditto	0,189	. .	Dry soap.
Ditto ditto ditto	0,218	0,646	Greased, and saturated with water.
Ditto ditto ditto	0,078	0,100	Tallow.
Ditto ditto ditto	0,075	. .	Hog's lard.
Ditto ditto ditto	0,075	0,100	Olive oil.
Ditto upon elm, ditto	0,077	. .	Tallow.
Ditto ditto ditto	0,061	. .	Olive oil.
Ditto ditto ditto	0,091	. .	Hog's lard and plumbago.
Ditto, ditto upon wrought iron	. .	0,100	Tallow.
Cast iron upon cast iron	0,314	. .	Water.
Ditto ditto	0,197	. .	Soap.

TABLE VIII.—*continued.*

SURFACES OF CONTACT.	FRICTION OF MOTION. Coefficient of Friction.	FRICTION OF QUIESCENCE. Coefficient of Friction.	UNGUENTS.
Cast iron upon cast iron . .	0,100	0,100	Tallow.
Ditto ditto . . .	0,070	0,100	Hogs' lard.
Ditto ditto . . .	0,064	. .	Olive oil.
Ditto ditto . . .	0,055	. .	Lard and plumbago.
Ditto upon brass . . .	0,103	. .	Tallow.
Ditto ditto . . .	0,075	. .	Hogs' lard.
Ditto ditto . . .	0,078	. .	Olive oil.
Copper upon oak, fibres parallel	0,069	0,100	Tallow.
Yellow copper upon cast iron .	0,072	0,103	Tallow.
Ditto ditto . . .	0,068	. .	Hogs' lard.
Ditto ditto . . .	0,066	. .	Olive oil.
Brass upon cast iron . . .	0,086	0,106	Tallow.
Ditto ditto . . .	0,077	. .	Olive oil.
Ditto upon wrought iron .	0,081	. .	Tallow.
Ditto ditto . . .	0,089	. .	Lard and plumbago.
Ditto ditto . . .	0,072	. .	Olive oil.
Ditto upon brass . . .	0,058	. .	Olive oil.
Steel upon cast iron . . .	0,105	0,108	Tallow.
Ditto ditto . . .	0,081	. .	Hogs' lard.
Ditto ditto . . .	0,079	. .	Olive oil.
Ditto upon wrought iron .	0,093	. .	Tallow.
Ditto ditto . . .	0,076	. .	Hogs' lard.
Ditto upon brass . . .	0,056	. .	Tallow.
Ditto ditto . . .	0,053	. .	Olive oil.
Ditto ditto . . .	0,067	. .	Lard and plumbago.
Tanned ox hide upon cast iron	0,365	. .	Greased, and saturated with water.
Ditto ditto . . .	0,159	. .	Tallow.
Ditto ditto . . .	0,133	0,122	Olive oil.
Ditto upon brass . . .	0,241	. .	Tallow.
Ditto ditto . . .	0,191	. .	Olive oil.
Ditto upon oak, . . .	0,29	0,79	Water.
Hempen fibres not twisted, moving upon oak, the fibres of the hemp being placed in a direction perpendicular to the direction of the motion, and those of the oak parallel to it . .	0,332	0,869	Greased, and saturated with water.
The same as above, moving upon cast iron	0,194	. .	Tallow.
Ditto	0,153	. .	Olive oil.
Soft calcareous stone of Jaumont upon the same, with a layer of mortar, of sand, and lime interposed, after from 10 to 15 minutes' contact.	. .	0,74	

TABLE IX.

FRICTION OF TRUNNIONS IN THEIR BOXES.—See § 361.

KINDS OF MATERIALS.	STATE OF SURFACES.	Ratio of friction to pressure when the unguent is renewed.	
		By the ordinary method.	Or, continuously.
Trunnions of cast iron and boxes of cast iron.	Unguents of olive oil, hogs' lard, and tallow	0,07 to 0,08	0,054
	The same unguents moistened with water	0,08	0,054
	Unguent of asphaltum	0,054	0,054
	Unctuous	0,14	. .
	Unctuous and moistened with water	0,14	. .
Trunnions of cast iron and boxes of brass.	Unguents of olive oil, hogs' lard, and tallow	0,07 to 0,08	0,054
	Unctuous	0,16	. .
	Unctuous and moistened with water	0,16	. .
	Very slightly unctuous	0,19	. .
Trunnions of cast iron and boxes of lignum-vitæ.	Without unguents	0,18	. .
	Unguents of olive oil and hogs' lard	. .	0,090
	Unctuous with oil and hogs' lard	0,10	. .
	Unctuous with a mixture of hogs' lard and plumbago	0,14	. .
Trunnions of wrought iron and boxes of cast iron.	Unguents of olive oil, tallow, and hogs' lard	0,07 to 0,08	0,054
Trunnions of wrought iron and boxes of brass.	Unguents of olive oil, hogs' lard, and tallow	0,07 to 0,08	0,054
	Old unguents hardened	0,09	. .
	Unctuous and moistened with water	0,19	. .
	Very slightly unctuous	0,25	. .
Trunnions of wrought iron and boxes of lignum-vitæ.	Unguents of oil or hogs' lard	0,11	. .
	Unctuous	0,19	. .
Trunnions of brass and boxes of brass.	Unguent of oil	0,10	. .
	Unguent of hogs' lard	0,09	. .
Trunnions of brass and boxes of cast iron.	Unguents of tallow or of olive oil.	. .	0,045 to 0,052
Trunnions of lignum-vitæ and boxes of cast iron.	Unguents of hogs' lard	0,12	. .
	Unctuous	0,15	. .
Trunnions of lignum-vitæ and boxes of lignum-vitæ.	Unguent of hogs' lard	. .	0,07

TABLE X.

OF WEIGHTS NECESSARY TO BEND DIFFERENT ROPES AROUND A WHEEL ONE FOOT IN DIAMETER.—See § 857.

No. 1. White Ropes—new and dry.

Stiffness proportional to the square of the diameter.

Diameter of rope in inches.	Natural stiffness, or value of *K*.	Stiffness for load of 1 lb., or value of *I*.
	lbs.	*lbs.*
0,39	0,4024	0,0079877
0,79	1,6097	0,0319501
1,57	6,4389	0,1278019
3,15	25,7553	0,5112019

No. 2. White Ropes—new and moistened with water.

Stiffness proportional to square of diameter.

Diameter of rope in inches.	Natural stiffness, or value of *K*.	Stiffness for load of 1 lb., or value of *I*.
	lbs.	*lbs.*
0,39	0,8048	0,0079877
0,79	3,2194	0,0319501
1,57	12,8772	0,1278019
3,15	51,5111	0,5112019

No. 3. White Ropes—half worn and dry.

Stiffness proportional to the square root of the cube of the diameter.

Diameter of rope in inches.	Natural Stiffness, or value of *K*.	Stiffness for load of 1 lb., or value of *I*.
	lbs.	*lbs.*
0,39	0,40243	0,0079877
0,79	1,13801	0,0525889
1,57	3,21844	0,0638794
3,15	9,10150	0,1806573

No. 4. White Ropes—half worn and moistened with water.

Stiffness proportional to the square root of the cube of the diameter.

Diameter of rope in inches.	Natural Stiffness, or value of *K*.	Stiffness for load of 1 lb., or value of *I*.
	lbs.	*lbs.*
0,39	0,8048	0,0079877
0,79	2,2761	0,0525889
1,57	6,4324	0,0638794
3,15	18,2037	0,1806573

Squares of the ratios of diameter, or values of d^2.	
Ratios d.	Squares d^2.
1,00	1,00
1,10	1,21
1,20	1,44
1,30	1,69
1,40	1,96
1,50	2,25
1,60	2,56
1,70	2,89
1,80	3,24
1,90	3,61
2,00	4,00

Square roots of the cubes of the ratios of diameter, or values of $d^{\frac{3}{2}}$.	
Ratios or d.	Power $\frac{3}{2}$, or $d^{\frac{3}{2}}$.
1,00	1,000
1,10	1,154
1,20	1,315
1,30	1,482
1,40	1,657
1,50	1,837
1,60	2,024
1,70	2,217
1,80	2,415
1,90	2,619
,00	2,828

TABLE X—*continued.*

No. 5. Tarred Ropes.

Stiffness proportional to the number of yarns.

[These ropes are usually made of three strands twisted around each other, each strand being composed of a certain number of yarns, also twisted about each other in the same manner.]

No. of yarns.	Weight of 1 foot in length of rope.	Natural stiffness, or value of K.	Stiffness for load of 1 lb., or value of I.
	lbs.	*lbs.*	*lbs.*
6	0,0211	0,1534	0,0085198
15	0,0497	0,7664	0,0198796
30	1,0137	2,5297	0,0411799

APPENDIX.

No. I.

Take the usual formulas for the transformation of co-ordinates from one system to another, both being rectangular, viz:

$$\left.\begin{aligned} x &= a x' + b y' + c z', \\ y &= a' x' + b' y' + c' z', \\ z &= a'' x' + b'' y' + c'' z'; \end{aligned}\right\} \quad \cdots \cdots \quad (1)$$

in which a, b, &c., denote the cosines of the angles which the axes of the same name as the co-ordinates into which they are respectively multiplied make with the axis of the variable in the first member. And hence,

$$\left.\begin{aligned} x' &= a x + a' y + a'' z, \\ y' &= b x + b' y + b'' z, \\ z' &= c x + c' y + c'' z; \end{aligned}\right\} \quad \cdots \cdots \quad (2)$$

Multiply the first of (2) by b, the second by a, and take the difference of the products; we get

$$b x' - a y' = y (a' b - a b') + z (a'' b - a b''); \quad \cdots \quad (3)$$

again, multiply the first by c, the third by a, and take the difference of products; we have

$$c x' - a z' = y (a' c - a c') + z (a'' c - a c'') \quad \cdots \quad (4)$$

Find the value of y in (4), substitute in (3), and reduce, we find

$$A z = (b c' - b' c) x' + (a' c - a c') y' + (a b' - a' b) z',$$

in which

$$A = c\,(a'\,b'' - a''\,b') + c'\,(a''\,b - a\,b'') + c''\,(a\,b' - a'\,b),$$

dividing by A, and subtracting the result from the third of Eqs. (1) we have

$$\left(a'' - \frac{b\,c' - b'\,c}{A}\right)x' + \left(b'' - \frac{a'\,c - a\,c'}{A}\right)y' + \left(c'' - \frac{ab' - a'\,b}{A}\right)z' = 0,$$

and since x', y' and z' are wholly arbitrary, we have

$$a'' - \frac{b\,c' - b'\,c}{A} = 0;\; b'' - \frac{a'\,c - a\,c'}{A} = 0;\; c'' - \frac{a\,b' - a'\,b}{A} = 0; \quad (5)$$

transposing, clearing the fraction, squaring, adding, collecting the co-efficients of c'^2, b'^2, a'^2, and reducing by the relations

$$a^2 + b^2 + c^2 = 1;\; a'^2 + b'^2 + c'^2 = 1;$$
$$a^2 + b^2 = 1 - c^2;\; c^2 + b^2 = 1 - a^2;\; a^2 + c^2 = 1 - b^2,$$

there will result

$$A^2 = 1 - (a\,a' + b\,b' + c\,c')^2.$$

But

$$a\,a' + b\,b' + c\,c' = 0,$$

whence $A = 1$, and, Eqs. (5),

$$a'' = b\,c' - b'\,c;\; b'' = a'\,c - a\,c';\; c'' = a\,b' - a'\,b.$$

No. II.

To find the radius of curvature of any curve, and its inclination to the co-ordinate axes.

Take the centre of curvature as the centre of a sphere of which the radius is unity. Through the same point draw the line $O\,X$, parallel to the axis x, and another $O\,T$, parallel to the tangent to the arc $M\,N$, of osculation. The planes of these lines and of the radius of curvature will cut from the sphere the spherical triangle $A\,B\,C$, of which the side $B\,C$ is 90°, $A\,C$ the angle which the radius of curvature makes with the axis x, and $A\,B$ the angle which the tangent to the curve makes with the same axis. Make

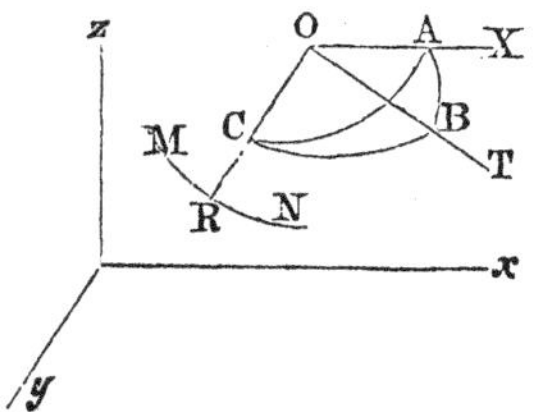

$$\rho = O\,R = \text{radius of curvature},$$
$$\theta' = A\,C;\; c = A\,B;\; C = A\,C\,B.$$

Then will

$$\cos c = \frac{d\,x}{d\,s} = \sin \theta'.\cos C;$$

differentiating, and regarding C as constant,

$$d\frac{d\,x}{d\,s} = \cos \theta'.\,d\,\theta'.\cos C;$$

but $d\,\theta'.\cos C$ is the projection of the arc $d\,\theta'$ on the osculatory plane, whence

$$d\,\theta'.\cos C = \frac{d\,s}{\rho}.$$

Substituting this above, we find

$$\cos \theta' = \rho \cdot \frac{d\dfrac{d\,x}{d\,s}}{d\,s};$$

and denoting by θ'' and θ''', the angles which the radius makes with the axes y and z, respectively, we may write

$$\cos \theta' = \rho \cdot \frac{d\dfrac{d\,x}{d\,s}}{d\,s}; \quad \cos \theta'' = \rho \cdot \frac{d\dfrac{d\,y}{d\,s}}{d\,s}; \quad \cos \theta''' = \rho \cdot \frac{d\dfrac{d\,z}{d\,s}}{d\,s} \cdot \cdot \cdot (1)$$

Squaring, adding and reducing by the relation,

$$\cos^2 \theta' + \cos^2 \theta'' + \cos^2 \theta''' = 1,$$

we have

$$\rho = \frac{d\,s}{\sqrt{\left(d\dfrac{dx}{d\,s}\right)^2 + \left(d\dfrac{d\,y}{d\,s}\right)^2 + \left(d\dfrac{d\,z}{d\,s}\right)^2}},$$

performing the operations indicated under the radical sign, and reducing by the relations

$$d\,s^2 = d\,x^2 + d\,y^2 + d\,z^2,$$
$$d^2\,s\,d\,s = d^2\,x\,d\,x + d^2\,y\,d\,y + d^2\,z\,d\,z,$$

we find

$$\rho = \frac{d\,s^2}{\sqrt{(d^2\,x)^2 + (d^2\,y)^2 + (d^2\,z)^2 - (d^2\,s)^2}}; \cdot \cdot \cdot \cdot (2)$$

If s be taken as the independent variable, then will $d^2\,s = 0$, and Eqs. (1) and (2) become

$$\cos \theta' = \rho \cdot \frac{d^2\,x}{d\,s^2}; \quad \cos \theta'' = \rho \cdot \frac{d^2\,y}{d\,s^2}; \quad \cos \theta''' = \rho \cdot \frac{d^2\,z}{d\,s^2}; \cdot \cdot \quad (3)$$

$$\rho = \frac{d\,s^2}{\sqrt{(d^2\,x)^2 + (d^2\,y)^2 + (d^2\,z)^2}}; \cdot \quad \cdot \cdot \cdot \cdot (4)$$

No. III

To integrate the partial differential equation

$$D \cdot \frac{d q}{d D} + \gamma \cdot p \cdot \frac{d q}{d p} = 0,$$

transpose and divide by D, and we have

$$\frac{d q}{d D} = -\gamma \cdot \frac{p}{D} \cdot \frac{d q}{d p};$$

and because q is a function of p and D, we have

$$d q = \frac{d q}{d D} \cdot d D + \frac{d q}{d p} \cdot d p;$$

and substituting the value of $\frac{d q}{d D}$,

$$d q = \frac{d q}{d p} \cdot \frac{D \cdot d p - \gamma \cdot p \cdot d D}{D};$$

multiplying and dividing by $\gamma \cdot D \cdot p^{\frac{1}{\gamma} - 1}$,

$$d q = \frac{d q}{d p} \cdot \frac{\gamma \cdot D}{p^{\frac{1}{\gamma} - 1}} \cdot \frac{D \cdot \frac{1}{\gamma} \cdot p^{\frac{1}{\gamma} - 1} \cdot d p - p^{\frac{1}{\gamma}} \cdot d D}{D^2};$$

but

$$\frac{D \cdot \frac{1}{\gamma} \cdot p^{\frac{1}{\gamma} - 1} \cdot d p - p^{\frac{1}{\gamma}} \cdot d D}{D^2} = d \left(\frac{p^{\frac{1}{\gamma}}}{D}\right),$$

and making

$$\frac{d q}{d p} \cdot \frac{\gamma \cdot D}{p^{\frac{1}{\gamma} - 1}} = F' \left(\frac{p^{\frac{1}{\gamma}}}{D}\right);$$

we may write

$$d q = F' \left(\frac{p^{\frac{1}{\gamma}}}{D}\right) \cdot d \left(\frac{p^{\frac{1}{\gamma}}}{D}\right);$$

and by integrating

$$q = F \left(\frac{p^{\frac{1}{\gamma}}}{D}\right).$$

in which F, denotes any arbitrary function.

No. IV.

To integrate Equation (414)′ of the text, add to both members

$$a \cdot \frac{d^2 r \varphi}{d t \,.\, d r},$$

and we have

$$\frac{1}{d t} \cdot d \left[\frac{d r \varphi}{d t} + a \frac{d r \varphi}{d r}\right] = \frac{a}{d r} \cdot d \left[\frac{d r \varphi}{d t} + a \frac{d r \varphi}{d r}\right];$$

and making

$$\frac{d r \varphi}{d t} + a \cdot \frac{d r \varphi}{d r} = V,$$

the above may be written,

$$\frac{d V}{d t} = a \cdot \frac{d V}{d r};$$

and V being a function of r and t, we have

$$d V = \frac{d V}{d r} \cdot d r + \frac{d V}{d t} \cdot d t;$$

or, by substitution for $\frac{d V}{d t}$ its value above,

$$d V = \frac{d V}{d r} \cdot (d r + a\, d t) = \frac{d V}{d r} \cdot d\,(r + a t),$$

and by integration,

$$V = \frac{d r \varphi}{d t} + a \cdot \frac{d r \varphi}{d r} = F'(r + a t);$$

in which F' is any arbitrary function. In like manner, by subtracting,

$$a \cdot \frac{d^2 r \varphi}{d r \,.\, d t},$$

from both members of Equation (414)′, we find

$$V' = \frac{d r \varphi}{d t} - a \cdot \frac{d r \varphi}{d r} = f'(r - a t);$$

in which f' denotes any arbitrary function. Whence, by addition,

$$V + V' = \frac{d r \varphi}{d t} = \tfrac{1}{2} F'(r + a t) + \tfrac{1}{2} f'(r - a t);$$

and by subtraction,

$$V - V' = \frac{d\,r\,\varphi}{d\,r} = \frac{1}{2\,a} \cdot F'\,(r + a\,t) - \frac{1}{2\,a} f'\,(r - a\,t).$$

But

$$d\,r\,\varphi = \frac{d\,r\,\varphi}{d\,t} \cdot d\,t + \frac{d\,r\,\varphi}{d\,r} \cdot d\,r.$$

Whence,

$$d\,r\,\varphi = \frac{1}{2\,a} \cdot F''\,(r + a\,t) \,.\, d\,(r + a\,t) - \frac{1}{2\,a} \cdot f' \,.\, (r - a\,t)\,d\,(r - a\,t)\,;$$

and, by integration,

$$r\,\varphi = F\,(r + a\,t) + f\,(r - a\,t),$$

in which F and f denote the primitive functions of which F' and f' are the derived.

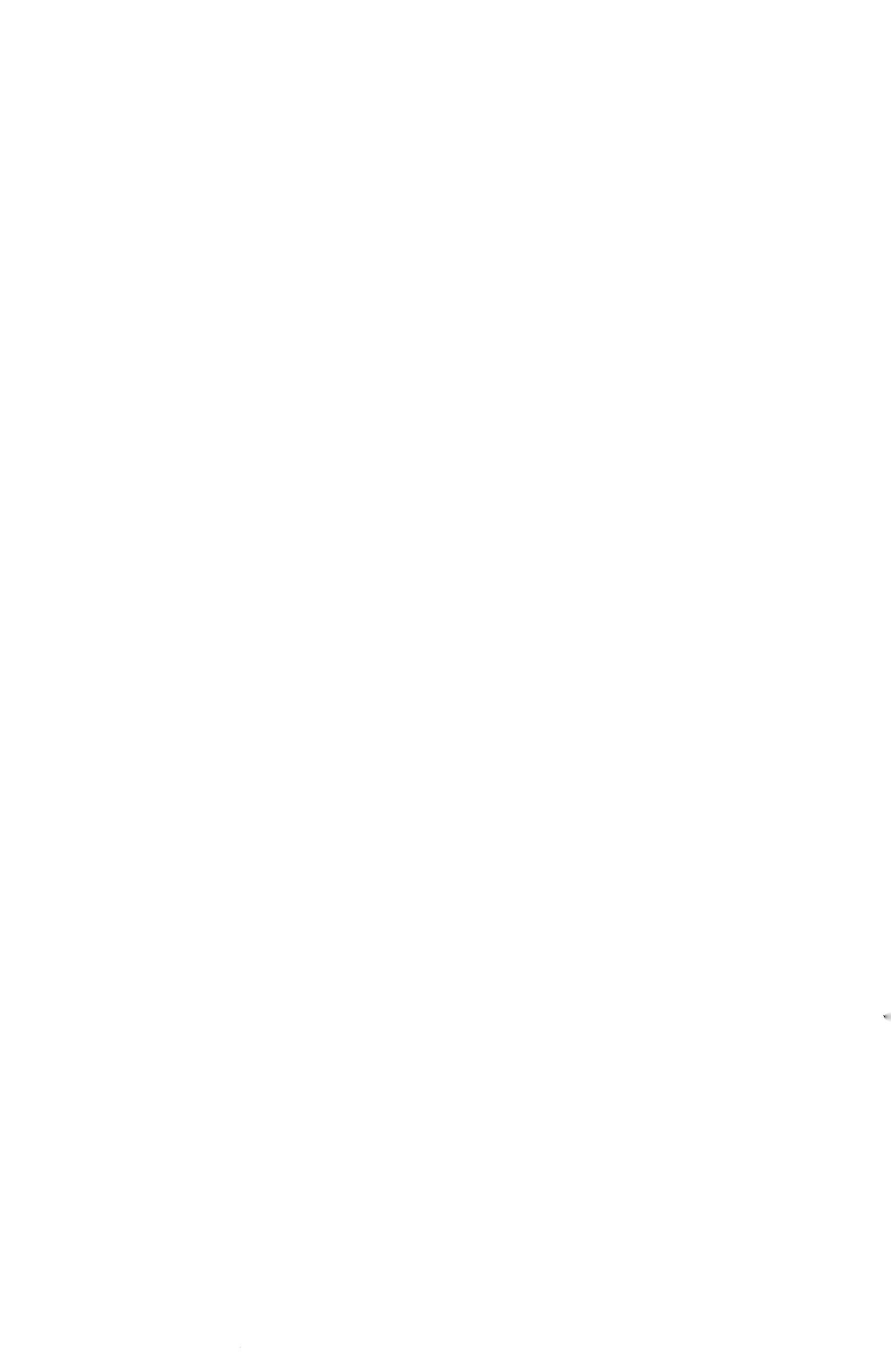

www.ingramcontent.com/pod-product-compliance
Lightning Source LLC
LaVergne TN
LVHW021316110826
845150LV00003B/588

* 9 7 8 1 4 2 5 5 5 7 7 9 9 *